The Study of Indigenous Landscape and Seascape Stewardship on the Central California Coast:

The Findings of a Collaborative Eco-Archaeological Investigation

Kent G. Lightfoot, Michael A. Grone, and Gabriel M. Sanchez

© 2025 Regents of the University of California
Published by eScholarship, Berkeley, CA
1st edition

PDF: 978-0-9982460-9-3
POD: 978-0-9890022-3-3

Available open access at: https://escholarship.org/uc/item/3rb3t6d7
Supplementary materials available at: https://escholarship.org/uc/item/3rb3t6d#supplemental

Publication of this book was made possible by funding from
UC Berkeley Library and Berkeley Research Impact Initiative Funds.

Cover photo by Matt Walker
Production: Westwood Press, www.westwoodpress.com
Design: Morgane Leoni

Contents

List of Figures | vii

List of Tables | xi

List of Appendices | xv

Chapter 1   Introduction to Volume | 1
             Kent G. Lightfoot, Michael A. Grone, and Gabriel M. Sanchez

Chapter 2   Collaborative Eco-Archaeological Investigations of
             Landscape and Seascape Stewardship Practices on
             the Central California Coast | 21
             Kent G. Lightfoot, Valentin Lopez, Mark G. Hylkema, Rob Q. Cuthrell,
             Michael A. Grone, Gabriel M. Sanchez, Peter A. Nelson, Roberta A. Jewett,
             Diane Gifford-Gonzalez, Paul V. A. Fine, Alec J. Apodaca, and Alexii Sigona

Chapter 3   The Study of Indigenous Landscape and Seascape Stewardship
             Practices: Taking it to the Next Step | 45
             Valentin Lopez, Carolyn T. Rodriguez, and K. Michelle Glowa

Chapter 4   The Eco-Archaeological Investigation of Indigenous
             Sites on the Santa Cruz Coast Using a Fine-Grained,
             Low-Impact Methodology | 55
             Kent G. Lightfoot, Valentin Lopez, Mark G. Hylkema, Rob Q. Cuthrell,
             Michael A. Grone, Gabriel M. Sanchez, Peter A. Nelson, Roberta A. Jewett,
             Diane Gifford-Gonzalez, Alec J. Apodaca, Alexii Sigona, and
             Ariadna Gonzalez

Chapter 5   Ground-Penetrating-Radar Survey on the Santa Cruz Coast | 91
             Peter A. Nelson

Chapter 6   Chronology of CA-SCR-7, CA-SCR-10, CA-SCR-14,
             AND CA-SCR-15 | 117
             Kent G. Lightfoot, Rob Q. Cuthrell, Gabriel M. Sanchez, and
             Michael A. Grone

Chapter 7   The Analysis of Artifacts from CA-SCR-7, CA-SCR-10,
             CA-SCR-14, and CA-SCR-15 | 121
             Ariadna Gonzalez and Kathryn Field

Chapter 8   Archaeobotanical Research at Nine Archaeological Sites
             West of the Santa Cruz Mountains: Implications for
             Subsistence Practices and Anthropogenic Burning | 143
             Rob Q. Cuthrell

Chapter 9   Ancient Mussel Bed Harvesting and Indigenous Stewardship
             on the Central California Coast | 199
             Michael A. Grone

Chapter 10   Seasonality of Mussel Harvesting at Three Holocene Sites
             on the Santa Cruz Coast: Insights from Isotopic Variation
             in Marine Mollusks | 229
             Alec J. Apodaca, Jordan F. Brown, and Michael A. Grone

Chapter 11   Middle and Late Holocene Fisheries of the Santa Cruz
             County Coast | 253
             Gabriel M. Sanchez

Chapter 12   Archaeofaunal Evidence for Landscape Management in Cotoni-
             Quiroste: Assessing Rodent Species Abundance as a Proxy | 297
             Diane Gifford-Gonzalez

Chapter 13   Ancient DNA from Voles from Archaeological Sites in Central
             California Reveals Population Continuity over the Last 6,000
             Years: Implications for Past Land Management Practices | 339
             *Paul V. A. Fine, Chris J. Conroy, Cameron Shard Milne, Beth Shapiro, Diane
             Gifford-Gonzalez, Gabriel M. Sanchez, and Kent G. Lightfoot*

Chapter 14   "Returning to the Path of our Ancestors": Using Collaborative
             Eco-Archaeology to Support Contemporary Indigenous
             Landscape and Seascape Stewardship | 355
             Alexii Sigona, Alec J. Apodaca, and Valentin Lopez

Chapter 15   The Findings from the Eco-Archaeological Study of the Central
             California Coast: New Insights on the Timing, Development,
             Scale, and Relevance of Indigenous Stewardship Practices | 371
             Kent G. Lightfoot, Valentin Lopez, Mark G. Hylkema, Rob Q. Cuthrell,
             Michael A. Grone, Gabriel M. Sanchez, Peter A. Nelson, Roberta A. Jewett,
             Diane Gifford-Gonzalez, Paul V. A. Fine, Alec J. Apodaca, Alexii Sigona,
             Jordan F. Brown, Ariadna Gonzalez, and Kathryn Field

List of Contributors | 403

Acknowledgments | 405

List of Figures

Chapter 1:

Figure 1.1.   Area of Conservation and Stewardship of the Amah Mutsun Land Trust within the Ancestral Territory of the Awaswas, Mutsun, and Chalon-Speaking Peoples.

Figure 1.2.   Map of Broader Study Area on the Santa Cruz Coast Showing Quiroste Valley Cultural Preserve and CA-SMA-113, as well as CA-SMA-113, CA-SCR-7, CA-SCR-10, CA-SCR-14, and CA-SCR-15.

Chapter 4:

Figure 4.1.   Location of Datums, Column Units, Auger Units, Excavation Units, and Surface Collection Units at CA-SCR-7.

Figure 4.2.   Northwest Profile of Upper Midden for CA-SCR-7.

Figure 4.3.   Location of Datums, Auger Units, Excavation Unit, and Surface Collection Units at CA-SCR-10.

Figure 4.4.   South Wall Profile of EU 1 for CA-SCR-10.

Figure 4.5.   North Wall Profile of EU 1 for CA-SCR-10.

Figure 4.6.   Location of Datums, Auger Units, Excavation Units, and Surface Collection Units at CA-SCR-14.

Figure 4.7.   East Wall Profile of EU 1 for CA-SCR-14.

Figure 4.8.   North Wall Profile of EU 1 for CA-SCR-14.

Figure 4.9.   Plan Map of Rock Concentration in Level 2, EU 1, CA-SCR-14.

Figure 4.10.   Plan Map of the Upper Layer of Fire-Cracked Rocks in Level 4, EU 1, CA-SCR-14.

Figure 4.11.   Plan Map of the Lower Layer of Fire-Cracked Rocks in Level 4, EU 1, CA-SCR-14.

Figure 4.12.   East Wall Profile of EU 2 for CA-SCR-14.

Figure 4.13.   North Wall Profile of EU 2 for CA-SCR-14.

Figure 4.14.   Location of Datums, Excavation Units, and Surface Collection Units at CA-SCR-15.

Figure 4.15.   East Wall Profile of EU 1 for CA-SCR-15.

Figure 4.16.   North Wall Profile of EU 1 for CA-SCR-15.

Figure 4.17.   East Wall Profile of EU 3 for CA-SCR-15.

Figure 4.18.   West Wall Profile of EU 3 for CA-SCR-15.

Figure 4.19.   Plan Map of Fire-Cracked Rocks in Level 5, EU 3, CA-SCR-15.

Chapter 5:

Figure 5.1.   GPR Profile of CA-SCR-7, File 239.

Figure 5.2.   GPR Profile of CA-SCR-7, File 235.

Figure 5.3.   GPR Profile of CA-SCR-7, File 230.

Figure 5.4.   GPR Profile of CA-SCR-7, File 241.

Figure 5.5.   GPR Profile of CA-SCR-10, Grid 1, Block 1, File 258.

Figure 5.6.   GPR Profile of CA-SCR-10, Grid 1, Block 1, File 262.

Figure 5.7.   GPR Planview of CA-SCR-10, Grid 1, Block 1.

Figure 5.8.   GPR Profile of CA-SCR-10, Grid 2, Block 1.

Figure 5.9.   GPR Profile of CA-SCR-10, Grid 2, Block 1.

Figure 5.10.   GPR Profile of CA-SCR-10, Grid 2, Block 2.

Figure 5.11.   GPR Profile of CA-SCR-10, Grid 2, Block 2.

Figure 5.12.   GPR Profile of CA-SCR-10, Grid 2, Block 3.

Figure 5.13.   GPR Planview of CA-SCR-10, Grid 2, Showing Plow Scars.

Figure 5.14.   GPR Profile of CA-SCR-14.

Figure 5.15.   GPR Profile of CA-SCR-14.

Figure 5.16.   GPR Profile of CA-SCR-15, Grid 1, Block 1.

Figure 5.17.   GPR Planview of CA-SCR-15, Grid 1, Block 1.

Figure 5.18.   GPR Profile of CA-SCR-15.

Chapter 7:

Figure 7.1.   Images of Core (1), Projectile Point (2), Biface (3), and Uniface (4).

Figure 7.2.   Images of Flake Shatter (1) and Angular Shatter (2).

Figure 7.3.   Images of Complete Flake (1) and Proximal Flake (2).

Figure 7.4.   Images of Hammerstone (1) and Pestle Fragment (2).

Figure 7.5.   Images of Fire-Cracked Rock.

Figure 7.6.   Pie Chart Showing Lithic Classification Percentages for CA-SCR-7.

Figure 7.7.   Pie Chart Showing Lithic Classification Percentages for CA-SCR-10.

Figure 7.8.   Pie Chart Showing Lithic Classification Percentages for CA-SCR-14.

Figure 7.9.   Pie Chart Showing Lithic Classification Percentages for CA-SCR-15.

Chapter 8:

Figure 8.1.   Map of Archaeological Sites Analyzed and Referenced in this Study.

Figure 8.2.   Soil Phytolith Content Expressed as a Percentage of Soil Dry Weight.

Figure 8.3.   Density of Diagnostic Grass Short Cell Phytoliths (Rondel, Bilobate, and Crenate Morphotypes) among Soil Samples Analyzed in this Project.

Figure 8.4.   Box Plot Inter-site Comparison of: A – Wood Charcoal Density (g/l); B – Edible Nutshell Density (g/l); and C – Edible Seed Density (n/l).

Figure 8.5.   Top: Inter-site Comparison of Percentages of Edible Nuts (Using Number of Nutshell Fragments), Edible Seeds, and Edible Nut Plus Seeds. Bottom: Inter-site Comparison of Ratio of Edible Seeds to Edible Nuts (Using Number of Nutshell Fragments).

Figure 8.6.   Inter-site Comparison of Selected Taxon Percentages.

Figure 8.7.   Inter-site Comparison of Edible Seed Densities by Taxon for Grasses (Poaceae), Panicled bulrush (*Scirpus microcarpus*), Tarweed (*Madia* sp.), and Clover (*Trifolium* sp.).

Figure 8.8.   A: Two-way Hierarchical Cluster Analysis of Edible Seed Density by Site/Component. B: Two-way Hierarchical Cluster Analysis of Edible Seed Density by Site/Component.

Figure 8.9.   Simpson's Diversity Index ("Simpson's D") of Food Plants and
                  Potential Food Plants at Inland Sites.
Figure 8.10.   Inter-site Comparison of Percentages of Wood Charcoal Taxa.
Figure 8.11.   Map of 1935 Vegetation in the Vicinity of Año Nuevo Point.

Chapter 9:
Figure 9.1.   Map of Study Area and Sites Sampled.
Figure 9.2.   Graph Depicting Regression Formula for Predicting Total Mussel
                  Length from Umbo Thickness.
Figure 9.3.   CA-SCR-7 Column 1; 3775–2200 BCE; n total = 603.
Figure 9.4.   CA-SCR-7 Column 2; AMS Date Range: 3940–2420 BCE;
                  n total = 520.
Figure 9.5.   CA-SCR-7 Column 7; AMS Date Range: 4785-4555 BCE;
                  n total = 316.
Figure 9.6.   CA-SCR-14 Excavation Unit 2; AMS Range 1160–1920 CE;
                  n total = 696.
Figure 9.7.   CA-SMA-216 Area 1 and 2; AMS Range 1300–1640 CE;
                  n total = 796.
Figure 9.8.   Comparison of Sites.

Chapter 10:
Figure 10.1.   Schematic of Spot Sampling Strategy.
Figure 10.2.   Probability of Harvest by Month per Archaeological Specimen.
Figure 10.3.   Harvesting Activity Averages per Month.
Figure 10.4.   Probability of Harvesting Month per Archaeological Specimen with
                  Effect of Differential Growth Rate.
Figure 10.5.   Harvesting Activity per Month with Effect of
                  Differential Growth Rate.
Figure 10.6.   Harvesting Activity Averages per Month with Effect Differential
                  SST Location and Dates.
Figure 10.7.   Seasonality Reconstruction by Individual Specimen.
Figure 10.8.   Diachronic Harvesting Activity by Chronological Period.

Chapter 11:
Figure 11.1.   An Overview Map of the California Coast with an Inset Map
                  Showing the Sites Discussed.
Figure 11.2.   Shannon-Weiner Index for Diversity and Evenness by Site and
                  Recovery Method.
Figure 11.3.   Results of the χ^2 Test of Hook-and-Line (H & N) and Net
                  Based Fishing.
Figure 11.4.   Density (Class) NISP/l for >3.2 mm and >2 mm Recovery Methods at
                  CA-SCR-10, CA-SCR-14, and CA-SCR-15.

Chapter 12:

Figure 12.1.   NMDS Distribution of Rodents from Reid's Four Habitat Live-trapping Samples, Counted as the Number of Individuals, plus CA-SMA-18, CA-SMA-113, and CA-SCR-7 Archaeofaunas counted as NISP.

Figure 12.2.   CA-SMA-18 and CA-SMA-113: Ratios of Thermally Altered Bone by the Taxonomic Group for the Two Sites.

Chapter 13:

Figure 13.1.   Haplotype Network Showing the Relationship among Northern California Vole Haplotypes.

Figure 13.2.   Map of the Vole Populations from the Complete Dataset.

Chapter 14:

Figure 14.1.   Culturally Important Plant Tarweed (*Madia sativa*) Growing in Quiroste Valley Preserve.

Figure 14.2.   Amah Mutsun Land Trust Native Stewards Participating in a Prescribed Burn at Cascade Field, San Mateo County.

Figure 14.3.   Huckleberries Provided by AMLT Youth Camp Boxes.

List of Tables

Chapter 5:

Table 5.1.   Survey and Grid Block Specifications for Geophysical Survey.

Chapter 7:

Table 7.1.   CA-SCR-7 Raw Material.

Table 7.2.   CA-SCR-7 Lithic Class.

Table 7.3.   CA-SCR-10 Raw Material Type.

Table 7.4.   CA-SCR-10 Lithic Class.

Table 7.5.   CA-SCR-14 Raw Material Type.

Table 7.6.   CA-SCR-14 Lithic Class.

Table 7.7.   CA-SCR-15 Raw Material Type.

Table 7.8.   CA-SCR-15 Lithic Class.

Chapter 8:

Table 8.1.   Summary Botanical Data Categories by Site.

Table 8.2.   Results of Anthracological Analysis—Counts and
                  Percentages of Taxa.

Table 8.3.   Estimates of Percentage Phytolith Content in Soils Near
                  Archaeological Sites.

Table 8.4.   Estimates of Grass Short Cell Phytolith and Diatom Densities in Soils
                  Near Archaeological Sites.

Table 8.5.   Percentages of Edible Nuts at Each Site/Component,
                  Quantified by Count.

Table 8.6.   Pairwise Statistical Significance Results of Chi-square Test (p values)
                  on Edible Nutshell and Edible Seed Counts between Archaeological
                  Sites Analyzed.

Table 8.7.   Results of Tukey HSD Test on Simpson's Diversity Index of Food
                  Plants and Potential Food Plants from Inland Sites.

Chapter 9:

Table 9.1.   Table Outlining Measurable Expectations for Detecting
                  Harvesting Practices.

Table 9.2.   Table Outlining Most Abundant Shellfish Taxa by Site.

Table 9.3.   Average Sizes of Mussel from Column 1 at CA-SCR-7.

Table 9.4.   Average Sizes of Mussel from Column 2 at CA-SCR-7.

Table 9.5.   Average Sizes of Mussel from Column 7 at CA-SCR-7.

Table 9.6.   Average Sizes of Mussel from Excavation Unit 2 at CA-SCR-14.

Table 9.7.   Average Sizes of Mussel from Area 1 and 2 at CA-SMA-216.

Table 9.8.   Average Sizes of Mussels from all Three Sites.

Table 9.9.   Interpretations of Harvesting Practices from All Sites Based on Size,
                  Seasonality, and Ridealong presence.

Chapter 10:

Table 10.1.   Two Examples of Seasonal Classification by Temperature Value and Temperature Trend, Generalized from the Specific Methods Described in Jew et al. (2013) and Jones et al. (2008).

Table 10.2.   Estimated (Most Likely) Month of Harvest by Specimen, Listed with Site and Associated Radiometric Dating Information.

Chapter 11:

Table 11.1.   Summary of Fish Remains from Previous Excavations from the Santa Cruz Coast.

Table 11.2.   Site, Volume of Flotation Samples, and Dry-screen Samples from Santa Cruz Archaeological Sites.

Table 11.3.   Faunal Analysis Results with NISP by Site, Context, Screen Size, and Taxon Total Across Sites.

Table 11.4.   Results of Fish Analysis for CA-SCR-7 from the >2 mm Recovery Method with NISP by Context.

Table 11.5.   Results of Fish Analysis for CA-SCR-10 from the >2 mm Recovery Method with NISP by Context.

Table 11.6.   Results of Fish Analysis for CA-SCR-10 from the >3.2 mm Recovery Method with NISP.

Table 11.7.   Results of Fish Analysis for CA-SCR-14 from the >2 mm Recovery Method with NISP.

Table 11.8.   Results of Fish Analysis for CA-SCR-14 from the >3.2 mm Recovery Method with NISP.

Table 11.9.   Results of Fish Analysis for CA-SCR-15 from the >2 mm Recovery Method with NISP.

Table 11.10.   Results of Fish Analysis for CA-SCR-15 from the >3.2 Recovery Method with NISP.

Table 11.11.   Surfperch NISP by Site, their Relative Abundance, Rank, the NISP of All Other Fishes, and Fish Rank.

Table 11.12.   Rockfish NISP by Site, their Relative Abundance, Rank, the NISP of All Other Fishes, and Fish Rank.

Table 11.13.   Results of the $\chi2$ Test by Site and Capture Method.

Chapter 12

Table 12.1.   Ancient DNA Species Identifications from 24 Non-thermally Altered Samples from CA-SMA-13, Submitted to Peter Heintzmann (Formerly of the University of California, Santa Cruz, now of Arctic University Museum of Norway, Tromsø) in 2012.

Table 12.2.   Reid's Live-trapped Rodent Census for Four Habitat Zones in the Año Nuevo Area.

Table 12.3.a.   Identifiable Rodent Taxa from CA-SMA-113, CA-SMA-18, and, for Information only, CA-SCR-7, Including *Thomomys bottae*.

Table 12.3.b.   Identifiable Rodent Taxa from CA-SMA-113, CA-SMA-18, and, for Information Only, CA-SCR-7, excluding *Thomomys bottae*.

Table 12.4.   CA-SMA-18: By-species Ratios of Thermally Altered Specimens,
                   Subdivided into Elements of Skull, Axial, and Extremity
                   Body Segments.

Table 12.5.   CA-SMA-113: By-species Ratios of Thermally Altered Specimens,
                   Subdivided into Elements of Skull, Axial, and Extremity
                   Body Segments.

Table 12.6.   CA-SMA-18 and CA-SMA-113: Values of Pairwise Two-tailed
                   Fisher's Exact Test for Between-site, By-taxon Comparisons of
                   Thermal Alteration to Bone.

Table 12.7.   CA-SMA-18: Values of Pairwise Two-tailed Fisher's Exact test
                   for Within Site, Taxon-to-Taxon Comparisons of Thermal
                   Alteration to Bone.

Table 12.8.   CA-SMA-113: Values of Pairwise Two-tailed Fisher's Exact
                   test for Within Site, Taxon-to-Taxon Comparisons of Thermal
                   Alteration to Bone.

Table 12.9.   Comparison of the Number of Rodent Specimens (NSP) Recovered,
                   Liters of Site Sediment Processed, and Rodent Recovery Rates per
                   Liter for CA-SMA-18, CA-SMA-113, and Cotoni sites of CA-
                   SCR-7, CA-SCR-10, CA-SCR-14, CA-SCR-15.

Chapter 13:

Table 13.1.   Site Information, aDNA Analysis Results, Zooarchaeological Species
                   Identification, Element, and Site Age Derived from Calibrated
                   Radiocarbon Dates on Charred Botanicals.

Table 13.2.   Conventional and Calibrated AMS ^{14}C Dates on Vole Bone from the
                   Central California Coast.

Chapter 15:

Table 15.1.   Densities of Archaeological Materials from Surface and Subsurface
                   Contexts for the Four Santa Cruz Sites.

List of Appendices

Available for download at: https://escholarship.org/uc/item/3rb3t6d#supplemental

Appendix 4.1.   Archaeological Materials Recorded in Surface Collection Units (Tables A1–A4).

Appendix 6.1.   Chronological Data (AMS Dates) for CA-SCR-7, CA-SCR-10, CA-SCR-14, and CA-SCR-15.

Appendix 8.1.   Archaeobotanical, Anthracological, and Phytolith Research Methods in Detail.

Appendix 8.2.   Archaeobotanical Data Tables.

Appendix 8.3.   Results of Macrobotanical, Anthracological, and Phytolith Analysis by Archaeological Site.

Appendix 11.1.   Fish Remains Previously Reported for the Santa Cruz Coast with NISP and Relative Abundance in Parentheses.

Appendix 12.1.   Rodents and Site Formation: Species Liable to Incorporation by Several Processes.

Appendix 12.2.   Other Methods and Considerations.

Appendix 12.3.   Table 1. Shannon Diversity Indices, Richness, and Evenness Statistics and Average Population Size from Reid's Live-Trapping Samples and the Archaeofaunal Samples from CA-SMA-18, CA-SMA-113, and CA-SCR-7.

Appendix 13.1.   Table 1. The Results of Paired Radiocarbon Dated Vole and Charred Botanical Chronological Modeling, Including 95% Probability Ranges.

Appendix 13.2.   Figure 1. A Maximum Parsimony Tree Based on Partial and Complete Mitochondrial Cytochrome B Sequences. In Relation to Figure 13.1, this Figure Shows Relationships Among Individuals that Were Classified in the Network Analysis. All Unlabeled Tips Were Classified to Be Included in the Large Haplotype A Group.

CHAPTER 1

Introduction to Volume

KENT G. LIGHTFOOT, MICHAEL A. GRONE, AND GABRIEL M. SANCHEZ

This volume presents the results of a collaborative eco-archaeological project that examines evidence for Indigenous landscape and seascape stewardship practices over 7,000 years on the Central California coast. The goal of this work is to develop a better understanding of practices employed by local Tribes to enhance the diversity, productivity, and sustainability of culturally important plants and animals in Tribal lands and waters. The centerpiece of the project is the Amah Mutsun Tribal Band (AMTB) whose members are Indigenous to the area and descend from survivors of the historic Franciscan missions Santa Cruz and San Juan Bautista.

They are working on developing a better understanding of their ancestral stewardship activities that may serve as a historical baseline for revitalizing Indigenous practices on the Central California coast today.

Unfortunately, Amah Mutsun knowledge about their ancestral stewardship practices suffered greatly after two and half centuries of ruthless, successive entanglements with Spanish, Mexican, and American intruders who attempted to remove them from their homes, curtail them from tending their lands, and deny them access to crucial Indigenous foods, medicines, and crafting materials (Lopez 2013, 2022; Rizzo-Martinez 2022; Chapter 3, this volume). As discussed in more detail in the next chapter, the combination of fire suppression policies (particularly in the American period), free-range livestock grazing, industrialized timber exploitation, misguided conservation practices, and the cessation of Indigenous stewardship activities resulted in a radically altered environment characterized by endangered Native ecosystems, perilously high fuel loads, and an increasing number of catastrophic wildfires. The once open landscape dotted with fertile coastal prairies and other Native habitats maintained by Indigenous people disappeared across much of the Central California coast.

Under the leadership of Chairman Valentin Lopez, the AMTB has embarked on an ambitious path to heal Mother Earth after centuries of neglect and mistreatment: it is a path that will honor and improve the livelihood of all creatures and plants in Tribal lands and waters (Lopez 2013; Chapter 3, this volume). In following this path, Tribal members are documenting stewardship practices used by their ancestors to enhance the health and vitality of culturally important terrestrial and marine species. The AMTB is actively employing this knowledge about their traditional resource management and Indigenous stewardship practices to help restore Native habitats and revitalize waterways along the Central California coast today (Sigona et al. 2021: 208).

In pursuing their goal of healing the environment, the Tribe founded its own land trust—the Amah Mutsun Land Trust (AMLT)—dedicated to restoring traditional Indigenous knowledge and revitalizing cultural and natural resources within the ancestral territories of the Awaswas, Mutsun, and Chalon-speaking peoples (Figure 1.1). They created the Native Stewardship Corps (NSC) as a workforce of young adult Tribe members to carry out a range of natural resource enhancement projects (see Chapters 3 and 14, this volume). The Tribe employs knowledge from elders, tribal oral traditions, as well as pertinent ethnohistoric observations and ethnographic accounts to develop a pathway forward in bringing Indigenous stewardship practices back to ancestral lands and waters.

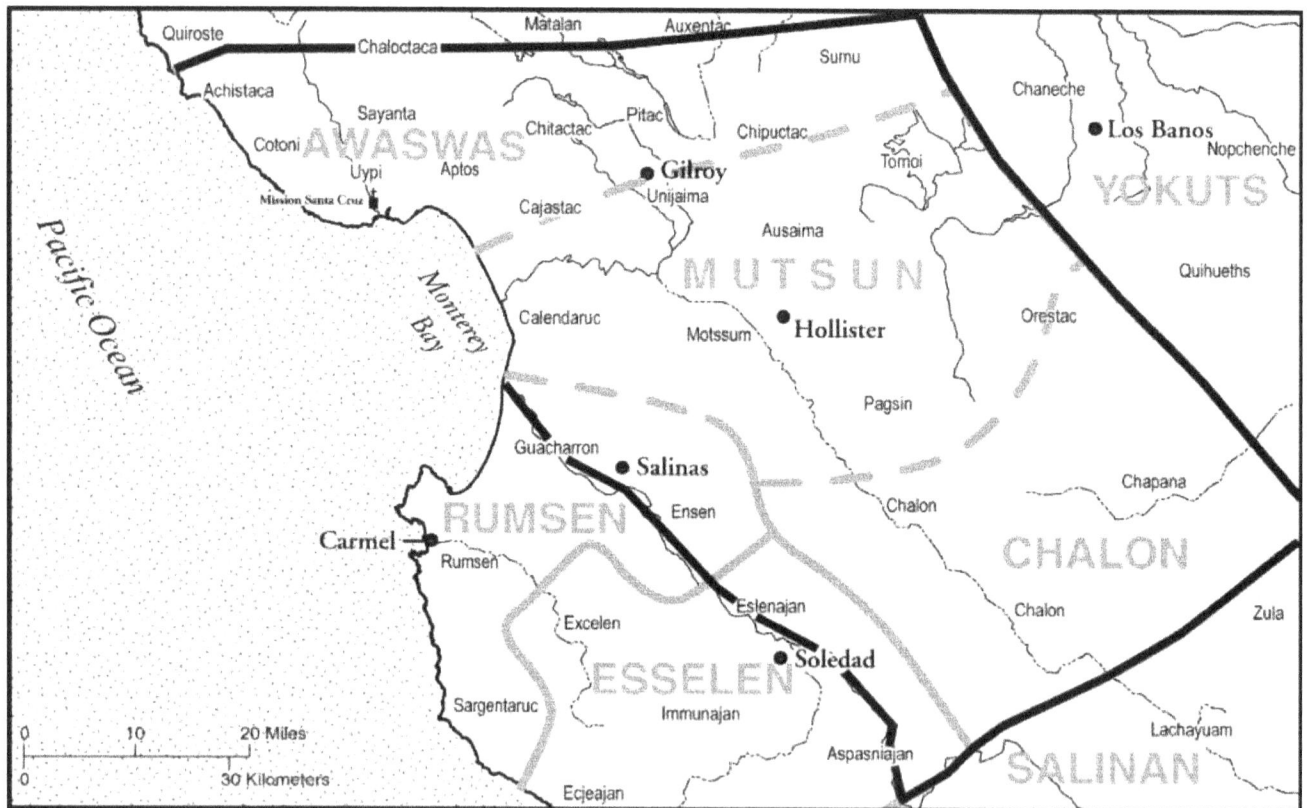

Figure 1.1. Area of Conservation and Stewardship of the Amah Mutsun Land Trust within the Ancestral Territory of the Awaswas, Mutsun, and Chalon-Speaking Peoples (AMLT Website 2020).

For example, they are working with the copious notes of the Smithsonian ethnographer John P. Harrington, who recorded a wealth of information about Amah Mutsun cultural practices and use of plants from Ascención Solórsano and other notable elders in the late 1920s and early 1930s (see Chapter 3, this volume; and Rizzo-Martinez 2022).

Tribal members also welcome collaboration with pertinent scholars who can help fill gaps in their knowledge about ancestral stewardship practices that have gone dormant after more than two centuries of colonialism. Our collaborative eco-archaeological project—the subject of this volume—is an outgrowth of this quest to document past stewardship practices. A team of scholars from the University of California, Berkeley (UC Berkeley), the University of California, Santa Cruz (UC Santa Cruz), and the California State Parks are working with the AMTB and AMLT to help shed light on these past practices. The first phase of this collaborative project, which focused on the Quiroste Valley Cultural Preserve in Año Nuevo State Park from 2007 to 2012, developed an eco-archaeological approach funded by the National Science Foundation (BCS-0912162) for examining past Indigenous landscape burning. The work demonstrated that local Native groups employed frequent cultural burns by at least 1000 CE to maintain highly productive grasslands interspersed with forest/savanna communities containing hazel and oak trees (Cuthrell 2013a, 2013b; Lightfoot et al. 2013a; and Chapter 2, this volume).

This volume reports on the second phase of this collaborative venture undertaken from 2016 to 2021, which was funded by a second National Science Foundation Grant (BCS-1523648). As outlined in more detail in Chapter 2, this work addresses four major goals concerning Indigenous landscape and seascape stewardship practices on the Central California coast. The first goal is to examine when Indigenous people may have first initiated sustained cultural burning and seascape stewardship practices in this region. The second goal is to examine how Indigenous stewardship practices may have changed over time in relation to climate change, sociopolitical and technological transformations, human population growth, and colonialism. The third goal is to evaluate the geographic scale and significance of Indigenous landscape and seascape stewardship practices on the Central California coast.

The fourth goal is to apply lessons from the past to better manage the lands and waters of the region. The AMLT is working closely with California State Parks to facilitate the restoration of Aboriginal habitats and terrestrial and marine species in state park holdings. The state agency is exploring new ways of managing public lands on the Central California coast that are rooted in the deep history of Tribal stewardship practices. As discussed in this volume and elsewhere (Lightfoot et al. 2021a; Sigona et al. 2021), some of the stewardship practices being re-implemented by the Amah Mutsun on state park lands have been informed from our collaborative eco-archaeological work, providing an example of applying information from the past to create collective benefits in our contemporary world (see Atalay et al. 2014: 7–8).

We chose the greater Santa Cruz coast as the study area for addressing the four goals of the project because this region contains a remarkable and diverse archaeological record spanning at least 7,000 years (Hylkema 1991, 2002; Hylkema and Cuthrell 2013). This volume presents the findings from our recent surface and subsurface investigation of four sites, CA-SCR-7, CA-SCR-10, CA-SCR-14, and CA-SCR-15, which are situated in the rural coastal area between the modern town/city of Davenport and Santa Cruz in central Santa Cruz County (Figure 1.2). Fieldwork also commenced at a fifth site, CA-SCR-38/123. However, due to significant bioturbation and sediment disturbances observed in our subsurface investigation, we were unable to detect any intact deposits with chronological integrity that we could employ in this current study of Indigenous stewardship practices. Consequently, we did not include the findings from CA-SCR-38/123 in this volume as they are presented elsewhere (Lightfoot et al. 2021b). While these four sites comprise the core of the investigation reported in this volume, the chapters herein also integrate prior results from other sites in Santa Cruz and San Mateo Counties where Indigenous landscape and seascape stewardship along the broader Santa Cruz coast were examined.

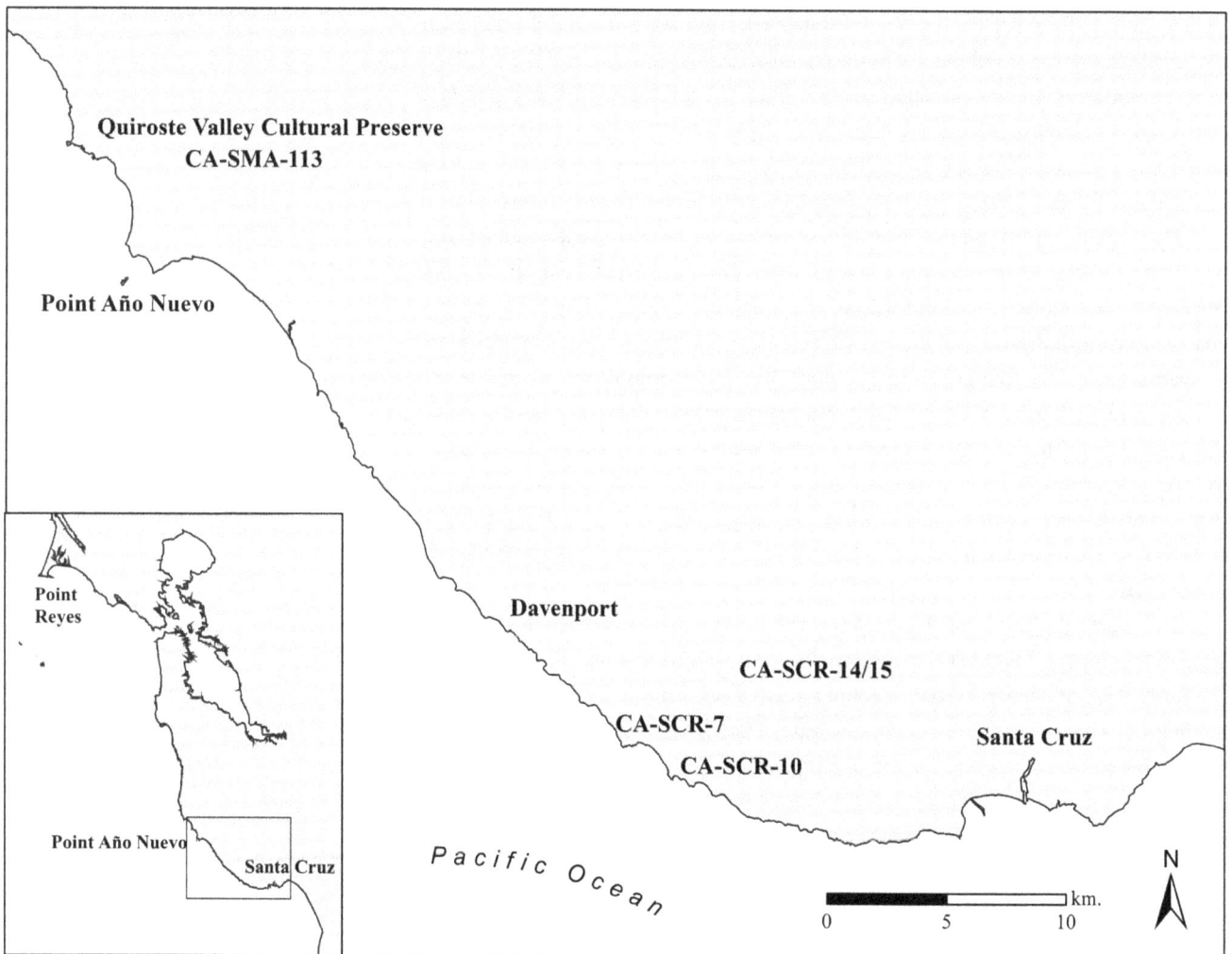

Figure 1.2. Map of Broader Study Area on the Santa Cruz Coast Showing Quiroste Valley Cultural Preserve and CA-SMA-113, as well as CA-SCR-7, CA-SCR-10, CA-SCR-14, and CA-SCR-15 (Map by Alec Apodaca).

Indigenous Archaeology?

One might ask, as a peer reviewer for this volume did, how our collaborative eco-archaeology approach relates to the contemporary practice of "Indigenous archaeology" and how it differs from traditional archaeology undertaken on the Central California coast. Indigenous archaeology refers to various forms of archaeology "done with, for, and by" Indigenous people; it is a field that has become an increasingly important component of North American archaeology since about 2000 (Nicholas and Andrews 1997: 3; see also Atalay 2006, 2008; Silliman 2008; Silliman and Ferguson 2010). Indigenous archaeology encompasses more than the work of archaeologists simply collaborating with Indigenous communities in the pursuit of similar goals. Rather, it involves integrating Indigenous values, worldviews, sensibilities, and social justice into the practice of archaeology and the interpretation of archaeological materials (see Atalay 2012: 39; Colwell-Chanthaphonh et al. 2010: 229; Laluk et al. 2022; Nicholas 2016: 10–12).

An important component of Indigenous knowledge that can inform archaeological investigations of human-land relationships and stewardship practices is Traditional Ecological Knowledge (TEK) (see discussion in Sigona et al. 2021; Chapter 14, this volume). TEK is defined in this volume as "a cumulative body of knowledge, practice, and belief, evolving by adaptive processes and handed down through generations of cultural transmissions, about the relationship of living beings (including humans) with one another and with their environment" (Berkes 2018: 18). TEK is place-based learning that is grounded on intimate observations of the world, generated over many centuries and guided by spiritual beliefs, the concept of reciprocity, and Tribal traditions (Berkes et al. 2000: 1251; Kimmerer 2011: 257–259; Mason et al. 2012: 187; Middleton 2015: 561). Regarding the stewardship of Tribal lands and waters, Indigenous knowledge is deeply contextual and derived from many generations of everyday, lived experiences and countless trials and errors about the appropriate ways to tend culturally important plants and animals in dynamic and ever-changing ecosystems (see Goode et al. 2022; Hankins 2021; Kimmerer 2011, 2013; Lake and Christianson 2019; Long et al. 2021; Smith et al. 2023).

There are aspects of Indigenous archaeology involved in our collaborative work with the Amah Mutsun. Team members from the AMTB and other California Tribes bring Indigenous values and knowledge to the project (e.g., Lopez 2013; Nelson 2020; Sigona et al. 2021), yet the majority of our research team is composed of Euro-American and, increasingly, Chicano/a researchers who are trained in academia in the method and theory of archaeology, anthropology, geography, ecology, and genetics. Consequently, we are uncomfortable in defining much of our work as truly "Indigenous." The senior author, for example, grew up in a white suburb of Santa Rosa, California, and received his initial training in the unabashedly scientific approach of processual archaeology in the 1970s when collaboration with Tribes and stakeholders rarely took place in North American archaeology (Lightfoot 2005). While he has now had the honor of collaborating with California Tribes for many years and learned much from Tribal partners, it seems somewhat disingenuous to claim he is undertaking Indigenous archaeology, which is fundamentally grounded in Indigenous knowledge, values, and beliefs handed down by Native peoples over many generations. As Liboiron (2021: 124) succinctly concluded after a lengthy discussion of similar issues concerning his research team, "Indigenous sciences are done by Indigenous peoples, full stop."

Following Two Paths

We recognize that Indigenous knowledge is a different way of knowing about the world than that based on Western science or Scientific Ecological Knowledge (SEK) (see Sigona et al. 2021; Chapter 14, this volume). We also recognize there are different perspectives about how research teams may integrate, or "braid" together, knowledge from both sources of information to facilitate the needs of Tribes (e.g., Atalay 2020; Lake et al. 2018). Given the diverse composition of our research team, rather than claiming that the work we do is fully grounded in Indigenous knowledge, values, and sensibilities, we employ an eco-archaeological approach that follows two separate but parallel paths that include pursuing the best of both TEK and SEK (Kimmerer 2011; Liboiron 2021: 127; Mason et al. 2012). Defined by Indigenous scholars as "Two-Eyed Seeing," this perspective allows us to employ knowledge from both TEK and SEK, which can provide a better understanding about Indigenous stewardship practices on the Central California coast than that afforded by either perspective alone (see Denny and Fanning 2016; Reid et al. 2021; Smith et al. 2023). Viewed in this manner, it is the collaborative relationship between Indigenous and non-Indigenous scholars that is so powerful—each can contribute much from different perspectives and vantage points to better address issues that concern Tribes today (see also Atalay 2020).

In keeping the Amah Mutsun as the centerpiece of the project, we are careful not to assimilate or extract Indigenous information into a broader SEK framework that can be harmful to our Indigenous partners. We view the AMTB and AMLT as the keepers of TEK for the Tribal territories they are working to revitalize in the ancestral lands of the Awaswas, Mutsun, and Chalon-speaking peoples. They control knowledge concerning traditional ecological practices and share it with us as appropriate. The eco-archaeological methods we employ, largely developed within an academic SEK framework, are used to support the Tribe in its recovery of information about dormant stewardship practices (Sigona et al. 2021: 209). We give this information back to our Tribal partners—they can then choose to incorporate it into their TEK as part of their re-implementation of Indigenous stewardship practices and the restoration of local environments after many decades of mistreatment under colonialism. This volume describes the goals, methods, and results of our collaborative eco-archaeological work undertaken to recover information about pertinent past Tribal practices that we were able to give back to the Amah Mutsun and its land trust.

Collaborative Eco-Archaeology

We define collaborative eco-archaeology as an approach that involves working with community partners in constructing a robust perspective about human relations with the environment over time by employing multiple ecological and archaeological data sets. For example, in the first phase of our project, we employed methods in zooarchaeology, archaeobotany, dendroecology, palynology, geomorphology, and plant population genetics derived largely from Western scientific knowledge to generate information about fire histories, fire regimes, faunal and floral populations, and vegetation conversions in Quiroste Valley over time (Lightfoot and Lopez 2013; Lightfoot et al. 2013a). Our approach employs the tenets of historical ecology that consider human agency at the landscape scale in examining how people engaged with local environments over multiple generations (Balée 1998; Balée and Erickson 2006; Crumley 1994). This approach recognizes from the outset that all human societies impact their local environments, that both agrarian and non-agrarian societies engage in the domestication of landscapes through various actions, including burning, transplanting, pruning, mound construction, and the like, and that human interactions with local environments may be beneficial or detrimental for the overall health and sustainability of biological communities when viewed in the longue durée (Balée 2006; Balée and Erickson 2006; Erickson 2006; Erickson and Balée 2006).

Our eco-archaeological program implements fine-grained methods designed specifically to recover artifacts, archaeobotanical remains, zooarchaeological specimens, sediment samples, and other relevant materials from archaeological contexts that provide insights about the cultural practices of past people and their relationships with local environments. Recovering the physical remains of plants and animals that were harvested and possibly tended by Native peoples is an important component of eco-archaeological investigations of past stewardship practices. Our eco-archaeological program also utilizes pertinent ecological data sets collected from non-archaeological contexts that offer insights about local soils, climate conditions, vegetation patterns, faunal populations, fire regimes, and water conditions (e.g., Klimaszewski-Patterson et al. 2018, 2021; Stephens and Fry 2005). Using radiocarbon chronologies, these different data sets can be integrated into regionally specific, historical baselines that represent a wealth of information about the temporal relationships of Native cultural practices, settlement systems, and the structure and composition of local environments. These eco-archaeological baselines provide important sources of information for Tribes and resource agencies in making decisions about revitalizing degraded environments, restoring plants and animals impacted by colonialism, and implementing Indigenous stewardship practices (see also Braje and Rick 2013; Braje et al. 2021; Hayashida 2005; Lightfoot et al. 2013b: 287–291; Lyman 2006; Rick and Lockwood 2012; Scharf 2014; Wolverton and Lyman 2012).

So how does our collaborative eco-archaeological program differ from traditional archaeology undertaken on the Central California coast? The basic SEK methods we use were developed and refined through many decades of archaeological and ecological research and are shared with many other scholars working in California and beyond. What makes our approach innovative is how we modified and enhanced these SEK methods specifically for doing collaborative work with Tribes. Interactions with Tribal partners have greatly influenced how we practice our eco-archaeological program.

We have created a low-impact archaeological methodology designed explicitly to minimize impacts to archaeological remains and to avoid disturbances to burials and other sacred Indigenous materials, while accentuating the recovery of useful information about Tribal histories and cultural resources. The methodology emphasizes close collaboration with Tribal partners with minimal destruction of the archaeological record. Researchers at UC Berkeley have been refining the methods of low-impact archaeology for the study of Indigenous sites in California over the last two decades (e.g., Byram et al. 2018; Byram and Sunseri 2021; Gonzalez et al. 2006; Gonzalez 2011, 2016; Lightfoot et al. 2001; Lightfoot 2005, 2005/2006, 2006a, 2006b, 2006c, 2008; Lightfoot and Gonzalez 2018a, 2018b; Lightfoot et al. 2021a; Nelson 2017, 2020, 2021; Parrish et al. 2000; Sanchez et al. 2021; Schneider 2010; Sunseri and Byram 2017).

Low-impact archaeology presents a fundamentally different approach for undertaking research on Indigenous sites than that employed by earlier generations of archaeologists in California. Previous archaeologists employed a series of high-impact exploratory field methods during their initial investigations of village sites to detect buried artifact concentrations, to locate features (e.g., house structures, hearths, earth ovens), and to determine the depth of archaeological deposits. These methods included broad-scale trenching (by hand or by backhoe), placing multiple test units across the site, or stripping large areal units by hand or machine. While California archaeologists still employ these methods today, as they can be effective for unearthing features and intact deposits, especially in buried or deeply stratified sites, they can come at a high cost. When used as initial search techniques, these high-impact methods often result in the destruction of significant components of archaeological sites before their adequate documentation, as demonstrated by Sunseri and Byram (2017). Unfortunately, the indiscriminate use of these high-impact search techniques over many decades led to the desecration of Indigenous burials and other sacred remains across the state without input from Tribal descendants.

Low-impact archeology, in contrast, employs a multi-phased field strategy that works to define the spatial patterning of artifacts, zooarchaeological and archaeobotanical remains, cultural features (e.g., house structures, hearths, underground ovens), discrete archaeological deposits, and other archaeological materials prior to any major subsurface disturbances. The successful implementation of this approach depends on collaborative partnerships with local Tribes and community members. At each phase of the multi-phased program, the members of the collaborative research team make decisions about the field strategies employed. The first phase involves the implementation of the lowest-impact methods designed explicitly to define site structures—here defined as the spatial distribution of diverse archaeological materials across site areas. In Chapters 4 and 5, we discuss the geophysical survey work and surface investigations involved in the first phases of study of sites on the Santa Cruz coast.

If the research team chooses to undertake additional phases of fieldwork with Tribal approval, then the results of the low-impact work inform subsequent field decisions that may involve subsurface testing. For example, based on our initial work, we were able to make enlightened decisions about where not to undertake subsurface investigations to protect and preserve potential burials or other sensitive remains on the Santa Cruz coast. We also generated plans about the precise placement of excavation units of specific sizes and shapes to investigate areas of the site of interest—such as potential house structures, hearths, pits, artifact concentrations, and so on. In Chapter 4, we discuss this process in more detail, stressing how all field decisions are undertaken in close consultation with Amah Mutsun partners, and describe the subsurface methods employed in the investigation of CA-SCR-7, CA-SCR-10, CA-SCR-14, and CA-SCR-15.

In undertaking subsurface testing programs as part of low-impact archaeology, we employ an intensive recovery methodology that carefully documents any of the archaeological deposits unearthed. This involves fine-screening (4 mm, 2 mm, 1 mm mesh), bulk soil processing, and the collection of archaeological sediment samples from which a broad suite of data classes can be recovered, including macro/micro artifacts, macro/micro faunal remains, macrobotanical remains, phytoliths, and radiocarbon samples (e.g., Lightfoot et al. 2021a, 2021b; see following chapters, this volume). As discussed by Reedy and Wohlgemuth (2016), California archaeologists have been slow to adopt these more intensive recovery methods.

Our collaborative work with the Amah Mutsun produced an incredible synergy resulting in considerable rethinking in how we undertake some basic archaeological practices. During our fieldwork on the Santa Cruz coast, the various members of our research team—undergraduate and graduate students, faculty, post-docs, AMLT stewards, Tribal elders, state park specialists, and assorted guests—resided in a field camp generously provided by California State Parks. The field camp involved tent living, communal dining with delicacies cooked by Indigenous chefs, lectures, workshops, talking circles, and nightly campfire gatherings. This provided an exceptional opportunity to talk about the work we were doing, the methods and goals of the project, the objectives and priorities of the Amah Mutsun, as well as many other topics.

The synergetic work reinforced to all team members the importance of respecting and honoring archaeological sites as places of ancestors. Tribal elders and stewards performed ceremonies, prayers, and smudging with the entire crew before any work took place at an ancestral place. These ceremonies highlighted the spiritual significance of these places and reified the need to work together to preserve and protect archaeological sites on the Santa Cruz coast.

Collaboration with the Amah Mutsun led to innovations in our low-impact archaeological program. As discussed in more detail in Chapter 4, systematic surface investigations involving the study of archaeological remains on the surface of sites can yield considerable information about past cultural practices, chronology, use of raw materials, and so on. The senior editor of this volume received training in this method from Fred Plog, one of its leading proponents, at Arizona State University when he was a graduate student in the 1970s. A series of collection units were laid out across sites and the materials systematically collected, provenienced, and taken back to the laboratory for analysis. Once we recorded and analyzed the materials in the laboratory, they were then permanently curated in state and federal repositories.

In the early 2000s, when the senior editor was co-directing archaeological research on Kashaya Pomo lands at Fort Ross State Historic Park with Otis Parrish, a Tribal elder and scholar, collaboration with members of the Kashaya Pomo made it very clear that they wanted to minimize the storage of archaeological remains in curation facilities. They referred to these facilities as "black holes" where materials went in and were never seen again. Our research team, comprised of some exceptional graduate students at the time, helped devise with Tribal members a "catch-and-release" methodology in which materials were collected, analyzed in the California Archaeology Laboratory at UC Berkeley, and then repatriated back to the ancestral places after analysis (see Gonzalez et al. 2006: 406–407; Gonzalez 2016: 543–545).

Collaborative interactions with the AMLT on the Santa Cruz coast led to further refinements of our "catch-and-release" methodology. The Amah Mutsun made it very clear that they wanted us to minimize not only the curation of materials from ancestral sites but also the removal of materials from these places. Consequently, we developed a field laboratory program for undertaking analysis of surface materials on site. As described in more detail in Chapter 4, this involved collecting materials systematically from the surface and training students and Tribal stewards how to identify, record, quantify, and weigh (using portable scales) archaeological materials in the field before they were carefully returned to the locations from which they were collected.

The synergetic work with the Amah Mutsun also resulted in modifications to methods involved in traditional archaeological surveys as conducted in California. Members of the AMLT Stewardship Corps and our archaeological staff jointly created the "integrated cultural resource survey" that records archaeological sites but also documents other tribally significant resources, including plants used for food, medicines, and crafting material, natural springs, caves, rock outcroppings, wildlife areas, and notable minerals (Sigona et al. 2021; Chapter 14, this volume). The Amah Mutsun and pertinent resource agencies can then employ information regarding culturally sensitive archaeological, botanical, and abiotic resources as components of "culturally significant landscapes" to make better-informed decisions about the placement of trails, parking lots, and development projects. This innovative approach also produces excellent information about remnant patches of culturally important plants that may be revived by the Amah Mutsun through various stewardship practices, as discussed in more detail in Chapter 14.

The synergistic work between Tribal stewards and archaeologists also created new perspectives about how to preserve and protect archaeological sites. Some sites that we recorded were associated with culturally significant plants, probably remnants of patches maintained by Indigenous ancestors who occupied these places. As discussed in Chapter 14, this provides an incentive for the AMTB to work with resource agencies to not only monitor these places for recent evidence of looting or damage due to erosion and other natural factors, but also to steward and revitalize the extant patches of plants so that they can be harvested for food, medicine, and basketry material. In cases where erosion is taking place on archaeological sites, Tribal members discussed the possibility of reintroducing culturally important plants that could not only stabilize the soil, but also engender harvestable patches of culturally important plants.

Organization of the Volume

The fifteen chapters in this volume detail the collaborative eco-archaeological approach we employed, the analysis of the diverse range of archaeological materials we uncovered, and our interpretations concerning Indigenous landscape and seascape stewardship practices in the Santa Cruz coast study area.

In Chapter 2, we outline the blueprint for the study of Indigenous stewardship practices that built upon previous work at Quiroste Valley in Año Nuevo State Park (Figure 1.1) and elsewhere. This chapter emphasizes why the study of Indigenous stewardship should matter to all Californians today. It also contextualizes the four major research goals that structured our investigation concerning the timing, diachronic development, scale, and contemporary relevance of Indigenous landscape and seascape stewardship on the Santa Cruz coast.

In Chapter 3, Lopez and colleagues present a Tribal perspective on collaborative eco-archaeology work. The authors articulate why the AMTB decided to participate in the project and what they hope to get out of it. Their goals concern the ecological revitalization of their Tribal territories, the restoration of Indigenous knowledge after centuries of brutal colonialism, and the healing of Mother Earth and all living things. They view collaborative eco-archaeology as a means of helping them "validate" various Indigenous stewardship practices that can be employed to reinvigorate their Tribal lands and waters.

Chapter 4 introduces CA-SCR-7, CA-SCR-10, CA-SCR-14, and CA-SCR-15 and details the field and laboratory methods employed in their study. In Chapter 5, Nelson describes the geophysical work undertaken at the four sites that delineated their spatial structures, including various subsurface features and archaeological deposits. The major findings from the ground penetrating radar inspections played a significant role in developing our plans for the subsurface investigation at each site.

Chapter 6 presents the chronological findings for CA-SCR-7, CA-SCR-10, CA-SCR-14, and CA-SCR-15 based on 60 radiocarbon assessments run by the Keck Carbon Cycle AMS Facility at the University of California, Irvine. In Chapter 7, Gonzalez and Field describe the results of the analysis of artifacts from the subsurface deposits excavated at the four sites. In Chapter 8, Cuthrell presents the findings from the archaeobotanical investigation of CA-SCR-7, CA-SCR-10, CA-SCR-14, and CA-SCR-14, along with five other sites situated in coastal and inland places west of the Santa Cruz Mountains. The study employs macrobotanical, anthracological (wood charcoal), and/or silica phytolith information to examine plant food consumption practices and to explore how Indigenous people used and managed plant resources through time with a particular focus on cultural burning systems.

Grone's study of invertebrate faunal remains in Chapter 9 examines the shellfish harvested by Indigenous people on the Santa Cruz coast spanning back 7,000 years. His detailed analysis of California mussel (*Mytilus californianus*) specimens explores different Indigenous harvesting methods (e.g., plucking and stripping) that may have been used to maintain productive mussel populations over time. In Chapter 10, Apodaca and colleagues present the findings from their isotopic analysis of a sample of California mussels from CA-SCR-7, CA-SCR-14, and CA-SMA-216. Their study provides estimates for the season(s) that Indigenous people harvested mussels and examines the timing of shellfish harvests on the Santa Cruz coast in the Middle and Late Holocene.

The archaeology of ancient and historic fisheries in Santa Cruz County is the topic of Chapter 11. Sanchez's investigation of fish remains from nine sites demonstrates the significant sampling biases that result in the study of fisheries when coarse-grained recovery methods (≥6.4 mm and 3.2 mm mesh sieves) are employed. He shows how fine-grained, eco-archaeological investigations, as highlighted in this volume, provide new insights about Indigenous fisheries, such as net-based mass capture fishing practices involving small schooling fish like Northern anchovies and herring.

In Chapter 12, Gifford-Gonzalez considers whether rodent relative taxonomic abundances or patterns of fire-related modifications from site archaeofauna may serve as an indirect proxy for vegetation and Indigenous landscape stewardship. In undertaking a detailed analysis of rodent remains from CA-SCR-7, CA-SCR-10, CA-SCR-14, CA-SCR-15, as well as other sites in San Mateo County, she demonstrates the challenges and potential equifinality problems of using rodent remains to make interpretations about past cultural burning and other landscape modifications. She also outlines recommendations for the types of sites and recovery methods that will optimize the recovery of evidence relevant to eco-archaeological research.

In Chapter 13, Fine and colleagues employ genetic data on contemporary and ancient California voles (*Microtus californicus*) to examine possible long-term Indigenous landscape management practices. They sequenced ancient vole DNA from the four Santa Cruz sites, along with other sites in Santa Cruz and San Mateo Counties, using a network analysis to investigate potential shifts in open coast prairie to closed chaparral and forest environments associated with Indigenous cultural burning, European colonization, and fire suppression. By integrating ancient samples into existing large-scale contemporary sampling of voles across their range, they examined population movement over millennia that indicate extensive grassland and coastal prairie existed on the Santa Cruz coast for at least 6,000 years.

In Chapter 14, Sigona and colleagues examine the relevance of collaborative eco-archaeology to current stewardship, cultural programming, and educational goals designed by and for the AMTB. Tribal members participating in the project are working to restore place-based ancestral practices, to create an innovative professional Tribal-driven archaeological resource management program, and to implement Amah Mutsun goals of cultural revitalization, land and seascape restoration, and healing. They detail how the results from eco-archaeological research are regularly integrated into cultural educational agendas and curriculum for youth camps.

The final chapter summarizes the results of the field and laboratory research at CA-SCR-7, CA-SCR-10, CA-SCR-14, and CA-SCR-15 and addresses the four primary goals for the project involving the timing, development, geographic scale, and contemporary relevance of Indigenous landscape and seascape stewardship practices on the greater Santa Cruz coast over the last 7,000 years. We evaluate the four goals using new information gleaned from the four sites, as well as by incorporating previous findings from other sites in Santa Cruz and San Mateo Counties. The chapter concludes with four major observations concerning our collaborative, fine-grained, low-impact, eco-archaeological study on the greater Santa Cruz coast.

References

Atalay, Sonya
2006 Indigenous Archaeology as Decolonizing Practice. *American Indian Quarterly* 30(2): 280–310.
2008 Multivocality and Indigenous Archaeologies. In *Evaluating Multiple Narratives: Beyond Nationalist, Colonialist, and Imperialist Archaeologies,* edited by Junko Habu, Claire Fawcett, and John M. Matsunaga, pp. 29–44. Springer, New York.
2012 *Community-Based Archaeology: Research With, By and For Indigenous and Local Communities.* University of California Press, Berkeley.
2020 Indigenous Science for a World in Crisis. *Public Archaeology* 19(1–4): 37–52.

Atalay, Sonya, Lee Rains Clauss, Randall McGuire, and John R. Welch
2014 Transforming Archaeology. In *Transforming Archaeology: Activist Practices and Prospects,* edited by Sonia Atalay, Lee Rains Clauss, Randall McGuire, and John R. Welch, pp. 7–28. Left Coast Press, Walnut Creek, California.

Balée, William
1998 Historical Ecology: Premises and Postulates. In *Advances in Historical Ecology,* edited by William Balée, pp. 13–29. Columbia University Press, New York.
2006 The Research Program of Historical Ecology. *Annual Review of Anthropology* 35: 75–98.

Balée, William, and Clark L. Erickson
2006 Time, Complexity, and Historical Ecology. In *Time and Complexity in Historical Ecology: Studies in the Neotropical Lowlands,* edited by William Balée and Clark L. Erickson, pp. 1–17. Columbia University Press, New York.

Berkes, Fikret
2018 *Sacred Ecology,* 4th ed. Routledge, London.

Berkes, Fikret, Johan Colding, and Carl Folke
2000 Rediscovery of Traditional Ecological Knowledge as Adaptive Management. *Ecological Applications* 10(5): 1251–1262.

Braje, Todd J., and Torben C. Rick
2013 From Forest Fires to Fisheries Management: Anthropology, Conservation Biology, and Historical Ecology. *Evolutionary Anthropology* 22: 303–311.

Braje, Todd J., Jon M. Erlandson, and Torben C. Rick
2021 *Islands Through Time: A Human and Ecological History of California's Northern Channel Islands.* Rowman and Littlefield, Lanham, Maryland.

Byram, Scott, and Jun Sunseri
2021   Principles and Practice of Investigating Buried Adobe Features with Ground-Penetrating Radar. *Remote Sensing* 12: 4980 https://doi.org/10.3390/rs13244980.

Byram, Scott, Kent G. Lightfoot, Peter Nelson, Jun Sunseri, Roberta A. Jewett, E. Breck Parkman, and Nicholas Tripcevich
2018   Geophysical Investigation of Mission San Francisco Solano, Sonoma, California. *Historical Archaeology* 52: 242–263.

Colwell-Chanthaphonh, Chip, T.J. Ferguson, Dorothy Lippert, Randall H. McGuire, George P. Nicholas, Joe E. Watkins, and Larry J. Zimmerman
2010   The Premise and Promise of Indigenous Archaeology *American Antiquity* 75(2): 228–238.

Crumley, Carole L.
1994   Historical Ecology: A Multidimensional Ecological Orientation. In *Historical Ecology: Cultural Knowledge and Changing Landscapes*, edited by Carole L. Crumley, pp. 1–16. School of American Research Press, Santa Fe, New Mexico.

Cuthrell, Rob Q.
2013a   Archaeobotanical Evidence for Indigenous Burning Practices and Foodways at CA-SMA-113. *California Archaeology* 5(2): 265–290.
2013b   An Eco-Archaeological Study of Late Holocene Indigenous Foodways and Landscape Management Practices at Quiroste Valley Cultural Preserve, San Mateo County, California. PhD Dissertation, Department of Anthropology, University of California, Berkeley.

Denny, Shelley K., and Lucia M. Fanning
2016   A Mi'kmaw Perspective on Advancing Salmon Governance in Nova Scotia, Canada: Setting the Stage for Collaborative Co-existence. *The International Indigenous Policy Journal* 7(3): 1–25.

Erickson, Clark L.
2006   The Domesticated Landscapes of the Bolivian Amazon. In *Time and Complexity in Historical Ecology: Studies in the Neotropical Lowlands*, edited by William Balée and Clark L. Erickson, pp. 235–278. Columbia University Press, New York.

Erickson, Clark L., and William Balée
2006   The Historical Ecology of a Complex Landscape in Bolivia. In *Time and Complexity in Historical Ecology: Studies in the Neotropical Lowlands*, edited by William Balée and Clark L. Erickson, pp. 187–233. Columbia University Press, New York.

Gonzalez, Sara L.

2011   Creating Trails from Tradition: The Kashaya Pomo Interpretive Trail at Fort Ross State Historic Park. PhD Dissertation, Department of Anthropology, University of California, Berkeley

2016   Indigenous Values and Methods in Archaeological Practice: Low-Impact Archaeology through the Kashaya Pomo Interpretive Trail Project. *American Antiquity* 81(3): 533–549.

Gonzalez, Sara L., Darren Modzelewski, Lee M. Panich, and Tsim D. Schneider

2006   Archaeology for the Seventh Generation. *American Indian Quarterly* 30(3/4): 388–415.

Goode, Ron W., Stephanie Farish Beard, and Christina Oraftik

2022   Putting Fire on the Land: The Indigenous People Spoke the Language of Ecology, and Understood the Connectedness and Relationship Between Land, Water, and Fire. *Journal of California and Great Basin Anthropology* 42(1): 85–95.

Hankins, Don

2021   Reading the Landscape for Fire. *Bay Nature* 21(1): 28–35.

Hayashida, Frances M.

2005   Archaeology, Ecological History, and Conservation. *Annual Review of Anthropology* 34: 43–65.

Hylkema, Mark G.

1991   Prehistoric Native American Adaptations along the Central California Coast of San Mateo and Santa Cruz Counties. MA Thesis, Department of Social Science, San Jose State University.

2002   Tidal Marsh, Oak Woodlands, and Cultural Florescence in the Southern San Francisco Bay Region. In *Catalysts to Complexity: The Late Holocene on the California Coast*, edited by Jon Erlandson and Terry Jones, pp. 233–262. Costen Institute of Archaeology, UCLA, Los Angeles, California.

Hylkema, 2013 Mark G., and Rob Q. Cuthrell

2013   An Archaeological and Historical View of Quiroste Tribal Genesis. *California Archaeology* 5(2): 225–245.

Kimmerer, Robin Wall

2011   Restoration and Reciprocity: The Contributions of Traditional Ecological Knowledge. In *Human Dimensions of Ecological Restoration Integrating Science, Nature, and Culture*, edited by Dave Egan, Evan E. Hjerpe, and Jesse Abrams, pp. 257–276. Island Press, Washington, DC.

2013   *Braiding Sweetgrass: Indigenous Wisdom, Scientific Knowledge, and the Teachings of Plants*. Milkweed Editions, Minneapolis, Minnesota.

Klimaszewski-Patterson, Anna, Peter J. Weisberg, Scott A. Mensing, and Robert M. Scheller
 2018 Using Paleolandscape Modeling to Investigate the Impact of Native American-Set Fires on Pre-Columbian Forests in the Southern Sierra Nevada, California, USA. *Annals of the American Association of Geographers* 108(6): 1635–1654.

Klimaszewski-Patterson, Anna, Christopher T. Morgan, and Scott A. Mensing
 2021 Identifying a Pre-Columbian Anthropocene in California. *Annals of the American Association of Geographers* 111(3): 784–794.

Lake, Frank K., and Amy C. Christianson
 2019 Indigenous Fire Stewardship. In *Encyclopedia of Wildfires and Wildland-Urban Interface (WUI) Fires*, edited by Samuel L. Manzello, pp. 1–9. Springer, New York.

Lake, Frank K., J. A. Parrotta, C.P. Giardian, I. Davidson-Hunt, and Y. Uprety
 2018 Integration of Traditional and Western Knowledge in Forest Landscape Restoration. In *Forest Landscape Restoration: Integrated Approaches to Support Effective Implementation*, edited by S. Mansourian and J. Parrotta, pp. 198–226. Routledge, New York.

Laluk, Nicholas C., Lindsay M. Montgomery, Rebecca Tsosie, Chistine McCleave, Rose Miron, Stephanie Russo Carroll, Joseph Aguilar, Ashleigh Big Wolf Thompson, Peter Nelson, Jun Sunseri, Isabel Trujillo, GeorgeAnn M DeAntoni, Greg Castro, and Tsim D. Schneider
 2022 Archaeology and Social Justice in Native America. *American Antiquity* 87(4): 659–682.

Liboiron, Max
 2021 *Pollution is Colonialism.* Duke University Press, Durham, North Carolina.

Lightfoot, Kent G.
 2005 Archaeology and Indians: Thawing an Icy Relationship. *News from Native California* 19(37–39).
 2005/2006 Collaboration: The Future of the Study of the Past. *News from Native California* 19(2): 28–31.
 2006a Experimenting with Low-Impact Field Methods. *News from Native California* 19(4): 16–19.
 2006b Rethinking Archaeological Field Methods. *News from Native California* 19(2): 21–24.
 2006c Sensing Archaeology Under Our Feet: The Potential for Geophysical Survey in Native California. *News from Native California* 20(1): 29–31, 37.
 2008 Collaborative Research Programs: Implications for the Practice of North American Archaeology. In *Collaborating at the Trowel's Edge: Teaching and Learning in Indigenous Archaeology*, edited by Stephen W. Silliman, pp. 211–227. University of Arizona Press, Tucson.

Lightfoot, Kent G., and Sara L. Gonzalez
   2018a   The Study of Sustained Colonialism: An Example from the Kashaya Pomo
   Homeland in Northern California. *American Antiquity* 83(3): 427–443.
   2018b   *Metini Village: An Archaeological Study of Sustained Colonialism. Volume 3,
   The Archaeology and Ethnohistory of Fort Ross, California.* Contributions of the
   University of California, Archaeological Facility Archaeological Research Facility,
   No. 69. Berkeley, California.

Lightfoot, Kent G., and Valentin Lopez
   2013   The Study of Indigenous Management Practices in California: An
   Introduction. *California Archaeology* 5(2): 209–219.

Lightfoot, Kent G., Otis Parrish, Roberta A. Jewett, E. Breck Parkman, and
Daniel F. Murley
   2001   The Metini Village Project: Collaborative Research in the Fort Ross State
   Historic Park. *Society for California Archaeology Newsletter* 35(2): 1, 23–26.

Lightfoot, Kent G., Rob Q. Cuthrell, Chuck J. Striplen, and Mark G. Hylkema
   2013a   Anthropogenic Burning on the Central California Coast in Late Holocene
   and Early Historical Times: Findings, Implications, and Future Directions.
   American Antiquity 78(2): 285–301.

Lightfoot, Kent G., Rob Q. Cuthrell, Cristie M. Boone, Roger Byrne, Andrea
B. Chavez, Laurel Collins, Alicia Cowart, Rand R. Evett, Fine V. A. Paul, Diane
Gifford-Gonzalez, Mark G. Hylkema, Valentin Lopez, Tracy M. Misiewicz, and
Rachel E. B. Reid
   2013b   Rethinking the Study of Landscape Management Practices Among
   Hunter-Gatherers in North America. *California Archaeology* 5(2): 371–390.

Lightfoot, Kent G., Rob Q. Cuthrell, Mark G. Hylkema, Valentin Lopez, Diane
Gifford-Gonzalez, Roberta A. Jewett, Michael A. Grone, Gabriel M. Sanchez, Peter
A. Nelson, Alec J. Apodaca, Ariadna Gonzalez, Kathryn Field, Jordan F. Brown, Alexii
Sigona, Paul V. A. Fine
   2021a   The Eco-Archaeological Investigation of Indigenous Stewardship
   Practices on the Santa Cruz Coast. *Journal of California and Great Basin
   Anthropology* 41(2): 185–204.

Lightfoot, Kent G., Rob Q. Cuthrell, Mark G. Hylkema, Valentin Lopez, Diane
Gifford-Gonzalez, Roberta A. Jewett, Michael A. Grone, Gabriel M. Sanchez, Peter A.
Nelson, Alec J. Apodaca, Ariadna Gonzalez, Kathryn Field, and Alexii Sigona
   2021b   *The Study of Indigenous Landscape and Seascape Stewardship Practices on the
   Santa Cruz Coast: A Collaborative Eco-Archaeological Approach.* Report Prepared
   for California Department of Parks and Recreation, Santa Cruz District,
   Archaeological Research Facility, University of California, Berkeley.

Long, Jonathan W., Frank K. Lake, and Ron W. Goode
2021   The Importance of Indigenous Cultural Burning in Forested Regions of the Pacific West, USA. *Forest Ecology and Management* 500: https://doi.org/10.1016/j.foreco.2021.119597.

Lopez, Valentin
2013   The Amah Mutsun Band: Reflections on Collaborative Archaeology. *California Archaeology* 5(2): 221–223.
2022   Forward. In *We Are Not Animals: Indigenous Politics of Survival, Rebellion, and Reconstitution in Nineteenth Century California*, by Martin Rizzo-Martinez, pp. xiii–xv. University of Nebraska Press, Lincoln, Nebraska.

Lyman, Lee
2006   Paleozoology in the Service of Conservation Biology. *Evolutionary Anthropology* 15: 11–19.

Mason, Larry, Germaine White, Gary Morishima, Ernesto Alvarado, Louise Andrew, Fred Clark, Mike Sr. Durglo, Jim Durglo, John Eneas, Jim Erickson, Margaret Friedlander, Kathy Hamel, Colin Hardy, Tony Harwood, Faline Haven, Everett Isaac, Laurel James, Robert Kenning, Adrian Leighton, Pat Pierre, Carol Raish, Bodie Shaw, Steven Smallsalmon, Vernon Stearns, Howard Teasley, Matt Weingart, and Spus Wilder
2012   Listening and Learning from Traditional Knowledge and Western Science: A Dialogue on Contemporary Challenges of Forest Health and Wildfire. *Journal of Forestry* 110(4): 187–193.

Middleton, Beth Rose
2015   Jahát Jatítotòdom*: Toward an Indigenous Political Ecology. In *The International Handbook of Political Ecology*, edited by Raymond L. Bryant, pp. 561–576. Edward Elgar Publishing, Northampton, Massachusetts.

Nelson, Peter A.
2017   Indigenous Archaeology at Tolay Lake: Responsive Research and the Empowered Tribal Management of a Sacred Landscape. PhD Dissertation, Department of Anthropology, University of California, Berkeley.
2020   Refusing Settler Epistemologies and Maintaining an Indigenous Future for Tolay Lake, Sonoma County, California. *American Indian Quarterly* 44(2): 221–242.
2021   The Role of GPR in Community-Driven Compliance Archaeology with Tribal and Non-tribal Communities in Central California. *Advances in Archaeological Practice* 9(3): 215–225.

Nicholas, George P.
2016   Introduction In *Being and Becoming Indigenous Archaeologists*, edited by George P. Nicholas and Claire Smith, pp. 9–18. Taylor & Francis Group, Milton Park, Oxfordshire.

Nicholas, George P., and Thomas D. Andrews
1997   Indigenous Archaeology in the Post-Modem World. In *At a Crossroads: Archaeology and First Peoples in Canada*, edited by George P. Nicholas and Thomas D. Andrews, pp. 1–18. Archaeology Press, Simon Fraser University, Burnaby.

Parrish, Otis, Daniel Murley, Roberta Jewett, and Kent Lightfoot
2000   The Science of Archaeology and the Response from Within Native California: The Archaeology and Ethnohistory of Metini Village in the Fort Ross State Historic Park. *Proceedings of the Society for California Archaeology* 13: 84–87.

Reddy, Seetha N., and Eric Wohlgemuth
2016   Introduction: Plant Use by Complex Hunter-Gatherers: Paleoethnobotanical Studies in California. *Journal of California and Great Basin Anthropology* 36(1): 1–4.

Reid, Andrea, Laruen E. Eckert, John-Francis Lane, Nathan Younr, Scott G. Hinch, Chris T. Darimont, Steven J. Cooke, Natalie C. Ban, and Albert Marshall
2021   "Two-Eyed Seeing": An Indigenous Framework to Transform Fisheries Research and Management. *Fish and Fisheries* 22: 243–261.

Rick, Torben C., and Bowman Lockwood
2012   Integrating Paleobiology, Archaeology, and History to Inform Biological Conservation. *Conservation Biology* 27(1): 45–54.

Rizzo-Martinez, Martin
2022   *We Are Not Animals: Indigenous Politics of Survival, Rebellion, and Reconstitution in Nineteenth Century California.* University of Nebraska Press, Lincoln, Nebraska.

Sanchez, Gabriel M., Michael A. Grone, Alec J. Apodaca, Scott Byram, Valentin Lopez, and Roberta A. Jewett
2021   Sensing the Past: Perspectives on Collaborative Archaeology and Ground Penetrating Radar Techniques from Central California. *Remote Sensing* 13 (285 https://doi.org/10.3390/rs13020285).

Scharf, Elizabeth A.
2014   Deep Time: The Emerging Role of Archaeology in Landscape Ecology. *Landscape Ecology* 29: 563–569.

Schneider, Tsim D.

2010   Placing Refuge: Shell Mounds and the Archaeology of Colonial Encounters in the San Francisco Bay Area, California. PhD Dissertation, Department of Anthropology, University of California, Berkeley.

Sigona, Alexii, Alec J. Apodaca, and Valentin Lopez

2021   Supporting Cultural Obligations: Using Eco-Archaeology to Inform Native Eco-Cultural Revitalization. *Journal of California and Great Basin Anthropology* 41(2): 207–229.

Silliman, Stephen W. (editor)

2008   *Collaborating at the Trowel's Edge: Teaching and Learning in Indigenous Archaeology*. Amerind Studies in Archaeology. University of Arizona Press, Tucson.

Silliman, Stephen W., and T. J. Ferguson

2010   Consultation and Collaboration with Descendant Communities. In *Voices in American Archaeology*, edited by Wendy Ashmore, Dorothy T. Lippert, and Barbara J. Mills, pp. 48–72. SAA Press, Washington, DC.

Smith, Carolyn, Sibyl Diver, and Ron Reed

2023   Advancing Indigenous Futures with Two-Eyed Seeing: Strategies for Restoration and Repair through Collaborative Research. *EPF: Philosophy, Theory, Models, Methods and Practice* 2(1–2): 121–143, https://doi.org/110.1177/26349825221142292.

Stephens, Scott L., and Danny L. Fry

2005   Fire History in Coast Redwood Stands in the Northeastern Santa Cruz Mountains, California. *Fire Ecology* 1(1): 2–19.

Sunseri, Jun U., and Scott Byram

2017   Site Interiography and Geophysical Scanning: Interpreting the Texture and Form of Archaeological Deposits with Ground-Penetrating Radar. *Journal of Archaeological Method and Theory* 24: 1400–1424.

Wolverton, Steve, and Lee R. Lyman

2012   *Conservation Biology and Applied Zooarchaeology*. University of Arizona Press, Tucson.

CHAPTER 2

Collaborative Eco-Archaeological Investigations of Landscape and Seascape Stewardship Practices on the Central California Coast

KENT G. LIGHTFOOT, VALENTIN LOPEZ, MARK G. HYLKEMA, ROB Q. CUTHRELL, MICHAEL A. GRONE, GABRIEL M. SANCHEZ, PETER A. NELSON, ROBERTA A. JEWETT, DIANE GIFFORD-GONZALEZ, PAUL V. A. FINE, ALEC J. APODACA, AND ALEXII SIGONA

This chapter outlines the research design we employed in the study of Indigenous landscape and seascape stewardship practices on the Central California coast undertaken by our collaborative research team. The study of Indigenous stewardship practices should matter to all Californians who are increasingly contending with catastrophic wildfires, deteriorating wildland environments, fishery collapses, and recurrent invasions of intrusive species. Indigenous communities learned to steward culturally important plants and animals in their territories over many centuries of intimate interactions with their environments. These cultural practices provide a deep pool of Traditional Ecological Knowledge (TEK) for rethinking how we manage our contemporary wildlands and seashores, such as enhancing the richness and diversity of native species, improving the overall health of biological communities, and minimizing the risks of catastrophic fires (Anderson 2018: 392–395; Goode et al. 2022; Hankins 2021; Long et al. 2020a; Long et al. 2020b).

Eco-archaeological research undertaken in partnership with local Tribes and resource agencies provides a unique perspective about how Indigenous people interacted with local environments over time. In recovering the physical evidence of plants, animals, and artifacts from the archaeological and sedimentary records, as discussed in Chapter 1, eco-archaeology complements other important sources of information (e.g., Tribal oral traditions/histories, ethnohistoric observations, ethnographic studies) for broadening our understanding of Indigenous practices employed in local regions and for giving back information about TEK to Tribal partners. Specifically, collaborative eco-archaeological research can contribute to the construction of regionally specific, historical baselines of past ecosystems and cultural practices that Tribes and resource agencies can employ in making decisions about the restoration of degraded landscapes and seascapes, for contemplating what plants and animals once flourished in Tribal territories, and for revitalizing Indigenous stewardship practices severely impacted by colonialism.

This chapter begins with a brief introduction about Indigenous stewardship practices in California and the devastating impacts that colonialism had on the Tribal peoples and the environments of California. We examine the many challenges that the Amah Mutsun Tribal Band (AMTB) faced in their entanglements with Spanish, Mexican, and American colonists. The chapter then discusses how the Amah Mutsun and other California Indian communities have emerged from this dark age as reinvigorated polities who are initiating Tribal revitalization programs that foster Native language retention, traditional crafts, wellness, spirituality, sovereignty, and environmental restoration. We outline how our eco-archaeological program is supporting the AMTB in recovering traditional ecological information about dormant stewardship practices that may be used to reinvigorate ancestral lands and waters degraded after years of settler colonialism. The final section presents background information on our collaborative eco-archaeological program. Here, we discuss our previous research findings at Quiroste Valley Cultural Preserve in Año Nuevo State Park and introduce the research design for undertaking fieldwork at CA-SCR-7, CA-SCR-10, CA-SCR-14, and CA-SCR-15 on the Santa Cruz coast. We conclude by discussing four major goals concerning the timing, development, geographic scale, and contemporary relevance of Indigenous landscape and seascape stewardship that directed our research.

Indigenous Stewardship Practices in California

Much has been written about the cultural practices that Native Californians employ in stewarding their lands and waters to enhance the productivity, diversity, and availability of culturally important plants and animals in their traditional territories (Anderson 2005; Blackburn and Anderson 1993; Lewis 1973; Lightfoot and Parrish 2009; Stewart 2002). Our purpose here is to highlight three observations gleaned from this extensive literature that structured our research program.

First, California is characterized by incredible environmental diversity that varies with topographic relief, the maritime conditions of the Pacific Ocean, and latitude (Lightfoot and Parrish 2009: 50–70). Tribal groups separated by relatively short distances often interacted with distinctive microclimates and habitats that provided opportunities for stewarding a diverse array of plant communities and animal species. It is important to recognize that both cultural and environmental differences in both time and space would have shaped TEK and the specific kinds of stewardship practices employed by Tribal groups across California. We advocate for eco-archaeological studies of Indigenous stewardship practices that take a long-term diachronic approach, are regionally focused, and that partner with local Tribes.

Second, extensive literature documents a diverse range of stewardship practices that Indigenous communities employed in the construction of productive landscapes, including tillage, pruning, seed broadcasting, weeding, irrigation, and cultural burning (e.g., Anderson 2005: 135; Blackburn and Anderson 1993; Lewis 1973; Stewart 2002). One of their primary stewardship tools was the use of "good" fires for clearing undergrowth, controlling insect infestations, promoting hunting, and encouraging the growth and prosperity of plants and animals used for foods, raw materials, and medicines (Anderson 2018; Lewis 1985). The specific characteristics of Tribal cultural burning varied across California, given the local ecology, the plants and animals being tended, and the frequency and timing of lightning strikes, which also contributed to local fire regimes. Yet most Indigenous communities appear to have instituted fire regimes characterized by frequent, relatively small, low-intensity surface burns (Anderson 2018; Goode et al. 2022; Hankins 2021; Lake and Christianson 2019; Lewis 1973; Lightfoot and Parrish 2009: 97–100; Long et al. 2021).

The ignition of relatively small patches of land, where fire coverage could be constrained by landscape features, such as streams, ridges, rock outcrops, and past cultural burns, would have created a rich mosaic of habitats (Hallam 1985; Keeley 2002; Lewis 1973). The cultural burning of small patches in a rotational pattern across Tribal territories would have produced a patchwork of vegetation stands at different stages of succession. By using fire to control vegetation successions over time, fire tenders could enhance the productivity and availability of grasslands, tubers, berries, nuts, and other important resources (Lightfoot and Parrish 2009: 100–102). Recent fires also produced excellent forage in successive years that attracted and supported a greater quantity of deer, antelope, tule elk, rabbits, and other favored game (Bean and Lawton 1976: 39; Lewis 1973; Keeley 2002; Shaffer and Laudenslayer 2006).

The consequences of Indigenous stewards burning small patches of land in a multiyear rotation across Tribal territories are threefold. First, it increased the productivity and sustainability of key economic resources used by people for foods, medicines, and raw materials (e.g., clothing, baskets, houses, dance regalia). By controlling the timing between cultural burns, Indigenous stewards maintained some control over the successional pattern of vegetation in their territories. They could choose to burn specific patches at fire return intervals that would sustain productive grasslands, berry patches, and oak woodlands that might otherwise have been engulfed by a succession of other plants, such as shrubs and conifers.

Second, the creation of a patchwork mosaic of vegetation communities at different stages of succession diversified the availability of a broad mix of resources across Tribal lands. This practice provided flexibility and various choices for Native peoples who could harvest alternative crops if one or more other resources failed, such as acorn harvests that can vary greatly in yields from year to year (Baumhoff 1963; Koenig et al. 1994; McCarthy 1993). It would also prove beneficial for buffering periods of environmental perturbations (see Lightfoot and Parrish 2009: 124–140). Finally, in creating a patchwork mosaic of diverse biological communities via cultural burning, Indigenous communities would have reduced fuel loads and constructed sizeable fuel breaks after recent burns. If this patchwork mosaic was maintained over a series of years, then we believe it may have served to minimize the size and number of catastrophic firestorms that swept across Tribal territories (see Goode et al. 2022: 89; Lake and Christianson 2019: 2; Long et al. 2021: 10).

Third, a growing literature also documents how Indigenous people stewarded seascapes by initiating cultural practices that nurtured and supported vertebrate and invertebrate fisheries among other creatures of the sea. In the Pacific Northwest, they fabricated engineered coastal landscapes through the creation of clam beds, stream scraping, holding ponds, fish weirs, and the use of shell middens as construction materials (Caldwell et al. 2012; Cannon and Burchell 2009; Grier 2014; Groesbeck et al. 2014; Lepofsky and Caldwell 2013; Lepofsky et al. 2015). While Indigenous seascape stewardship practices are not as well documented in California, there are ethnographic observations of clam gardens in Tomales Bay (Baker 1992; Grone 2020). There are also archaeological findings that some Native peoples harvested large quantities of shellfish and other coastal resources in ways that sustained these populations over thousands of years by removing predators, shifting village locations, "fishing up the food web," and employing plucking and stripping methods of gathering shellfish (Braje et al. 2009; Erlandson et al. 2008; Erlandson et al. 2009; Whitaker 2008). Recent eco-archaeological fieldwork in Central California suggests considerable promise in the study of shoreline practices affecting intertidal and wetland resources that may have involved the stewardship of fisheries and shellfish populations, as will be discussed later in this volume.

The Consequences of Colonialism

Colonialism altered the close-knit relationship that many Indigenous people in California maintained with their local environments. The intrusion of Spanish, Russian, Mexican, and later American settlers into Indigenous lands touched off a terrible dark age for Native communities who experienced a devastating genocide, chilling anti-Indian policies, blatant discrimination, pervasive land appropriations, and loss of homelands (Castillo 1978; Heizer and Almquist 1971; Hurtado 1988; Lightfoot 2005: 210–233; Lindsay 2012; Madley 2016). Settler colonists enacted policies that inhibited Indigenous people from implementing traditional landscape and seascape stewardship practices. These prohibitions began with the arrival of the first settlers, who made a concerted effort to stop Indigenous cultural burning. Early colonial settlements, including the huge mission complexes established by the Franciscan missionaries and the Russian settlement of Colony Ross in northern California, were extensive agrarian enterprises that involved crop production and thousands of free-range livestock (Lightfoot 2005). The last thing these colonists wanted were Natives in the hinterland setting fires that could devastate herds and fields of wheat, barley, and other crops. Consequently, in 1793, the governor of Spanish Alta California, José Joaquín de Arrillaga, issued a proclamation that prohibited cultural burning by both "Christian and Gentile Indians," an act that could result in "severe punishment" for those who continued to tend the land (Timbrook et al. 1993: 129–133).

The founding of the early Spanish, Mexican, and Russian agrarian settlements along the coasts of Southern and Central California not only prohibited Indigenous burning but also hindered the ability of Native communities to tend landscapes and to practice traditional subsistence pursuits on land that had been besieged with free-range livestock, intrusive weeds, and hostile settlers (Anderson 2005: 76–77; Crosby 2004: 152–154; Dartt-Newton and Erlandson 2006; Lightfoot 2005: 86–87; Rizzo-Martinez 2022: 55–58). Furthermore, the displacement of Indigenous settlements from productive coastal places by colonial enterprises began the long-term process of limiting Native peoples' access to littoral resources and seascapes, which continues to this day.

Indigenous relationships with their traditional landscapes and seascapes degenerated even further during the American period (post-1846) when thousands of settler colonists flooded into the state, unleashing a plethora of atrocities that fueled the genocide, seizures of Native lands, displacement of Tribes, the enactment of flagrant discrimination policies, and wholesale environmental degradation (see Anderson 2005: 62–121; Castillo 1978; Lindsay 2012; Madley 2016). As nicely summarized by Anderson, the combination of commercial fishing, logging, mining, ranching, farming, and urbanism, particularly in the American period,

> caused major declines in dozens of bird and mammal populations, denuded entire landscapes, caused major changes in the state's hydrological processes, vastly accelerated rates of erosion, destroyed countless acres of productive wildlife and plant habitats, radically altered much of the state's vegetation, and decimated Indian populations and destroyed their cultures. (Anderson 2005: 120)

A pertinent example of how American settler colonialism negatively affected California's environment is fire suppression. The establishment of large-scale forest reserves in late 1800s and early 1900s and the founding of the United States Forest Service in 1905 led to the creation of an explicit fire suppression program, which had major implications for federal and state lands (Stephens and Sugihara 2018: 400). Unlike the previous Spanish, Russian, and Mexican colonial regimes that forbade Tribes from practicing cultural burning, the American policy not only prohibited Indigenous fire stewardship but also attempted to put out any fires ignited by humans or nature (e.g., lightning). Essentially, the goal was to remove wildfires from the California ecosystem. This led to the 10:00 am policy, a concerted effort to put out any wildfire by 10:00 am the next morning, employing a large labor force of firefighters, improved access to wildlands, better documentation of fire occurrences, and innovations in firefighting equipment (Stephens and Sugihara 2018: 401). More than a century later, the upshot of these fire suppression activities has been the emergence of a radically transformed environment involving a shift from fire-adapted to fire-averse plants, many of which are foreign intruders. Once kept open by cultural burning, productive coastal prairies have been engulfed by scrublands and conifers, magnificent oak woodlands have become entangled with undergrowth, and open forestlands have become dense, overcrowded masses with less species diversity (Cuthrell 2013a: 266–269; Johnson 2014; Stephens et al. 2015).

We emphasize that the complex history of colonialism led to distinctive challenges and consequences for the Tribes of California today. Colonialism did not take place as a unified process across the state but varied greatly from Tribe to Tribe depending on different chronologies of encounters and colonial succession patterns. California Tribes have had to contend with distinctive colonial enterprises that often invaded their lands for very different reasons (Lightfoot and Gonzalez 2018; Lightfoot et al. 2021). In some areas of California, such as the Northwest Coast and Sierra Nevada Mountains, sustained colonial engagements with Tribes took place in the mid-1800s with the Gold Rush, a watershed moment that unleashed a deluge of foreigners into Tribal lands, resulting in massacres, land displacements, and other atrocities. In other areas, particularly coastal Central and Southern California, Tribes have grappled with sustained settler colonialism for almost 250 years since the establishment of the Franciscan missions and Spanish presidios and pueblos in the late 1700s. These Tribes had to struggle not only with many years of colonial repression, but also with a succession of different colonial agents from Spain, Russia, Mexico, and the United States who enacted divergent policies and practices in dealing with Indigenous communities (Lightfoot 2005).

The Amah Mutsun Tribal Band

The AMTB exemplify the significant challenges that Tribal people faced with the colonization of their lands on the Central California coast. Rizzo-Martinez's (2022) superb study details Indigenous entanglements with Spanish and Mexican colonists with the founding and occupation of Mission Santa Cruz and Villa de Branciforte beginning in the late 1700s and continuing through the 1830s. Mission Santa Cruz became home to hundreds of people from at least 27 Tribal polities who spoke Awaswas, Mutsun, and Yokuts languages from the greater Santa Cruz region and the northern San Joaquin area (Rizzo-Martinez 2022: Tables 6 and 22). Despite their facing many decades of discrimination, attempted subjugation and enculturation, deadly diseases, sexual abuses, corporal punishments, land appropriations, malnutrition, and armed violence, they were able to sustain their Indigenous cultures and languages. Indigenous survival took place in spite of the abysmal mortality of Native family members, particularly women and children, through various actions involving rebellions, assassinations, fugitivism, mobility, and most importantly, maintaining viable Indigenous communities (see Rizzo-Martinez 2022: 25–107).

With the secularization of Mission Santa Cruz from 1834 to 1840, survivors of the mission obtained small parcels of land where they built Indigenous communities, which offered sanctuaries for gatherings, feasts, dances, sweathouse ceremonies, and the retainment of Native languages and various cultural practices (Rizzo-Martinez 2022: 179–187). During the mission and early post-mission years, local Natives maintained some of their stewardship practices that involved harvesting and tending to both terrestrial and maritime resources. Archaeological investigations of mission sites—including Mission Santa Cruz—have recovered fish, shellfish, terrestrial game, and various seeds and nuts in mission Indian housing that they obtained either via trade with outlying Native groups or through their own acquisition from approved mission furloughs or clandestine visits to their homelands (Allen 1998: 55–61; Lightfoot 2005: 97–99; Panich and Schneider 2014; Schneider 2021: 65–70). While free-range livestock—including thousands of cattle, horses, and pigs, and other intrusive species—took their toll on the local environment (see Rizzo-Martinez 2022: 55–58, 70–76), it appears that at least some Tribally important plants and animals persevered in the Indigenous hinterland of Mission Santa Cruz.

The Santa Cruz Mountains appear to have served as a refuge for Native peoples fleeing the mission or attempting to keep a safe distance from it (Rizzo-Martinez and Hylkema 2021; see also Schneider 2015, 2021: 65–75) and provided a haven to tend and harvest resources beyond the prying eyes of the colonists. Spanish and Mexican settlers had little reason to venture into the redwood forests and oak woodland, as their economic livelihood focused on the agrarian fields and pasturelands in the coastal terraces and foothills (Rizzo-Martinez and Hylkema 2021). While Indigenous laborers procured timber for various colonial endeavors, the colonists lacked the skill and technology to mill lumber on a large scale and relied heavily on adobe as their preferred building material (Meniketti 2016: 38). We believe the sparse number of colonists, coupled with their infrequent use of some stretches of the coast and uplands of the Santa Cruz Mountains, allowed Indigenous people to continue to undertake landscape and seascape stewardship practices in secret through the early 1840s, despite the prohibitions of the Spanish and Mexican colonial administrations.

The situation changed dramatically in the American period when it became almost impossible for Indigenous people along the Santa Cruz coast and mountains to implement any sustained landscape and seascape stewardship practices. The reasons are threefold.

First, beginning with California statehood in 1850, a substantial number of non-Indigenous people thrust their way into the Santa Cruz vicinity, while the population of Native peoples declined rapidly due to high mortality rates and their movement elsewhere. Rizzo-Martinez (2022: 204, Table 27 and 29) shows that while Indians comprised a significant percentage (27%) of the area's population toward the end of the Mexican period (1845), their numbers dropped precipitously over time: from 1852 with 9% Indigenous (out of a total population of 1,219), to 1860 with 4.4% Indigenous (out of a total of population of 4,944), to 1880 with 1% Indigenous (out of a total population of 12,802), to 1900 with .03% Indigenous (out of a total population of 21,512).

The onslaught of settler colonists in the American period pushed Indigenous people to the margins of society in rapidly growing coastal cities such as Santa Cruz (Rizzo-Martinez 2022: 230). By the 1880s, settlers had driven Indigenous people out of their homes and land parcels, forcing them to find housing elsewhere. This process of relocation and resettlement involved some Native peoples becoming laborers in settler homes and businesses in town or serving as workers on nearby ranches. Others blended into Californio homes and neighborhoods and/or moved into the less populated areas of Pescadero, Aptos, and Watsonville (Rizzo-Martinez 2022: 230, 248). Still other Indigenous people relocated to the traditional lands of Yokuts-speaking Tribes in the San Joaquin Valley (e.g., Los Banos, Madera) where they remain today. Indigenous depopulation on the Santa Cruz coast would have made it more challenging to implement stewardship practices that involved sustained group efforts. Furthermore, the major influx of non-Indigenous people residing along the coast would have made it more difficult to undertake these activities in secret or without interference.

Second, the heavily forested Santa Cruz Mountains, which served as a refuge for Indigenous people during the Spanish and Mexican periods, became a major hub of economic activity for American entrepreneurs who terminated Indigenous occupation and stewardship practices—particularly, cultural burning. It is estimated that at least 52% of Santa Cruz County was covered by valued redwood forests in the 1800s (Payne 1978). While the Spanish and Mexican colonists harvested some lumber in the 1790s and early 1800s, the timber industry in the Santa Cruz Mountains did not really take off until the 1850s–1880s. Innovative technologies for tree cutting and milling, in combination with the creation of transportation systems for getting the wood to markets, made the exploitation of redwoods and other timber products highly profitable. This resulted in the clearcutting of forests, the construction of miles of roads and rail lines, and the building of dams, mills, and logging camps across much of the heavily forested areas of the Santa Cruz Mountains (see Lasnier 2022; Meniketti 2016: 37–41; Payne 1978). These commercial activities produced a transformed, industrialized landscape associated with "colossal environmental damage," characterized by flooding, massive erosion, extensive road cuts, changes in hydrology due to dams, mill ponds, and dumping tons of sawdust into streams, as well as the widespread stripping of redwoods, firs, oaks (for firewood), and tanoaks from the land (see Meniketti 2016: 35–37).

While the timber industry provided lumber for the growing cities of California, it also removed forest coverage, which enabled settlers to move into some of the less rugged terrain to farm and ranch (Meniketti 2016: 41; Payne 1978)—an action that further pushed Native peoples off their lands. The existence of valued timberlands in the Santa Cruz Mountains, across California, and beyond was a significant factor in the development and support of American fire suppression policies. Timber companies did not want to see their investments in land, labor, and equipment disappear in a cloud of smoke. While these companies employed fire to burn slash and downed wood, they barred Indigenous people from cultural burning on their properties. Although the timber industry may have employed some Tribal members as laborers, it appears that by the late 1850s and 1860s, Indigenous people practicing TEK were not welcome across much of the industrialized landscape of the Santa Cruz Mountains.

Third, the conservation movement of the late 1800s and early 1900s essentially banished Tribal peoples from lands that would become parklands and preserves in the Santa Cruz area and across California—a legacy that continued for many decades. The creation of California's first state park in 1902, now known as Big Basin Redwoods State Park, exemplifies this point. The current 18,000-acre (73 km²) park, situated in the ancestral homelands of the Awaswas-speaking Cotoni and Quiroste Tribes, is in the heart of the Santa Cruz Mountains. Bliss (2021) presents a masterful account of the politics and people involved in the development of the park and in saving remnant patches of old growth redwoods from destruction by the timber industry. She details the establishment of the Sempervirens Club in the early 1900s that provided much of the funding and political will to persuade the state legislature to purchase Big Basin as a state park. While John Muir was not directly involved in these efforts, his previous endeavors with the Sierra Club in preserving remnant giant Sequoias stands in the Sierra Nevada Mountains were very influential. In advocating for the preservation of "pristine wilderness" lands in Sequoia and Yosemite, supposedly untouched by humankind, Muir believed that Indigenous people had no role in their stewardship or upkeep—either in the past or in the future as part of the conservation movement. His brand of environmental preservation and racism proved detrimental for several reasons: it excluded Native peoples from using or stewarding parks and wilderness areas; it rejected in principle human management of Native ecosystems; and it strongly promoted forest protection and fire suppression measures to eliminate the risk of wildfires (Brune 2020; Grad 2020; Johnson 2014).

Muir's imprint is evident in the establishment of Big Basin Redwoods State Park. Bliss (2021: Chapter 8) documents how members of the Sempervirens Club strongly advocated that the forest should "be preserved in its primeval state." Her book recounts the negative reaction that resulted when the park warden attempted to extensively clear underbrush, remove old logs, and cut fire-damaged trees after a major wildfire in 1904. Preservationists strongly objected to this "rape of the redwoods," claiming that it would result in the maintenance of the pristine forest as a "city park" (see Bliss 2021: Chapter 8). While strongly objecting to human management of the park's natural resources, members of the Sempervirens Club and the Sierra Club utilized the area for camping trips, fishing, and so on. Although they supported recreational use of the lands—primarily for white families who could afford to visit—they made no efforts to include local Indigenous people in the use or stewardship of the park. Unfortunately, the early movement to maintain parklands as "pristine wildernesses" that minimized any kind of Indigenous stewardship continued with the creation of other state and federal parks in California, a legacy that has had longstanding implications for Indigenous sovereignty and access to public lands by the AMTB and other California Tribes.

Tribal Revitalization in California

Despite the many years of challenges from colonial invaders, Indigenous communities across the state have emerged from this dark period as reinvigorated entities who are initiating Tribal revitalization programs to foster Native language retention, traditional crafts, wellness, spirituality, and sovereignty. A significant component of this renaissance involves their reengagement with stewardship practices that were prohibited or made inaccessible by settler colonialism. Much of this work is aimed at reversing the process of environmental degradation and fire suppression that has been inflicted on their ancestral landscapes and seascapes. Many Tribes across the state are now working to bring cultural burning back to their lands, which have not been tended for many decades (Codero-Lamb et al. 2018; Goode 2015; Goode et al. 2022; Hankins 2021; Lake et al. 2017, 2018; Long et al. 2020a, 2020b; Long et al. 2021; Marks-Block et al. 2019, 2021). This initiative is taking place while state and federal resource agencies and parks, as well as private landholding conservation organizations, are rethinking fire suppression policies and actively working to bring "good" fires back to their properties (see Stephens and Sugihara 2018).

The Central California coast is a case example of this development. After facing countless hardships that forced them from their homes and denied them access to resources in their ancestral lands and waters, the Amah Mutsun and other Indigenous groups are now following a path of Tribal revitalization that is reenergizing their ceremonies, dances, language fluency, health, and cultural practices. As Rizzo-Martinez (2022) details, many Tribal cultural practices persevered through the maintenance of Tribal communities where Tribal oral histories were passed down through families, ethnohistoric accounts, and early ethnographers such as John P. Harrington who documented a wealth of their linguistic, foodway, ceremonial, and ecological knowledge in the late 1920s and early 1930s. While this Tribal revitalization is taking place, California State Parks, the National Park Service, the Bureau of Land Management, and other local and private conservation groups are actively working with the Amah Mutsun and other Tribes to allow them access to culturally important plants and animals for tending and harvesting for food, medicine, and raw materials.

Unfortunately, the repressive colonial policies by a succession of Spanish, Mexican, and American settler colonists prohibiting Indigenous stewardship of terrestrial and maritime resources over many generations had a huge impact on local Tribes. As noted above, Indigenous people on the Santa Cruz Mountains and coast have probably not been involved in any sustained cultural burning or other major stewardship practices since at least the 1850s. The deprivation of these activities for over 170 years has understandably led to lapses or gaps in the memories of how such practices were enacted (Lopez 2013; Chapter 3, this volume).

Collaborative Eco-Archaeology on the Central California Coast

We have implemented a collaborative eco-archaeological program with the AMTB to help bring some of these Indigenous landscape and seascape stewardship practices out of dormancy (Sigona et al. 2021; Chapter 14, this volume). The program is designed to examine the timing, development, scale, and contemporary relevance of Indigenous landscape and seascape maintenance on the Central California coast. We are collaborating closely with the Amah Mutsun Land Trust (AMLT), dedicated to conserving and restoring cultural and natural resources within the ancestral territory of the Awaswas, Mutsun, and Chalon-speaking peoples (Figure 1.1). The AMLT is working with various state, federal, and private land-owning entities across this broad area to help steward extensive wildland tracts. The land trust provides opportunities for Tribal members to tend their ancestral landscapes and seascapes using traditional stewardship practices in combination with modern management approaches to increase the diversity and availability of native fauna and flora, remove harmful invasive species, protect sacred sites, and preserve important cultural landscapes (see Sigona et al. 2021; Chapters 3 and 14, this volume). The AMLT created the Native Stewardship Corps (NSC) as an entity that employs and trains young Tribal members to serve as environmental stewards who provide the boots on the ground for undertaking landscape and seascape revitalization projects.

First Phase of Collaborative Eco-Archaeological Research

Our eco-archaeological program was founded in the early 2000s when a collective of scholars from the AMTB, California State Parks, and the University of California campuses at Berkeley and Santa Cruz initiated the study of past Indigenous landscape stewardship practices on the Central California coast. Our first phase of research focused on developing an eco-archaeological approach for better understanding how Indigenous people implemented cultural burning in ancient and historical times. This approach involved the investigation of multiple lines of evidence that would shed light on past fire regimes, fire return intervals, and Indigenous harvesting of fire-enhanced plants and animals. Our collaborative research team undertook the following work:

1. We collected and analyzed information on cultural burning from relevant Tribal histories, ethnohistorical observations, and ethnographic sources (Cuthrell 2013b);

2. We implemented archaeological fieldwork that examined transformations in ancient and historical cultural practices and community organizations, and documented changes in floral and faunal resource use over time. The fieldwork employed an intensive sampling program to recover rich and robust assemblages of archaeobotanical and zooarchaeological remains from well-dated archaeological contexts (Cuthrell 2013a, 2013b; Gifford-Gonzalez et al. 2013);

3. We investigated silica phytoliths from archaeological features that provided insights about the spatial patterning of plant processing activity areas, while the measurement of phytolith content in landscape soils represented an important data source for evaluating the existence of past grassland habitats (Cuthrell 2013b; Evett and Cuthrell 2013);

4. We collected and analyzed sediment cores from two wetlands that afforded a relatively high-resolution data set for examining diachronic changes in pollen counts and frequencies, as well as the accumulation rates of charcoal as proxies for past fire events (Cowart 2014; Cowart and Byrne 2013);

5. We undertook the synthesis of past fire ecology studies and initiated our own study of fire scars from coast redwood (*Sequoia sempervirens*) trunk sections, which provided a method for directly reconstructing fire return intervals on the Central California coast in Late Holocene and Historic times (Stephens and Fry 2005; Striplen 2014);

6. We experimented with the population genetics of plants and animals that may have been impacted by Indigenous stewardship practices. For example, we examined the genetic variability of modern California hazelnuts (*Corylus cornuta* var. *californica*) across space to evaluate if pervasive fire suppression associated with colonialism may have reduced the occurrence of this culturally important plant that is commonly found in some archaeobotanical assemblages (Fine et al. 2013).

To make a strong case for cultural burning, we argued that some level of temporal concordance should exist across the different data sets. For example, archaeological evidence for fire enhancement strategies, such as notable increases in the exploitation of fire-following or fire-enhanced species, should be temporally associated with changes in fire regimes and vegetation successions, which deviate significantly from baseline predictions for natural fire regimes based on lightning strikes alone (Lightfoot and Lopez 2013: 216–218).

The results of our eco-archaeological research at Quiroste Valley in Año Nuevo State Park (Figure 1.2) demonstrated that Native peoples employed fire to maintain highly productive grasslands interspersed with forest/savanna communities containing hazelnut and oak trees from at least 1000–1300 cal CE. Excavations at CA-SMA-113, a Late Holocene/Historic village, unearthed evidence that Indigenous people harvested a diverse range of food plants that thrived with frequent fires, including seeds of grasses (Poaceae), panicled bulrush (*Scirpus microcarpus*), coast tarweed (*Madia* spp.), clover

(*Trifolium* spp.), composites (Asteraceae), California hazelnut, tanoak (*Notholithocarpus densiflorus*), and other plants (Cuthrell 2013a, 2013b). The anthracological investigation of charcoal from woody species showed that they used primarily fire-enhanced species—e.g., coast redwood and California lilac (*Ceanothus thyrsiflorus*)—as sources of fuel and raw materials (Cuthrell 2013a: 279–283, 2013b). The zooarchaeological study indicated that they hunted or trapped a substantial number of mule deer (*Odocoileus hemionus*), various lagomorphs, and voles (*Microtus* sp.), which would have thrived in the mixed mosaic of grassland and open woodland environment (Gifford-Gonzalez et al. 2013).

The high proportions of grassland-associated food plants, high density of hazelnut remains, high ratio of voles (specially adapted to grassland habitats) to wood rats (*Neotoma* sp., which prefer dense woodland or forest environments), and the dominance of fire-enhanced shrubs and trees analyzed from archaeological contexts stand in sharp contrast to the fire-intolerant vegetation that populates Quiroste Valley today: this indicates that more open-forest and fire-adapted species flourished here in the past. In synthesizing the results of multiple lines of evidence drawn from relevant ethnohistorical, archaeological, phytolith, pollen/charcoal, dendroecological, and genetic data sets, we concluded that local groups employed frequent, low-intensity fires to convert north coast scrub and Douglas fir forests into extensive grasslands in Late Holocene times. This greatly enhanced the diversity, quantity, and predictability of plant and animal resources for human subsistence. We argued that lightning ignitions alone (with fire return intervals of 50–100 years) were insufficient to sustain these coastal prairies. We showed that the long-term maintenance of coastal grasslands in Central California instead required a sub-decadal fire return interval, with fires probably set at intervals of one to five years (Cuthrell 2013a, 2013b; Lightfoot et al. 2013).

Second Phase of Eco-archaeological Research

We built upon our earlier work in designing our next stage of research that addressed four major goals concerning the timing, development, geographic scale, and contemporary relevancy of Indigenous landscape and seascape stewardship practices on the greater Santa Cruz coast, which is the topic of this volume.

The first goal is to examine when people may have first initiated sustained cultural burning and seascape stewardship practices in the broader region. Expanding our work beyond Quiroste Valley and the archaeological investigation of CA-SMA-113, we sought to extend the time-depth of our investigation before and after 1000–1300 CE. We are particularly interested in examining sites on the greater Santa Cruz coast with occupations dating to the Middle Holocene (6000–3000 BP), Late Holocene (3000–500 BP), and Historic periods (500–100 BP).

The second goal is to examine whether and how local Indigenous communities may have modified stewardship practices over time in relation to climate change, sociopolitical and technological transformations, human population growth, and colonialism. Long-term developments in anthropogenic fire regimes are poorly known in California. We are particularly interested in examining how Native peoples may have altered stewardship practices during past episodes of climate change. Our work at Quiroste Valley indicates strong evidence for frequent fires and fire-enhanced species during the period from

1000 to 1300 CE. This timespan falls within the Medieval Climatic Anomaly (MCA), a climatic regime from about 900 to 1300 CE marked by prolonged intervals of decreased precipitation, "epic" droughts, and warmer summer temperatures (Jones et al. 1999; Klimaszewski-Patterson et al. 2018, 2021). As Klimaszewski-Patterson and colleagues (2021: 2) note, we should expect more evidence for frequent fires during the warmer, drier climate of the MCA and a greater propensity for fire-enhanced vegetation consequently. It raises the question of how much of the eco-archaeological pattern we observed at Quiroste Valley is due primarily to climate and, to a lesser extent, to people.

Therefore, in evaluating whether people employed sustained cultural burning for stewarding terrestrial landscapes in ancient times, we are interested in examining eco-archaeological evidence for Indigenous harvesting practices before and after the onset of the MCA, when fire-adapted resources are expected to be less common (see Klimaszewski-Patterson et al. 2021). We are particularly interested in the Little Ice Age (LIA), from about 1350 to 1850 CE, to explore how overall cooler temperatures may have structured fire regimes and Indigenous stewardship practices in Late Holocene times. If Indigenous harvesting practices in the MCA were influenced primarily by climate, then we would expect to see a decline in the frequency of fire-enhanced species in the archaeological record in the cooler LIA. On the other hand, if Indigenous stewardship practices and cultural burning were well integrated into the lifeways of local communities in MCA times, then we would expect to see the tending and harvesting of fire-adapted species continue into the LIA.

We are also interested in examining how Indigenous stewardship practices may have been transformed over time with sociopolitical and technological developments, human demography, and colonialism. We are particularly attentive to how the process of resource intensification that may be associated with human population increase and technological innovations may have modified Indigenous stewardship practices in the longue durée (e.g., Broughton 1994, 2002). Likewise, we consider the timing and development of seascape stewardship practices in the region in relation to climate change and societal transformations, as well as how these practices may have overlapped with terrestrial stewardship activities.

The third goal is to evaluate the geographic scale of Indigenous landscape and seascape stewardship practices on the Central California coast. There is considerable debate about the degree of landscape modifications undertaken by Native Californians. While some scholars believe cultural burning was employed widely in California and the American West, others contend that the magnitude of these practices has been greatly exaggerated in the anthropological literature. The latter scholars contend that climate was the primary factor influencing fire regimes in California and that recent changes in vegetation patterns, which are attributed to the abolition of Indigenous burning practices, have more to do with state and federal fire suppression policies and subsequent changes in fire regimes across the state (Barrett et al. 2005; Bendix 2002; Parker 2002; Vale 1998, 2002). Eco-archaeological studies evaluate this issue by examining the geographic scale of landscape and seascape stewardship practices: for example, were they highly localized and employed only in a few places? Or is there evidence for their widespread use among Tribes in different regions of California? In expanding our study to include the greater Santa Cruz coast, we evaluate whether cultural burning at Quiroste Valley was an isolated case—or part of a broader pattern of landscape stewardship instituted by multiple Tribes in Central California.

Our fourth goal is to give back information about TEK concerning landscape and seascape stewardship practices to the AMTB and AMLT. This information may be of assistance to them as they collaborate with California State Parks and other resource agencies to develop new protocols for the contemporary management of public spaces rooted in the deep history of Tribal practices. California State Parks have extensive holdings, from Point Año Nuevo to just north of Santa Cruz, which are the focus of this study. They are actively working with the Tribe to re-implement Indigenous stewardship practices that will enhance the richness and diversity of native species and thus improve the health of biological communities, reduce fuel loads, and minimize the risk of catastrophic fires.

The Santa Cruz Coast

We address these four research goals using a fine-grained, low-impact, eco-archaeological approach designed to minimize impacts on ancestral places and to avoid disturbances to burials and other sacred remains, while maximizing the recovery of useful information about Tribal histories and cultural resources. We chose to examine sites south of Año Nuevo State Park that had chronological components spanning the Middle Holocene (6000–3000 BP), Late Holocene (3000–500 BP), and Historic periods (500–100 BP). We defined our study area as the Santa Cruz coast situated between the modern communities of Davenport and Santa Cruz in central Santa Cruz County. Here, we investigated four sites (CA-SCR-7, CA-SCR-10, CA-SCR-14, and CA-SCR-15) that are the focus of this volume (Figure 1.2). In Chapter 4, we describe our study area in more detail, introduce the four sites, and outline the field investigations implemented at each site.

Conclusion

This chapter serves as an introduction to our study of Indigenous stewardship practices on the Santa Cruz coast. We provide background information on the literature on Indigenous stewardship practices in California and describe the devastating impacts that colonialism had on the Tribal peoples and the environments of California. We focus our discussion on the many challenges that the AMTB and other Indigenous people faced in their engagements with Spanish, Mexican, and American colonists on the Central California coast. The chapter then highlights how California Indians—with particular emphasis on the AMTB—have emerged from this dark age as motivated agents of change, who are initiating Tribal revitalization programs that foster Native language retention, traditional crafts, wellness, spirituality, sovereignty, and environmental restoration. We argue that eco-archaeological research undertaken in partnership with Tribes and resources agencies offers an important avenue for revitalizing Indigenous stewardship practices that can further resuscitate ancestral lands and waters degraded after years of settler colonialism.

In the final section, we introduce the first phase of eco-archaeological research that we implemented on the Central California coast and describe the research design for the second phase of research, which is the focus of this volume. In the following chapters, our collaborative research team discusses the findings from our eco-archaeological investigation of CA-SCR-7, CA-SCR-10, CA-SCR-14, and CA-SCR-15, as well as other relevant sites, which are employed to address four goals concerning the timing, development, geographic scale, and contemporary relevance of Indigenous stewardship practices on the Santa Cruz coast.

References

Allen, Rebecca
1998   Native Americans at Mission Santa Cruz, 1791–1834: Interpreting the Archaeological Record. *Perspectives in California Archaeology*, vol. 5. Institute of Archaeology, University of California, Los Angeles.

Anderson, M. Kat
2005   *Tending the Wild: Native American Knowledge and the Management of California's Natural Resources.* University of California Press, Berkeley.
2018   The Use of Fire by Native Americans in California. In *Fire in California's Ecosystems,* edited by Jan W. Van Wagtendonk, Neil G. Sugihara, Scott L. Stephens, Andrea E. Thode, Kevin E. Shaffer, and Joann Fites-Kaufman, pp. 381–397. University of California Press, Berkeley.

Baker, Rob
1992   The Clam "Gardens" of Tomales Bay. *News from Native California* 6(2): 28–29.

Barrett, Stephen W., Thomas Swetnam, and William L. Baker
2005   Indian Fire Use: Deflating the Legend. *Fire Management Today* 65(3): 31–34.

Baumhoff, Martin A.
1963   Ecological Determination of Aboriginal California Populations. *University of California Publications in American Archaeology and Ethnology* 49(2): 155–236.

Bean, Lowell John, and Harry Lawton
1976   Some Explanations for the Rise of Cultural Complexity in Native California with Comments on Proto-Agriculture and Agriculture. In *Native Californians: A Theoretical Retrospective*, edited by Lowell John Bean and Thomas C. Blackburn, pp. 19–48. Ballena Press, Menlo Park, California.

Bendix, Jacob
2002   Pre-European Fire in California Chaparral. In *Fire, Native Peoples, and the Natural Landscape*, edited by Thomas R. Vale, pp. 269–293. Island Press, Covelo, California.

Blackburn, Thomas C., and Kat Anderson (editors)
1993   *Before the Wilderness: Environmental Management by Native Californians.* Ballena Press, Menlo Park, California.

Bliss, Traci
2021   *Big Basin Redwood Forest: California's Oldest State Park.* Arcadia Publishing, Mount Pleasant, South Carolina.

Braje, Todd J., Jon M. Erlandson, Torben C. Rick, Paul K. Dayton, and
Marco B.A. Hartch
    2009   Fishing from Past to Present: Continuity and Resilience of Red Abalone
    Fisheries on the Channel Islands, California. *Ecological Applications* 19(4): 906–919.

Broughton, Jack M.
    1994   Declines in Mammalian Foraging Efficiency during the Late
    Holocene, San Francisco Bay, California. *Journal of Anthropological Archaeology*
    13(4): 371–401.
    2002   Pristine Benchmarks and Indigenous Conservation? Implications from
    California Zooarchaeology. In *The Future from the Past: Archaeozoology in Wildlife*
    Conservation and Heritage Management, edited by Roel C. Lauwerier and Plug Ina,
    pp. 6–18. Oxbow Books, Oxford, England.

Brune, Michael
    2020   Pulling Down Our Monuments. *Sierra Club Newsletter.* https://www.
    sierraclub.org/michael-brune/2020/07/john-muir-early-history-sierra-club.

Caldwell, Megan E., Dana Lepofsky, Georgia Combes, Michelle Washington, John R.
Welch, and John Harper
    2012   A Bird's Eye View of Northern Coast Salish Intertidal Resource
    Management Features, Southern British Columbia, Canada
    Canada Journal of Island and Coastal Archaeology 7(2): 219–233.

Cannon, Audrey, and Meghan Burchell
    2009   Clam-Growth Stage Profiles as a Measure of Harvest Intensity and
    Resource Management on the Central Coast of British Columbia. *Journal of*
    Archaeological Science 36: 1050–1060.

Castillo, Edward D.
    1978   The Impact of Euro-American Exploration and Settlement. In *Handbook*
    of North American Indians: Volume 8, California, edited by Robert F. Heizer, pp.
    99–127. Smithsonian Institution, Washington, DC.

Codero-Lamb, Julie, Jared Dahl Aldern, and Teresa Romero
    2018   Bring Back the Good Fires. *News from Native California* 31(3): 14–17.

Cowart, Alicia
    2014   Paleoenvironmental Change in Central California in the Late Pleistocene
    and Holocene: Impacts of Climate Change and Human Land Use on Vegetation
    and Fire Regimes. PhD Dissertation, Department of Geography, University of
    California at Berkeley.

Cowart, Alicia, and Roger Byrne

2013 A Paleolimnological Record of Late Holocene Vegetation Change from the Central California Coast. *California Archaeology* 5(2): 337–352.

Crosby, Alfred W.

2004 *Ecological Imperialism: The Biological Expansion of Europe, 900–1900,* 2nd ed. Cambridge University Press, Cambridge.

Cuthrell, Rob Q.

2013a Archaeobotanical Evidence for Indigenous Burning Practices and Foodways at CA-SMA-113. *California Archaeology* 5(2): 265–290.

2013b An Eco-Archaeological Study of Late Holocene Indigenous Foodways and Landscape Management Practices at Quiroste Valley Cultural Preserve, San Mateo County, California. PhD Dissertation, Department of Anthropology, University of California, Berkeley.

Dartt-Newton, Deana, and Jon M. Erlandson

2006 Little Choice for the Chumash: Colonialism, Cattle, and Coercion in Mission Period California. *American Indian Quarterly* 30(3–4): 416–430.

Erlandson, Jon M., Torben C. Rick, and Todd J. Braje

2009 Fishing Up the Food Web? 12,000 Years of Maritime Subsistence and Adaptive Adjustments on California's Channel Islands. *Pacific Science* 63: 711–724.

Erlandson, Jon M., Torben C. Rick, Todd J. Braje, Alexis Steinberg, and Rene L. Vellanoweth

2008 Human Impacts on Ancient Shellfish: A 10,000 Year Record from San Miguel Island, California. *Journal of Archaeological Science* 35: 2144–2152.

Evett, Rand R., and Rob Q. Cuthrell

2013 Phytolith Evidence for a Grass-Dominated Prairie Landscape at Quiroste Valley on the Central Coast of California. *California Archaeology* 5(2): 319–335.

Fine, Paul V. A., Tracy M. Misiewicz, Andrea B. Chavez, and Rob Q. Cuthrell

2013 Population Genetic Structure of California Hazelnut, an Important Food Source for People in Quiroste Valley in the Late Holocene. *California Archaeology* 5(2): 353–370.

Gifford-Gonzalez, Diane, Cristie M. Boone, and Rachel E. B. Reid

2013 The Fauna From Quiroste: Insights into Indigenous Foodways, Culture, and Land Modification. *California Archaeology* 5(2): 291–317.

Goode, Ron W.

2015 Tribal-Traditional Ecological Knowledge. *News from Native California* (Spring): 23–28.

Goode, Ron W., Stephanie Farish Beard, and Christina Oraftik
　　2022　Putting Fire on the Land: The Indigenous People Spoke the Language of Ecology, and Understood the Connectedness and Relationship Between Land, Water, and Fire. *Journal of California and Great Basin Anthropology* 42(1): 85–95.

Grad, Shelby
　　2020　Sierra Club Calls Out the Racism of John Muir. *Los Angeles Times*, July 22, 2020. https://www.latimes.com/california/story/2020-07-22/sierra-club-calls-out-the-racism-of-john-muir.

Grier, Colin
　　2014　Landscape Construction, Ownership and Social Change in the Southern Gulf Islands of British Columbia. *Canadian Journal of Archaeology* 38(211–249).

Groesbeck, Amy S., Kirsten Rowell, Dana Lepofsky, and Anne K. Salomon
　　2014　Ancient Clam Gardens Increased Shellfish Production: Adaptive Strategies from the Past Can Inform Food Security Today. *pLoS One* 9(3): e91235. https://doi.org/10.1371/journal.pone.0091235.

Grone, Michael A.
　　2020　Of Molluscs and Middens: Historical Ecology of Indigenous Shoreline Stewardship along the Central Coast of California. PhD Dissertation, Department of Anthropology, University of California, Berkeley.

Hallam, S. J.
　　1985　The History of Aboriginal Firing. In *Fire Ecology and Management in Western Australian Ecosystems*, edited by J. R. Ford, pp. 7–20. Western Australian Institute of Technology. WAIT Environmental Studies Group Report No. 14, Perth, Australia.

Hankins, Don
　　2021　Reading the Landscape for Fire. *Bay Nature* 21(1): 28–35.

Heizer, Robert F., and Alan F. Almquist
　　1971　*The Other Californians: Prejudice and Discrimination under Spain, Mexico, and the United States to 1920.* University of California Press, Berkeley.

Hurtado, Albert L.
　　1988　*Indian Survival on the California Frontier.* Yale University Press, New Haven.

Johnson, Eric Michael
　　2014　Fire Over Ahwahnee: John Muir and the Decline of Yosemite. *Scientific American.* http://blogs.scientificamierican.com/primate-diaries/201408/13/fire-over-ahwahnee-john-muir-and-the-decline-of-yosemite/.

Jones, Terry L., Gary M. Brown, L. Mark Raab, Janet L. McVickar, W. Geoffrey Spaulding, Douglas J. Kennett, Andrew York, and Phillip L. Walker
> 1999   Environmental Imperatives Reconsidered: Demographic Crises in Western North America during the Medieval Climatic Anomaly. *Current Anthropology* 40(2): 137–170.

Keeley, Jon E.
> 2002   Native American Impacts on Fire Regimes of the California Coastal Ranges. *Journal of Biogeography* 29: 303–320.

Klimaszewski-Patterson, Anna, Peter J. Weisberg, Scott A. Mensing, and Robert M. Scheller
> 2018   Using Paleolandscape Modeling to Investigate the Impact of Native American-Set Fires on Pre-Columbian Forests in the Southern Sierra Nevada, California, USA. *Annals of the American Association of Geographers* 108(6): 1635–1654.

Klimaszewski-Patterson, Anna, Christopher T. Morgan, and Scott A. Mensing
> 2021   Identifying a Pre-Columbian Anthropocene in California. *Annals of the American Association of Geographers* 111(3): 784–794.

Koenig, Walter D., Ronald L. Mumme, William J. Carmen, and Mark T. Stanback
> 1994   Acorn Production by Oaks in Central Coastal California: Variation Within and Among Years. *Ecology* 75(1): 99–109.

Lake, Frank K., and Amy C. Christianson
> 2019   Indigenous Fire Stewardship. In *Encyclopedia of Wildfires and Wildland-Urban Interface (WUI) Fires*, edited by Samuel L. Manzello. Springer, New York. https://doi.org/10.1007/978-3-319-51727-8_225-1.

Lake, Frank K., Vita Wright, Penelope Morgan, Mary McFadzen, Dave McWethy, and Camille Stevens-Rumann
> 2017   Returning Fire to the Land: Celebrating Traditional Knowledge and Fire. *Journal of Forestry* 115(5): 343–353.

Lake, Frank K., J. A. Parrotta, C. P. Giardian, I. Davidson-Hunt, and Y. Uprety
> 2018   Integration of Traditional and Western Knowledge in Forest Landscape Restoration. In *Forest Landscape Restoration: Integrated Approaches to Support Effective Implementation*, edited by S. Mansourian and J. Parrotta, pp. 198–226. Routledge, New York.

Lasnier, Guy
> 2022   San Vicente Redwoods: A History of Industry, a Future of Conservation. Santa Cruz Mountains Trail Stewardship. September 22, 2022. https://santacruztrails.org/blog/san-vicente-redwoods-a-history-of-industry-a-future-of-conservation.

Lepofsky, Dana, and Megan E. Caldwell
2013   Indigenous Marine Resource Management on the Northwest Coast of North America. *Ecological Processes* 2(12): 1–12.

Lepofsky, Dana, Nicole F. Smith, Nathan Cardinal, John Harper, Mary Morris, Gitla Eloy White, Randy Bouchard, Dorothy I. D. Kennedy, Anne K. Salomon, Michelle Puckett, and Kirsten Rowell
2015   Ancient Shellfish Mariculture on the Northwest Coast of North America. *American Antiquity* 80(2): 236–259.

Lewis, Henry T.
1973   *Patterns of Indian Burning in California: Ecology and Ethnohistory.* Ballena Press, Ramona, New Mexico.
1985   Why Indians Burned: Specific Versus General Reasons. In *Proceedings: Symposium and Workshop on Wilderness Fire*, edited by J. E. Lotan, B. M. Kilgore, W. C. Fischer, and W. R. Mutsch, pp. 75–80. Intermountain Forest and Range Experiment Station, USDA Forest Service, Ogden, Utah.

Lightfoot, Kent G.
2005   *Indians, Missionaries, and Merchants: The Legacy of Colonial Encounters on the California Frontiers.* University of California Press, Berkeley.

Lightfoot, Kent G., and Otis Parrish
2009   *California Indians and their Environment: An Introduction.* University of California Press, Berkeley.

Lightfoot, Kent G., and Valentin Lopez
2013   The Study of Indigenous Management Practices in California: An Introduction. *California Archaeology* 5(2): 209–219.

Lightfoot, Kent G., and Sara L. Gonzalez
2018   The Study of Sustained Colonialism: An Example from the Kashaya Pomo Homeland in Northern California. *American Antiquity* 83(3): 427–443.

Lightfoot, Kent G., Rob Q. Cuthrell, Cristie M. Boone, Roger Byrne, Andrea B. Chavez, Laurel Collins, Alicia Cowart, Rand R. Evett, Fine V. A. Paul, Diane Gifford-Gonzalez, Mark G. Hylkema, Valentin Lopez, Tracy M. Misiewicz, and Rachel E. B. Reid
2013   Anthropogenic Burning on the Central California Coast in Late Holocene and Early Historical Times: Findings, Implications, and Future Directions. *California Archaeology* 5(2): 371–390.

Lightfoot, Kent G., Peter A. Nelson, Michael A. Grone, and Alec J. Apodaca
2021   Pathways to Persistence: Divergent Native Engagements with Sustained Colonial Permutations in North America. In *The Routledge Handbook of the Archaeology Indigenous-Colonial Interaction in the Americas*, edited by Lee M. Panich and Sara L. Gonzalez, pp. 129–145. Routledge, New York.

Lindsay, Brendan C.
   2012   *Murder State: California Native American Genocide 1846–1873.* University of Nebraska Press, Lincoln, Nebraska.

Long, Jonathan W., Frank K. Lake, and Ron W. Goode
   2021   The Importance of Indigenous Cultural Burning in Forested Regions of the Pacific West, USA. *Forest Ecology and Management* 500: https://doi.org/10.1016/j.foreco.2021.119597.

Long, Jonathan W., Frank K. Lake, Ron W. Goode, and Benrita Mae Burnette
   2020b   How Traditional Tribal Perspectives Influence Ecosystem Restoration. *Ecopsychology* 12(2): 71–82.

Long, Jonathan W., Ron W. Goode, and Frank K. Lake
   2020a   Recentering Ecological Restoration with Tribal Perspectives. *Fremontia* 48(1): 14–19.

Lopez, Valentin
   2013   The Amah Mutsun Band: Reflections on Collaborative Archaeology. *California Archaeology* 5(2): 221–223.

Madley, Benjamin
   2016   *An American Genocide: The United States and the California Indian Catastrophe.* Yale University Press, New Haven.

Marks-Block, Tony, Frank K. Lake, and Lisa M. Curran
   2019   Effects of Understory Fire Management Treatments on California Hazelnut, An Ecocultural Resource of the Karuk and Yurok Indians in the Pacific Northwest. *Forest Ecology and Management* 450: 1–12.

Marks-Block, Tony, Frank K. Lake, Rebecca Bliege Bird, and Lisa M. Curran
   2021   Revitalizing Karuk and Yurok Cultural Burning to Enhance California Hazelnut for Basketweaving in Northwestern California, USA. *Fire Ecology* 17(6): https://doi.org/10.1186/s42408-42021-00092-42406.

McCarthy, Helen
   1993   Managing Oaks and the Acorn Crop. In *Before the Wilderness: Environmental Management by Native Californians*, edited by Thomas C. Blackburn and Kat Anderson, pp. 213–228. Ballena Press, Menlo Park, California.

Meniketti, Marco G.
   2016   The Timber Industry of the Early San Francisco Bay Region. *The Journal of the Society for Industrial Archaeology* 42(2): 35–54.

Panich, Lee M., and Tsim D. Schneider (editors)
    2014   *Indigenous Landscapes and Spanish Missions: New Perspectives from Archaeology and Ethnohistory.* University of Arizona Press, Tucson.

Parker, Albert J.
    2002   Fire in Sierra Nevada Forests: Evaluating the Ecological Impact of Burning by Native Americans. In *Fire, Native Peoples, and the Natural Landscape*, edited by Thomas R. Vale, pp. 233–267. Island Press, Covelo, California.

Payne, Stephen Michael
    1978   *A Howling Wilderness: A History of the Summit Road Area of the Santa Cruz Mountains 1850–1906.* Loma Prieta Publishing, Santa Cruz, California.

Rizzo-Martinez, Martin
    2022   *We Are Not Animals: Indigenous Politics of Survival, Rebellion, and Reconstitution in Nineteenth Century California.* University of Nebraska Press, Lincoln, Nebraska.

Rizzo-Martinez, Martin, and Mark G. Hylkema
    2021   Forward. In *Big Basin Redwood Forest: California's Oldest State Park*, edited by Traci Bliss, pp. 1–2. Arcadia Publishing, Mount Pleasant, South Carolina.

Schneider, Tsim D.
    2015   Placing Refuge and the Archaeology of Indigenous Hinterlands in Colonial California. *American Antiquity* 80(4): 695–713.
    2021   *The Archaeology of Refuge and Recourse: Coast Miwok Resilience and Indigenous Hinterlands of Colonial California.* University of Arizona Press, Tucson.

Shaffer, Kevin E., and William F. Laudenslayer
    2006   Fire and Animal Interactions. In *Fires in California's Ecosystems*, edited by Neil G. Sugihara, Jan W. van Wagtendonk, Kevin E. Shaffer, Joann Fites-Kaufman, and Andrea E. Thode, pp. 118–144. University of California Press, Berkeley.

Sigona, Alexii, Alec J. Apodaca, and Valentin Lopez
    2021   Supporting Cultural Obligations: Using Eco-Archaeology to Inform Native Eco-Cultural Revitalization. *Journal of California and Great Basin Anthropology* 41(2): 207–229.

Stephens, Scott L., and Danny L. Fry
    2005   Fire History in Coast Redwood Stands in the Northeastern Santa Cruz Mountains, California. *Fire Ecology* 1(1): 2–19.

Stephens, Scott L., and Neil G. Sugihara

2018   Fire Management and Policy since European Settlement
In *Fire in California's Ecosystems,* edited by Jan W. Van Wagtendonk, Neil G.
Sugihara, Scott L. Stephens, Andrea E. Thode, Kevin E. Shaffer, and Joann Fites-
Kaufman, pp. 399–410. University of California Press, Berkeley.

Stephens, Scott L., Jamie M. Lydersen, Brandon M. Collins, Danny L. Fry, and
Marc D. Meyer

2015   Historical and Current Landscape-Scale Ponderosa Pine and Mixed
Conifer Forest Structure in the Southern Sierra Nevada. *Ecosphere* 6(5): 1–20.

Stewart, Omer C.

2002   *Forgotten Fires: Native Americans and the Transient Wilderness.* University of
Oklahoma Press, Norman.

Striplen, Chuck J.

2014   A Dendroecology-Based Fire History of Coast Redwoods (*Sequoia
sempervirens*) In Central Coastal California. PhD Dissertation, Department
of Environmental Science, Policy and Management, University of
California, Berkeley.

Timbrook, Jan, John R. Johnson, and David D. Earle

1993   Vegetation Burning by the Chumash. In *Before the Wilderness:
Environmental Management by Native Californians*, edited by Thomas C. Blackburn
and Kat Anderson, pp. 117–149. Ballena Press, Menlo Park, California.

Vale, Thomas R.

1998   The Myth of the Humanized Landscape: An Example from Yosemite
National Park. *Natural Areas Journal* 18(3): 231–236.
2002   The Pre-European Landscape of the United States: Pristine or
Humanized? In *Fire, Native Peoples, and the Natural Landscape*, edited by Thomas R.
Vale, pp. 1–39. Island Press, Covelo, California.

Whitaker, Adrian R.

2008   Incipient Aquaculture in Prehistoric California?: Long-term Productivity
and Sustainability vs. Immediate Returns for the Harvest of Marine Invertebrates.
Journal of Archaeological Science 35(4): 1114–1123.

CHAPTER 3

The Study of Indigenous Landscape and Seascape Stewardship Practices: Taking It to the Next Step

VALENTIN LOPEZ, CAROLYN T. RODRIGUEZ, AND K. MICHELLE GLOWA

Originating from when our People were first taken to the missions, there is a vital teaching from a Mutsun elder. Upon arrival at the mission, one of our elders said that our People would suffer for seven generations. With the seventh generation, he said, things will get better. I (Valentin Lopez) represent that seventh generation. Our Tribe recognizes it is time for things to get better. The only way things are going to get better is if we heal from our historical trauma and return to the path of our ancestors so we can continue their journey and work to fulfill our sacred obligation to Creator.

Before we began our research partnership with the University of California, Berkeley (UC Berkeley), our Tribe had been on a path to reclaim and increase our understanding of the ancestral ways of land care and stewardship. We, unfortunately, lost much of our connection to our traditional knowledge and practices due to the three brutal periods of colonization. First, in the late 1700s, when Spanish missionaries enslaved our People at Missions San Juan Bautista and Santa Cruz. Second, in the 1800s, when our People worked as slaves on ranchos under a peonage system after Mexico won its independence and secularized the California missions. Lastly, in 1851, when California's first governor declared a war of extermination on California Natives until their extinction.

Nevertheless, our Tribe actively works to restore the knowledge of our ancestors—the knowledge learned over thousands of years, with hundreds of generations learning, knowing, and practicing our traditional ways. We listened to elders, talked with neighboring Tribes, and read whatever we could access, including notes at the Smithsonian from our ancestor Ascención Solórsano de Cervantes, a powerful healer and leader.

In 2006, Tribal elders attended an Elders Meeting and reminded us that Creator never rescinded our responsibility to take care of Mother Earth and all living things. These words were an unequivocal directive, and we knew we must find a way to return to our territory and fulfill our responsibility. As part of this work, our Tribe started the Amah Mutsun Land Trust (AMLT), committed to restoring Indigenous knowledge and practices to our traditional territories. Today, our restoration efforts are primarily for the purpose of relearning the Indigenous knowledge of our ancestors, so that we can continue on their path and fulfill our sacred obligation. For example, within AMLT, we started the Native Stewardship Corps (NSC), so Tribal members can return to our ancestral homelands to relearn and recover cultural traditions and traditional land stewardship.

Because we did not own any Tribal land in our territory, we could not imagine how we could return to our homelands and bring back our traditions. We just knew we had to find a way. Following the Elders Meeting, we prayed and asked Creator to help us find a way to return to our traditional territory. It was not long before we received a call from the new Superintendent of Pinnacles National Monument, Eric Brunnemann. Pinnacles, which is now a national park, is within the traditional Tribal territory of the Chalon Tribe who spoke the Mutsun language and were part of the greater Mutsun "Nation." He invited us to a meeting and told us that he wanted us to have a voice in the management of the park. Of course, we cautiously accepted.

Shortly after beginning our work with Pinnacles, researchers at UC Berkeley approached us. The university was developing an archaeological field research project about 25 miles north of Santa Cruz at a location now known as Quiroste Valley. California State Parks and the University of California, Santa Cruz (UC Santa Cruz), were also going to be important partners in this research. Quiroste Valley is within the boundaries of Año Nuevo State Park. It was believed that this location was a historic first-contact site between the Quiroste Tribe and the Portolá Expedition in 1769, the first overland expedition entering California from Mexico. The Quiroste Tribe spoke the Awaswas language and was part of the Awaswas Nation. Because our Tribe has lineages taken to Mission Santa Cruz, and because the Awaswas are a neighboring Tribe of the Amah Mutsun, we felt it was essential to work on Awaswas territory to restore the traditional landscape and to honor the Awaswas and ensure they are never forgotten. We have invited all neighboring Tribes to join us in this work. This work provides our Tribe with knowledge and understanding of the importance of bringing Indigenous management back to the Central Coast as we work to gain access to traditional Mutsun lands.

Through this research, we have discovered new paths of learning about ancestral practices and knowledge, which have greatly assisted us in our efforts to fulfill our sacred obligation. We have worked alongside these researchers, restoring traditional knowledge and recognizing the critical contribution of California Indians. By reasserting relationships of integrity with Mother Earth, we can address the monumental problems of climate change, unprecedented catastrophic fires, and rapid loss of biodiversity that all of us face today.

The Early Research

The research project with UC Berkeley is an integrated archaeological study that first needed to determine if the study site was indeed the Quiroste Village site. The study examined how the Native Peoples along the central coast of California lived and how they managed the landscapes. Several areas of particular interest to the Tribe and researchers included: Did the Tribes in this area intentionally burn the landscape, and, if so, with what frequency? Secondly, what foods did their diet consist of—specifically, how were these food resources sustained and stewarded? Thirdly, what did the landscape look like at first contact? And finally, how had it changed over time?

When we received the invitation to participate in this research, our Tribe's immediate response was to say "no." The history of our Tribe's relationship with archaeologists has almost always been contentious and full of distrust. Up to this time, we experienced the destruction of many of our cultural and spiritual sites, which includes disruption to our ancestors' afterlife. With no concern, input, or approval from the Tribe, archaeologists removed our ancestors from the ground and took them to labs, museums, or special collections. However, Mutsun member Chuck Striplen was pursuing his PhD at UC Berkeley and was an integral member of the research team. He talked to us and asked us to talk to the researchers to understand how their approach would have a very low impact. As a result, we agreed to meet with the project leaders.

The project leaders explained their geomorphology resistivity testing usage for trying to avoid disturbing human remains and other sensitive cultural or spiritual resources. Wanting to help our Tribal member, we agreed to conduct the study, provided our Tribe maintained a yes/no authority on all decisions related to any impacts on cultural resources—in particular, human remains. First, UC Berkeley agreed with this condition; next, we discussed how Tribal members could participate in the project as paid field staff.

From this time on, our Tribe participated in all field study work while a few of our members worked in the lab to process the collected materials. This relationship has been great for our Tribal members. This was an opportunity for them to work side-by-side with university students, who were eager to help our members learn. We could not be prouder of our Tribal members, seeing that they were working on the land and bringing back Indigenous knowledge. We constantly reminded them that if our traditional knowledge is to survive, they must be successful in their learning. At the same time, the university students got to work with our Tribal members, some of whom did not go to college or finish high school. Nevertheless, besides being hard workers, they shared with the students the history of our Tribe and the difficult struggles that our members faced at home. Thus, there was a reciprocal learning relationship; students and researchers learned from our Tribal members, while students taught our members how archaeology could help our Tribe restore our traditional Indigenous knowledge.

It is important to know that we were fully involved in the research. We worked side-by-side with the professors and students on every step of the research to understand what they were doing, why they were doing it, what they were hoping to learn, and finally, how to interpret the research results. In the evenings after field days, Tribal members and researchers would circle up at the fire to converse and share together. Our members would tell stories about their backgrounds and experiences, our history, our more recent elders who had passed, and how they lived their daily lives. The students would talk about school, their families, and their path to archaeology. Telling our stories together helped connect the students and researchers to our members in a way that helped both parties understand each other and see the world differently than either had ever seen it before. As a result, a critical part of the research process was developing friendships and mutual understandings. Tribal members and researchers established an appreciation for each other that has allowed our partnership to continue to grow and evolve.

Delving Into the Research Projects

The first test pit excavated was a cooking pit. The research team excavated a pit in the ground about the size of a shovel's length by width and sifted through the removed soil. We were able to see what the excavation process was going to look like and began to get a sense of the wide range of uncovered cultural materials. We quickly realized how this material could be very helpful to our Tribe's goal of restoring landscapes to their pre-contact condition. Despite the trees and forest-like conditions we see in today's landscape, our Tribe was beginning to understand how the central coast of California looked very different under Indigenous management—prior to colonization. The area around the Quiroste Village was once a very productive coastal prairie, part of one of the most biodiverse landscapes in North America. The more we learned, the more we realized that our Tribe needed to restore the native landscapes that existed before first contact.

In those initial digs, the research team, including our Tribal members, found grass seeds that indicated the areas that are now Douglas fir forests were most likely maintained by ancestors as more open grasslands. This finding was supported by the presence of vole bone fragments, an animal relative that lives in grasslands and non-dense stands of Douglas firs. Additionally, grass seeds accompanied a high number of hazelnut shells. We learned that these ancestors relied on the hazelnut as a dominant nut source—more than acorns. We also saw their use of redwood as firewood and that there were traces of tobacco. Tobacco does not grow along that portion of the coast: this means that these ancestors actively traded with more inland Tribes, which encouraged us to learn more about the trade routes and intertribal connections.

When our Tribe reviewed all that was discovered, we learned a lot about the ancestors who lived at Quiroste. They were actively managing the landscape. These ancestors were not hunters and gatherers, nor did they practice agriculture. Instead, they used frequent cultural burns to keep the coastal prairies open. They carefully tended both land and coastal resources such as foods, medicines, and crafting materials. Burning made sure there were adequate amounts of food for their People and all wildlife. Their land management practices created mosaics in the landscape that benefited humans and non-humans.

Cultural Burns

The Amah Mutsun recognizes fire as sacred and a gift from Creator. In 2020, we, unfortunately, saw catastrophic wildfires throughout California, including in the Santa Cruz Mountains and Diablo Range within the Amah Mutsun Tribal Band (AMTB) territory. The Santa Cruz fire, known as the CZU Lightning Complex fires, burned more than 86,500 acres, 1,400 structures, and 911 homes. These fires are a symptom of an unhealthy relationship between people and the land. Today, a long history of fire suppression, in concert with climate change, has resulted in an unsustainable relationship with wildfire, increasing risks to our lives and homes as time goes on.

For countless generations, the Native Peoples of California used cultural burning to maintain healthy and productive landscapes. The Spanish invaders first outlawed the cultural burning practices of California Indians in the late 1700s. This prohibition continued through the Mexican and American periods of colonization. Because the prairies were not being routinely burned as the ancestors had done, shrubs built up and the Douglas firs and redwoods encroached into what had previously been open grasslands. Fire is necessary for certain native plants to germinate, and these plants were significant food resources for wildlife. The diet of these ancestors was sometimes close to 40% seeds, so keeping up seed production in plants through fire kept them and many birds fed. Cultural burns maintained the biodiversity and health of the land and of many species.

This research partnership helped us to learn about the fire management practices of the Quiroste Valley ancestors. Hearing of the loss of biodiversity and changes in the landscape after colonization led us to understand that we have the responsibility and obligation to restore the landscape and the coastal prairie. We want to restore the native grasses that were there during the time of the ancestors, including the return of all the species that once lived there. We clearly understand that if our traditional approach to fire had continued, California would not be facing the catastrophic fires that are occurring today.

Before the 2020 fire, when we would talk about restoring Indigenous burns, community members would not listen; they were all afraid of fire. After the CZU fires, people understood that they had to do something different, and they are now listening to our talks about restoring cultural burns. Today, all of our AMLT Native Stewards are certified firefighters. We partner with California State Parks, CalFire, Prescribed Burn Associations, and other organizations to bring back cultural burns to this region. Particularly, we are working to conduct cultural burns in Quiroste Valley to restore the coastal prairie. Just this year, our Native Stewards created over 400 burn piles, which were burned in early spring. At the same time, we are working to develop a fire plan for a broadcast burn at Quiroste Valley.

Our Native Stewards have relearned cultural burning practices in many ways. For example, we worked with the TREX, Fire Training Exchanges, and the Yurok People from the Klamath River to help them conduct burns on Tribal lands. Our collaboration provided a vital opportunity to learn through practice, acquiring information and knowledge on cultural burning as we went. We have also learned from other Tribes about their cultural burning knowledge and practices. We combine this with the knowledge gained from the archaeological work to look for successful models to bring to Awaswas and Mutsun territories.

Building the Structure for Increasing Indigenous Stewardship

The more we learned, the more we felt obligated to find a way to restore Indigenous knowledge of stewardship and begin restoring traditional landscapes. Notably, in 2012, our AMTB established the AMLT, and in 2014, we became a 501c3 non-profit. Our Tribe wanted to create a vehicle and structure to support our work, increasing Mutsun stewardship of the land. We say that all restoration—whether of our language, knowledge of the environment, wildlife, and so on—must start with restoring the sacredness of the landscapes. Plants and animals are relatives. All come from the same father, Father Sky (Creator), and the same mother, Mother Earth. Furthermore, the Mutsun understand them as relatives, and therefore, we pray for them, talk to them, sing to them, and communicate with them.

The development of the AMLT supports our continued commitment to restoring this sacred relationship to our relatives and the land. The goals of the AMLT are to

- Protect and conserve cultural and sacred sites
- Conduct research to restore Indigenous knowledge and Tribal history
- Educate our members and the public regarding the importance of Indigenous land stewardship
- Establish an Amah Mutsun Stewardship Corp to restore traditional stewardship

The AMLT is always under the leadership of the Tribe. There is also a board of directors consisting of Tribal members, scholars, and researchers to oversee the range of initiatives. The NSC provides the space for Tribal members to engage in research, conservation, education, and professional development while restoring their relationship with Mother Earth. In most cases, the Native Stewards are the first generation of Tribal members to restore their intimate relationships with Mother Earth since first contact. Today, Native Stewards are upholding our sacred relationship and responsibility to care for Mother Earth. We are recovering our knowledge to protect the lands and all living things. The Stewards also have the role and responsibility to share the knowledge with the larger Tribal community. They must ensure the knowledge is passed on to future generations.

The Land Trust also developed a summer camp and internship program to forward cultural knowledge to young Tribal members. The AMLT Native Youth Stewardship Summer Camp lasts two weeks during the summer months and strives to teach our youth about the Amah Mutsun culture. Native youths, ages 4–17, attend camp to connect to Indigenous territories while learning culturally relevant environmental education. Camp curriculum focuses on land and coastal stewardship, conservation, environmental education, cultural relearning, and recreational activities. Youth learn about removing invasive species, cleaning the beach, archaeological sites, plant identification, and ethnobotany. They learn about Amah Mutsun history, ceremonial practices, and their spiritual connection to Mother Earth. Youth also hear stories and have the chance to learn and practice speaking the Mutsun language and singing traditional songs. At summer camp, the Native Stewards help youth build a sense of Tribal identity as they learn about the Mutsun culture and stewardship. After summer camp, members ages 16 and older can participate in the Internship program for four to six weeks. Interns receive a wage while working alongside the Stewards and learning more about land stewardship. The internship also helps them gain work experience.

Today, our Land Trust is very effectively using the knowledge we have learned from this research, in addition to the knowledge we possessed within the Tribe. We have also worked to develop relationships and sign MOUs with the National Park Service, Bureau of Land Management, California State Parks, open space districts, local land trusts, and private landowners. These MOUs allow us access to more than 140,000 acres.

In addition to our work of reestablishing cultural burning, we are actively reeducating our Tribal members about our ancestors' ethnobotanical relationships to food plants. We have worked with several partners, including the UC Santa Cruz Arboretum, California State Parks, Pie Ranch, and the Morgan Hill Museum, to develop native plant gardens that will help teach the broader public about the cultural importance of many plant species. In 2018, the AMLT worked with Sempervirens Fund, California Nativescapes, and Muwekma Ohlone Tribe members to design and install an ethnobotanical demonstration garden at Castle Rock State Park's new Robert C. Kirkwood Entrance. Together, we planted over 700 culturally significant native plants in the garden. We are working to restore plant communities, as well as our relationships to seeds, bulbs, corms, and nuts. Our work includes conducting ethnobotany surveys to identify patches of our traditional foods so we can grow the patch larger, collect seeds for planting, and gather plants for food now. In 2021, we also partnered with Pie Ranch and California State Parks to propagate over 120,000 native plants at Cascade Ranch, where the plants will be tended to and produce seeds that will be sown at Quiroste Valley Cultural Preserve. The seeds used for propagation are from plants found in Quiroste Valley that are now receiving cultural management to support their growth. In April 2021, we started transplanting these seedlings for the restoration of native coastal prairie landscapes.

Our AMLT is also actively participating in research and Tribal activities centered on coastal stewardship. Specifically, we are partnering with researchers from UC Berkeley to understand our traditional relationships and stewardship practices with sea mammals, shellfish, salmon, small fishes (such as sardines, anchovies, and smelt), and many other species of sea plants and crustaceans. As these were significant food resources for our ancestors and many others, they intentionally and effectively stewarded the environment for the well-being of these life forms. For the last several years, our Native Youth Stewardship camp has been able to come with researchers to see and experience the importance of the sea to our ancestors, tasting different kinds of seaweed and exploring along the coast. We are also one of the founding members of the Tribal Marine Stewardship Network in which we collaborate with other Tribes to restore knowledge and return stewardship to the coast.

Our Tribe participated in the removal of Mill Creek Dam on the San Vicente Creek in Santa Cruz. This dam removal project relates to our goal of returning Coho salmon to this creek. We are also conducting an eDNA monitoring project to observe the impacts of the dam removal project on the salmon and other species at this site. This eDNA project is in collaboration with the Center for Diverse Leadership in Science at the University of California, Los Angeles.

Finally, the AMLT provides Indigenous leadership to address the current climate change crisis. The research with our partners shows us how crucial traditional stewardship is in addressing climate change. We are working to incorporate adaptation strategies into our stewardship practices that will address climate change and promote resilience for humans and native species alike.

It is essential that the Tribe restores knowledge in ways that honor our ancestors and other California Native Peoples. First, we restore knowledge by learning from our elders. We talk to them to document the cultural knowledge that they hold. For example, from them, we have learned about medicinal plants and teas and traditional healing practices. The second way we restore knowledge is by learning from neighboring Tribes as we do in our Land Trust. For example, as part of the Tribal Marine Stewardship Network, we are working with four other California Tribes to learn how to restore and monitor traditional cultural resources. These Tribal communities shared information and were able to obtain a grant to enhance our stewardship collaboration. The third way we restore knowledge is through active research with partners—particularly, our partnerships with UC Berkeley, UC Santa Cruz, and other academic institutions. For instance, through research partnerships, we have begun transcribing John Peabody Harrington's notes on the Amah Mutsun.

In 1929–1930, Harrington, an ethnographer, linguist, and researcher from the Smithsonian Institute, worked with the Tribe's last Mutsun speaker, Ascención Solórsano, to document her knowledge. Ascención was a repository for Tribal history. Harrington interviewed Ascención near the end of her life. Together, they produced 78,000 pages of Tribal anthropological field notes on our history before first contact, the mission, and Mexican and early American periods. In addition, Ascención shared information on the Mutsun language, songs, ceremonies, native plants, medicinal plants, and much more. It is the Tribe's responsibility to continue going through the notes to restore her knowledge for transmission to future generations. There is much to learn from Ascención, our elders, and other California Tribes. Notably, the most critical responsibility of the Tribal Stewards and future research projects is to restore the sacredness of the lands and spiritual sites.

Moving Forward with Continued Partnership and Indigenous Leadership

Our participation in field research with UC Berkeley starting in 2007 took us down a path that we could not have imagined. Since that time, we have developed research that continues to unfold and deepen. In addition to working in Quiroste Valley, we have also participated in projects with UC Berkeley and our other partners at locations such as Año Nuevo, Wilder Ranch, and Pinnacles National Park. The restoration work that resulted from the research has allowed our members to return to traditional lands to practice and relearn ancestral ways of stewarding the land. This work has been essential in our Tribe's efforts to address the legacy of historical trauma that resulted from brutal colonization. It has brought us opportunities to restore our identity and self-esteem as we work to recuperate the Indigenous knowledge that allowed our ancestors to live in a sustainable way and have balance in their lives.

One problem that has made our Tribe's struggle harder is that when anthropologists first started documenting and researching our Mutsun ancestors, they looked for tangible objects to identify as cultural sites, such as middens, petroglyphs, or bedrock mortars. They very seldom documented spiritual sites or understood the spiritual practices of our ancestors. Because of this, today, Native American spirituality is not understood, respected, or valued. To this day, this failure is much to our great detriment. When we talk about our sacred relationship to Mother Earth and saving our spiritual sites, counties and governments ask to be shown the documentation demonstrating that the site in question is a spiritual site. Consequently, 1) without anthropologists' documentation and 2) with our Tribal knowledge being supported only by oral history (when our oral history is never given enough weight), we struggle to protect and gain access to sacred lands. Thus, not documenting these sites as sacred has seriously impacted our Tribe's capacity to protect our spiritual sites. For example, since 2015 we have faced this exact challenge as we try to protect our most sacred mountains and landscape, Juristac, from a proposed sand and gravel mine. This landscape has multiple cultural sites and was an important spiritual place where Tribes gathered for ceremony. Despite this, Juristac was not identified as a spiritual center by early anthropologists. Our Tribe continues to try to protect Juristac from development so we may develop a Tribal Park with the help of our partners where we could do research, bring back Indigenous stewardship and ceremony, and tell the true story of the Amah Mutsun people.

In doing collaborative research with UC Berkeley, we have a better understanding of the everyday lives of the Quiroste Tribe ancestors and their relationships to the coastal prairie, oceans, and plant and animal relatives. For us, this understanding connects the spiritual ways of these ancestors to what is today considered ecological knowledge. Furthermore, it helps us address the problem that our humanity, spirituality, environments, and knowledge have not been acknowledged or respected by governments to the present day.

Current land management in most research projects, government agencies, and even conservation organizations ignore the sacredness of the land. Mother Earth has a living sacredness. All people need to have a relationship with Mother Earth and all living things. Unfortunately, contemporary practices have eliminated this responsibility. As our increasing access to knowledge and tools returns us to traditional stewardship practices, we gain strength on our path of honoring our duty to Creator. Additionally, we are providing leadership with local conservation and land management efforts. There is much to be learned from the California Natives who lived sustainably with these landscapes for thousands of years. Today, we can work together, following Indigenous leadership to reestablish relationships with Mother Earth.

We are actively working to return to Mutsun sacred sites and lands. Because our creation story tells us that Creator gave us the responsibility to take care of Mother Earth, we recognize this responsibility as both a moral directive and moral authority. We must restore sacredness to these lands, and we must do so by restoring our relationship with landscapes, by stewarding with reciprocity, and by restoring our sacred ceremonies and prayers. We must begin teaching our children their obligations for how they are to live their lives and how they are responsible to Creator for taking care of Mother Earth. Protecting and regaining access to our sacred sites and landscapes will give our members a place where we can go and learn the Indigenous knowledge passed down by our ancestors for hundreds, or perhaps a thousand generations or more, and for thousands and thousands of years. We want to restore that knowledge and get back to taking care of the lands as our ancestors did.

Our Tribe no longer calls the work of UC Berkeley and other universities "research." We refer to their work as "validation studies" instead. The studies we conducted together have validated that the knowledge of our ancestors and their practices must be restored if we are to ever live in a sustainable way and successfully deal with climate change. Today, we are reclaiming the knowledge that colonization attempted to eliminate. We survived, and so has the knowledge of our ancestors. Today, we have an obligation to restore our traditional relationship with Mother Earth and all living things. It is essential that we bring back traditional stewardship if we are going to survive into the future and deal with the multiple crises we face.

CHAPTER 4

The Eco-Archaeological Investigation of Indigenous Sites on the Santa Cruz Coast Using a Fine-Grained, Low-Impact Methodology

KENT G. LIGHTFOOT, VALENTIN LOPEZ, MARK G. HYLKEMA, ROB Q. CUTHRELL, MICHAEL A. GRONE, GABRIEL M. SANCHEZ, PETER A. NELSON, ROBERTA A. JEWETT, DIANE GIFFORD-GONZALEZ, ALEC J. APODACA, ALEXII SIGONA, AND ARIADNA GONZALEZ

The purpose of this chapter is to describe the fine-grained, low-impact methodology employed in our collaborative, eco-archaeological investigation of Indigenous sites on the Santa Cruz coast. We designed this study to address four primary research goals concerning the timing, development, geographic scale, and contemporary relevance of Indigenous landscape and seascape stewardship practices on the Santa Cruz coast, as detailed in Chapter 2. The study area extends from the modern community of Davenport to the city of Santa Cruz in central Santa Cruz County (Figure 1.2). This chapter describes the eco-archaeological investigation of four sites from this study area—CA-SCR-7, CA-SCR-10, CA-SCR-14, and CA-SCR-15—as they offered an excellent opportunity to evaluate the above research goals.

The chapter begins with an introduction to the Santa Cruz coast study area and the four sites chosen for investigation. We then discuss our low-impact approach that minimizes impacts to ancestral places and avoids disturbing burials and other sacred remains while maximizing the recovery of useful information about Tribal histories and cultural resources. Our low-impact, eco-archaeological field methodology emphasizes the use of surface and near-surface prospection to obtain information about site structure, features, and cultural materials prior to any significant subsurface investigations (see

Chapters 1 and 5, this volume; Lightfoot 2008: 218–221; Sanchez et al. 2021). We then outline how this approach was implemented at CA-SCR-7, CA-SCR-10, CA-SCR-14, and CA-SCR-15.

Santa Cruz Coast Study Area

The research area lies along the north-central coast of California, encompassing today's central Santa Cruz County—the homeland of the historic Cotoni Tribe and a land characterized by varied shorelines and diverse vegetation mosaics. The shores of the study area comprise relatively narrow coastal plains on tectonically uplifted marine terraces. The region has a few broad, sandy beaches, but it consists predominantly of rocky shorelines with small, sandy coves, often backed by steep cliffs 45–60 meters high. While these plains today are under nearly continuous cultivation, it originally comprised a coastal prairie with abundant native grasses and forbs, the flowers of which so impressed early American Period travelers, such as John Muir. Immediately behind the coastal plain are the foothills of the Santa Cruz Mountains and the Santa Cruz Mountains themselves, which rise to 810 m above sea level. Today, they are covered by a diverse array of shrublands, as well as varied forest type characterized by oaks, bay, buckeye, hazelnut, and closed-cone conifer forest, with Douglas fir (*Pseudotsuga menziesii*) and coast redwoods (*Sequoia sempervirens*) as the dominant tree species. The Santa Cruz Mountains and coastal plains are transected by small and often deeply incised drainages, with larger streams debouching in small estuaries along the coast. Redwoods and riparian plant communities line these watercourses down to their lower reaches, where distinctive estuarine plant communities dominate.

Sites Selected for Investigation

The following section introduces the four sites selected for study. We present a brief description and discussion of previous research undertaken at CA-SCR-7, CA-SCR-10, CA-SCR-14, and CA-SCR-15 (see Figure 1.2).

CA-SCR-7 (Sand Hill Bluff Site)

Located 8 km north of Santa Cruz near the mouth of Laguna Creek, this extensive complex of sand dunes and archaeological deposits covers an estimated 8.3 hectares. Four major loci have been defined for the site complex. Locus 1 comprises the major component of the site and consists of an imposing sand dune mound rising 10.6 m above the coastal terrace that is interlaced with archaeological strata. Locus 2 is a disturbed midden deposit situated southeast of Locus 1 in an area heavily impacted by agriculture and an abandoned abalone farm. Locus 3 consists of a smaller dune complex to the north of Locus 1 that contains archaeological materials. Our fieldwork also involved the study of another related complex of archaeological materials on the leeward side of Locus 1. Defined here as Locus 4, this area consists of a broad scatter of artifacts found in adjacent agricultural fields once planted with Brussels sprouts. These former fields, walked by farmers and artifact collectors for years, have produced significant projectile point collections that are renowned in the local region. Some of these collections have recently been donated to California State Parks and analyzed by Mark Hylkema (2021).

CA-SCR-7 has been on the radar of California archaeologists for almost 150 years. The first known mention of the site was by A. E. Saxe (1875) at the 1873 meeting for the California Academy of Sciences where he described a 3.6–4.5 m thick shell midden deposit containing projectile points and chalcedony boulders. The site was first recorded by members of the University of California Archaeological Survey in 1950 (Archaeological Site Survey Record 10/6/50). Donald W. Lathrop and William J. Wallace described two components for the site: 1) SCR 7a, consisting of an eroding, large shell midden that covers a sand dune on a cliff covering an area of 22.9x122 m (75x400 ft), and 2) SCR 7b, described as an area of black soil with shell extending over a 15.2x67 m (50x220 ft) area. The recorders noted that numerous artifacts had been collected from the site in the past (the site record mentions some of the collectors by name) and that fields of Brussels sprouts were being raised on the leeward side of the sand dune. The next major activity at CA-SCR-7 took place in 1973–1974 by Victor Morejohn, a Biology Professor at San Jose State University, who collected faunal remains from the exposed shell mound. His study detected evidence of an extinct flightless duck, *Chendytes lawi*, and yielded radiocarbon dates from mussel shell for the upper and lower deposits of Locus 1 (3790 +/- 110 BP and 5390 +/- 100 BP, respectively) (Moratto 1984: 244–245). In the 1970s and 1980s, sporadic field investigations were undertaken at CA-SCR-7 by Cabrillo College and others; these enhanced the number of known artifact types and radiocarbon dates from Locus 1 (Jones and Hildebrandt 1990: 17).

The next major spurt of archaeological work took place in the late 1980s. This work was to mitigate the adverse effects resulting from the proposed construction of an abalone farm on a 1.5-hectare area south of Locus 1, near Locus 2. A three-phase data recovery plan was enacted that included an initial surface survey, initial test excavations (8 cubic meters excavated), and a broad-scale data recovery-mitigation excavation program (20.2 cubic meters excavated) (Jones and Hildebrandt 1990). The archaeological investigation yielded substantial information about Locus 1 and Locus 2. During the 1988 field season, archaeologists from Far Western Anthropological Research Group (FWARG) excavated eight units (three 1x1 m, four 1x2 m, and one .5x1 m sized units) along the lower intact shell midden at the base of Locus 1, which had been exposed through erosion of the upper dune and archaeological complexes. The maximum depth of the units ranged from 90 to 130 cm below surface and all materials were sifted through ¼" mesh, with the exception of the .5x1 m unit where 1/8" mesh was used. Given the disturbed context of Locus 2, the 1988 excavation strategy involved the surface investigation of eight 3x3 m units to a depth of 10 cm using ¼" mesh (Jones and Hildebrandt 1990).

The fieldwork at both loci resulted in the collection of a diverse assemblage of flaked stone tools and debitage, along with ground stone artifacts (hand stones, hammerstones, one possible pestle) and battered cobble tools. The quantity of vertebrate faunal remains recovered was relatively small (n=326 specimens) but included avifauna, terrestrial and marine mammals, and a few fish (Jones and Hildebrandt 1990). A 20x20x10 cm shell column sample revealed primarily mussel (94% by weight) and barnacles (5% by weight) with a few other species represented. A series of obsidian hydration dates were run from both loci, as well as one radiocarbon date from the base of the intact midden deposit in Locus 1 (Unit 9, 80–90 cm), which yielded a date of 5970 +/- 120 years.

The next major pulse of fieldwork took place in 2008 when Cabrillo College, in collaboration with California State Parks, excavated four 1x2 m units on the leeside of Locus 1 as part of the construction of an impressive boardwalk designed to keep hikers off the main portion of the site. One additional 1x2 m unit was also placed in Locus 3. While laboratory work is ongoing, the investigation has yielded a series of obsidian hydration dates, obsidian sourcing for 20 specimens, and four radiocarbon dates from intact deposits in Unit 2—at depths ranging from 90–100 cm to 150–160 cm, with dates extending from 3660–3490 cal BC to 3965–3750 cal BC (Schlagheck 2011).

The findings from previous investigations have contributed much to our understanding of the site structure and chronology of CA-SCR-7. Two distinct archaeological strata have been defined along the eroded, seaward face of the sand dune complex in Locus 1 (Hildebrandt et al. 2007). The first or lower stratum is a dark compacted deposit about one-meter-thick running along the base of the sand dune. Conducted as part of the mitigation of the abalone farm, excavations in the 1980s produced an extensive sample of archaeological materials from this lower stratum. The second or upper stratum, about one-meter-thick, caps the top of the sand dune. Limited investigations have taken place here, but this rich intact midden deposit contains ample shell, charcoal, artifacts, and vertebrate faunal remains.

Hildebrandt and colleagues (2007) employ chronological information to reconstruct how the archaeological strata and sand dune deposits in Locus 1 developed over time. Five known radiocarbon dates exist for the lower stratum, ranging from 4650–4665 cal BP to 5880–6410 cal BP, while three dates have been published for the upper stratum, ranging from 2830–3400 cal BP to 4710–5230 cal BP. They propose that at about 6000 BP, people began to use the exposed marine terrace at CA-SCR-7 for various hunting-gathering-fishing activities. Since the shoreline would have existed farther to the west at this time, it is possible that people were living on the leeward side of a sand dune that no longer exists. The initial occupation that produced the lower stratum terminated at about 4900 BP; and at this time, the current sand dune deposit began to take shape. The sand dune deposit is believed to have resulted from increased sediment loads in Laguna Creek and the formation of extensive beaches in the nearby environs, which supplied wind-blown sand that was then transported onto the marine terrace. Since the vertical distance between the lower and upper strata is about 11 meters, it appears that the sand dune deposit developed very quickly. The dating of the two archaeological strata at the base and top of the sand dune suggest the mound complex may have been created over a few centuries. A significant finding from this work is that archaeological strata in Locus 1 are largely intact and undisturbed—particularly, the lower stratum that was rapidly buried by the sand dune (Hildebrandt et al. 2007).

While Locus 2 is highly disturbed from various historic activities, field investigations yielded information on the archaeological assemblage and chronology. While no radiocarbon dates have yet been run, obsidian hydration dates and a Rossi Stemmed projectile point suggest a Middle Period age (2000–4000 BP). This suggests that the upper stratum of Locus 1 and the archaeological materials recovered from Locus 2 were probably contemporaneous (Jones and Hildebrandt 1990: 69–70). No known dates exist for Locus 3, but analysis of materials from a recent excavation unit is ongoing. Locus 4 encompassing the nearby off-mound agricultural fields appears to date to Middle Holocene and Late Holocene times, based on the projectile point types (Hylkema 2021).

CA-SCR-10

Situated 2.6 km southeast from CA-SCR-7, the majority of this site has been under active row crop cultivation for many years. First recorded in 1950 by John Costa, Donald Lathrop, and William Wallace of the University of California Archaeological Survey (Archaeological Site Survey Record 5/1/50), CA-SCR-10 was described as a large (60x213 m / 200x700 ft) "occupation site" with black soil and shell containing arrowheads, scrapers, and mortars. The sketch map showed its location west of nearby Baldwin Creek and to the north and east of Highway 1, although a portion of the site may have extended across the highway as well. The recorders noted that burials had been removed from the site and reburied in the adjacent hill. Much of the site was under active Brussels sprout cultivation.

In the early 1990s, an archaeological investigation of CA-SCR-10 took place to mitigate the adverse effects of a proposed buried water line by the Santa Cruz City Water Department along a 10.2 km section of Highway 1. The proposed Area of Potential Effect (APE) would cross the southwest periphery of the site along the northeast side of the highway. An initial surface examination of the site by field crews from FWARG described a well-developed midden deposit, measuring about 400 m N/S and 250 m E/W situated on a terrace overlooking Baldwin Creek (Jones and Hildebrandt 1994). A suite of 17 auger units, excavated to a depth of 18–100 cm using 1/8" mesh to screen sediments, was placed along the APE that yielded shell and debitage along the eastside of the highway, but sterile units were placed to the west of the highway. A series of 13 excavation units were then laid out in areas containing archaeological materials. The twelve 1x1 m and one 1.5x1 m units were excavated to depths of 40–70 cm with sediments screened through ¼" mesh. The fieldwork resulted in the collection of a diverse assemblage of flaked stone tools, including projectile points, drills, bifaces, core tools, expedient flaked tools, and considerable debitage from mostly Monterey chert—but also from some obsidian and Franciscan chert. Ground stone artifacts were identified as hand stones, pitted cobbles, cobble tools, battered cobbles, and one possible pestle fragment. A paucity of vertebrate faunal remains was recovered (25 specimens) representing large mammals, intrusive rodents, birds, and cow. A 20x20x10 cm shell column sample revealed primarily mussel (64.8% by weight) and barnacle (32.7% by weight) and a limited number of other invertebrate taxa. Unfortunately, much of this area of the site had been significantly disturbed by agricultural and road construction activities, and most of the archaeological contexts lacked stratigraphic integrity. No radiocarbon dates were run, but obsidian hydration analysis indicated a relatively late date (315–870 BP), while the three projectile points suggested an earlier chronology, ranging between 1000 to 5000 BP (Jones and Hildebrandt 1994).

In 2011, field crews from Cabrillo College and California State Parks completed the excavation of three 1x1 m units at CA-SCR-10. The analysis of material from these excavations is ongoing. The initial findings from one of the units (Unit 2) looked very promising. Situated on the southeast perimeter of the extant agricultural field, the investigation of this unit yielded a 1.7 m stratigraphic profile, with large quantities of shellfish remains and lithic artifacts. Two radiocarbon dates from the unit indicated ancient deposits. Shell from the 60–70 cm level yielded a date of 4425–4205 cal BP (Beta-429487); and from the 170–180 cm level, a date of 3545–2775 cal BP (Beta-429488) was assessed. Projectile points recovered from the site also indicate a Middle Holocene age (Mark Hylkema, personal communication).

CA-SCR-14

This upland site, located 91 m from the southeastern bank of Laguna Creek, was recorded in 1950 by Donald Lathrop and William Wallace of the University of California Archaeological Survey (Archaeological Site Survey Record 10/6/50). They described a large midden deposit, measuring 46x61 m (150x200 ft) with black soil and mussel shell. No previous excavation work had been undertaken at the site, but materials had been surface collected by local collectors, including "mortars and other objects." Some materials were in the Santa Cruz Museum (Pilkington Collection). A bedrock mortar was recorded on the site. The site form was updated in 1977 with additional information on the dispersal of shell fragments on the surface of the site not covered by dense vegetation.

CA-SCR-15

This upland site is situated about 15 m from the south bank of Laguna Creek near CA-SCR-14. Recorded in 1950 by Donald Lathrop and William Wallace of the University of California Archaeological Survey, CA-SCR-15 is described as a large midden site with black soil and mussel shell on a knoll near the creek. The area was formerly cultivated as a hayfield (Archaeological Site Survey Record 10/6/50). The site measured 61x76 m (200x270 ft) and once contained mortars that had been rolled down the slope when the area was plowed. The site record was updated in 1977 by Mark Valliar, who noted a dirt road running through the middle of this "large village midden." The site boundaries were difficult to estimate, but the midden was most visible on the large knoll adjacent to the creek. It was estimated to be 200x400 m in size and contained a hammerstone, side scraper, retouched flakes and debitage of Monterey chert, along with a wide variety of shellfish species.

On-Site Field Methodology

Eco-archaeological fieldwork at the four sites was conducted for two summer field seasons: May 23–June 17, 2016, and May 22–June 9, 2017. The research team included scholars from the Amah Mutsun Tribal Band (AMTB), the Native Stewardship Corps (NSC) of the Amah Mutsun Land Trust (AMLT), California State Parks, and the University of California campuses at Berkeley and Santa Cruz. A full list of the participants is presented in this volume's acknowledgments section.

We addressed the above four research goals by implementing a fine-grained, low-impact, eco-archaeological approach that emphasized the recovery of artifacts, ethnobotanical samples, and zooarchaeological remains and avoided disturbances to burials and other sacred Indigenous materials. Our field methodology involved four basic tasks: 1) systematic surface collection, 2) geophysical prospecting, 3) subsurface testing employing soil augers and small excavation units, and 4) collection of phytolith samples in the hinterland of several sites.

Systematic Surface Investigation

Archaeological remains were systematically recorded from across the site area using one or more collection transects. If site datums did not exist, we established them, along with one or more subdatums. Surface units (SU) were positioned along one or more transects radiating out from the datums to the boundary of the site area. While the primary transects were typically oriented N/S or E/W, the compass bearing varied from site to site, depending on the spatial configuration of surface materials, vegetation coverage, poison oak stands, and so on. Surface units were laid out at 5-meter, 10-meter, or 50-meter intervals, depending on the size of the site. The units were placed along the primary and secondary transects until surface artifacts disappeared, which marked the boundary of the site. Surface units were laid out using compass and tape or a GPS unit. Each surface unit was circular in shape (what is known as dog leash units). The units employed in the plowed field at CA-SCR-10 measured 40 cm in radius and comprised a 0.5 m² area. The surface units used at CA-SCR-7, CA-SCR-14, and CA-SCR-15 measured 28 cm in radius and comprised a .25 m² area. Units were numbered from 1 to n (e.g., SU 1, SU 2, SU 3). GPS coordinates were recorded for the site datum, subdatums, and all surface units. Crew members produced a map of the site showing the location of the site datum, subdatums, surface units, and subsurface units. Scale, north arrow, and a key to symbols used for surface units, excavation units, and cultural and natural features were included.

We used Surface Collection forms to record information on the soil and archaeological materials. Crew members used trowels to cut through the duff/grass roots to expose and collect materials from the surface of mineral soil. Except for CA-SCR-10, we screened near-surface sediments through 3.2 mm mesh. Information was recorded about the soil texture and color. All artifacts and faunal remains were recorded and provenienced by SU number. The catch-and-release methodology was employed so that no archaeological materials were collected from the surface units (see Chapter 1 for discussion on the development of this approach). Artifacts were counted and weighed using portable digital scales. The artifact categories include obsidian lithics, Monterey chert lithics, Franciscan chert lithics, other lithics, ground stone, fire-cracked rock (if bigger than fist), worked bone, shell beads, bottle glass, flat glass, worked glass, glass beads, ceramics, flat nails, round nails, and unidentified metal. The Surface Collection form also had space for adding other artifact types not listed above, which were also counted and weighed. Mammal, bird, and fish taxa were recorded as present or absent, and the total faunal remains were counted and weighed. Shellfish species (e.g., clam, mussel, oyster, barnacle) were noted as present if length was greater or equal to 1 cm. The total shellfish assemblage from the unit was then weighed, but not counted. Charred botanical remains were weighed, recorded as present or absent, but not counted.

Geophysical Prospecting

We employed a Ground Penetrating Radar (GPR) to inspect the subsurface deposits of site areas with minimal vegetation coverage (see Chapter 5). Using a GSSI SIR 3000 GPR unit with a 400 MHz antenna, one or more profiles or survey blocks were walked across the site area to record subsurface profiles. Careful analysis of the output can provide information on the depth of deposits and the detection of subsurface features. As outlined below and in Chapter 5, GPR surveys were conducted at all four sites.

Subsurface Units

We initiated subsurface testing at the four sites using three kinds of subsurface units (auger, column, and excavation). The selection of a specific type of unit was based on the depth of deposits, the context of the archaeological constituents, and our goal of minimizing subsurface impacts to sites. The placement of excavation units was typically based on the results of the GPR survey.

Soil Auger Units
Soil augers were utilized in situations involving deep deposits or during the initial testing of GPR anomalies. Each soil auger unit (AU) was numbered (e.g., AU 1, AU 2) and its GPS coordinates recorded. Sediments were collected in 20 cm levels, bagged separately, and provenienced by AU number and depth. An Auger Record form was recorded for each AU, which provides space for describing archaeological materials, sediments, charcoal, and so on, for each 20-cm level. Sediments collected from auger units were either dry screened using 3.2 mm mesh or floated using the method described below.

Column Units
We employed column units to excavate and record the exposed profiles at CA-SCR-7. Each column unit (CU) was numbered (e.g., CU 1, CU 2) and its GPS coordinates recorded. Sediments collected in 10-cm levels were bagged separately and provenienced by CU number and depth. The sediments from all column units were floated, as described below and in Chapter 8.

Excavation Units
Larger units were laid out and excavated in situations where enhanced subsurface samples were justified or warranted. Excavation units varied in size from 50x50 cm to 1x1 m. Each excavation unit (EU) was numbered (e.g., EU 1, EU 2) and its GPS coordinates (corners) marked. Excavations were undertaken in natural or arbitrary (10 cm) levels, depending on the nature of the stratigraphy. Column samples from each level were collected for flotation. The remainder of the sediments from specific levels were dry screened through 3.2 mm mesh.

Phytolith Survey

We also collected near-surface soils in the proximity of CA-SCR-7 and of CA-SCR-10 and between CA-SCR-14 and CA-SCR-15 for phytolith analysis.

On-Site Field Investigations and Findings

The specific suite of methods employed at each of the four sites are outlined below, along with a discussion of the findings from surface investigations conducted.

CA-SCR-7

In our study of CA-SCR-7, we employed a fine-grained, low-impact methodology that was designed to complement and enhance the findings from previous investigations. Our work focused on developing a better understanding of the site structure of the eroded, exposed sand dune complex of Locus 1, as well as obtaining information about archaeological deposits in the leeward side of the mound complex (Locus 4). Field crews initiated the following activities: examining the upper and lower midden deposits visible in Locus 1; searching for additional midden deposits buried in Locus 1; surveying the off-mound area in Locus 4; and undertaking a phytolith survey in the nearby environs. Four datums were established on the main dune (Locus 1) and one datum on the northern dune (Locus 3). As described below, the work involved the investigation of 7 column units, 3 auger units, 2 excavation units, and 76 surface units (see Figure 4.1).

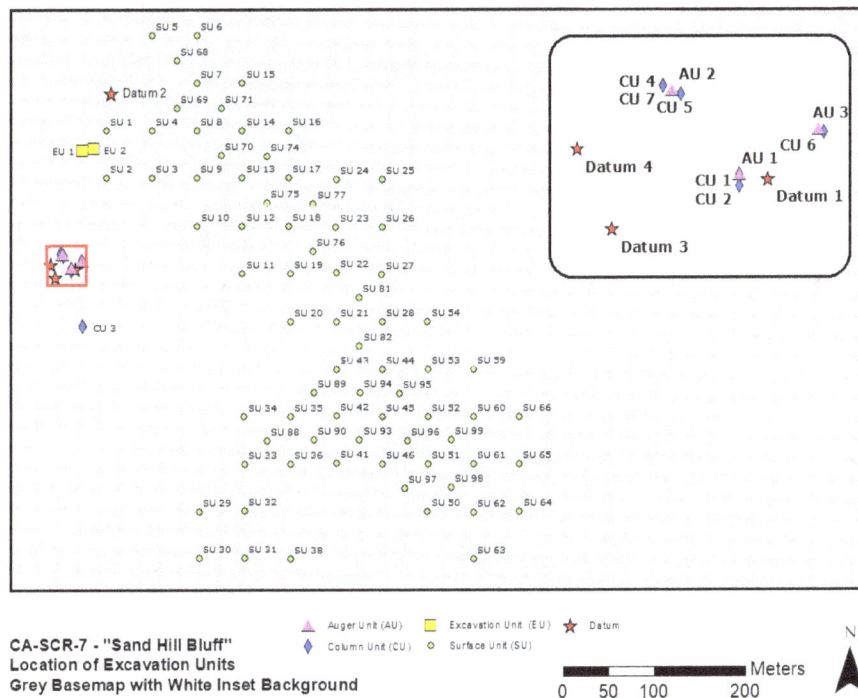

Figure 4.1. Location of Datums, Column Units, Auger Units, Excavation Units, and Surface Collection Units at CA-SCR-7. The data are schematically represented without reference to topographic features or geospatial landmarks following site stewardship protocols of California State Parks and the Amah Mutsun Tribal Band for publishing information on archaeological sites (Map by Rob Cuthrell and Alec Apodaca).

Upper Midden Deposit: CU 1 and CU 2

The first activity involved the recording and excavation of two column samples from the upper midden stratum in Locus 1 that caps much of the sand dune deposit. We exposed a 3.7-meter clean profile of the upper midden deposit (Northwest Profile) that exposed three stratigraphic layers to a depth of about 1 m (Figure 4.2). The uppermost layer (Stratum 1) consisted of a loose dark grayish brown (10YR 4/2) sand deposit that varied from ca. 1 to 10 cm in depth below profile datum. The second layer (Stratum II) was composed of dense fine shell in a black (10YR 2/1) matrix that varied in depth from ca. 1 to 20 cm in depth. The majority of the profile was comprised of the third layer (Stratum III) of very dense coarse shell in a very dark gray (10YR 3/1) matrix that extended from about 20 cm to about 1 m below profile datum. Wall fall mixed with brown sand (10 YR 3/2) appeared about 1 m below profile datum. Two locations were then chosen for column sampling that appeared to be relatively intact with minimal rodent disturbance or irregular faces. Column Unit 1 (CU 1) was established at 30–80 cm along the profile face (first set of dotted lines in Figure 4.2), while Column Unit 2 (CU 2) was situated at 230–285 cm along the profile (second set of dotted lines in Figure 4.2). The columns were created in a trapezoidal shape to avoid creating 90-degree corners more vulnerable to collapse. Sediments were excavated in 10-cm levels—generally 5–15 liters, depending on slope geometry—and collected in bulk for later flotation. We also collected microbotanical columns from the side walls of each column. This involved collecting about 100 g of soil at 10-cm increments beginning in the middle of each 10-cm level. Excavation notes were taken on unit forms.

Figure 4.2. Northwest Profile of Upper Midden for CA-SCR-7. Dudleya refers to the succulent plant drawn in the profile (Profile by Rob Cuthrell and Ariadna Gonzalez).

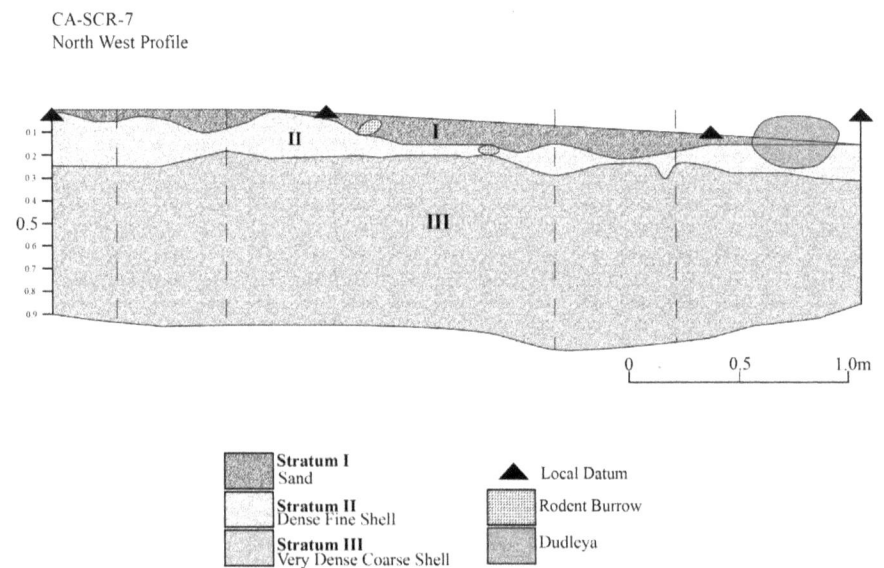

We excavated eleven levels from CU 1 to a depth of 110 cm. Most of the deposit consisted of the very dense coarse shell dominated by mussels and some large barnacles. Some occasional chert artifacts and faunal remains were reported and collected. The lower 100–110 cm consisted of a mixture of lighter sand and shell with fewer artifacts. Field crews working at CU 2 excavated nine levels to a depth of 90 cm. Similar to CU 1, much of the unit consisted of the very dense coarse shell layer full of mussels and very large barnacles. Some artifacts and faunal remains were found throughout the unit. Sand began to show up in increasing concentrations at about 65–70 cm below profile datum, although artifacts continued to be found. The unit was terminated at 90 cm.

Lower Midden Deposit: CU 3

The second activity was the excavation of a column unit from the lower midden stratum in Locus 1 at the base of the sand dune. Since the FWARG investigation in the 1980s sampled an extensive area of the lower stratum, we only placed one column sample here that would build upon previous findings. Our work provided the opportunity to recover flotation samples and to undertake more fine-grained recovery of materials from this archaeological deposit. In the summer of 2016, we observed a ca. 50 cm deposit of highly organic dark sand with shell and some artifacts that was still visible along the cliff face. However, placing a crew along the cliff face was untenable given the precarious situation. Consequently, we placed Column Unit 3 (CU 3) about three meters interior from the cliff face where it was safe to conduct our study.

CU 3 measured about .5x.2 m in size, and six levels were excavated to a depth of 60 cm. The surface of the unit appeared to be possibly deflated. Our excavations exposed the deposit visible on the cliff face, but it only extended from the surface to about 25 cm below surface, and shell was relatively sparse and small. Our findings suggest the deposit visible on the cliff face tapered off toward interior and the original extent of the deposit has mostly been scoured away. Sediments were excavated in 10-cm levels and collected in bulk for flotation.

Search for Additional Midden Strata: AU 1, AU 2, AU 3, CU 4, CU 5, CU 6, CU 7

The third activity considered whether other archaeological deposits may be found in Locus 1. Little is known about the existence of additional archaeological remains within this massive sand dune complex that may exist between the upper and lower midden deposits. To better understand how and when this area may have been used, we initiated GPR profiles to detect additional cultural deposits that may be interred in the sand dune sediments, as described in Chapter 5. Employing the findings of the geophysical survey, along with a careful inspection of the eroded, windward side of the mound, we searched for evidence of buried midden deposits in the mid-section of the mound. This activity involved the excavation of three auger units and four column units.

AU 1. The first auger unit was placed in the bottom of CU 1. The goal was to evaluate if archaeological remains were found directly below the upper midden deposit in the underlying sand dune. AU 1 was excavated in 15 levels of 20 cm to a depth of about 3 m. The first and second levels had moderate shell density, probably materials associated with the upper cultural deposit. Sediments from level 1 (10 YR 4/2) were collected for flotation, and sediments from level 2 (10 YR 5/4) were dry screened through 3.2 cm mesh, which yielded no artifacts. Sediments from subsequent levels were dry screened through 3.2 mesh but consisted primarily of pure sand with a few shell, which became sparse after the first meter or so. No artifacts were recovered. The shell may be from natural deposition with birds dropping shell into sand dune. The last level (15) from 280–300 cm from the bottom of CU 1 was collected as a bulk sample for flotation.

AU 2. We placed AU 2 about one-meter upslope of an exposure of cultural deposit in the lower sand dune deposit. Our goal was to determine if the deposit continued into the dune or represented a slump of material from the upper midden deposit. We excavated seven 20-cm levels to a depth below the starting point of about 120 cm. Cultural materials were recovered at the bottom of level 3 (40–60 cm), in level 4 (60–80 cm), and the top of level 5 (80–100 cm). It appears the cultural level is about 40–60 cm thick. We then excavated 40 cm below the cultural deposit (levels 5 and 6); and after determining that the lower deposits were almost pure sand, we closed the unit. We collected the sediments from levels 4 and 5 for flotation, while sediments from the other levels were dry screened through 3.2 cm mesh. It appears that a cultural deposit of about 40 cm in thickness was detected and sampled.

AU 3. Field crews established AU 3 on the upper surface of the dune where the results of the GPR survey indicated a possible cultural layer at an estimated depth of 1 m (see Chapter 5). AU 3 was situated along an open trail that people used in walking across the top of the dune. We excavated eight 20-cm levels to a depth of about 160 cm below surface. The first three levels (0–60 cm) were almost pure sand (10 YR 5/4) with a few shell inclusions. In the bottom of level 4 at about 65 cm, cultural materials were detected. The cultural deposit (10YR 3/2, 10YR 4/3) continued in level 5 (80–100 cm) and the upper 5 cm of level 6 (100–105cm). The remainder of level 6 (105–120cm), level 7 (120–140cm), and level 8 (140–160 cm) returned to almost pure sand (10YR 5/4, 10YR 6/4). Sediments excavated from levels 4, 5, and 6 that contained cultural materials were collected as bulk samples for flotation. Sediments from the other levels were dry screened through 3.2 mm mesh. Thus, AU 3 resulted in recovery of cultural deposit from 65 to about 105 cm depth below surface. Sediments from this very dark matrix, comprised of dense shell but in very small pieces, were floated and the resulting materials analyzed under laboratory conditions at UC Berkeley.

CU 4. This column unit was situated about two meters downslope of AU 2 to further test the area with the exposed cultural deposit in the lower sand dune deposit. CU 4 measured about 40–50 cm in width and involved the profiling and excavation of three 10-cm levels. Each level was excavated as a step to avoid destabilizing this area of the dune. Sediments from the three levels were collected for flotation. During the excavation, crew members noted that these samples may be compromised by slope wash from above.

CU 5. We placed this column unit about two meters upslope of AU 2 in our efforts to continue to document the area with the exposed cultural materials in the lower sand dune deposit. Crew members laid out CU 5, measuring 35 cm in width, and excavated two 10-cm levels to a depth of about 20 cm. Sediments contained dense shell with some larger pieces, but the matrix was not dark and organic—it was just slightly darker sand. Charcoal, clay, and a burned mammal bone were noted during excavation. It is possible that this deposit may represent old slope wash or material that had been buried and was now becoming exposed by erosional processes on the windward side of the dune. All sediments were collected as bulk samples and floated and earmarked for further laboratory analysis.

CU 6. Crew members established CU 6 just above AU 3 on the upper surface of the mound complex. Measuring 47 cm in width, the unit was excavated in four levels: level 1 (0–10cm), level 2 (10–20 cm), level 3 (20–35cm), and level 4 (35–45cm). The levels sampled a very dark organic matrix with moderate density of small shell pieces. It is possibly a continuation of the upper midden deposit profiled (CU 1 and CU 2) on the windward side of the dune. Sediments from all four levels were collected as bulk samples and floated.

CU 7. In locating this column unit near CU 4, we continued sampling this area with the purpose of excavating deeper and horizontally into the dune deposit so that cultural materials could be collected stratigraphically. The first level (4–16cm) was a dark cultural deposit with dense shell and chert artifacts. The second level (16–22 cm) was comprised of sand with moderate density of shell. The third level (22–29cm) unearthed the dark cultural deposit with dense shell, charcoal, and bone. The fourth level, extending the unit to 34 cm below surface, also contained dense shell but with a sand matrix. All sediments from the four levels were collected as bulk samples and floated.

Off-Mound Investigation

The fourth major activity involved fieldwork in adjacent off-mound locations where cultural materials were detected in former agricultural fields (Locus 4). The purpose was to obtain information on the diversity of cultural remains in the off-mound area and to define the broader spatial distribution of materials and boundaries of the site. We laid out a grid system oriented north/south with a series of east/west-oriented transects that extended from the leeward edge of the mound to the outlying landscape. Surface units were placed at 50-meter intervals in an eastern direction along the transects. Surface units were inspected every 50 m along the transect until cultural materials disappeared. We employed a GPS unit to locate each surface unit along eastern transects. Each surface unit was circular in shape (aka dog leash units), measured 28 cm in radius, and comprised a 0.25 m² area. For each surface unit, crew members excavated 30 liters of sediments, screening the contents though 3.2 mm mesh and then recording cultural materials using our survey form described above. While the quantity of sediments screened per surface unit was only 5 liters at the other sites, we decided to increase the volume of sediments examined at CA-SCR-7 to 30 liters to better define the spatial patterning of archaeological materials across the formerly plowed fields. Consequently, in calculating surface densities below, we used two measures: cultural materials per liter screened (n/l) and cultural material per area (n/m²). In calculating the latter measure, we divided the total number of remains per m² by 6 to make the densities for the surface units from CA-SCR-7, based on the screening of 30 liters of sediments, comparable with the other sites where we only screened 5 liters. A total of 76 surface units (28 cm radius) were investigated along the survey transects (see Figure 4.1).

Survey Results. The on-site surface investigation of the 76 units in the summer of 2016 revealed a relatively high density of lithic artifacts, shellfish and vertebrate faunal remains, and some charred botanical materials (Appendix 4.1, Table A1). We recorded 1,094 artifacts. The vast majority were identified as Monterey chert flakes and debitage, tallying 1,015 artifacts, followed by 20 Franciscan chert and one obsidian lithics. Then, 5 ground stone artifacts and 6 fire-cracked rocks were also noted, along with 12 other lithic artifacts. No worked bone, antler or shell artifacts were recorded. Euro-American artifacts included 6 sherds of bottle glass, 6 sherds of flat glass, 1 piece of worked glass, 1 glass bead, 1 ceramic artifact, 6 unidentified metal fragments, 8 plastic pieces, and 6 other artifacts. The mean number of artifacts per unit was 14.39 (SD = 24.94) or .47 artifacts/liter (SD=.83). This yielded a density (as calculated above) of 9.59 artifacts/m² (SD=16.6).

We recorded a total of 1,059 lithic artifacts in the surface units, with a mean number of 13.93 lithics per unit (SD=24.82) or .46 lithic/liter (SD=.83). This yielded a density of 9.33 lithics/m^2 (SD=16.55). The spatial distribution of the lithic artifacts suggests two areas of concentration in the former agricultural fields: in the northwest surface units between Locus 1 and Locus 3 and to the southeast of Locus 1 (see Lightfoot et al. 2021).

Then, 41 vertebrate faunal specimens, weighing 12.91 grams, were observed. These included mammal (from 10 units), bird (from 4 units) and fish (from 1 unit). The mean number of faunal specimens per unit was .54 (SD=2.68) or .02 elements/liter (SD=.09), and the mean weight of the faunal remains per unit was .17 grams (SD=.94). This yielded faunal densities for the surface units of .36 elements/m^2 (SD=1.79) and .11 grams/m^2 (SD=.63). The highest density of faunal remains was found between Locus 1 and 3, with only sporadic finds found elsewhere. An assortment of shellfish species, weighing 188.57 grams, was recorded for the units. The most common invertebrate remains recorded were mussel (43 units), followed by oyster (18 units), clam (16 units), turban snails (7 units), barnacle (5 units), abalone (1 unit), and other shell (2 units). The mean weight of shellfish per unit was 2.48 grams (SD=9.13) or .083 grams/liter (SD=.30), with a density of 1.65 grams/m^2 (SD=6.09). The spatial distribution of shell resembled that of the lithic artifacts: the greatest concentration of shell was detected between Locus 1 and 3, with some additional shell found to the northeast and southeast of Locus 1 (see Lightfoot et al. 2021). Eleven units contained charred botanical remains that weighed ca. 4.04 grams or about .05 grams per unit (SD=.205).

Notable cultural materials observed during the survey included flakes, thinning flakes, and shatter, and a Monterey chert (MC) projectile point (SU 2); a complete flake, thinning flakes, debitage, and a MC core (SU 4); a MC flake showing evidence of edge-modification, along with other flakes (SU 41, SU 46); possible worked glass and lithics (SU 63); a MC biface and other lithics (SU 94); the probable tip of a quartzite projectile point, obsidian flake, a MC core, and all stages of MC lithic manufacture (SU 69); and other units with MC lithics showing all stages of lithic manufacture (e.g., SU 42, SU 43, SU 89) (see Figure 4.1).

Exploratory GPR and Excavation Units. Based on survey results, we chose two areas in the summer of 2017 located between Locus 1 and Locus 3 where dense and diverse archaeological remains had been recorded. In each area, field crews laid out and surveyed a 20x20 m grid unit using the GSSI SRI-30000 GPR with 400 MHz antenna with survey wheel. In the first geophysical grid, which produced the most interpretable results, we placed two 50x50 cm excavation units (EU 1, EU 2) (Figure 4.1).

EU 1. This 50x50 cm excavation unit was excavated in 9 levels to a depth of about 90 cm. Much of the unit is characterized by disturbed plow zone that contained some shell and lithics. All sediments were dry screened through 3.2 mm mesh. In level 8, about 70 cm below datum, a possible intact shell lens was detected. A 5-liter bulk sample of this deposit was collected for flotation. In level 9, at about 90 cm below datum, the underlying clay horizon was reached, and the unit terminated.

EU 2. This second 50x50 cm excavation unit was set up near EU 1 but had to be shut down due to time constraints. Field crews observed lithics, historic glass, and shell in the upper levels of the unit.

Phytolith Survey

The fifth activity undertaken at CA-SCR-7 was a phytolith survey that sampled near-surface soils for phytolith remains in the proximity of the site. The location of the four phytolith samples are documented by Lightfoot and colleagues (2021).

Summary of Field Investigation at CA-SCR-7

In summary, our fieldwork supported previous work at CA-SCR-7, indicating it is a large sand dune mound sandwiched by archaeological deposits at its base and top. The geophysical survey and subsequent subsurface investigation show that the mound complex is also interlaced with strands of cultural layers. The sampling of these layers and subsequent analysis of materials from CU 1 to CU 7 and from AU 1 to AU 3 provided additional information about the chronology, cultural materials, and internal spatial structure of the mound complex. We collected sediments for flotation from selected levels of the auger units where archaeological materials were detected. These included level 1 (0–20 cm) and level 15 (280–300 cm) from AU 1, level 4 (60–80 cm) and level 5 (80–100 cm) from AU 2, and level 4 (60–80 cm), level 5 (80–100 cm), and level 6 (100–120 cm) from AU 3. We floated all the sediments excavated from the 7 column samples. These included 11 levels (0–110 cm) from CU 1, 9 levels (0–90 cm) from CU 2, 5 levels (0–50 cm) from CU 3, 3 levels (0–30 cm) from CU 4, 2 levels (0–20 cm) from CU 5, 4 levels (0–45 cm) from CU 6, and 4 levels (0–35 cm) from CU 7. We also floated the bulk sample from level 8 of EU 1 located between Loci 1 and 3. The light and heavy fractions from all these flotation samples, as well as the dry screened samples from EU 1, were subsequently analyzed under laboratory conditions at UC Berkeley and elsewhere. The findings from this work are discussed in subsequent chapters.

Our fieldwork also recorded the extensive surface distribution of materials that extends from the leeward edge of the mound complex (Loci 1–3) into nearby agricultural fields (Locus 4). The investigation of the surface units shows that a relatively dense and diverse range of archaeological materials are found in the agricultural fields near the mound complex. The survey work helped define the boundary of CA-SCR-7 more thoroughly.

CA-SCR-10

Our investigation consisted of three activities. The first was to excavate a 1x1 m unit in the southeast quadrant of the site. The second was to explore the central area of the site during the fallow season following the harvest of the Brussels sprout crop. During this short window of opportunity, we conducted a geophysical survey, initiated a limited subsurface investigation with auger units, and performed a surface survey. The third activity involved a phytolith survey in the nearby environs. Figure 4.3, the site map for CA-SCR-10, shows the location of the 2 datums, 1 excavation unit, 3 auger units, and 38 surface collection units.

Excavation Unit (EU 1)

The initial investigation of CA-SCR-10 focused on the southeastern edge of the agricultural field. As outlined in our research design (Chapter 2), we proposed to excavate

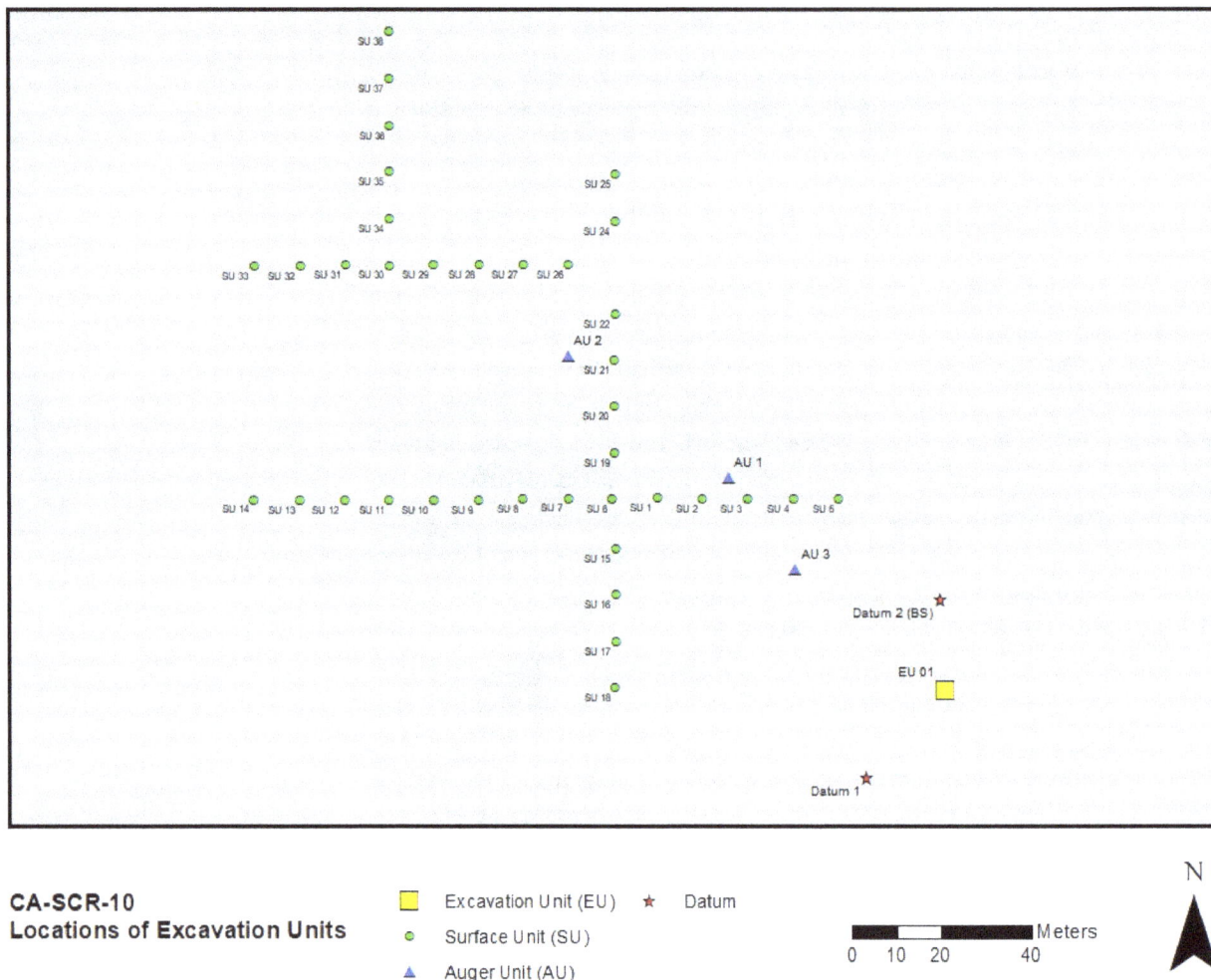

Figure 4.3. Location of Datums, Auger Units, Excavation Unit, and Surface Collection Units at CA-SCR-10. The data are schematically represented without reference to topographic features or geospatial landmarks following site stewardship protocols of California State Parks and the Amah Mutsun Tribal Band for publishing information on archaeological sites (Map by Rob Cuthrell and Alec Apodaca).

a 1x1 m unit near Unit 2, which had been excavated by Cabrillo College/California State Parks in 2011. Previous excavations in Unit 2 detected a dense assemblage of shell, vertebrate remains, and artifacts to a depth of 1.7 meters below surface. We recognized from the outset that the two radiocarbon dates from the unit were reversed with the 60–70 cm level dating to 4425–4205 cal BP, and the lowest level (170–180 cm) dating to 3545–2775 cal BP, possibly indicating some stratigraphic disturbance. Yet we reasoned that once we got below the plow zone and area of agricultural impacts, the archaeological materials should be relatively intact. Our purpose in opening this second unit, defined as EU 1, was to complement the results of the prior study by collecting more samples for fine-grained analysis that would yield paleoethnobotanical remains, micro artifacts, and small-bodied faunal specimens. We employed the GPR and magnetometry to help us pinpoint the placement of EU 1 near the original location of Unit 2 (Chapter 5). Our methodology involved excavating cultural/natural levels using shovels, trowels, and other hand tools. We employed arbitrary levels (10 cm) in contexts where cultural/natural stratigraphy was not visible. Bulk sediments were taken from most levels for flotation, along with separate 100-gram microbotanical samples, and all remaining sediments were dry screened through 3.2 mm mesh. The sediments from cultural features, such as hearths, ash lenses, ovens, and so on, were collected as bulk samples for flotation. We designated the northeast corner as the unit datum and all depth measurements taken in the unit refer to depth below datum.

2016 Field Season. As expected, when we initiated our subsurface investigation in late May and early June 2016, the excavation of the upper levels (0–70 cm) of EU 1 unearthed a diverse range of artifacts, shellfish, and faunal remains in a highly disturbed context. We completed level 1 (0–12 cm), level 2 (12–22 cm), level 3 (22–32 cm), level 4 (32–42 cm), level 5 (36–42 cm), level 6 (42–48 cm), level 7 (48–52 cm), level 8 (52–62 cm), and level 9 (62–72 cm). Excavators recovered Monterey chert flakes and debitage, some obsidian, Olivella shell fragments, and plentiful faunal remains, along with plastic bags, other historic materials, and evidence of plowing (plow-scarred rocks) and rodent burrows. Unfortunately, excavation of subsequent levels 10 (72–81 cm), 11 (81–92 cm), and 12 (92–102 cm) detected similar evidence of modern disturbance. Crew members were disappointed to be working in a highly impacted context more than a meter below surface. The last level excavated in the summer of 2016 finally brought some hope to the subsurface investigation. In level 13 (102–112 cm), excavators detected shell that appeared to be in intact, as well as an increasing number of fire-cracked rocks that may also have been in situ. While some plastic was also recovered, it appeared to be from an upper rodent burrow. Excavators closed the unit at the end of the 2016 field season.

2017 Field Season. Field work resumed at EU 1 with the reopening of the unit to the bottom of the previous level. Once the unit was cleaned (level 13/14), we began the excavation of level 14 (112–118 cm), which revealed a potential hearth area characterized by a concentration of intact fire-cracked rock surrounded by shell and fish and mammal bones. The feature was collected as a bulk sediment sample. Level 15 (112–120 cm) exposed charcoal, possible worked bone, and chert artifacts, along with intact shell, including burned mussel. In level 16 (120–130 cm), an intact ash lens was unearthed and a 5-liter bulk sample taken for flotation. While some plastic was also uncovered, it appeared to be from a distinct rodent burrow. Levels 17 (120–128 cm) and 18 (130–144 cm) continued with the exposure of the ash feature and intact fire-cracked rocks in the south and northeastern section of the unit. Bulk sediment samples were taken from the feature. Level 19 (132–135 cm) sampled an ash lens with charcoal and faunal remains in the northeastern corner of the unit. Level 20 (130–131 cm) sampled a similar feature in the southern boundary of the unit. Level 21 (129–136 cm) exposed another ash feature in the southwestern corner of the unit associated with intact shell and charcoal, while level 22 (136–142 cm) documented multiple ash lenses and intact shell midden in the southern and eastern boundaries of the unit. Finally, levels 23 (136 cm) and 24 (136–140 cm) sampled a dense shell lens in the southern section of the unit. Bulk sediment samples were taken for all the cultural features.

The unit was terminated at the end of level 24 at a depth below datum of 144 cm. We photographed and drew profiles of all four walls and completed a unit form. The South Wall profile (Figure 4.4) and North Wall profile (Figure 4.5) illustrate the disturbed stratigraphy of EU 1. For the South Wall profile, we defined three different strata for the upper deposits to a depth of about 80 to 110 cm: Stratum I as brown (10YR 3/2) clay loam; Stratum II as very dark gray (10YR 3/1) sandy loam; and Stratum III as black (10YR 2/1) clay loam. All three strata had been highly disturbed by rodent burrows and agricultural activities. Stratum IV represents the intact archaeological deposit in EU 1 for the South Wall. It consists of black (10YR 2/1) sandy loam that contains ash lenses and shell concentrations.

The North Wall profile documents a similar stratigraphic record. We defined six different strata in the upper 90–110 cm of the unit: Stratum I as a very dark gray (10YR 3/1) soft sandy deposit; Stratum II as a brown (10YR 3/2) clay loam; Stratum III as very dark gray (10YR 3/1) sandy clay loam; Stratum IV as very dark gray (10YR 3/1) clay loam; Stratum V as a small pocket of very dark gray (10YR 3/1) sandy clay loam; and Stratum VI as black (10YR 2/1) sandy clay loam. All these strata showed signs of bioturbation and agricultural disturbance. Stratum VII, defined as a black (10YR 2/1) clay loam, denotes the intact deposit for EU 1 for the North Wall. This stratum contained the ash lenses and shell midden pockets that were collected as bulk sediment samples.

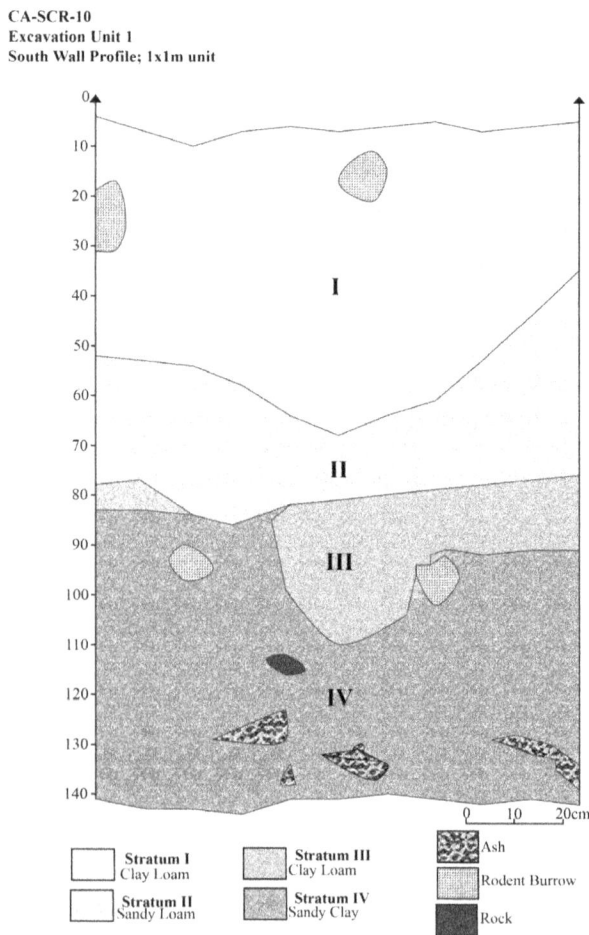

Figure 4.4. South Wall Profile of EU 1 for CA-SCR-10 (Profile by Ariadna Gonzalez).

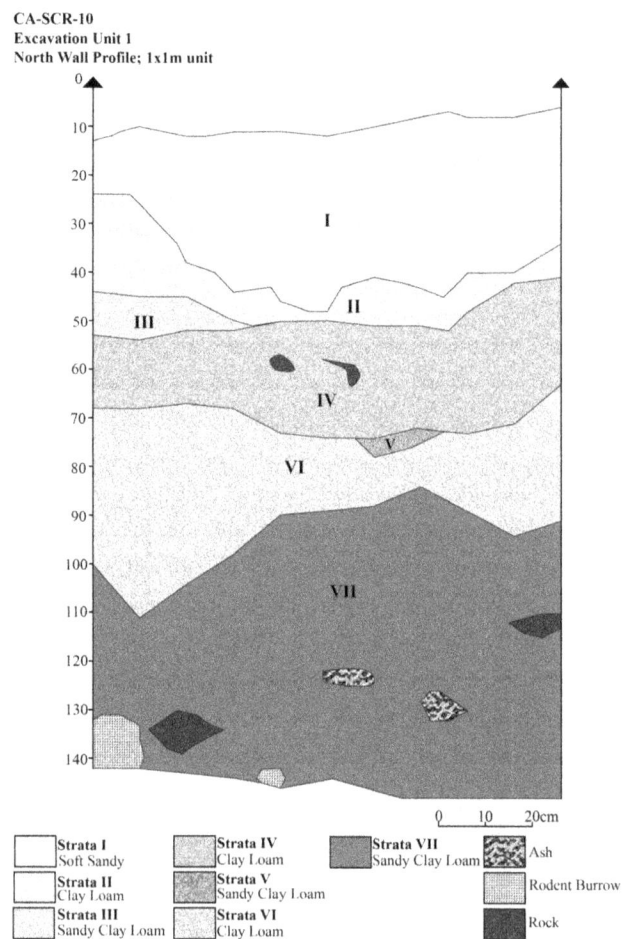

Figure 4.5. North Wall Profile of EU 1 for CA-SCR-10 (Profile by Ariadna Gonzalez).

Surface Investigation and Soil Augering

The second major activity at CA-SCR-10 involved recording and sampling the broader area of the site that was under cultivation. The opportunity to examine the central area of the site containing artifacts, shellfish, and faunal remains took place during a brief window (October 2016) when the site was left fallow. (Shortly after our fieldwork was completed and the results reported to California State Parks, the area of CA-SCR-10 was removed from agricultural production.) Despite many decades of plowing, our investigation documented an extensive mounded feature in the central area of the site. It is unclear what the topography of the site looked like prior to agricultural production, but the mounded contour of the site still existed. The fieldwork conducted during the fallow period of the year involved three tasks: geophysical survey, surface investigation, and soil augering.

Geophysical Survey. We established three 20x20 m blocks in the central area of the site. Field crews then employed the GSSI SIR-3000 GPR with 400 MHz antenna and survey wheel to complete GPR surveys of each grid. The findings from this work are detailed in Chapter 5.

Surface Investigation. The second task was to undertake a systematic surface investigation of archaeological materials across the central area of the fallow site. We established a series of north/south and east/west transects with surface units located about 10 m apart (Figure 4.3). Each surface unit measured 40 cm in radius, which comprised a 0.50 m^2 area. A GPS unit was used to locate and record the coordinates of each surface unit. We found the surface of the fallow site to be quite compact after the farmers had plowed the ground to bare earth. The hard pan earth was not suitable for screening. Two-person field crews used trowels to pick archaeological materials from the compact surface of each unit. Archaeological remains from a total of 38 surface units were recorded using the Surface Unit forms.

The on-site surface investigation of the 38 units recorded a total of 68 artifacts. Lithic flakes and debitage comprised the majority of the assemblage (n=55), followed by fire-cracked rock (n=12), and one bottle glass fragment (Appendix 4.1: Table A2). We did not differentiate Franciscan versus Monterey chert in this survey, but survey forms indicated most of the lithic flaked artifacts were Monterey chert. While no ground stone artifacts were recorded in the survey, our discussion with the people farming the property indicated that they removed any large stones, including fire-cracked rock and ground stone tools, from the plowed fields. This appears to be an activity that has gone on for many years, resulting in many ground stone tools on the periphery of the site. The mean number of artifacts per unit was 1.84 with a SD of 1.80. This yielded a density of 3.68 artifacts/m^2 (SD=3.6). A total of 67 lithic artifacts were recorded. One bottle glass sherd was also recorded. The mean number of lithics per unit was 1.81 (SD=1.82), which yielded a density of 3.62 lithics/m^2 (SD=3.64). The spatial patterning of the lithics indicate they are concentrated in the central mounded area of the fallow field (see Lightfoot et al. 2021). Three vertebrate faunal specimens were recorded in SU 3, SU 20, and SU 26 (Figure 4.4) and were all identified as mammal. No faunal weights were made. The mean number of faunal specimens per unit was .08 (SD=.28) or .16 elements/m^2 (SD=.56). Mussel shell was recorded in 30 units, barnacle in 21 units, clam shell in 3 units, and abalone in 3 units. No shellfish weights were made. No charred plant remains were recorded in the units.

Soil Augers. Employing the results of the GPR survey, we selected three locations in the central area of the site for subsurface investigation using soil augers (Figure 4.3). We excavated AU 1 in five 20-cm levels to a depth of about 100 cm. Shell and artifacts were observed in the first four levels; however, level 5 (80–100 cm) appeared to be relatively sterile. Sediments from all the levels were collected as bulk samples. We floated the sediments from level 4 (60–80 cm). AU 2 was excavated in three 20-cm levels to a depth of about 60 cm, when the frequency of shell and artifacts declined markedly and a reddish colored subsoil was observed. Sediments from all levels were collected as bulk samples, and we floated level 3 (40–60 cm). Field crews excavated four 20-cm levels for AU 3 and observed artifacts and large shell pieces. At the bottom of level 4 (60–80 cm), cultural material declined markedly, and gravel and underlying bedrock was exposed. Sediments from all levels were collected as bulk samples, and we floated the contents of level 4.

Phytolith Survey
The third major activity involved the collection of samples from near-surface soils for phytolith analysis. Three phytolith samples were collected from the nearby environs of CA-SCR-10 (see Lightfoot et al. 2021).

Summary of Field Investigation at CA-SCR-10

In summary, the investigation of EU 1 indicated that archaeological materials continued to a depth of about 1.44 m below surface on the southern periphery of the site. Our study observed evidence of recent historical disturbance to a depth of about 1 m in this area, as evidenced by plastic wrappers and other contemporary materials. We suspect this was a consequence of the combination of agricultural activities and burrowing animals. In the summer of 2017, we detected undisturbed deposits containing archaeological materials at a depth of about 112 cm to 144 cm below surface (Stratum IV in the South Wall profile and Stratum VII in the North Wall profile, Figures 4.4 and 4.5, respectively). We unearthed a series of intact cultural features. The shell lenses, concentrations of fire-cracked rocks, and ash lenses were collected as bulk samples for flotation and subsequent analysis.

Two types of samples from EU 1 were prepared for subsequent laboratory research. The 3.2 mm dry-screened samples from the 24 levels were transported to the California Archaeology Laboratory at UC Berkeley for analysis. The bulk sediment samples from the EU 1 levels and cultural features were floated in the field and the light and heavy fractions subsequently analyzed under laboratory conditions at UC Berkeley and elsewhere.

Our investigation of the central area of the site during the fallow season indicates the surface distribution of cultural materials in the plowed area covers about 140 m (n/s) by 130 m (e/w). We detected primarily lithic flakes, debitage, and fire-cracked rocks. The density of vertebrate faunal remains was low, although shellfish was found in almost half of the units. The low density of material recorded in the surface investigation may be a product of our methodology: we did not do any screening of sediments (as in our other surface investigations), given the hard pan surface of the site created by recent tractor activity. Another reason for the low density of cultural material is probably due to the long-term practice of moving larger artifacts and rocks from the plowed field area. The results of the three auger units indicate that the central area of the site contains archaeological deposits to a depth of about 60–80 cm below surface. Sediment samples from the three auger units were floated and the light and heavy fractions analyzed under laboratory conditions. They included sediments from level 4 (60–80 cm) of AU 1, level 3 (40–60 cm) of AU 2, and level 4 (60–80 cm) of AU 3.

The archaeological deposits in the central area of the site were considerably shallower than what we expected given the depth of cultural materials found along the southern periphery of the site (previously excavated Unit 2=1.7 m b.s.; EU 1=1.44 m b.s.). We believe this situation is a product of many decades of agrarian practice. Materials from the central mounded area of the site have been flattened and pushed over time to the periphery of the site. Thus, while the central area of the site has been deflated over time, the periphery area has been built up with the redeposition of deposits from the mound area (see Chapter 5, this volume). Our investigation suggests that only about a 30 cm deposit on the periphery area of the site was intact (112–144 cm), while the upper levels appear to have been largely redeposited from earlier agrarian activities.

CA-SCR-14

Field crews initiated a diverse range of activities in the summer of 2017 including geophysical survey, systematic surface investigation, the excavation of soil auger units and excavation units, and a phytolith survey. The work involved establishing 2 datums, 5 auger units, 35 surface collection units, and 2 excavation units (Figure 4.6).

Geophysical Survey
The study of CA-SCR-14 began with some vegetation removal to clear five profiles using the GSSI SIR 3000 GPR unit with a 400 MHz antenna. The findings indicated anomalies that might represent intact subsurface cultural deposits (0–50 cm in depth) and features (see Chapter 5, this volume).

Surface Investigation
Field crews laid out several survey transects across the site and collected 35 surface units, each measuring 28 cm in radius (.25 m^2 area) and spaced 5 m apart (Figure 4.6). Then, 5 liters of sediments were excavated from each unit and screened through 3.2 mm mesh. A total of 150 artifacts were recorded, including 75 Monterey chert flakes and debitage pieces, 3 other flaked stone artifacts (1 basalt and 2 "overheated chert"), 1 ground stone (possible pestle fragment), and 62 pieces of fire-cracked rock. Euro-American manufactured materials consisted of 4 bottle glass sherds, 2 flat glass fragments, 2 ceramic sherds, and 1 unidentified metal piece (Appendix 4.1: Table A3). The mean number of artifacts per unit was 4.28 (SD=3.59), yielding a density of 17.12 artifacts/m^2 (SD=14.36). The 141 lithic artifacts yielded a mean number of lithics per unit of 4.03 (SD=3.23) or 16.12 lithics/m^2 (SD=12.92). The spatial distribution of lithic artifacts suggests three major loci of cultural materials: one concentration in the northcentral area, a second in the southcentral area, and a third trailing down the slope to Laguna Creek on the west side of the site (Lightfoot et al. 2021).

In total, 77 vertebrate faunal specimens, weighing 44.06 grams, were identified. These included mammals (from 26 units) and fish (from 1 unit). The mean number of

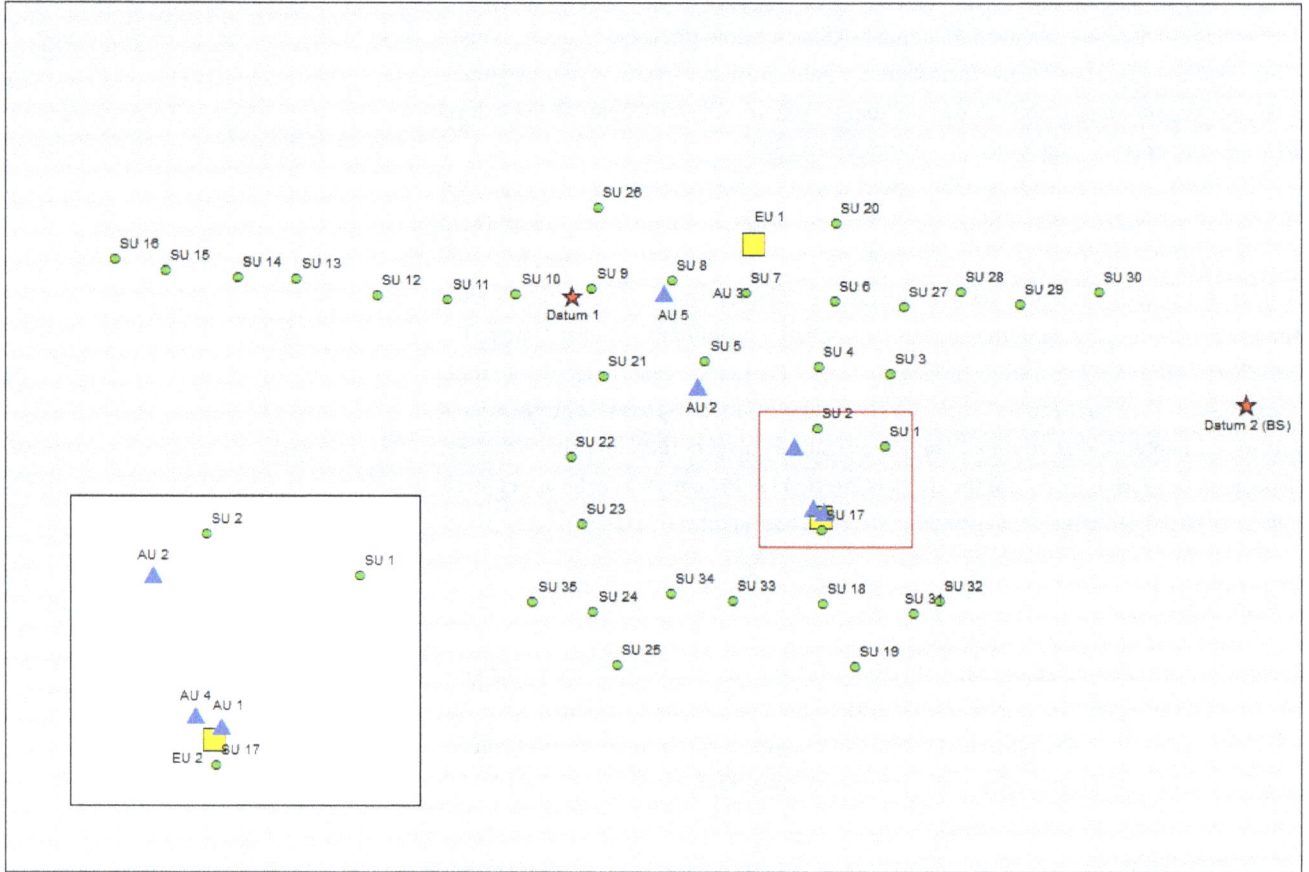

CA-SCR-14
Locations of Excavation Units
Grey Basemap with White Inset Background

- Surface Unit (SU)
- Auger Unit (AU)
- Datum
- Excavation Unit (EU)

Meters
0 2.5 5   10

N

Figure 4.6. Location of Datums, Auger Units, Excavation Units, and Surface Collection Units at CA-SCR-14. The data are schematically represented without reference to topographic features or geospatial landmarks following site stewardship protocols of California State Parks and the Amah Mutsun Tribal Band for publishing information on archaeological sites (Map by Rob Cuthrell and Alec Apodaca).

faunal specimens per unit was 2.2 (SD=2.59), and the mean weight of the faunal remains per unit was 1.26 grams (SD=1.96). This yielded faunal densities for the surface units of 8.8 elements/m^2 (SD=10.36) and 5.04 grams/m^2 (SD=7.84). The spatial distribution of the faunal remains follows a similar pattern as that of the lithic artifacts, with three concentrations in the northcentral, southcentral, and western areas of the site (see Lightfoot et al. 2021). A diverse range of shellfish species, weighing 744.81 grams, was recorded for the units. They included mussel (32 units), barnacle (27 units), clam (10 units), oyster (8 units), chiton (4 units), abalone (1 unit), turban snail (1 unit), and other shell (1 unit). The mean weight of shellfish per unit was 21.28 grams (SD=28.1) with a density of 85.12 grams/m^2 (SD=112.4). The spatial patterning of the shellfish remains is also similar to the lithic artifacts and faunal remains (see Lightfoot et al. 2021). Finally, 18 units contained charred botanical remains that weighed ca. 11.69 grams or about .33 grams per unit (SD=.81).

Subsurface Investigation

Field crews initiated the subsurface study through the excavation of five auger units to test geophysical anomalies and obtain information on the depth and kinds of deposits at the site (Figure 4.6). The excavation of AU 1 involved two levels to a depth of 30 cm when rocks were encountered. AU 2 comprised two levels with sterile deposits observed 35–36 cm below surface. Three levels were excavated from AU 3 until sterile sediments were hit at 50 cm. AU 4 was prematurely terminated with the detection of large rocks about 15 cm below surface. One 20-cm level and half of a second level were completed at AU 5 before sterile sediments were recorded at 30 cm below surface. The findings from the auger units indicated that CA-SCR-14 consisted of a rich midden deposit with shellfish, vertebrate faunal remains, and artifacts that were relatively shallow—about 30 to 50 cm in depth. Employing the results of the geophysical survey, auger units, and surface units, we staked out two 50x50 cm excavation units (EU 1, EU 2) in the southcentral and northcentral areas of the site, respectively (Figure 4.6).

EU 1. We placed this excavation unit near the location of AU 1 and AU 4 to evaluate if the auger units may have detected a possible rock feature about 15–30 cm below surface. The northeast corner served as the datum unit, and all measurements were tied into this point. The excavation involved trowels, the collection of bulk sediment samples, and the dry screening of all remaining sediments through 3.2 mm mesh. Excavations terminated at about 45 cm below datum at which time we drew profiles of the East Wall (Figure 4.7) and North Wall (Figure 4.8). We defined three soil strata in the wall profiles: Stratum I consisted of black (10YR 2/1) loam with sparse shell; Stratum II was black (10YR 2/1) loam with dense shell; and Stratum III comprised the very dark gray/dark grayish-brown (10YR 3/1–4/2) clay sub soil.

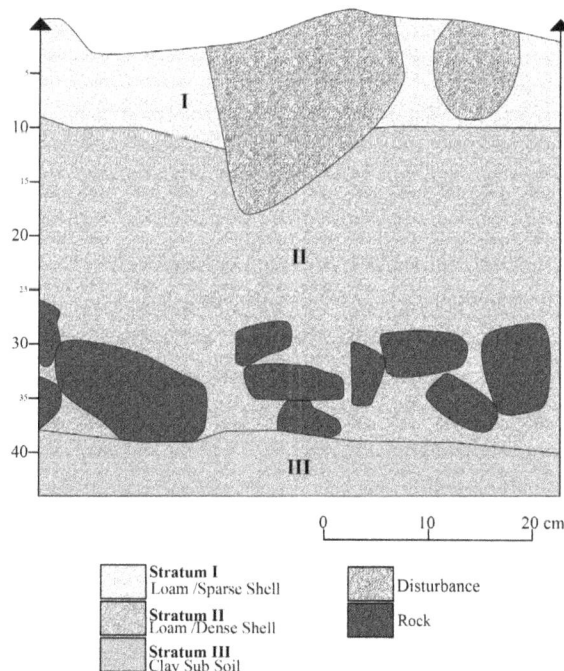

Figure 4.7. East Wall Profile of EU 1 for CA-SCR-14 (Profile by Ariadna Gonzelez).

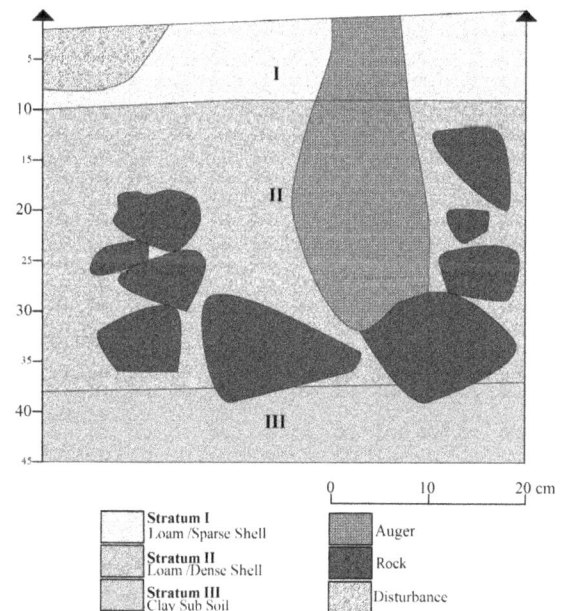

Figure 4.8. North Wall Profile of EU 1 for CA-SCR-14 (Profile by Ariadna Gonzelez).

We excavated five levels in EU 1. Level 1 (0–10 cm) consisted of Stratum I midden with sparse shell that contained chert artifacts and faunal bones. Level 2 (10–20 cm) yielded the Stratum II loam and a cluster of six rocks, several measuring 10–15 cm in size, in the center of the unit (Figure 4.9). Level 3 (20–28 cm) exposed more of the Stratum II deposit that contained fire-cracked rocks, shell, chert artifacts, and faunal remains. Level 4 (28–40 cm) revealed more of the Stratum II loam and a concentration of fire-cracked rocks across the entire unit. Two layers of rocks, cracked and broken from probable burning, were mapped, photographed, and removed from the level. The plan maps for the upper and lower layers of fire-cracked rocks are presented in Figures 4.10 and 4.11, respectively. Some of the rocks were identified as granite and measured about 10–15 cm in size. Charcoal, chert artifacts, shell, and other faunal remains were associated with the two layers of rocks. Level 5 (40–45 cm) exposed the remainder of the fire-cracked rock feature and terminated in the clay subsoil (Stratum III), about 45 cm below datum.

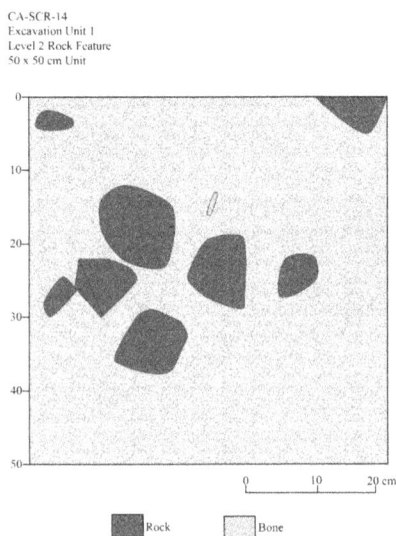

Figure 4.9. Plan Map of Rock Concentration in Level 2, EU 1, CA-SCR-14 (Map by Ariadna Gonzalez).

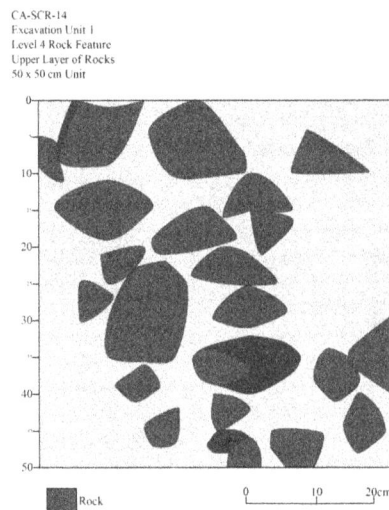

Figure 4.10. Plan Map of the Upper Layer of Fire-Cracked Rocks in Level 4, EU 1, CA-SCR-14 (Map by Ariadna Gonzalez).

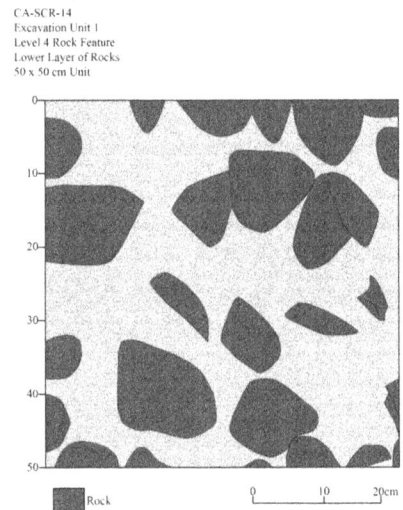

Figure 4.11. Plan Map of the Lower Layer of Fire-Cracked Rocks in level 4, EU 1, CA-SCR-14 (Map by Ariadna Gonzalez).

EU 2. We placed this unit in the northcentral area of the site where relatively dense concentrations of lithics, faunal remains, and shellfish were recorded in our surface investigation. The unit is situated on a gentle slope that drains to Laguna Creek. The southeast corner served as the datum unit, and all measurements were tied into this point. The excavation involved trowels, the collection of bulk sediment samples, and the dry screening of all remaining sediments through 3.2 mm mesh. The unit was excavated to a depth of 55 cm below datum. The excavation revealed three major stratigraphic layers, as illustrated in the wall profiles. Figure 4.12 shows the East Wall profile, while Figure 4.13 illustrates the North Wall profile. Stratum I consists of black (10YR 2/1) midden that is relatively homogenous with fine shell (less than 1 cm in size). Stratum II is a black to dark grayish-brown (10YR 2/1–4/2) midden that is heterogeneous in consistency and contains shell up to 5 cm in size. We unearthed several intact ash and ashy lenses in this stratum as illustrated in the wall profiles and detailed below. Stratum III is defined as the transitional zone to the brown (10YR 3/2–4/3) clay subsoil.

We excavated seven levels in EU 2. Level 1 (0–15 cm) consisted of the Stratum I midden deposit comprised of fine shell, including abalone, along with faunal bones, chert artifacts, fire-cracked rock, and bottle glass. Level 2 (16–25 cm) continued to reveal the Stratum I midden deposit with the addition of ash concentrations. Some bioturbation due to rodent activity was also noted. Level 3 (30–34 cm) unearthed the Stratum II midden deposit that contained almost pure ash with some charcoal and small pieces of burned shell. The entire level was collected as a bulk sediment sample. Level 4 (30–40 cm) consisted of more diffuse shell-rich ash Stratum II deposits that may be the consequence of multiple dumping events, particularly in the northwestern section of the unit. Level 5 (40–43 cm) in the Stratum II deposit unearthed shell concentrations containing ash probably from discrete dumping episodes in the northwest section of the unit. Level 6 (43–51 cm) completed the excavation of Stratum II and revealed shell, chert artifacts, and faunal bones, along with shell and ash-rich deposits that had been disturbed by rodent burrows. Level 7 (51–55 cm) marked the transition of the shell-rich deposit to the lighter brown subsoil clay horizon (Stratum III), where the unit was terminated after scraping the entire unit to the underlying clay.

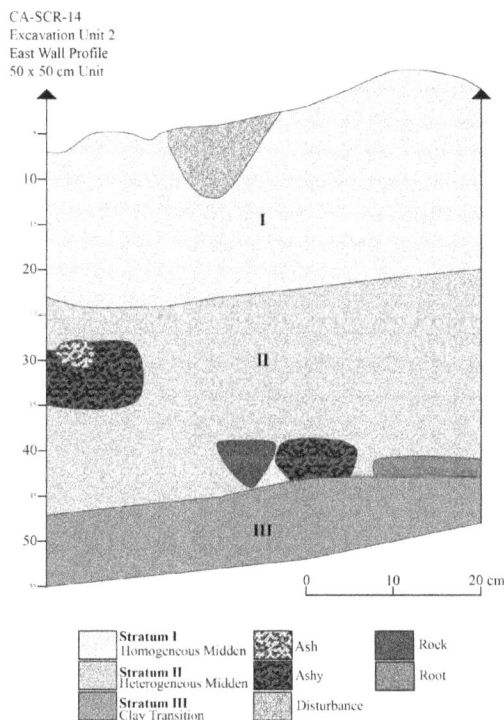

Figure 4.12. East Wall Profile of EU 2 for CA-SCR-14 (Profile by Ariadna Gonzalez).

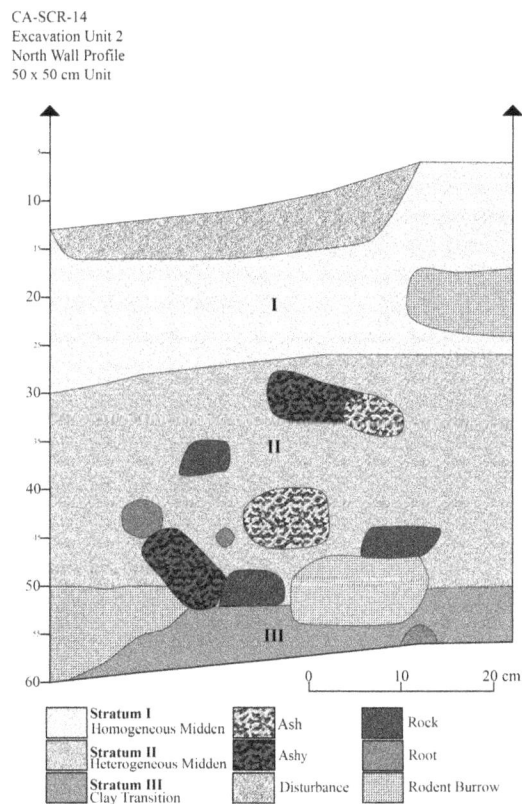

Figure 4.13. North Wall Profile of EU 2 for CA-SCR-14 (Profile by Ariadna Gonzalez).

Phytolith Survey

The final major activity at CA-SCR-14 involved the collection of samples from near-surface soils for phytolith analysis. Three phytolith samples were collected from the nearby environs of CA-SCR-14 (see Lightfoot et al. 2021).

Summary of Field Investigation at CA-SCR-14

The study of CA-SCR-14 revealed a moderate-sized site along Laguna Creek characterized by a diverse and dense assemblage of artifacts, shellfish, faunal remains, and subsurface cultural features. The surface investigation revealed Monterey chert flakes and debitage, fire-cracked rocks, and various shellfish and vertebrate faunal remains. Some modern Euro-American artifacts were also recorded, probably the consequence of a known artist who lived on the site in recent years. The geophysical survey and auger units indicated that subsurface features may be found on the site. The auger units also indicated a fairly shallow depth to the site, ranging from 30 to 50 cm or so. Excavation crews unearthed an intact fire-cracked rock feature in EU 1. It may represent a rock surface used for pit cooking or rocks discarded from cooking, sweats, or other nearby activities. The excavation of EU 2 revealed an area probably used for dumping refuse on the periphery of the site, as evidenced by multiple, small, discrete deposits of ashy, eco-fact rich materials.

Two types of samples from EU 1 and EU 2 were prepared for subsequent laboratory research. The 3.2 mm dry-screened samples from the two units were transported to the California Archaeology Laboratory at UC Berkeley for analysis. The bulk sediment samples from the five levels from EU 1 and the seven levels from EU 2 were floated in the field and the light and heavy fractions subsequently analyzed under laboratory conditions at UC Berkeley and elsewhere.

CA-SCR-15

In the summer of 2016, field crews initiated four primary activities at CA-SCR-15: geophysical survey, systematic surface collection, the excavation of two 50x50 cm units, and a phytolith survey. Field crews returned in the summer of 2017 to better define the boundary of the site through the placement of additional surface units on the periphery of the site. Both seasons of fieldwork involved the placement of 2 datums and the investigation of 39 surface units and 2 excavation units (Figure 4.14).

Geophysical Survey

We instigated a magnetometry and GPR survey on the western knoll overlooking Laguna Creek. The survey of a 20-meter square area (Block 1) using the GSSI SIR 3000 GPR unit with a 400 MHz antenna unit revealed a clear demarcation between the midden deposit and subsoil, as well as a series of potential features, including a possible house floor (see Chapter 5 for details). Additional GPR profiles also suggested intact features, as unearthed in EU 3 below.

Surface Investigation

Field crews laid out a primary east/west transect that extended from the western knoll across the asphalt road and through the swale and upper area in the eastern section of the site. A series of small north/south transects were then extended from the primary transect to cover other areas of the site. Surface units, measuring .28 m in radius, were placed 10 m apart. Sediments from the upper 5 cm of the units were removed and dry screened through 3.2 mm mesh. A total of 24 surface units were collected in the summer of 2016 (Figure 4.14). In the summer of 2017, 15 additional surface units were collected to help define the outer site boundaries.

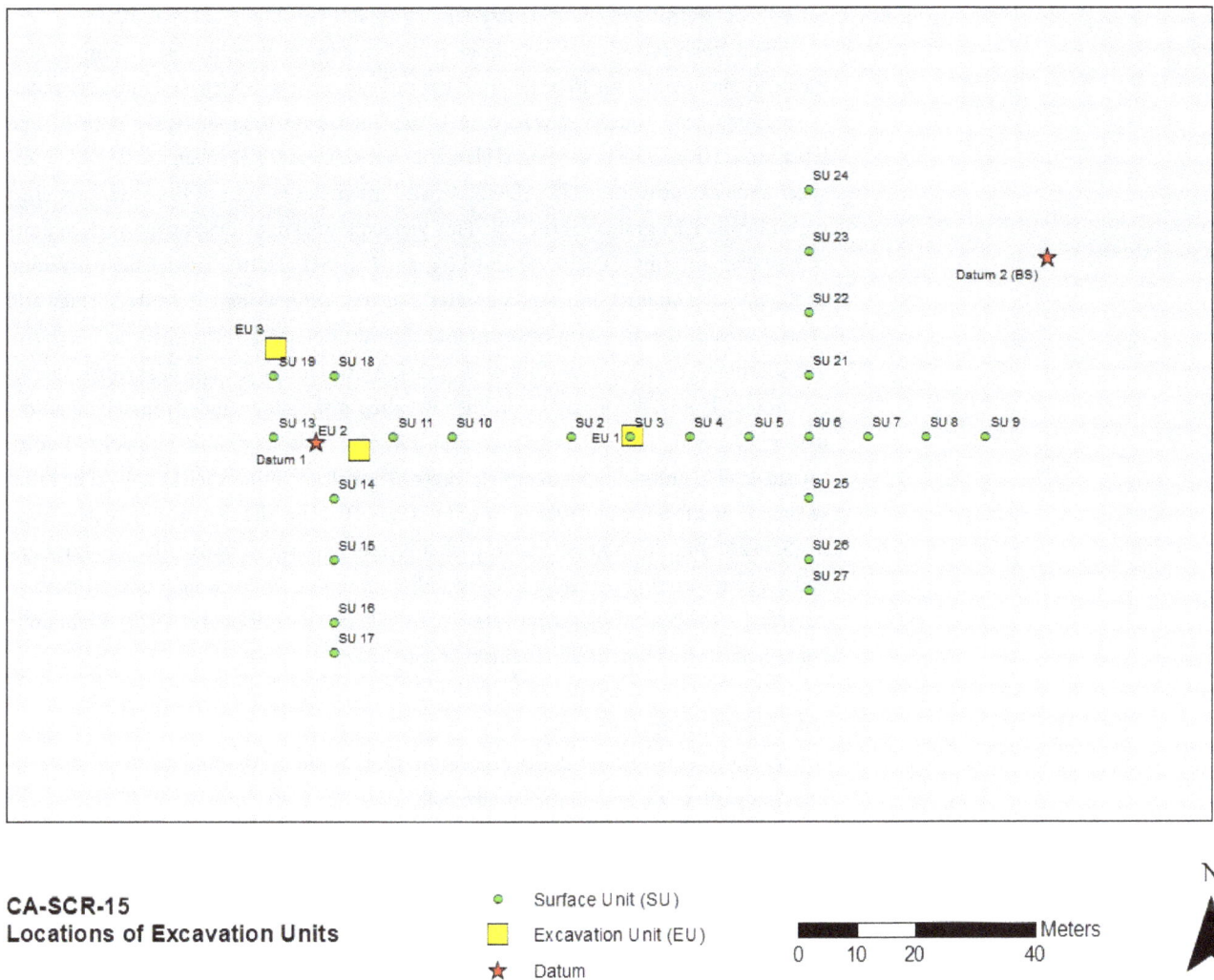

CA-SCR-15
Locations of Excavation Units

- Surface Unit (SU)
- Excavation Unit (EU)
- Datum

0 10 20 40 Meters

N

Figure 4.14. Location of Datums, Excavation Units, and Surface Collection Units at CA-SCR-15. The data are schematically represented without reference to topographic features or geospatial landmarks following site stewardship protocols of California State Parks and the Amah Mutsun Tribal Band for publishing information on archaeological sites (Map by Rob Cuthrell and Alec Apodaca).

The surface investigation of the 39 units revealed a total of 605 artifacts, including 181 Monterey chert flakes and debitage pieces, 8 other pieces of debitage of unknown material, 2 ground stone fragments, 413 pieces of fire-cracked rock, and 1 fragment of a shell bead (Appendix 4.1: Table A4). Notable observations included an unfinished projectile point; chipped stone material exhibiting primary, secondary, and tertiary decortication; and indications that some of the fire-cracked rock may have been reused ground stone artifact fragments. No Euro-American manufactured materials were recorded. The mean number of lithic and shell artifacts per unit was 15.51 (SD=18.94), yielding a density of 62.04 artifacts/m^2 (SD=75.76). The mean number of Monterey chert flake stone artifacts was 4.64 (SD=5.72) per unit or 18.56 lithics/m^2 (SD=22.88). The spatial distribution of lithic artifacts across the site area is relatively continuous from the western ridge overlooking Laguna Creek to the east along the main collection transect, along with high numbers along the north/south collection transects. The high artifact densities and extensive patterning of cultural materials are notable (see Lightfoot et al. 2021).

Field crews recorded 69 vertebrate faunal specimens, weighing 48.7 grams. These included mammal (from 18 units), bird (from 3 units), and fish (from 2 units). Some of the mammal bones observed were quite large. One bat ray fragment was identified, and it was noted that some of the faunal elements were charred. The mean number of faunal specimens per unit was 1.77 (SD=2.74), and the mean weight of the faunal remains per unit was 1.25 grams (SD=2.79). This yielded faunal densities for the surface units of 7.08 elements/m^2 (SD=10.96) and 5.0 grams/m^2 (SD=11.16). The spatial patterning of the faunal elements indicates a greater concentration along the western ridge and central area of the site with fewer faunal remains detected in the eastern, northern, and southern areas (see Lightfoot et al. 2021). A diverse assemblage of shellfish species was noted. The 1209.5 grams of shellfish weighed at the site included mussel (30 units), oyster (12 units), clam (19 units), barnacle (29 units), chiton (10 units), abalone (1 unit), turban snail (5 units), limpet (4 units), whelk (6 units), and other shell (1 unit). The mean weight of shellfish per unit was 31.01 grams (SD= 60.36) with a density of 124.04 grams/m^2 (SD=241.44). The highest density of shellfish remains was found on the western ridge and the southern slope of Laguna Creek, although an impressive weight of shell was also recorded in the central and eastern areas of the site (see Lightfoot et al. 2021). Eight units contained charred botanical remains that weighed ca. 1.4 grams.

Subsurface Investigation

We located the placement of the excavation units based on the findings from the geophysical survey and systematic surface collection. EU 1 was located east of the asphalt road in the location of SU 3 where a dense assemblage of cultural materials was recorded. We established a second excavation unit (EU 2) on the western knoll in a promising area based on the magnetometry survey. Unfortunately, in the upper level of this unit, human remains were detected. After a smudging ceremony involving the reburial of the ancestral remains, the unit was shut down. We employed the results of the GPR survey to select another area for investigation away from the human remains. In a location containing anomalies that might represent intact subsurface cultural features, we laid out another excavation unit (EU 3) on the western knoll of the site (see Figure 4.14).

EU 1. The first excavation unit, located at SU 3 in the eastern section of the site, sampled an area with a dense concentration of surface materials. In establishing the 50x50 cm unit, we designated the southeast corner as the unit datum from which measurements were taken. Then, 5-liter bulk sediment samples were collected from each level, along with separate 100-gram microbotanical samples. Additional sediments from each level were dry screened through 3.2 mm mesh. Our excavation crew described two major strata in the profiles of the East Wall (Figure 4.15) and North Wall (Figure 4.16). The upper levels were characterized by a black or very dark gray (10YR 2/1–3/1) silty clay loam that contained concentrations of shell in some places. The lower levels (50–60 cm below datum) transitioned into a dark gray clay-rich (10YR 3/1) loam.

CA-SCR-15
Excavation Unit 1
East Wall Profile
50 x 50 cm Unit

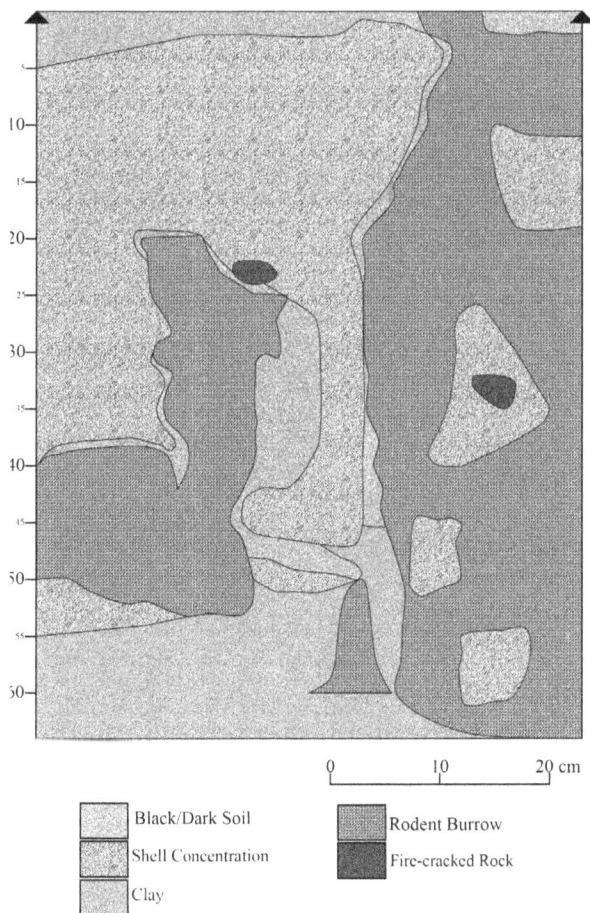

CA-SCR-15
Excavation Unit 1
North Wall Profile
50 x 50 cm Unit

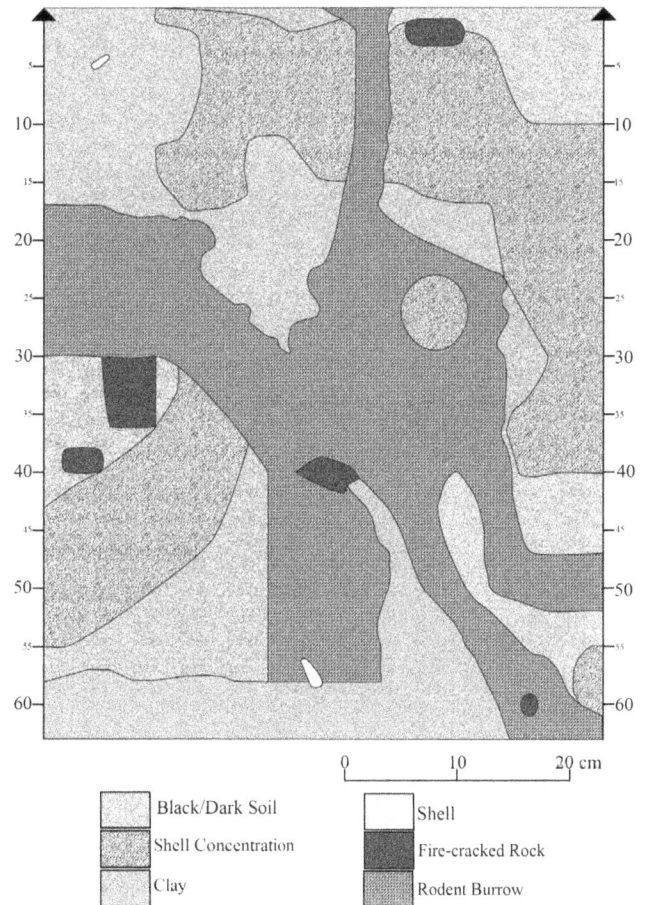

Black/Dark Soil

Shell Concentration

Clay

Rodent Burrow

Fire-cracked Rock

Black/Dark Soil

Shell Concentration

Clay

Shell

Fire-cracked Rock

Rodent Burrow

Figure 4.15. East Wall Profile of EU 1 for CA-SCR-15 (Profile by Ariadna Gonzalez).

Figure 4.16. North Wall Profile of EU 1 for CA-SCR-15 (Profile by Ariadna Gonzalez).

We excavated seven levels in EU 1. The dark silty clay loam in Levels 1 (0–10 cm), 2 (10–20 cm), 3 (20–30), and 4 (30–40 cm) contained a diverse range of shellfish (mussel, barnacle, etc.), vertebrate faunal remains, chert artifacts, and fire-cracked rock. However, there was also considerable evidence for bioturbation from rodents (Figures 4.15 and 4.16). Toward the bottom of level 4, the excavators detected a large fire-cracked rock.

In level 5 (38–45 cm), a concentration of fire-cracked rocks was unearthed that was associated with shellfish fragments. Sediments from the fire-cracked rock feature were collected as a bulk sample. While sediments below the feature remained dark (10YR 3/1), the amount of clay increased. Level 6 (50–60 cm), below the fire-cracked rock feature, exposed dark clay rich sediments, along with mussel, clam, and oyster shells, lithic artifacts, and an abalone shell bead blank. Soils became increasingly compact and difficult to dig. The excavation of level 7 (60–70 cm) continued to reveal more compact, dark clay-rich sediments. While some shell fragments and small lithics were found, the density of cultural material dropped off significantly. We then placed a soil auger unit (AU 1) in the bottom of EU 1 to evaluate the sediments from 70 to 90 cm below datum. AU 1 revealed a dark silty clay stratum (70–80 cm) that transitioned into sterile, yellow (2.5YR 5/4) clay about 88 cm below datum. The auger unit was terminated at 90 cm below datum in the underlying clay stratum.

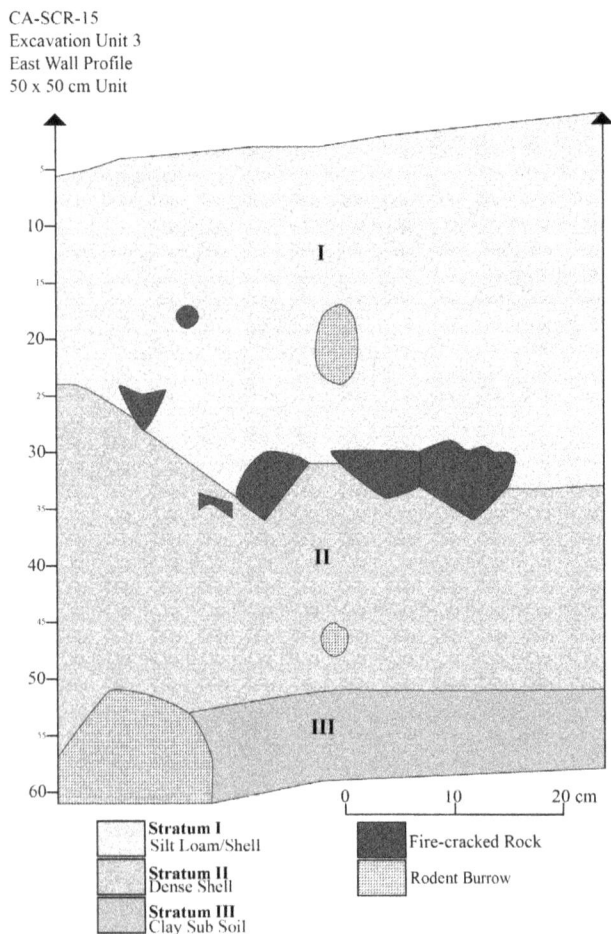

Figure 4.17. East Wall Profile of EU 3 for CA-SCR-15 (Profile by Ariadna Gonzalez).

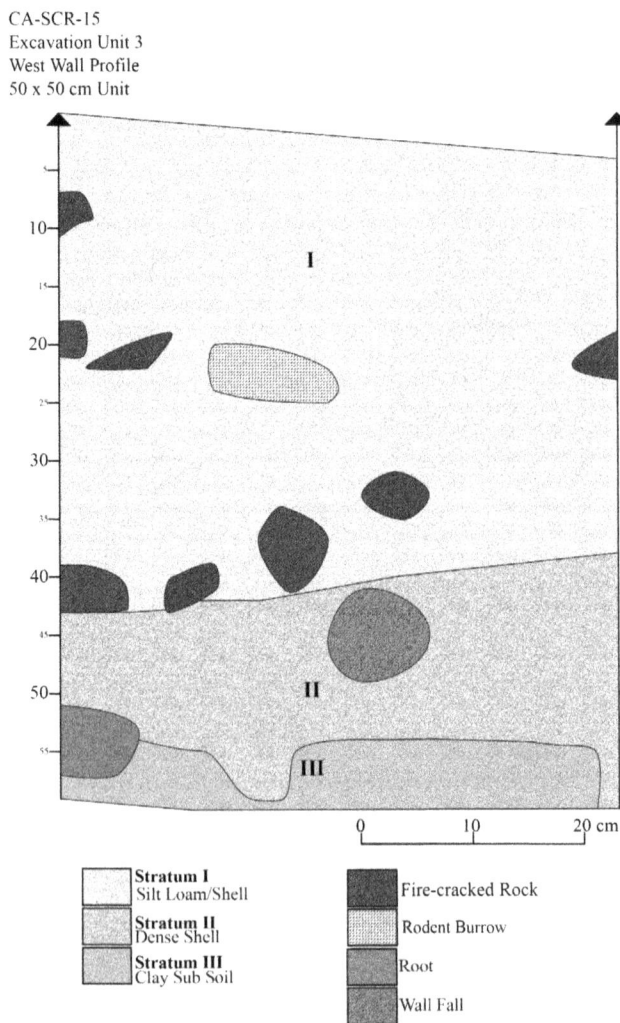

Figure 4.18. West Wall Profile of EU 3 for CA-SCR-15 (Profile by Ariadna Gonzalez).

EU 3. This 50x50 cm unit was situated in the western knoll area of the site. The southeast corner was designated the unit datum and all measurements taken from this point. Then, 5-liter bulk sediment samples were collected from each level, along with separate 100-gram microbotanical samples. Additional sediments from each level were then dry screened through 3.2 mm mesh. The excavation revealed three major soil strata as illustrated in the East Wall profile (Figure 4.17) and in the West Wall profile (Figure 4.18). Stratum I consisted of a dark gray (10YR 3/1) silt loam with some shell. Stratum II was comprised of black to dark gray (10YR 2/1–3/1) silt loam with dense shell. Stratum III was the sterile yellow (2.5YR 5/4) clay subsoil.

Our field crew excavated seven levels in EU 3. The upper 40 cm of the unit, including level 1 (0–14 cm), level 2 (14–24 cm), level 3 (24–34 cm), and level 4 (34–40 cm), consisted of the dark silt loam with a high density of shellfish, as well vertebrate faunal remains, chert artifacts, and fire-cracked rock. There was also evidence of bioturbation created by rodent burrows. In level 5 (40–44 cm), excavators detected a cultural feature comprised of a concentration of fire-cracked rock associated with a high density of shellfish and other materials. A 10-liter bulk sample was taken from the feature. The feature was embedded in dark (10YR 3/1) organic-rich sediments at the base of Stratum I (see Figures 4.17 and 4.18). The plan map of the fire-cracked rocks unearthed in level 5 is presented in Figure 4.19. Level 6 (44–54 cm) continued to expose fire-cracked rock and the dark shell-rich sediments in Stratum II, along with increasing amounts of clay. In level 7 (54–64 cm), more of the clay-rich soil was exposed, and the substratum of sterile clay (Stratum III) was reached at about 60–64 cm.

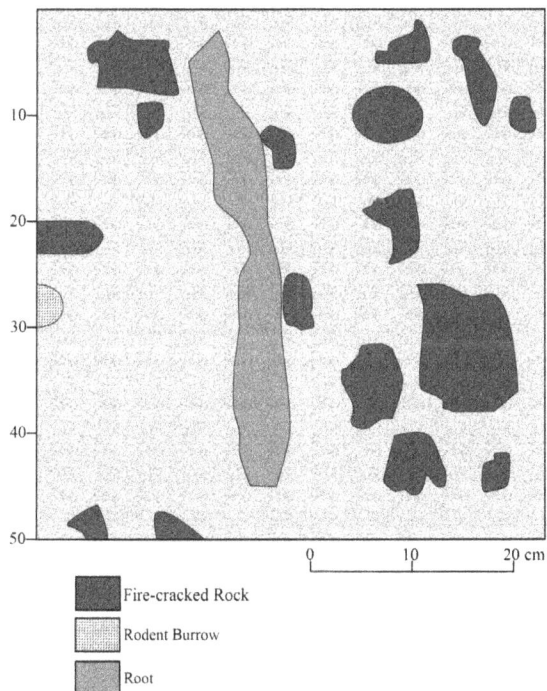

Figure 4.19. Plan Map of Fire-Cracked Rocks in Level 5, EU 3, CA-SCR-15 (Map by Ariadna Gonzalez).

Phytolith Survey

The fourth major activity at CA-SCR-15 involved the collection of samples from near-surface soils for phytolith analysis. Three phytolith samples were collected from the nearby environs (see Lightfoot et al. 2021).

Summary of Field Investigation at CA-SCR-15

The findings from the geophysical survey, surface units, and subsurface investigations indicate that CA-SCR-15 is an extensive site with a dense and diverse range of cultural materials.

The detection of GPR and magnetometry anomalies suggests a high potential for intact subsurface features. The surface investigation of units spaced ten meters apart documented a diverse assemblage of shellfish, vertebrate faunal remains, and artifacts dispersed across a ca. 130x60 m area. The excavation of two small units (with an additional auger unit) unearthed a rich midden deposit characterized by a relatively homogenous, dark silty loam with shell, which transitioned into an underlying clay substratum about 60 cm below surface. Both units reported extensive bioturbation from burrowing animals. Yet both units also detected fire-cracked rock features associated with high densities of cultural material. A dense shell lens (Stratum II) was profiled below the cultural feature in EU 3, while it appears the underlying deposit below the feature in EU 1 consisted of clay-rich sediments.

The materials collected during the subsurface investigation of CA-SCR-15 were prepared for subsequent laboratory analysis. This included both the bulk sediment and dry-screened samples from EU 1 (seven levels) and EU 3 (seven levels), as well as the bulk sediment sample from AU 1 (70–90cm). All bulk sediment samples were floated in the field. The resulting light fraction and heavy fraction samples, along with the dry-screened materials recovered in the 3.2 mm mesh, were transported to the California Archaeology Laboratory at UC Berkeley for detailed analysis.

Conclusion

This chapter described the low-impact methodology employed in our collaborative, eco-archaeological investigation of four sites on the Santa Cruz coast. Building on previous work in Quiroste Valley and elsewhere on the Central California coast, we designed this study to address four primary research goals concerning the timing, development, geographic scale, and contemporary relevance of Indigenous landscape and seascape stewardship practices, as detailed in Chapter 2. We examined sites that had chronological components spanning Middle Holocene (6000–3000 BP), Late Holocene (3000–500 BP), and Historic (500–100 BP) periods in a study area north of the city of Santa Cruz (Figure 1.2). This chapter describes the eco-archaeological investigation of CA-SCR-7, CA-SCR-10, CA-SCR-14, and CA-SCR-15 from this study area, which offered an excellent opportunity to evaluate the above research goals.

Our eco-archaeological approach emphasized the recovery of artifacts, ethnobotanical samples, and zooarchaeological remains, as well as a low-impact methodology designed specifically to minimize disturbances to ancestral places. Tribal members collaborated in making all field decisions employed in the study. The field methodology involved four basic tasks: 1) systematic surface collection, 2) geophysical prospecting, 3) subsurface testing employing soil augers and small excavation units, and 4) the collection of phytolith samples in the hinterland of several sites. This chapter discusses the findings from the surface investigation for each of the four sites. Subsequent chapters describe the findings of the analysis of various archaeological materials recovered from subsurface contexts.

The laboratory analysis of the subsurface materials from auger units, column units, and excavation units involved working with both flotation and dry-screen samples. All dry-screened samples had been sieved in the field using 3.2 mm mesh. Rob Cuthrell directed the flotation of sediment samples in the field using two SMAP-type flotation tanks that he constructed. A full description of the flotation methodology is presented by Cuthrell in Chapter 8. After flotation was completed, the heavy and light fraction materials were bagged separately and transported to UC Berkeley for analysis along with the dry-screen samples.

Our initial laboratory investigation involved sorting the dry-screened and floated samples into various classes of archaeological materials. A full discussion of the treatment of the light fraction samples is presented in Chapter 8, which describes the analysis of the archaeobotanical and phytolith samples. The heavy-fraction samples from flotation were subsequently screened in the California Archaeology Laboratory through 4 mm and 2 mm sieves, as well as 1 mm sieves for some additional samples. The dry-screened samples and the screened heavy-fraction samples were then separated into broad artifact types, archaeofaunal groups, and archaeobotanical taxa, as outlined in subsequent chapters. The macrobotanical assemblage also provided ephemeral plant materials suitable for submitting radiocarbon dates. Radiocarbon assessments for the four sites were undertaken at the Keck Carbon Cycle AMS Facility at the University of California, Irvine, as outlined in Chapter 6.

References

Hildebrandt, William R., Deborah A. Jones, and Mark G. Hylkema
  2007   *Sand Hill Bluff Site: CA-SCR-7*, National Register of Historic Places
          Registration Form. On file, California Department of Parks and Recreation,
          Sacramento, California.

Hylkema, Mark G.
  2021   Middle Holocene Projectile Points from the Santa Cruz Coast of Northern
          Monterey Bay. Paper Presented at the 2021 Annual Meeting for the Society for
          California Archaeology, A Virtual Event. Paper Archived by the Society for
          California Archaeology.

Jones, Deborah A., and William R. Hildebrandt
  1990   *Archaeological Excavation at Sand Hill Bluff: Portions of a Prehistoric Site
          CA-SCR-7, Santa Cruz County, California*. Far Western Anthropological Research
          Group, Inc., Davis, California.
  1994   *Archaeological Investigations at Sites CA-SCR-10, CA-SCR-17, CA-SCR-304, and
          CA-SCR-38/123 for the North Coast Treated Water Main Project, Santa Cruz County,
          California*. Far Western Anthropological Research Group, Inc., Davis, California.

Lightfoot, Kent G.
  2008   Collaborative Research Programs: Implications for the Practice of North
          American Archaeology. In *Collaborating at the Trowel's Edge: Teaching and Learning
          in Indigenous Archaeology*, edited by Stephen W. Silliman, pp. 211–227. University
          of Arizona Press, Tucson.

Lightfoot, Kent G., Rob Q. Cuthrell, Mark G. Hylkema, Valentin Lopez, Diane
Gifford-Gonzalez, Roberta A. Jewett, Michael A. Grone, Gabriel M. Sanchez, Peter A.
Nelson, Alec J. Apodaca, Ariadna Gonzalez, Kathryn Field, and Alexii Sigona
  2021   *The Study of Indigenous Landscape and Seascape Stewardship Practices on the
          Santa Cruz Coast: A Collaborative Eco-Archaeological Approach*. Report Prepared
          for California Department of Parks and Recreation, Santa Cruz District.
          Archaeological Research Facility, University of California, Berkeley.

Moratto, Michael J.
  1984   *California Archaeology*. Academic Press, Orlando.

Sanchez, Gabriel M., Michael A. Grone, Alec J. Apodaca, Scott Byram, Valentin Lopez,
and Roberta A. Jewett
  2021   Sensing the Past: Perspectives on Collaborative Archaeology and Ground
          Penetrating Radar Techniques from Central California. *Remote Sensing* 13(2): 285
          https://doi.org/10.3390/rs13020285.

Saxe, A. W.

1875   Verbal Remarks on a Mound Near Santa Cruz, California. *Proceedings of the California Academy of Sciences* 5: 157.

Schlagheck, John

2011   Obsidian Trade at Sand Hill Bluff. *Society for California Archaeology Proceedings* 25: 1–17.

CHAPTER 5

Ground-Penetrating-Radar Survey on the Santa Cruz Coast

PETER A. NELSON

Introduction

Ground-penetrating radar (GPR) is a premier geophysical technology that can aid in the non-invasive or low-impact study of near-surface cultural resources and anthropogenic landscapes. The utility and versatility of this technology was proven in several diverse contexts along the Central Coast of California during a University of California, Berkeley (UC Berkeley), field project in 2016. The GPR survey included the investigation of four sites in Santa Cruz County, the ancestral homeland of the Amah Mutsun Tribal Band (AMTB), in a collaborative project between UC Berkeley, the Tribe, and California State Parks. The goals of this collaborative project were 1) to examine the antiquity of steward-ship practices such as burning, 2) to investigate evidence for changes in anthropogenic fire regimes, and 3) to evaluate the geographic scale of Indigenous stewardship. Given these research objectives, GPR was strategically deployed to evaluate the formation processes operating at these sites, the relative chronology of components for radiocarbon dating (if discernable), intact versus disturbed contexts, and discrete features that would likely produce abundant botanical and faunal materials, such as domestic areas and hearths or earth ovens.

This chapter will provide an overview of the results from the GPR survey and how interpretations of these data were utilized independently from, and in tandem with, other data to interpret aspects of these resources as well as to guide excavations and aid in archaeological data collection within the project. Recommendations will also be provided for future survey at these sites or in similar contexts considering the challenges that many coastal and remote sites pose for researchers and land managers. Overall, GPR technology has been extremely beneficial in guiding the work of archaeologists on the California coast. GPR surveys also support efforts in California to respond to the impacts of climate change on cultural heritage, especially along the coast where winter storms cause accelerated erosion. The information from GPR can be very beneficial for Tribes, State Parks, and other land managers seeking to evaluate potential impacts, conduct research, and protect and preserve these invaluable resources from further damage.

Background to Ground-Penetrating Radar in California

Geophysical surveys of archaeological sites and Tribal cultural resources have been conducted in California for several decades with technologies such as magnetometry and resistivity (e.g., Allan 1997; Cuthrell 2013; Gonzalez 2011; Grenda et al. 1998; Lightfoot 2006, 2008; Lightfoot and Gonzalez 2018; Lightfoot et al. 2013; Nelson 2017; Schneider 2010; Silliman et al. 2000; Somers 2008; Tschan 1997). However, GPR surveys from the early decades of geophysical archaeology in California were very rare in the 1990s and 2000s (e.g., Arnold et al. 1997). Fortunately, GPR surveys have become increasingly more common in California in the past decade, fueled by collaborative research with Native American Tribes (e.g., Byram et al. 2018; Nelson 2017, 2020, 2021; Sunseri and Byram 2017). This integration of geophysical archaeology into these archaeological projects originates theoretically and methodologically in collaborative and Indigenous archaeologies. These methods have been an essential component of how these projects reduce the impacts of archaeological research on Tribal cultural heritage (Gonzalez 2016; Lightfoot 2008; Nelson 2020).

Though magnetometry and resistivity can be very productive in a suite of low-impact methodologies and continue to be used in this project along the Santa Cruz coast and in others, GPR has become the preferred technology for use in most contexts among the UC Berkeley collaborative research team. The reasoning behind the preference for using GPR over other techniques is its unique ability to quickly produce two-dimensional profiles with exact depths of features. These profiles can also be post-processed together for three-dimensional data analysis without any additional re-survey.

GPR, broadly defined, is a method of geophysical prospection that uses high-frequency radar pulses transmitted from an antenna to detect and map natural geological formations, as well as anthropogenic modifications to the natural world (Conyers 2004: 11–12; 2012; Conyers and Goodman 1997). GPR has been used to identify architectural features, such as walls and foundations, as well as other anthropogenic features such as hearths, pits, and floors, and historic graves in cemeteries. Much of the success of GPR in identifying subsurface features depends on how different the materials in the ground are from surrounding soils. Much of what we see in the GPR data reflects the difference between materials and transitions from one material to another. Thus, some features may not be visible in GPR because there is not enough of a difference between the feature and the surrounding substrate. However, there are some features or stratigraphic changes that may not be visible to the human eye but may be detected by the GPR. For instance, trace amounts of moisture and water retention in or on top of some constituents may create very highly reflective features (Conyers 2012: 37). Thus, GPR results can guide interpretations or further investigation plans for a site beyond the physical limits of other methods.

Methods

Instrument

The GPR used in the 2016 survey was a GSSI SIR-3000 with a survey wheel setup. The survey wheel setup—with only one wheel trailing behind the antenna bin—was more advantageous for surveying through the "rough" terrain of coastal California with chaparral, other dense brush, and undulating ground surface topography. Other configurations, such as the cart or carriage with three or four wheels on the sides of the antenna bin, make the setup wider and less able to navigate through brush, which could not be trimmed for various reasons (e.g., sensitive dune plants preventing erosion). The survey wheel has also proved to offer improved flexibility as it moves with undulating ground topography, allowing the antenna to remain coupled with the ground much more than the cart or carriage configuration would (in this latter case, the movement of the antenna bin is more restricted). The 2016 survey was conducted with a 400 MHz antenna. This decision was predicated on previous archaeological work and reports and primary records that indicated that most of these sites ranged in depth from 0.5 to 2 meters—with one exception at CA-SCR-7. This assessment was confirmed by the GPR survey in the field and the reevaluation of our strategy after operating the instrument at each site. The particularities of CA-SCR-7 will be discussed further in the results section, but they did not hinder the acquisition of useful data to address our goals during the 2016 survey with the 400 MHz antenna. Thus, we completed the survey solely using the 400 MHz antenna.

Grid System

The 2016 GPR survey included five sites along the coast of Santa Cruz County. All four of these sites contained dark, organic-rich, midden soils deposited by Awaswas ancestors of the AMTB throughout hundreds or thousands of years before contact with Europeans. However, the physical settings, site formation processes, post-contact disturbances, etc., for each site differed from one to another. Since the survey conditions at each site were diverse, the survey strategies were tailored to address the particularities and challenges encountered at each site. In general, two strategies were employed to gather information about the sites—individual profiles and formal survey blocks. Table 5.1 summarizes the areas surveyed at each site.

Individual profiles are composed of one continuous string of GPR data that forms a two-dimensional vertical profile with length and depth being the spatial units represented. Formal survey blocks are composed of several parallel data profiles that are post-processed

Table 5.1. Survey Grid and Block Specifications for Geophysical Survey.

Site Name	Grid Number	Block Number	Block Dimensions
CA-SCR-7	--	--	Individual profiles
CA-SCR-10	--	--	Individual profiles
CA-SCR-10	1	1	3.5m x 27m
CA-SCR-10	2	1	20m x 20m
CA-SCR-10	2	2	20m x 20m
CA-SCR-10	2	3	20m x 20m
CA-SCR-14	--	--	Individual profiles
CA-SCR-15	--	--	Individual profiles
CA-SCR-15	1	1	20m x 20m

together to generate several two-dimensional horizontal plan view maps with length and width represented at different depths or three-dimensional representations of the entire block. At the beginning of every survey, individual profiles are collected (although not always saved) while the GPR operator assesses site conditions and configures the GPR settings for the survey wheel, dielectric permittivity, depth of range in nanoseconds, and so on. During this initial phase of the survey, the GPR operator may also gain a better sense for where GPR survey will be most productive and whether or not to continue collecting individual profiles or advance to a more systematic and intensive survey strategy such as formal survey blocks. Individual profiles were collected when site condition posed obstructions, such as heavy brush, that were not conducive to surveying in formal blocks. While more time-intensive to collect, formal survey blocks offered much more resolution within a survey area and were employed when there was a large unobstructed space (or when one could be created through brush clearing, etc.) within a site.

For the purposes of this survey, three terms—grid, block, and unit—denoted different scales and orientations of space. A grid is defined as the entire surveyable area aligned with a chosen orientation. An orientation could be defined by real-world coordinates and true north or by an arbitrarily designated coordinate system and orientation. A grid system aligned to true north that utilizes real world coordinates is of course preferable, but it is not always practical within the constraints of time, obstructions, and the potential orientation of features within the ground. Thus, arbitrary grids targeting specific areas can prove to be very useful. During the 2016 survey, all grids were arbitrary and placed judgmentally with the aid of a Sokkia SET 530R3 total station, measuring tapes, and compasses, while considering the factors of time, obstructions, and potential feature orientation. Within a grid, surveyed areas were divided up into blocks. A block is defined as any delineated area within a grid that is surveyed with contiguous adjacent profiles. These adjacent profiles can be post-processed together to form a three-dimensional representation of data or two-dimensional slices of data in plan view. The real-world coordinates of the four corners of all blocks within these grids were recorded with a Trimble GeoXH6000. Units refer to any designation of space within a grid and block that were investigated through ground disturbing activities such as catch-and-release surface collection, augering, shovel test pits, or other mode of excavation.

Post-processing Software

After GPR data was collected, it was downloaded and post-processed using the GPR Viewer and GPR Process programs developed by Larry Conyers and Jeff Lucius. GPR Viewer was used to perform a "remove background" function to eliminate artificial, horizontal banding in the data. This software was also used to increase or decrease the "range gain" to improve data visibility and facilitate visual analysis of the profiles. This software also facilitated the correction of the depth of the ground surface and features within the profiles. GPR Process was used to process several GPR profiles together to create a text file with x and y spatial coordinates as well as z reflection data (rather than elevation), which was then imported into Golden Software's Surfer 8 program. Surfer 8 allows for x, y, z data to be interpolated and gridded to produce a contour map. These contour maps were then analyzed and compared to the profile data to confirm the shape and orientation of features.

Results

CA-SCR-7

This site is situated within the sand dune system along the Santa Cruz coast. Visually above ground, the dune has the shape of an oblong hill about 225 meters in length and between 80 to 100 meters in width. The main longitudinal axis of the dune has a northwest to southeast orientation with most wind exposure and surf found along the northwest end and half of the south side closest to this northwest end. In this exposed northwest end, the cultural components of the site were very visible on the surface. These cultural components are organic-rich, dark midden soils, shell, bone, charred botanical materials, fire-cracked rock, and chipped and ground stone materials. The sand and midden soils at this site were very conducive to GPR survey, allowing for very deep penetration of the GPR signal and very clear features throughout the profiles. However, this site was particularly challenging to survey because there were steep slopes and dense brush, grass, and other vegetation covering the majority of the site (with the exception of the exposed northwestern end). Adding to these challenges, the dune sands were actively eroding, and California State Parks, justifiably so, required that the brush remain intact for stabilizing the dune and the cultural site.

The strategy for GPR survey at this site relied on individual, non-contiguous profiles to collect data in several opportunistic bare-earth areas on the northwest end of the dune. The northwest end was the most exposed to wind and winter storms. The rest of the dune, being more protected from these elements, was covered too densely with vegetation to survey effectively without vegetation removal. Even though the profiles were all collected at the northwest end of the dune, the profiles were collected in a great diversity of locations along the base of the dune on the north and south sides, in front of, and on top of it. The profiles also traversed in several directions parallel to the longitudinal axis of the dune as well as crossed the dune perpendicularly to the longitudinal axis. Though not all of these profiles will be presented in this discussion, the examples given will explore this range of diversity.

GPR survey was able to explore the structure of the dune system in which CA-SCR-7 was embedded. As is apparent from the two GPR profiles in Figures 5.1 and 5.2, the dune is composed of many overlapping layers of stratigraphy. The profile in Figure 5.1 is an example of what these layers of deposition and dune formation look like in a cross section parallel to the longitudinal axis of the dune. The GPR profile begins at the base of the dune about 20 meters away from the visual rise of the dune above ground and gradually climbs up the initial rise of the dune from about 20 meters to the end of the profile at 30 meters. The layers of dune sand in the profile are layered diagonally and terminate at the surface where the GPR was surveying. There is some skewing of the shape of these layers because the profile image is presented as though there is no contour to the ground. However, the image does give an accurate depiction of the unbroken, regular layers of geological stratigraphy in this area. These layers appear as low-, medium-, and high-amplitude reflections that terminate at their intersection with a very high-amplitude planar reflection ranging in depth from 25 to 50 nanoseconds or 1.5–2.8 meters below the ground surface. This high-amplitude reflection also includes very large hyperbolas or point reflections and indicates the interface with bedrock material underneath the dune sands. Below the initial high-amplitude interface between sand and bedrock, the GPR signal attenuates quickly, which appears "fuzzy" or blurred like the signal from a bad channel on an old analog television. While the bedrock interface is very clearly defined from 0 to 18 meters of horizontal distance in the profile, this interface becomes less well-defined from 18 to 30 meters as the GPR gains elevation up the slope of the dune. The diagonal layers of dune sand stratigraphy in this area also begin to trail down to depths of 50–60 nanoseconds or between 2.8 to 3.2 meters at the bottom of the viewable area in the profile.

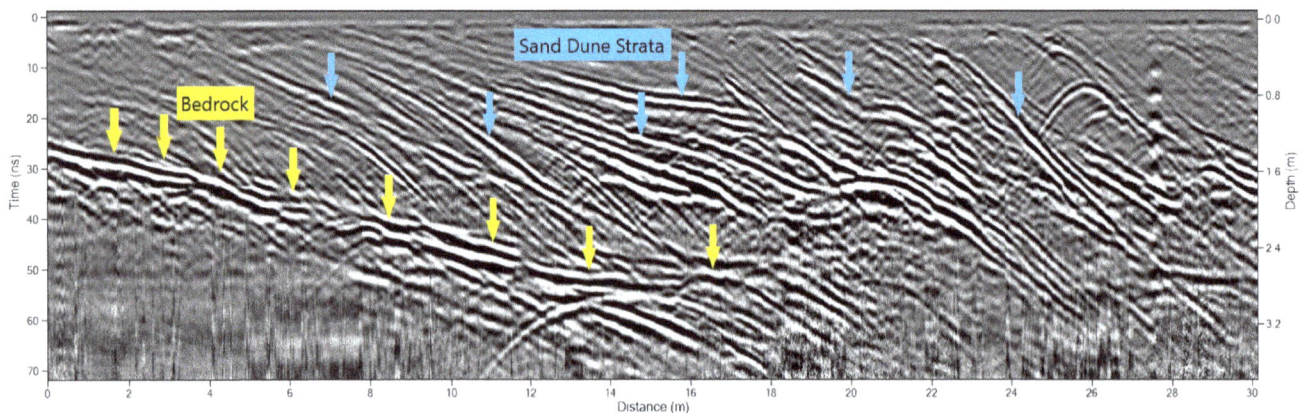

Figure 5.1. GPR Profile of CA-SCR-7, File 239.

The sand dune stratigraphy is also viewable in a cross-section perpendicular to the longitudinal axis of the dune. The profile in Figure 5.2 is an example of such a cross-section (heading from northeast to southwest) across the exposed northwest end of the dune. There is one planar reflection representing a layer of stratigraphy that begins at 25 nanoseconds of depth at the 0-meter mark in the profile. This layer of stratigraphy rises in a gentle arc to the ground surface at the top of the profile at about the 7-meter mark and begins descending again at the 11-meter mark. It continues down to 60 nanoseconds of depth or about 3.2 meters below the ground surface at the 26-meter mark. This layer is completely continuous and unbroken across the entire profile. Underneath this unbroken layer of stratigraphy, the GPR data reveals that this arc is composed of several overlapping smaller arcs and flat horizontal layers. In this profile, bedrock appears at about 55 nanoseconds or about 3 meters below the ground surface.

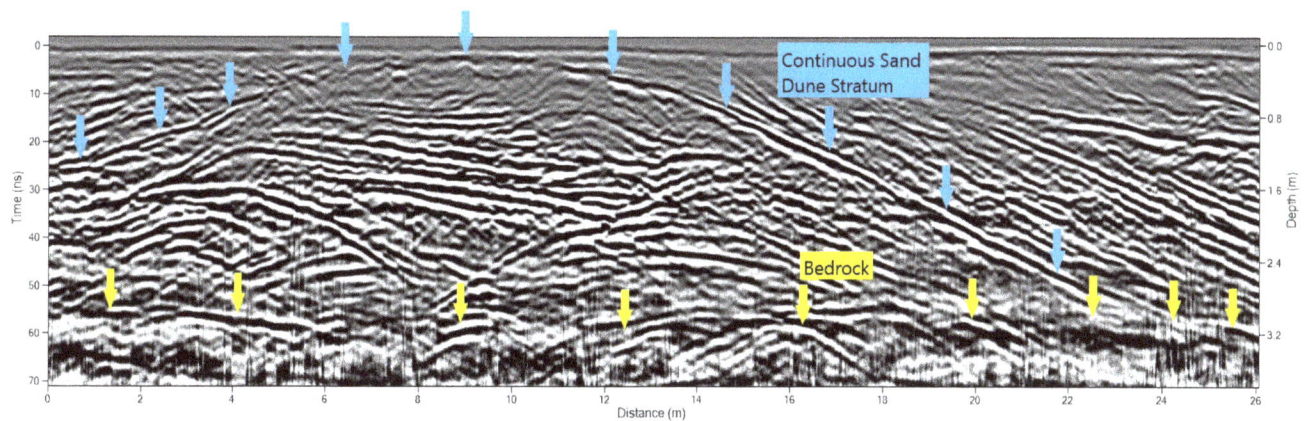

Figure 5.2. GPR Profile of CA-SCR-7, File 235.

These two images provide a good baseline for understanding the development of the sand dune independent from other areas where this development occurs in tandem with human activities and materials that become incorporated into the structure of the dune. This is important for understanding what no discernable or material human activity in an area looks like compared with areas containing midden soils and other cultural materials.

Midden soils and other cultural materials were sporadically exposed to the surface across the dune, so it was also helpful to survey one of the larger exposed midden areas to create a baseline for how these features appeared to be incorporated within the dune system. The profile in Figure 5.3 is an example of an exposed midden deposit that extends further underground than the visible boundary of midden on the surface of the site. This profile begins at the top of the northwest end of the dune where midden is exposed and continues southeast back toward the middle of the dune. The profile shows midden at the ground surface from the 0- to 5-meter marks. The bottom of the midden soils is represented by a continuous, high-amplitude planar reflection that extends from 0 nanoseconds at the 0-meter mark to 24 nanoseconds or 1.4 meters of depth at the 12.5-meter mark. The upper limit of the midden soil, also represented by a high-amplitude planar reflection, dips below the surface in the profile at about the 5-meter mark and continues to descend

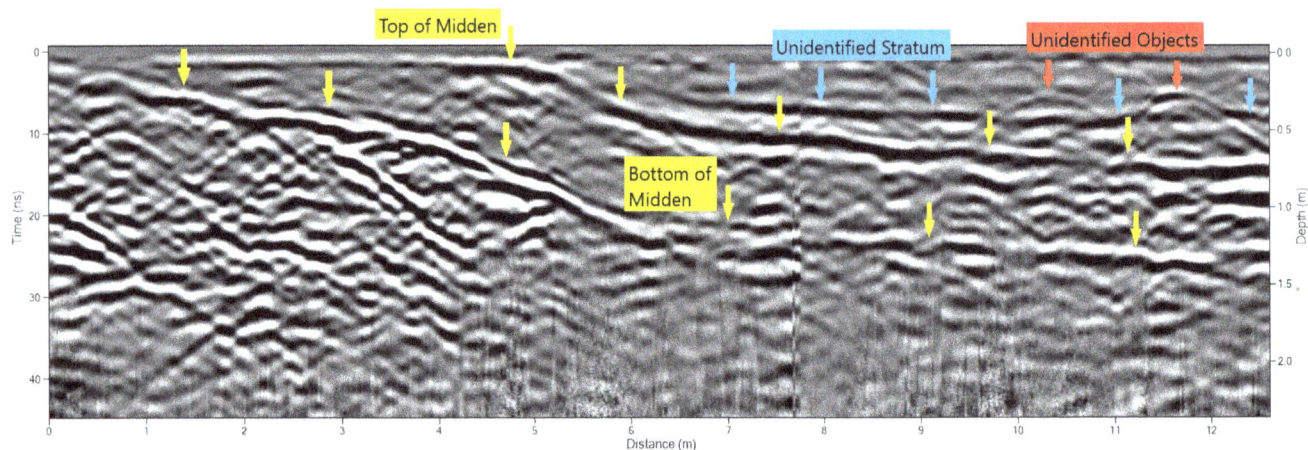

Figure 5.3. GPR Profile of CA-SCR-7, File 230.

to about 15 nanoseconds or 0.75 meters of depth. Between the 5- and 12.5-meter marks, the midden soil maintains a thickness of about 12 nanoseconds or 0.65 meters.

There is also a fairly consistent planar reflection above the 0.65-meter-thick midden soils from the 6- to 12.5-meter marks at about 0.4 meters of depth. This represents some sort of additional layer of stratigraphy above the midden. This could be additional midden that is slightly different, perhaps mixed with sand, or it could be a sand layer that is somewhat different than the layers of sand above it. There are two objects resting on this surface at the 10.3- and 11.8-meter marks, which seem consistent with it being an old surface or substantial layer of stratigraphy.

Investigating these exposed midden areas allowed us to determine the depth and thickness of these deposits and explore some of the extent of their underground components. The profile was terminated at 12.5 meters of distance because the brush on top of the dune was too thick to continue surveying any further. Even with such a small surveyable area, the profile allowed us to better understand how these midden soils fit within the structure of the dune, and confirmed that there were much more extensive portions of the midden that were not visible. This finding was important for setting our expectations about how completely buried midden soils may appear in profile without any visual indication of their presence on the surface.

With expectations set for both non-anthropogenic and anthropogenic formation of dune components, more profiles were collected in areas with no visual signs of midden soils. One area that produced cultural features in the GPR data is shown in figure 5.4. This profile runs along the north side of the dune in between dense brush, beginning at the northwest end of the dune heading to the southeast. Within the profile, there are several diagonal layers of sand, such as between the 0- to 4-meter marks from about 20 to 40 nanoseconds or 1 to 2 meters of depth. Resting on some of these diagonal sand layers is a layered mound feature from the 5- to 12-meter marks, at about 27 to 40 nanoseconds or 1.4 to 2 meters of depth. This feature is oriented and shaped very differently from the diagonal sand layers. There is also another one of these features from the 13- to 16-meter marks at about 11 to 20 nanoseconds or 0.6 to 1.1 meters of depth.

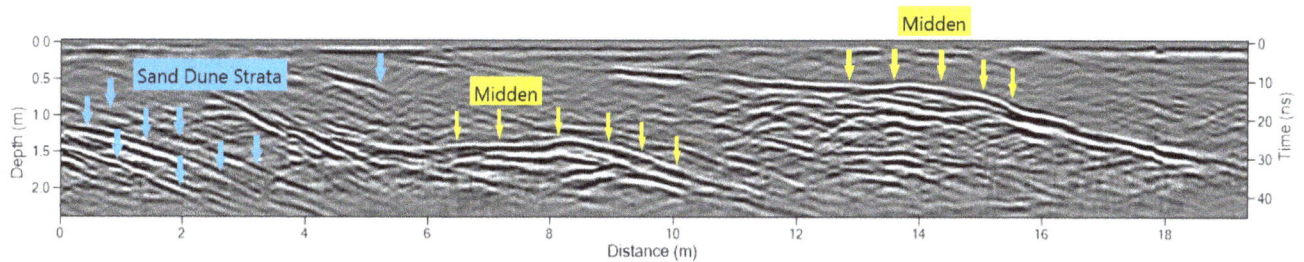

Figure 5.4. GPR Profile of CA-SCR-7, File 241.

Given the great potential of the features in figure 5.4 to be midden soils, an auger unit (AU 3, see Chapter 4) was placed (at about the 14-meter mark in the GPR profile) to investigate one of these features. The auger confirmed that the first 0 to 0.6 meters below surface was almost pure sand with a few shell inclusions. Midden soil was detected at about 0.65 meters below the surface and continued until about 1.05 meters below the surface, where it returned to almost pure sand for the remaining 0.55 meters of the auger unit. Given these findings, there is great potential for more such lenses and isolated deposits of midden representing potentially very different and discrete episodes of occupation.

CA-SCR-10

This site is situated completely within an agricultural field surrounded by a dirt service road, drainage ditches, and a steep drop off in topography down to a creek along the northeastern side. The site has very clearly been flattened from years of plowing, and the midden soils have most likely been spread out horizontally from the original boundaries of the site. Knowing that there were some major impacts to the near-surface components of this site, it was hoped that GPR, magnetometry, and excavation could detect and investigate intact features in the layers of the site below the plow zone.

The field in which this site was situated was actively being used to grow Brussels sprouts during the summer of 2016. These fields were being watered with sprinklers, saturating the ground, and the Brussels sprouts were planted in rows between furrows. These furrows were unfortunately too narrow to consider surveying because it would risk damaging the farmer's crop. Thus, the only option for this first component of the GPR survey at CA-SCR-10 was to construct one long grid (grid 1, block 1) in the middle of the dirt service road at the southeast end of the site. At this end, midden soils and cultural materials such as shell, chipped and ground stone, faunal bone, etc., were present. After a reevaluation of the GPR data alongside the excavation data, it was determined that there was very little intact midden at this end of the site and that the best course of action was to revisit the site in the fall after the harvest, when the middle of the field would be open for study. This second component of the survey at CA-SCR-10 was composed of three blocks (grid 2, blocks 1, 2, and 3) within the center of the site and completed in the fall of 2016. The ground had been plowed to bare earth, destroying the rows and furrows. However, the plow scars remained, oriented in a northwest-southeast direction, and the ground was bumpy with clods of dirt and debris from the Brussels sprouts. The uneven nature of the soil in this area (as opposed to the smooth compact soil in the road area) resulted in more instances where the antenna decoupled from the ground and disrupted or obscured some portions of the data, though not enough to impact broader interpretations of the site.

The profiles in Figures 5.5 and 5.6 represent the typical stratigraphy found throughout grid 1, block 1 in the area of the service road. There are several parallel low-, medium-, and high-amplitude planar reflections or interfaces between soil horizons. These very regular horizons remain even after a background removal process is performed to remove artificial banding in the data. There are also several small, medium, and large point reflections or objects resting on these various soil horizons, indicating that they are relatively intact layers of stratigraphy.

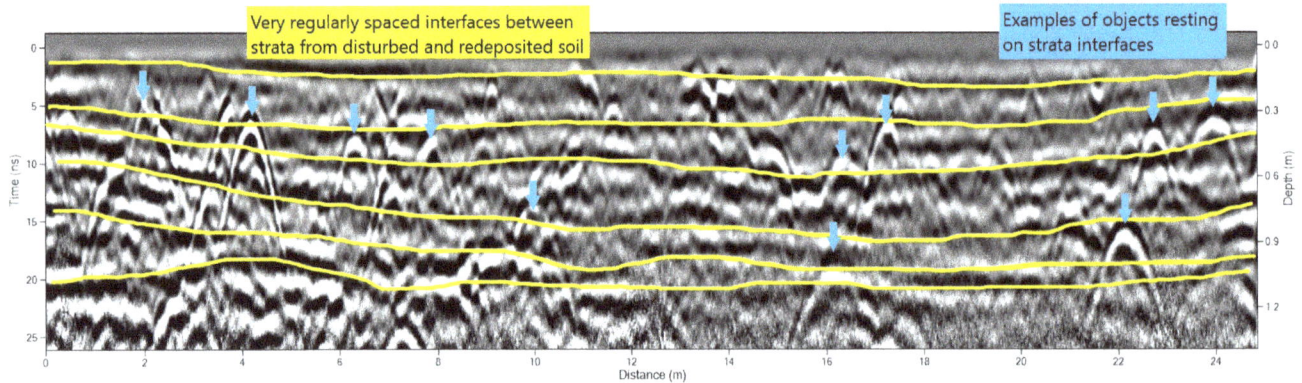

Figure 5.5. GPR Profile of CA-SCR-10, Grid 1, Block 1, File 258.

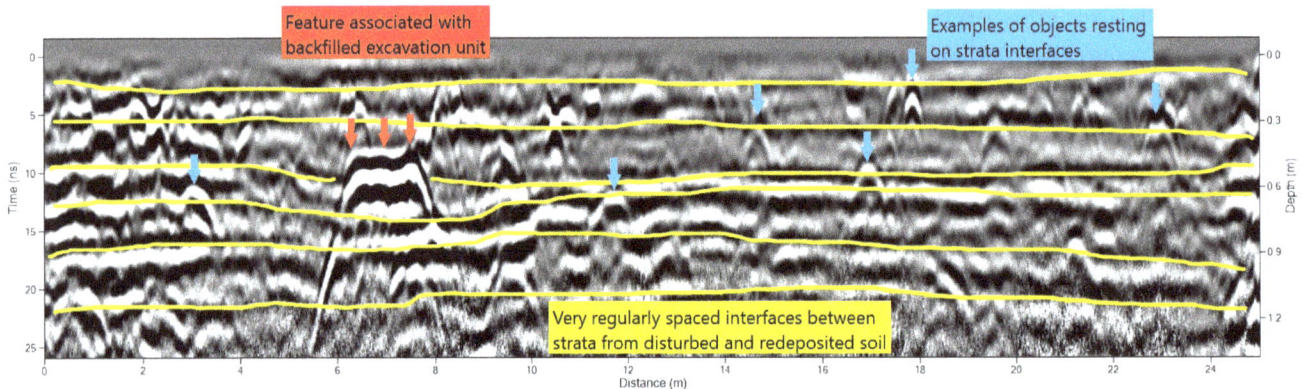

Figure 5.6. GPR Profile of CA-SCR-10, Grid 1, Block 1, File 262.

However, excavation of a 1x1 m unit (EU 1, see Chapter 4) revealed deep plow scars down to about 0.5 or 0.6 meters and modern trash as deep as 1 meter. Excavators continued work in this unit during a second field season in 2017 and uncovered intact midden soils between 1.12 to 1.4 meters below the surface. This assessment aligns with previous studies of CA-SCR-10 along the highway that mention heavy disturbance from farming equipment and road building activities, such as plowing and bull-dozing for agricultural fields, a concrete-lined ditch, a series of telephone poles, and highway and farm roads, leaving homogenous soils with little stratigraphic differentiation and broken chunks of bedrock mixed throughout the deposits (Jones and Hildebrandt 1994:

1, 42, 45, 69). Though grid 1, block 1 was placed as far from the highway as possible at the southern end of this site, it appears that the survey was not able to escape the amount of disturbance done to this site.

Given the feedback from excavation of a very disturbed site, how might the regular horizons in the GPR be explained? Jones and Hildebrandt (1994: 45) noted that the soil was relatively homogenous without much stratigraphic differentiation. However, if bulldozing the road and regular plowing of the agricultural field are flattening the site overall, these activities, done on yearly, seasonal cycles, may be causing regular erosional events that produce perfectly layered stratigraphy at the extreme edges of the site. These layers are mixed contexts and would look homogenous to the eye of a human excavator, but the GPR is able to differentiate between these events. Thus, the GPR profile shows intact horizons of disturbance events that flattened and shaped the modern agricultural field. Unfortunately, these results indicate that very little of the site along its margin may remain intact. However, if erosion is covering intact soils below, these intact midden soils would be protected from other subsequent disturbance, whereas the center of the site would continue to experience cutting, flattening, and mixing from the continuance of these activities into the future.

One additional feature to note in Figures 5.6 and 5.7 is a perfectly rectangular planar reflection between the 6- and 8-meter marks at about 10 nanoseconds or 0.55 meters of depth. The dimensions of this feature are approximately 1 by 2 meters, and it appears in the 10–15 and 15–20 nanosecond slices shown in Figure 5.7. In these slices, the feature appears between 3 and 4 meters on the x-axis and 16 and 18 meters on the y-axis. In addition to the planar reflection, there appear to be additional diagonal or point reflections above the feature at both ends, perhaps representing the sidewalls of a vault or cut in the ground. This feature overall appears to be an old archaeological excavation unit (Unit 2, see Chapter 4), most likely from the study in this area of the site led by State Parks archaeologist Mark Hylkema in coordination with Cabrillo College in July of 2008. Since Hylkema's excavation units did not stop at 0.55 meters of depth, the planar reflection in the GPR data may be some kind of backfilling event in the middle of their unit. Often, GPR reflections are most pronounced when there are differentials in water retention between materials or the compaction of materials. Thus, the GPR reflection may be more related to the unintentional circumstances of closing the unit rather than a planned or intentional action by the excavators at this depth.

Since the only areas of CA-SCR-10 that were accessible for survey during summer 2016 were heavily disturbed, three additional blocks in a second grid were placed in the middle of the site in the fall of 2016. This follow-up survey in the middle of the briefly fallow agricultural field produced very different results from the dirt road area on the margins of the site. Many of the profiles from these three blocks are difficult to interpret, though this area did produce a few promising areas for future investigation.

Figure 5.8 is an example of a typical profile from these blocks exhibiting a "choppiness" in the data due to a dense presence of overlapping point and planar reflections as well as artificial features from obstructions. Examples of these artificial features are present between the 18- and 18.5-meter marks where the features appear dragged and blurred, ending abruptly at an artificially straight vertical line, after which the following data is disjunct or vertically shifted from the position of corresponding features before this line.

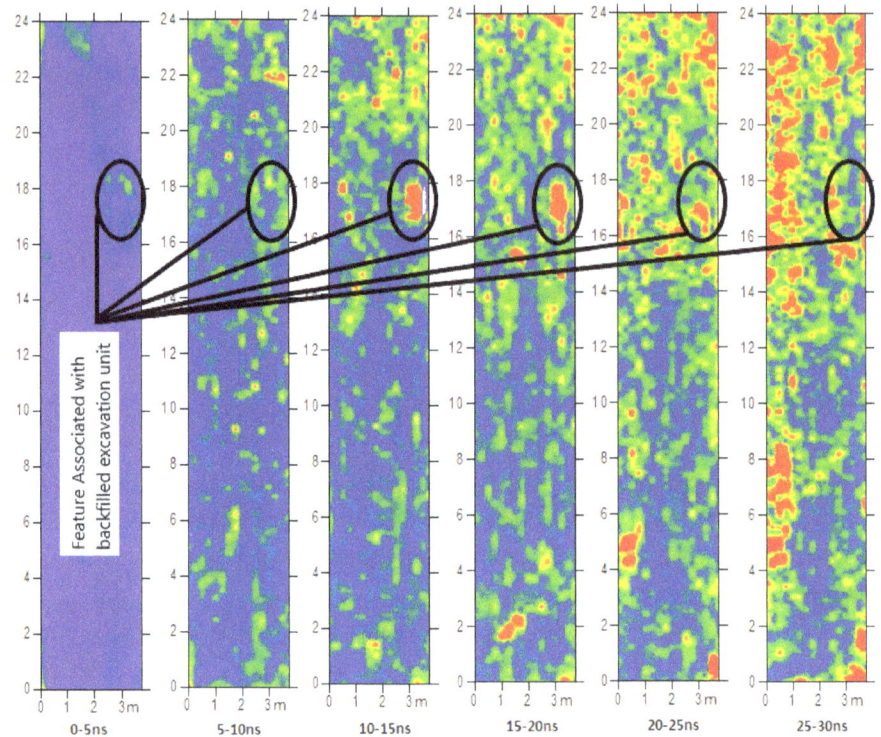

Figure 5.7. GPR Planview of CA-SCR-10, Grid 1, Block 1.

The appearance of "dragging" is the result of the survey wheel not making full contact with the ground, resulting in more data collected per unit of distance in the profile. The abrupt disjuncture is caused by an abrupt decoupling of the antenna from the ground when it rolls over or is interrupted in its path by a rock or other object, uneven ground, and void space, etc. These antenna decouplings can also appear blurred, or not as sharply contrasted as the areas around it, when they occur more gradually or the obstruction is large. In profiles such as these, it is difficult to interpret patterns between individual point reflections or see contiguous planar reflections of any significant length.

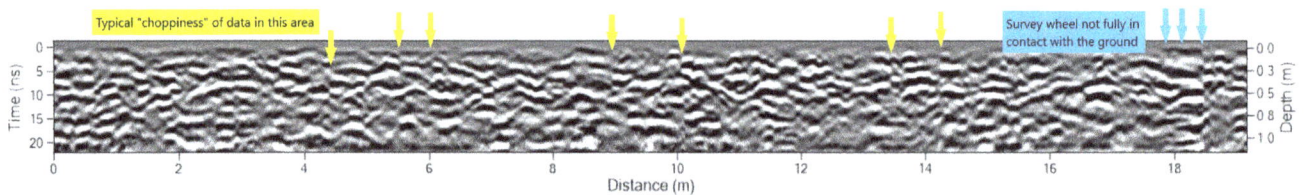

Figure 5.8. GPR Profile of CA-SCR-10, Grid 2, Block 1.

That being said, the profile in Figure 5.9 shows a possible surface or downward sloping layer of stratigraphy at about 7 nanoseconds or 0.4 meters of depth at the 12-meter mark to about 19 nanoseconds or 1 meter of depth at the 17-meter mark. Similarly, there is another example of a smaller planar reflection in the profile in Figure 5.10. This planar reflection appears at about 11 nanoseconds or 0.6 meters of depth between the 11.7- and

14-meter marks. This planar reflection has a "bowl" or downturned lenticular shape with raised ends. This feature is consistent with the shape of a potential house floor. The center of this feature is also of interest because it becomes a little bit diffuse, indicating there may be a partial gap in the floor, potentially a central hearth pit. One last planar feature of note in Figure 5.11 is a "hill" or upturned lenticular shape at 13 nanoseconds or 0.7 meters of depth between the 13.5- and 16.5-meter marks. Both ends of this feature terminate at a depth of about 19 nanoseconds or 1 meter of depth. This planar reflection is somewhat discontinuous with three 1-meter sections, and so despite the appearance of a potential surface or surfaces, it could also represent an accumulation of smaller objects that appear as surfaces when detected by the 400 MHz antenna as opposed to the 900 MHz antenna. An example of a feature that could potentially appear as this does is a larger flat accumulation of rocks from an earthen oven. Without further survey with a more precise, higher frequency antenna or confirmation through excavation of what this feature may be, there is not enough confidence to merit a more specific interpretation.

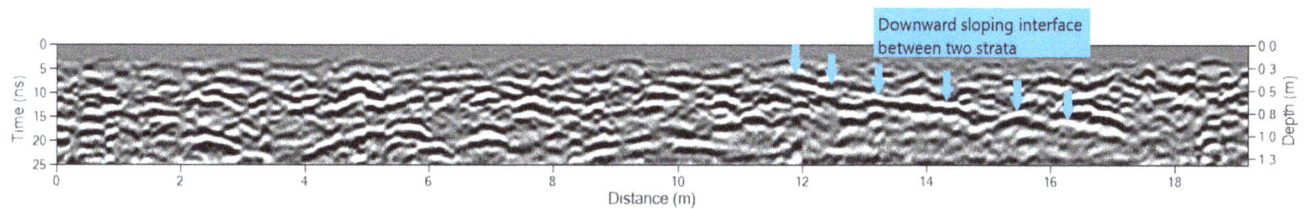

Figure 5.9. GPR Profile of CA-SCR-10, Grid 2, Block 1.

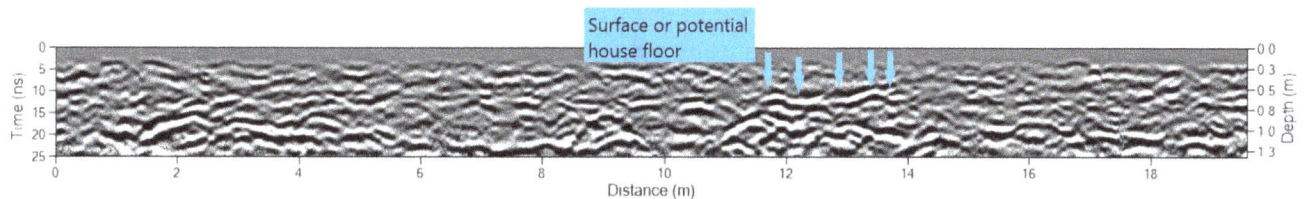

Figure 5.10. GPR Profile of CA-SCR-10, Grid 2, Block 2.

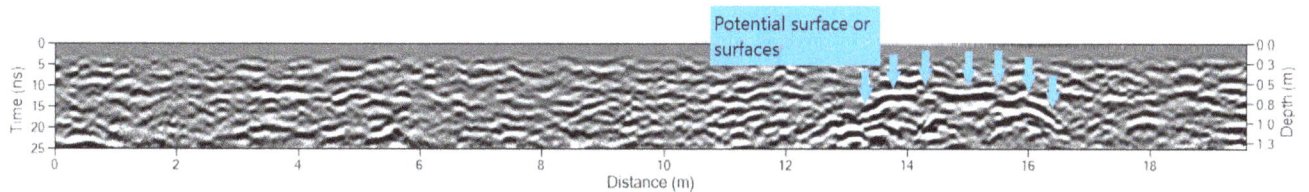

Figure 5.11. GPR Profile of CA-SCR-10, Grid 2, Block 2.

The profile in Figure 5.12 is a rare, pristine example of the GPR data detecting "intact" soil horizons within the profile. Even though these horizons were not detected in many other areas, this profile is fairly representative of the overall stratigraphy in this section of the site. Within this profile, there are two noteworthy planar reflections that extend throughout the entire length of the profile. One of these planar reflections appears at a depth of about 5 nanoseconds or 0.3 meters. This reflection is almost completely parallel with the ground surface, and is marked by several small point reflections. This feature is consistent with the bottom of the active plow zone in this area that would usually end between 30 to 40 cm below the surface of an agricultural field. The many point reflections are also consistent with rocks being churned up vertically and resting on the bottom of the plow zone. Some of the choppiness along the bottom of the plow zone as well as "V" shaped features may also be plow scars dipping below the average bottom of the plow zone. The disturbed plow zone soil and intact soil below are essentially composed of the same midden soil. The difference between these two areas producing this GPR reflection may reflect different levels of compaction and water retention between the two levels as well as fertilizer and other botanical debris from agricultural crops that decompose in the plow zone area every year.

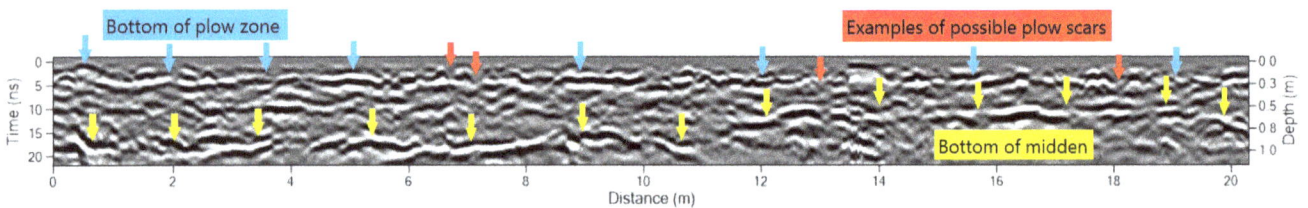

Figure 5.12. GPR Profile of CA-SCR-10, Grid 2, Block 3.

The second planar reflection in Figure 5.12 begins at a depth of 20 nanoseconds or 1 meter at the 0-meter mark, starts to become shallower around the 12-meter mark, and ends at a depth of about 13 nanoseconds or 0.7 meters at the 20-meter mark. The shallower depth is consistent with the auger data (AU 1, see Chapter 4) collected from this area. Within the first 30–40 cm of the auger, tiny broken pieces of shell were observed in the midden soils, which is consistent with this range of depth, representing the active plow zone of the field. After about 40 cm in depth, there were much larger pieces of shell in the midden soil from the auger. The midden soils continued to a depth of about 70–80 cm, where these soils terminated in subsoil with very little or no other cultural material. Given the findings from the GPR profile, there may be shallower and deeper sections of this site that are still as of yet unstudied and could potentially be older than the shallow portions of the site that were investigated with the auger.

One final unanticipated result from the survey was the appearance of several perfectly straight and parallel, high-amplitude linear features in the plan view slice maps of the blocks in Figure 5.13. These lines are oriented in a northwest-southeast direction and are consistent with the orientation of the rows and furrows created by farmers in the plowed field after the harvest in October 2016. These linear features are most apparent in the first two slices, which are the average of data between 0 to 5 nanoseconds (0–0.3 meters) and 5 to 10 nanoseconds (0.3–0.5 meters), respectively. The appearance of these linear features in both of the first two slices is consistent with the known location of the plow zone, because it terminates at about 30–40 cm below the ground surface and straddles the range of depth just above and below the boundary between these two slices. These linear features are parallel with the largest dimension or length of the field. However, the rows and furrows along which Brussels sprouts were planted and growing earlier in the summer were perpendicular to the way this field was tilled after the harvest. The features observed by the GPR may be very ephemeral and recent and could be erased by the next plowing and planting event. It is common practice for farmers to plow and till the ground after a harvest, reincorporating leftover pieces of the crops back into the ground to provide additional nutrients to the soil for the next season's crop. The post-harvest plowing was most likely completed in a parallel orientation to the length of the field to save time and labor. There are many factors that could be causing these linear features in the GPR data including a reorientation of subsurface objects parallel with the rows of the plow, cuts and voids along the bottom of the plow zone in the plowing direction, and variable accumulations of moisture and water retention along loosened rows of soil.

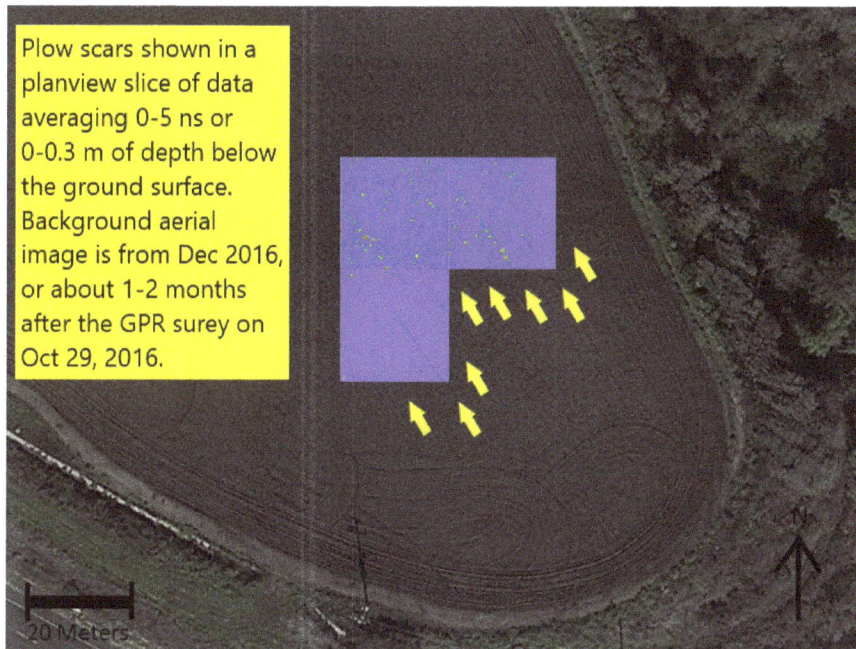

Figure 5.13. GPR Planview of CA-SCR-10, Grid 2, Showing Plow Scars.

CA-SCR-14

This site was characterized by a gently sloping topography and was partially covered by trees and dense brushy vegetation. Due to the demands on crew for excavation at other sites as well as limited availability of surveyable area, the GPR survey was conducted by collecting individual profiles in tandem with a GPS tracklog rather than contiguous profiles in a block. Even with these few profiles, the GPR survey was able to provide valuable information about the structure, features, and depth of the site.

The most striking feature in both Figures 5.14 and 5.15 is a pervasive, high-amplitude planar reflection that runs continuously throughout the entire profile. This planar reflection is more clearly present (higher amplitude) in some sections and subtler (lower amplitude or absent) in other sections of the profile. It appears in the profile between 6 and 10 nanoseconds or approximately 30–50 cm below the ground surface. Above this continuous planar reflection are other smaller medium- and high-amplitude point and planar reflections, indicating that the soil between roughly 0 and 50 cm is very conducive to GPR survey. This soil matrix between 0 and 50 cm likely represents midden soils, which have produced similar results in other studies in Central California (Nelson 2017). Below the continuous planar reflection, the profile becomes blurry and no more distinct features are apparent, indicating that the GPR signal quickly attenuates at this depth in the ground. This portion of the profile below about 50 cm represents subsoil that is not conducive to GPR survey. Thus, the continuous planar reflection between the midden soils and subsoil represents the interface between these two soil matrices. This transition was apparent throughout the entire survey of this site and clearly represented in every profile collected with the GPR.

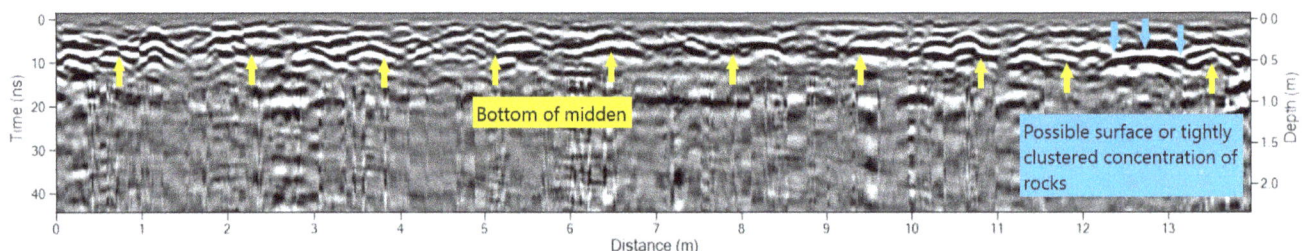

Figure 5.14. GPR Profile of CA-SCR-14.

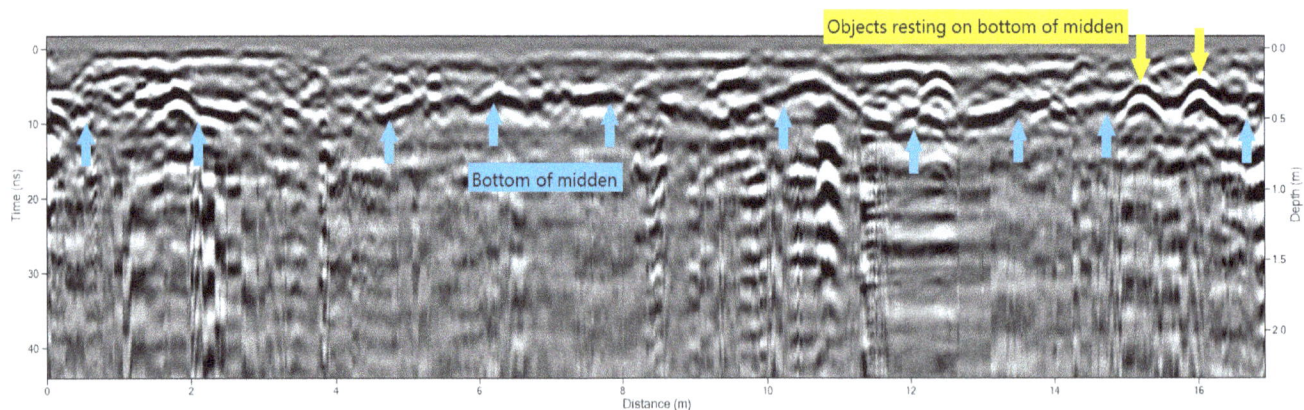

Figure 5.15. GPR Profile of CA-SCR-14.

One feature of note in Figure 5.14, located between the 12- and 13-meter marks at a depth of about 50 cm, is associated with or resting on top of the bottom of the midden soils. This feature is a very high-amplitude planar reflection that quickly attenuates the signal below, indicating a hard or compact surface, potentially a living/working space or a very tightly clustered discrete concentration of rocks from a hearth or earth oven. Given that the survey was conducted with a 400 MHz antenna, it is very possible that this feature could be composed of tightly packed rocks, making it appear as a reflective surface, whereas a higher frequency antenna (such as 900 MHz) may be able to provide greater resolution and distinguish between a truly continuous surface or several tightly-packed individual objects.

While the features in these profiles were not tested through excavation, test units in other areas of the site confirmed that the site was very shallow (between 15 and 50 cm), and features such as tight clusters of rocks from earth ovens or cooking activities were present in EU 1. Rock clusters may be the most likely interpretation for many of the point reflections in the GPR profile within the limits of the midden soils.

CA-SCR-15

This site, composed of midden soil, fire-cracked rock, ground stone, lithics, bone, and shell, was very similar to and more extensive than CA-SCR-14. CA-SCR-15 was also situated in the middle of a grassy meadow area surrounded by dense trees and vegetation with a paved one-lane road running through the middle of the site. The meadow area was kept open and clear of encroaching shrubs and forest by the landowners, so there was very good access to surveyable areas within the site despite the disturbance from the road. One other notable obstruction or disturbance in the ground was a large metal water pipe (30 cm in diameter) that serves as a water pipe for the city of Santa Cruz. The pipe is visible above ground down near the creek to the north of the site but dives into the ground and runs close to the paved road. Also present near the pipe is a very subtle berm that is most likely the back dirt dug from the trench that was excavated for this water pipe. Given these parameters, and time and crew capacity for geophysical survey in the overall project, the crew surveyed both non-contiguous, individual GPR profiles with a GPS tracklog and contiguous profiles within a formal block area at this site.

Block 1 at CA-SCR-15 was a 20-meter square area chosen based on the surface level presence of ashy midden and shell. This block was also surveyed with magnetometry/gradiometry. Within Block 1, several results became immediately apparent about the site's structure in the GPR data. In Figure 5.16, the interface between the midden soils and subsoil appears in the profiles between 6 and 15 nanoseconds or about 40–60 cm in depth. In Figure 5.16, this interface is a long, undulating, medium- to high-amplitude planar reflection that stretches the entire length of the profile. It is choppy in some places and there are also point reflections or objects in the ground that are "resting" on top of this interface. This interface and the depth of the site overall is shallower in the southern portion of Block 1 and deeper in the north and east end of Block 1. There is some variation in the depth in the center of this interface because it is not completely uniform. Thus, the interface appears somewhat irregular in the plan view slice maps in Figure 5.17.

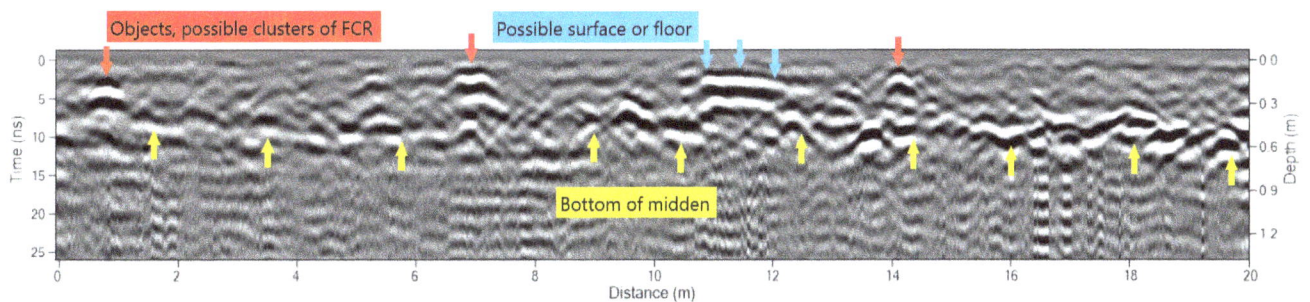

Figure 5.16. GPR Profile of CA-SCR-15, Grid 1, Block 1.

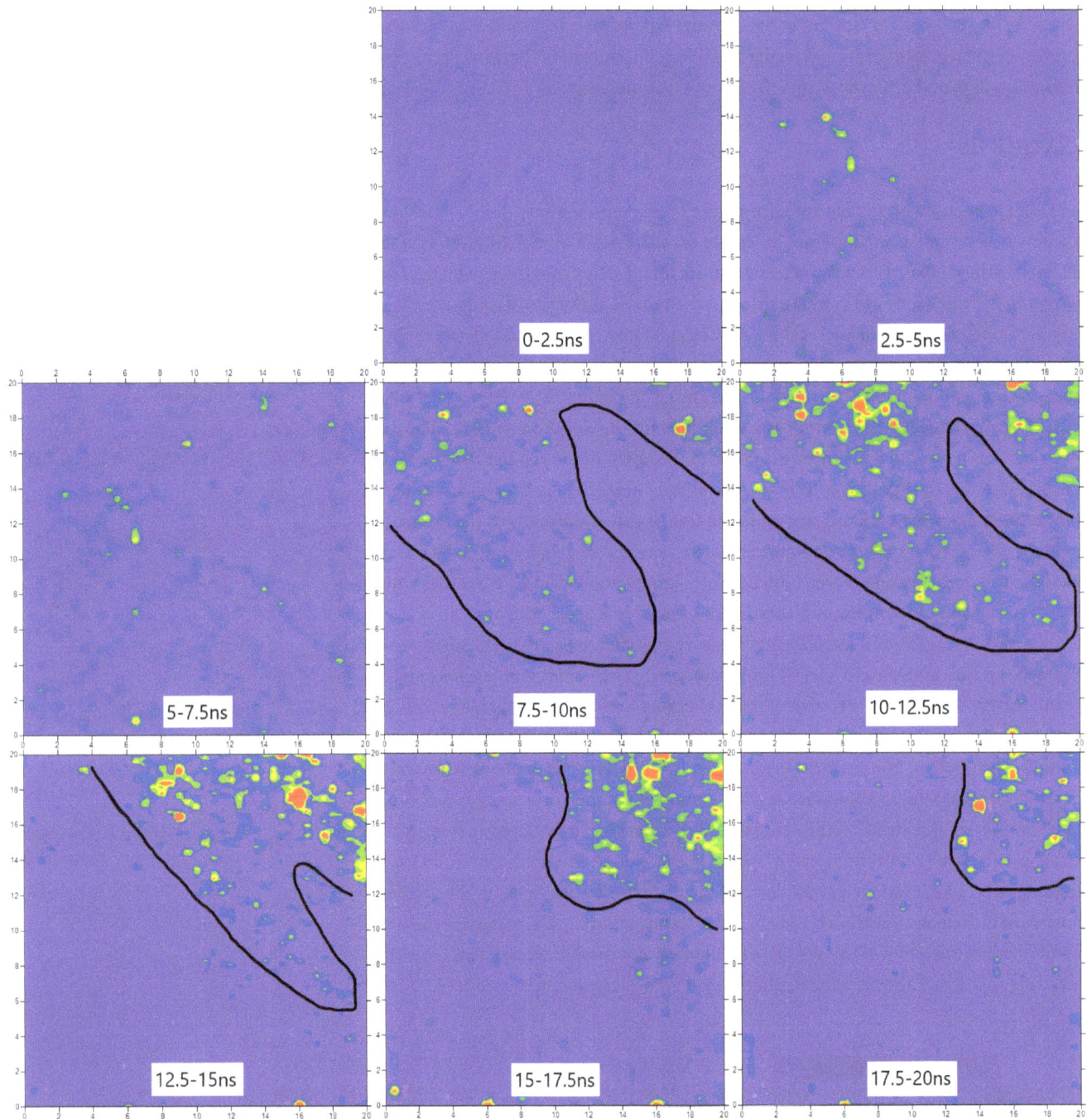

Figure 5.17. GPR Planview of CA-SCR-15, Grid 1, Block 1.

In general, the interface between midden and subsoil appears as clusters of medium- and high-amplitude reflections in the plan view slice maps in Figure 5.17. These reflections are represented in the image by clusters of yellow, green, and red blotches of color. Each slice contains 2.5 nanoseconds of vertical data that is averaged together, meaning that the data may not appear as precisely as the profiles in the direction of vertical depth. The interface between midden and subsoil appears primarily in these slices between

7.5 and 17.5 nanoseconds covering a little more range of depth than the profiles, but these results are consistent with the range of depth in the profiles from 6 to 15 nanoseconds when considering that the slices were cut and averaged across this range.

The interface between midden and subsoil begins to appear in the upper left or northwest corner of the fourth slice (7.5–10ns) in Figure 5.17, expands throughout the entire upper right or northeast half of the fifth (10–12.5ns) and sixth (12.5–15ns) slices, and recedes toward the northeast corner where it largely recedes in the seventh slice (15–17.5ns). The homogenous, purple area devoid of point reflections in the southwest corner of the fourth through sixth slices and the majority of the seventh and eighth slices is subsoil below the midden site where the GPR signal attenuated. In Figure 5.16, this area underneath the interface between midden and subsoil can be viewed in profile and appears "fuzzy" or blurred compared to the well-defined features above it. The continuation of some medium- or high-amplitude reflections in the deepest plan view slice may represent artificial continuations of reflection trails rather than the center of the midden-subsoil interface.

In the profiles from this site, there are clear signatures of features within the midden soils. In Figure 5.16, there are high-amplitude point reflections or objects between 3 and 5 nanoseconds or 20–30 cm in depth at 1, 7, and 14 meters of horizontal distance in the profile. These features may represent large, individual objects such as rocks or groundstone or clusters of smaller fire- cracked rocks from cooking or thermal features such as hearths. There is also a high-amplitude planar reflection from 10.5 to 12 meters at about 3 nanoseconds or 20 cm of depth. This 1.5-meter-long feature is dimly present in the profiles adjacent to it, which indicates that it may be a surface with a diameter of 1.5 meters. This feature may likely be a small structure floor. However, a higher resolution GPR antenna, such as 900 MHz, or excavation would need to be employed in order to strengthen or confirm this interpretation.

Sensitive materials were encountered in an excavation unit from this area where magnetometry/gradiometry grids 1–3 and GPR Block 1 were placed. Thus, the decision was made to relocate further excavations to another location northwest of this gridded area. GPR was used as the primary instrument to guide the placement of the next excavation unit. Figure 5.18 is an individual, non-contiguous profile showing the feature targeted in this area. The feature in this profile is located 8–10.5 meters horizontal distance and 6–10 nanoseconds or 30–50cm below the ground surface. This feature is composed of two large, high-amplitude point reflections, which appear to be resting on a medium-to

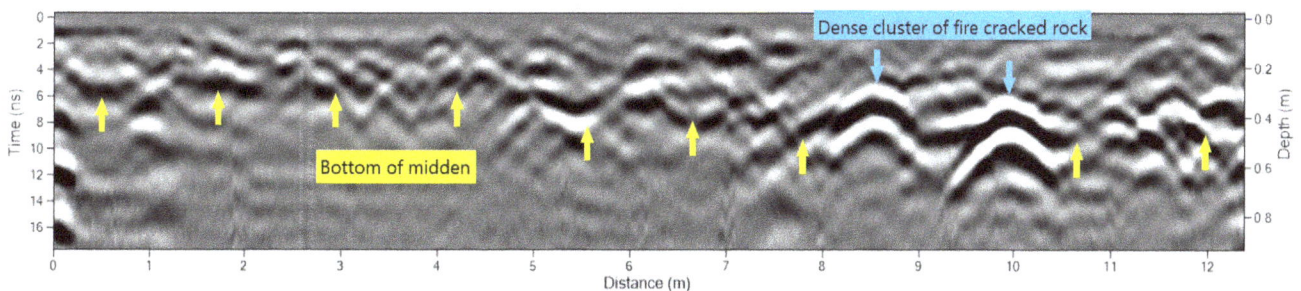

Figure 5.18. GPR Profile of CA-SCR-15.

high-amplitude planar reflection. Initial field interpretations of this feature were a possible floor or surface with objects resting on top of it. Further investigation with the excavation of EU 3 (see Chapter 4) revealed that this feature was a dense cluster of fire-cracked rock and cultural materials at about 40–44cm depth below the ground surface. Excavators also encountered the interface between midden soils and clay subsoil at about 60 cm below the ground surface. Both of the true depths confirmed by excavation for the fire-cracked rock feature and the beginning of the subsoil are also consistent with the depths for these components at 10 meters horizontal distance in the GPR profile.

Discussion

Despite a variety of challenges, the GPR survey was able to produce information at every site within the study. Much of the flexibility in the GPR strategy can be attributed to the instrument's ability to survey individual noncontiguous profiles for analysis of two-dimensional profiles, as well as whole blocks of contiguous profiles for analysis of two-dimensional profiles and plan view slice maps or three-dimensional blocks. No matter whether individual profiles or whole blocks of data were collected, the GPR detected accurate depths of features, as well as the bottom depth of midden soils in all cases. These data allow for a comparison between the four sites, which were studied in terms of their depth and composition. These data also allowed for a comparison of the relative intactness of each site.

In terms of the lowest depth of midden soils recorded at each site, there was a great range of diversity. CA-SCR-7 contained the deepest midden at around 2 meters, while CA-SCR-14 and CA-SCR-15 contained the shallowest midden soils at about 0.5 meters. CA-SCR-10 was comparatively somewhere in the middle with its deepest midden at 1.4 meters. All of the midden components detected with GPR at CA-SCR-7 were 0.6–0.7 meters thick, whereas CA-SCR-14 and CA-SCR-15 were slightly thinner at about 0.5 meters thick. The center of midden soils at CA-SCR-10 was about 0.7–0.8 meters thick, while the margin of the site was only about 0.3 meters thick. This may not represent the maximum thickness of this site due to the impacts from agriculture and other disturbances.

CA-SCR-7 is particularly interesting, because GPR was able to identify three discrete components of midden soils representing three separate occupations of this site along the northwestern end of the sand dune. These components of midden shown in Figures 5.3 and 5.4 were buried under sand from 0.75 to 1.4 meters (0.65 meters thick), 1.4 to 2 meters (0.6 meters thick), and 0.6 to 1.1 meters (0.7 meters thick). These components may represent discrete contemporaneous occupations of this site and/or decisions to move camps and village sites sequentially through time to different locations as the dune itself continued to shift and evolve. These decisions could have been part of a rotational preference or generational cycle of occupying different areas of the dune and/or pragmatic considerations of weather exposure and wind breaks, viewsheds, and so on.

Based on the very limited GPR study of CA-SCR-7, as well as previous archaeological excavations in other areas of the site (Jones and Hildebrandt 1990), it is very likely that this site is extremely complex stratigraphically, with more midden components throughout the site similar to the three identified in the GPR study. Rather than one large component of midden embedded in or on top of the dune in a classic "layer cake" model of stratigraphy, there are lenses of midden threading through the sand at different depths and locations. Given the ability of the GPR to detect and distinguish both environmental and anthropogenic depositional events, it may be easier and more feasible to disentangle the structure of, and relative ages within, this complex stratigraphy than researchers thought possible three decades ago. However, this close analysis of stratigraphy would require full GPR survey coverage of the site to establish more of these relationships between different components.

Without full GPR survey coverage at CA-SCR-7, it is unclear from the GPR data alone whether the midden components in Figures 5.3 and 5.4 are contemporaneous. However, two components from Figure 4 appear in the same profile, and so it is possible to comment on the relative ages of these two. Since the stratigraphic layers of the shallower midden deposit overlap the stratigraphic layers of the deeper deposit at around the 10-meter mark and appear unbroken, the shallower deposit is younger in age than the deeper deposit. Augering alone would not have been able to definitively confirm the relative age of these components without radiocarbon dating, making GPR a very useful tool for field exploration and for guiding excavation in the moment. GPR data can also be used to interpret site structure and guide the selection of radiocarbon dates from discrete stratigraphic components, which may produce a more informed and cost-effective sampling strategy than judgmentally/systematically sampling these dates based on auger information alone.

Despite differences in depth and thickness of midden components of the four sites studied with GPR, many of the types of features detected by the GPR were similar. The most prevalent type of feature throughout the sites was point reflections and small planar reflections that represent objects or clusters of objects in the ground. Some of these—such as the feature at CA-SCR-15 in Figure 5.18—represent clusters of fire-cracked rock from earth ovens and other cooking or fire-related activities. There were also a fair number of stratigraphic layers representing the upper and lower extent of midden soils, as well as a couple of small- to medium-sized planar reflections in the range of 1–2 meters in length, which could represent surfaces or house floors, such as the one at CA-SCR-10 in Figure 10. None of the small- to medium-sized planar reflection features were excavated. These were rare compared to other feature types, and some of these may represent larger fire-cracked rock accumulations, which the 400 MHz antenna represents as a planar reflection because the rocks are so tightly clustered together.

GPR was also able to help assess the relative intactness of various components at each site. At CA-SCR-10, the GPR distinguished between relatively intact midden from 0.3 to 0.8 meters in the middle of the site, even though more than a meter of the midden was disturbed on the margin of the site. The GPR was also able to detect completely intact and buried components of CA-SCR-7 at three different locations. CA-SCR-14 and CA-SCR-15 were interpreted to be intact even though they were very shallow sites. However, all sites contained some intact components and contributed to the study. And any impacts or disturbances to the midden soils does not diminish the importance of any of these sites, because these materials contain traces of ancestral activity and contribute to the significance of the larger cultural landscape.

Conclusion

Given the time constraints to complete surveys at the four sites before excavation began and, in some cases, limitations on clearing brush from more sensitive sites, full GPR survey coverage was not conceivable during the 2016 GPR field season. GPR survey focused on areas of the sites that the crew were able to access. Thus, most interpretations based on GPR data alone about the horizontal extent of midden components at these sites were either not feasible or too incomplete to merit discussion. However, these constraints did allow for an exploration of the depth, extent, and intactness of various site components, which provided excellent information that guided excavation and interpretations of these sites.

More extensive future survey could fill in the gaps in the GPR data in order to answer questions about the extent and structure of these sites that are still unresolved. In considering future research, CA-SCR-14 and CA-SCR-15 should be resurveyed with a 900 MHz antenna in order to obtain greater near-surface resolution because of the shallow depth of these sites. Since CA-SCR-7 is primarily composed of sand that is very conducive to GPR, it should be resurveyed with a 200 MHz antenna to extend the range of depth at which the antenna is able to detect features within the ground. The 400 MHz GPR survey at CA-SCR-10 could benefit from much more horizontal coverage and include the entire agricultural field. Since the field is not under cultivation in the fall after the harvest and before the rains arrive, full coverage order would not be difficult to obtain at this site. Full coverage would help to confidently interpret the site's structure and potential features such as house floors or surfaces, shell lenses, hearths, and other features embedded in the intact portions of the site. This extended survey would also provide further resolution as to where the intact and eroded midden soils interface with one another, as in Figure 5.12.

Overall, there was a very high success rate of identifying productive features and site components from the 2016 GPR data—of which some were excavated—that led to strong interpretations of site structure. This high success rate of identifying features and interpreting a site's structure with more precision and no disturbance to the site from the GPR survey itself makes GPR a fantastic tool when working with sensitive sites where non-invasive or low-impact strategies are necessary. This is especially useful in cases such as the 2016 survey where the AMTB, like many Tribal communities engaging in the study and management of Tribal cultural resources, wished to minimize damage to these ancestral places. When it is carefully considered, GPR is a very useful tool to better understand the site structure and to guide the excavation strategy. The broader the survey, the better the information produced, though individual profiles can also produce great information—and very expediently.

References

Allan, James M.
  1997   Searching for California's First Shipyard: Remote Sensing Surveys at Fort Ross. *Kroeber Anthropological Society Papers* 81: 50–83.

Arnold, Jeanne E., Elizabeth L. Ambos, and Daniel O. Larson
  1997   Geophysical Surveys of Stratigraphically Complex Island California Sites: New Implications for Household Archaeology. *Antiquity* 71: 157–168.

Byram, Scott, Kent G. Lightfoot, Rob Q. Cuthrell, Peter Nelson, Jun Sunseri, Roberta A. Jewett, E. Breck Parkman, and Nicholas Tripcevich
  2018   Geophysical Investigation of Mission San Francisco Solano, Sonoma, California. *Historical Archaeology* 52: 242–263.

Conyers, Lawrence B.
  2004   *Ground-Penetrating Radar for Archaeology.* Geophysical Methods for Archaeology. AltaMira Press, Walnut Creek, California.
  2012   *Interpreting Ground-Penetrating Radar for Archaeology.* Left Coast Press, Inc., Walnut Creek, California.

Conyers, Lawrence B., and Dean Goodman
  1997   *Ground-Penetrating Radar: An Introduction for Archaeologists.* AltaMira Press, Walnut Creek, California.

Cuthrell, Rob Q.
  2013   An Eco-Archaeological Study of Late Holocene Indigenous Foodways and Landscape Management Practices at Quiroste Valley Cultural Preserve, San Mateo County, California. PhD Dissertation, Department of Anthropology, University of California, Berkeley, Berkeley, California.

Gonzalez, Sara Lynae
  2011   Creating Trails from Traditions: The Kashaya Pomo Interpretive Trail at Fort Ross State Historic Park. PhD Dissertation, Department of Anthropology, University of California, Berkeley, Berkeley, CA.
  2016   Indigenous Values and Methods in Archaeological Practice: Low-Impact Archaeology through the Kashaya Pomo Interpretive Trail Project. *American Antiquity* 81(3): 533–549.

Grenda, Donn R., Christopher J. Doolittle, and Jeffrey H. Altschul
  1998   *House Pits and Middens: A Methodological Study of Site Structure and Formation Processes at CA-ORA-116, Newport Bay, Orange County, California.* Statistical Research Technical Series. Statistical Research, Inc., Tucson, Arizona.

Jones, Deborah A., and William Hildebrandt

1990   *Archaeological Excavation at Sand Hill Bluff: Portions of a Prehistoric Site CA-SCR-7, Santa Cruz County, California.* Far Western Anthropological Research Group, Davis, California.

1994   *Archaeological Investigations at Sites CA-SCR-10, CA-SCR-17, CA-SCR-304, and CA-SCR-38/123 for the North Coast Treated Water Main Project, Santa Cruz County, California.* Far Western Anthropological Research Group, Inc., Davis, California.

Lightfoot, Kent G.

2006   Experimenting with Low-Impact Field Methods. *News from Native California* 19(4): 16–19.

2008   Collaborative Research Programs: Implications for the Practice of North American Archaeology. In *Collaborating at the Trowel's Edge: Teaching and Learning in Indigenous Archaeology*, edited by S. Silliman, pp. 211–227. University of Arizona Press, Tucson.

Lightfoot, Kent G., and Sara L. Gonzalez

2018   *Metini Village: An Archaeological Study of Sustained Colonialism in Northern California.* Contributions of the University of California Archaeological Research Facility No. 69. Archaeological Research Facility, University of California, Berkeley.

Lightfoot, Kent G., Peter A. Nelson, Roberta A. Jewett, Rob Q. Cuthrell, Paul Mondragon, Nicholas Tripcevich, and Sara Gonzalez

2013   *The Archaeological Investigation of McCabe Canyon Pinnacles National Park.* Archaeological Research Facility Field Reports. Archaeological Research Facility, University of California, Berkeley.

Nelson, Peter A.

2017   Indigenous Archaeology at Tolay Lake: Responsive Research and the Empowered Tribal Management of a Sacred Landscape. PhD Dissertation, Department of Anthropology, University of California, Berkeley.

2020   Refusing Settler Epistemologies and Maintaining an Indigenous Future for Tolay Lake, Sonoma County, California. *American Indian Quarterly* 44(2): 221–242.

2021   The Role of GPR in Community-Driven Compliance Archaeology with Tribal and Non-Tribal Communities in Central California. *Advances in Archaeological Practice* 9(3): 215–225.

Schneider, Tsim D.

2010   Placing Refuge: Shell Mounds and the Archaeology of Colonial Encounters in the San Francisco Bay Area, California. PhD Dissertation, Department of Anthropology, University of California, Berkeley.

Silliman, Stephen W., Paul Farnsworth, and Kent G. Lightfoot

2000   Magnetometer Prospecting in Historical Archaeology: Evaluating Survey Options at a 19th Century Rancho Site in California. *Historical Archaeology* 34(2): 89–109.

Somers, Lewis

2008   Resistivity and Magnetometry Field Gradient Survey. In *The Creekside Village Archaeological Testing Program, Santa Rosa, Sonoma County, California*, edited by William Roop, and Emily Wick, pp. 112–126. Archaeological Resource Service, Rohnert Park, CA. On file with the Northwest Information Center, Sonoma State University.

Sunseri, Jun U., and Scott Byram

2017   Site Interiography and Geophysical Scanning: Interpreting the Texture and Form of Archaeological Deposits with Ground-Penetrating Radar. *Journal of Archaeological Method and Theory*: 1–25.

Tschan, Andre P.

1997   Sensing the Past and the Remoteness of the Future: A Soil Resistivity Survey at the Native Alaskan Village Site. In *The Archaeology and Ethnohistory of Fort Ross, California, Volume 2 The Native Alaskan Neighborhood: A Multiethnic Community at Colony Ross*, edited by Kent G. Lightfoot, Ann M. Schiff, and Thomas A. Wake, pp. 107–128. Contributions of the University of California Archaeological Research Facility No. 55. Archaeological Research Facility, University of California, Berkeley.

CHAPTER 6

Chronology of CA-SCR-7, CA-SCR-10, CA-SCR-14, and CA-SCR-15

KENT G. LIGHTFOOT, ROB Q. CUTHRELL, GABRIEL M. SANCHEZ, AND MICHAEL A. GRONE

The macrobotanical assemblages recovered from intact archaeological deposits (see Chapter 8) at CA-SCR-7, CA-SCR-10, CA-SCR-14, and CA-SCR-15 provided ephemeral plant remains suitable for submitting radiocarbon dates. Rob Cuthrell identified terrestrial paleoethnobotanical remains for radiocarbon dating. Rhytidome and parenchymous tissue of terrestrial vegetation was selected to avoid biases or the "old wood" effect (e.g., Ashmore 1999; Schiffer 1986; Stuiver et al. 1986). At least one basal and one upper deposit radiocarbon sample were selected from each archaeological site and context. We submitted a total of 60 radiocarbon samples from the four sites to the Keck Carbon Cycle AMS Facility at the University of California, Irvine (UC Irvine), for Accelerator Mass Spectrometry (AMS) dating (Appendix 6.1). The calibrated AMS age (2-sigma) for the 60 radiocarbon assessments was based on the IntCal13 Radiocarbon Age Calibration Curve (Reimer et al. 2013).

Results

CA-SCR-7

The 27 AMS dates calibrated for the 3 auger units and 7 column units in Locus 1 indicate a Middle Holocene chronology ranging from ca. 4800 to 2200 cal BCE (Appendix 6.1). The relatively tight sequence of dates indicates the deposits have high integrity.

The radiocarbon dates indicate two distinct temporal components: Component A, dating to ca. 4800–3600 cal BCE, and Component B, dating to ca. 2700–2200 cal BCE (see Cuthrell 2021). Component A includes deposits associated with AU 1, AU 2, AU 3, CU 3, CU 4, CU 5, and CU 7. Component A also includes levels 8–11 in CU 1 and levels 8–9 in CU 2. Component B is represented by the deposits associated with CU 6, as well as levels 1–7 in CU 1 and CU 2.

Two radiocarbon samples collected from EU 1 in Locus 4 revealed ambiguous results. The two AMS dates from level 8 (60–70 cm below datum) ranged from 1400–1430 cal CE to 4040–3610 cal BCE. These results suggest the possibility of either sample contamination or bioturbation.

CA-SCR-10

For this site, 14 AMS dates were obtained from CA-SCR-10, 3 from the auger units and 11 from the excavation unit. The three radiocarbon samples from the basal levels of AU 1, AU 2, and AU 3 returned dates from ca. 3800 to 3200 cal BCE. As outlined in Chapter 4, the excavation of EU 1 revealed considerable evidence for rodent bioturbation, agricultural impacts, and modern material culture (e.g., plastic) in the upper levels of the unit (levels 1–12). Beginning with level 13 and continuing to basal level 24, we observed evidence for intact deposits and features (e.g., fire-cracked rock concentrations, ash lenses, shell deposits) and minimal indications of bioturbation or modern impacts. Three AMS assessments from the upper levels revealed a diverse range of dates: 810–795 cal BCE from level 4, 1225–1270 cal CE from level 9, and 875–970 cal CE from level 12. We interpret these dates as further evidence for the disturbed nature of these upper deposits that involved the mixing of older and younger aged materials.

The radiocarbon assessments from the lower levels of EU 1 indicated a much tighter age range. Eight AMS dates from levels 14, 16, 18, 20, 22, 23 and 24 revealed dates from 680 to 880 cal CE. These findings further suggest that the lower deposits of EU 1 show a high degree of integrity with materials still intact and relatively undisturbed.

CA-SCR-14

We submitted eight radiocarbon assessments to the Keck Carbon Cycle AMS Facility from EU 1 and EU 2. The dates suggested a Late Holocene occupation ca. 1000–1700 cal CE. However, the upper levels (level 2) of both EU 1 and EU 2 revealed multiple calibrated temporal intervals including a short span in the late seventeenth to early eighteenth centuries and a longer span in the early nineteenth to early twentieth centuries. We recognize that the samples from level 2 may be from potentially disturbed near-surface contexts or indicate an occupation into the early eighteenth century (see Cuthrell 2021). If we exclude these two dates, the site occupation range is ca. 1000–1510 cal CE.

CA-SCR-15

We submitted nine radiocarbon assessments from EU 1 and EU 3. Seven of the dates revealed a relatively tight age range from ca. 1050 to 1400 cal CE. The other two radiocarbon samples returned dates before and after this range. One of these dates (EU 1 level 6; ca. 800–900 cal CE) may be from a stratum that involved the mixing of cultural deposits and subsoil and may not represent site occupation. The other date—from EU 1, level 2—may indicate site occupation into the 1500s to early 1600s CE; but like the upper levels of CA-SCR-14, this date may also reflect near-surface disturbance and later historical activities on the site (see Cuthrell 2021).

Conclusion

The 60 AMS assessments for the four Santa Cruz coast sites indicate a diverse range of Middle Holocene and Late Holocene dates extending back almost 7,000 years. We defined two temporal components for CA-SCR-7: Component A from ca. 4800 to 3600 cal BCE and Component B from ca. 2700 to 2200 cal BCE. The one radiocarbon assessment from the previous excavation in Locus 1 by Jones and Hildebrandt (1990: 55) in their Unit 9 (near our CU 3) produced an uncalibrated date of 5970 $^{+/-}$ 120 BP, which appears to fall within the Component A date range. We defined two temporal components for CA-SCR-10. The basal deposits of the central area of the site sampled by the three auger units appear to date from ca. 3800 to 3200 cal BCE. The lower, intact deposits of EU 1 on the southern edge of the site revealed a later occupation from ca. 680 to 880 cal CE. The disturbed, upper levels of the unit appear to be the product of mixed deposits of varying ages due to agricultural activities and bioturbation (see discussion in Chapter 5). The two radiocarbon assessments from the 2011 excavation of Unit 2 by Cabrillo College (near EU 1) returned a date of 4425–4205 cal BP from the 60–70 cm level and a date of 3545–2775 cal BP from the 170–180 cm level. These results suggest that older deposits from the central area of the site may also be found in the site's periphery, but we did not encounter any evidence of these ancient remains within intact deposits in EU 1. We feel confident in dating the main occupations of CA-SCR-14 and CA-SCR-15 from ca. 1000 to 1510 cal CE and ca. 1050 to 1400 cal CE, respectively, with the possibility that later occupations may also be represented as well.

References

Ashmore, Patrick J.
    1999   Radiocarbon Dating: Avoiding Errors by Avoiding Mixed Samples. *Antiquity* 73(279): 124–130.

Cuthrell, Rob Q.
    2021   Archaeobotanical Research at Eight Archaeological Sites West of the Santa Cruz Mountains: Implications for Subsistence Practices and Anthropogenic Burning. In *The Study of Indigenous Landscape and Seascape Stewardship Practices on the Santa Cruz Coast: A Collaborative Eco-Archaeological Approach*, edited by Kent G. Lightfoot, Rob Q. Cuthrell, Mark G. Hylkema, Valentin Lopez, Diane Gifford-Gonzalez, Roberta A. Jewett, Michael A. Grone, Gabriel M. Sanchez, Peter A. Nelson, Alec J. Apodaca, Ariadna Gonzalez, Kathryn Field, and Alexii Sigona. Report Prepared for California Department of Parks and Recreation, Santa Cruz District, Archaeological Research Facility, University of California, Berkeley.

Reimer, Paula, Edouard Bard, Alex Bayliss, J. Warren Beck, Paul Blackwell, et al.
    2013   IntCal13 and Marine13 Radiocarbon Age Calibration Curves 0–50,000 Years cal BP. *Radiocarbon* 55(4): 1869–1887, ff1810.2458/azu_js_rc.1855.16947ff. ffhal-02470111f.

Schiffer, Michal B.
    1986   Radiocarbon Dating and the "Old Wood" Problem: The Case of Hohokam Chronology. *Journal of Archaeological Science* 13(1): 13–30.

Stuiver, Minze, Gordon W. Pearson, and Tom Braziunas
    1986   Radiocarbon Age Calibration of Marine Samples Back to 9000 cal yr BP. *Radiocarbon* 28(2): 980–1021.

CHAPTER 7

The Analysis of Artifacts from CA-SCR-7, CA-SCR-10, CA-SCR-14, and CA-SCR-15

ARIADNA GONZALEZ AND AND KATHRYN FIELD

Introduction

Our artifact analysis from four Santa Cruz sites (CA-SCR-7, CA-SCR-10, CA-SCR-14, CA-SCR-15) focuses primarily on the lithic assemblages. Through a combination of dry-screen and fine-grain recovery methods, our analysis of the lithic material has allowed for a broader understanding of tool production and food processing on the Santa Cruz coastline. In the sections to follow, we present a brief background, describe our methodology and results, and provide interpretations gathered from the lithic assemblage.

Background

The Santa Cruz coastline's environment is diverse with its north-west/south-east trending Santa Cruz Mountains, which reach elevations of 915 meters. These mountains stretch to meet grasslands that press against eroding cliff sides, rocky shorelines, and sandy beaches (Jones et al. 2008). The four sites of our analysis reside on a diverse selection of coastal habitats including two coastal sites (CA-SCR-7 and CA-SCR-10) and two upland valley sites (CA-SCR-14 and CA-SCR-15). The vegetation communities surrounding this broader study area include highly productive northern seashore habitats, coastal scrub and prairie mosaics, redwood forests, mixed hardwood forests, as well as coastal cypress and pine forests (Hylkema 1991). This highly productive landscape supports an abundance of terrestrial and marine creatures available for human exploitation. Faunal resources that were available for exploitation along the coastline were ocean mollusks, terrestrial herbivores, and marine mammals, all of which were found at the four sites within our analysis. Locally available Monterey chert is also found in abundance throughout the broader study area in the form of formal stone tools and stone tool debitage.

Any study of California coastal settlements must consider the effect of sea-level rise, which is a critical component to the transformation and disappearance of coastal sites throughout the world. Rising sea levels and coastal erosion have dramatically shaped the California coast over the last 15,000 years, which has put a huge toll on the archaeological record (Hylkema 1991).

The Santa Cruz coastline contains stretches of open coast where the sand dunes are continually eroding. One significant site within our study, the Sand Hill Bluff site (CA-SCR-7), is increasingly at risk of eventually fully eroding into the ocean due to sea level rise. Mitigating these losses includes performing salvage archaeology, where excavations include a full recovery of material before it is churned into the sea. Additionally, providing mitigations to help decrease the effects of erosion is critical, such as applying fencing or planting native coastal plants to secure the integrity of the midden deposits. Working collaboratively with Native communities is integral to understanding the management and preservation of sites. It is vital for eco-archaeological programs to undertake partnership with Tribes and agencies that can enable the revitalization of Indigenous stewardship practices in the lands and waters of California (Lightfoot et al. 2021).

At the core of this project is the collaborative relationship with the Amah Mutsun Tribal Band (AMTB) and Native Stewardship Corps (NSC). Members of the Stewardship Corps continue to be a part of every process, including excavation and initial laboratory processes. Information gathered through archaeological survey, excavation, and laboratory investigations can provide additional knowledge contributing toward the restoration and revitalization of Traditional Ecological Knowledge (TEK).

Methods

Our lithic and ground stone analysis of CA-SCR-7, CA-SCR-10, CA-SCR-14, and CA-SCR-15 includes the artifacts from the dry-screen and bulk sediment samples recovered from subsurface units including column, auger, and excavation units. The following outlines field and laboratory methods used to collect, process, and further analyze lithic archaeological samples. The field excavation and initial laboratory processes were performed in collaboration with the Amah Mutsun Land Trust's NSC.

Flotation Methodology

Bulk sediment samples excavated were processed via soil flotation. As outlined in Chapter 8, flotation utilizes the flow of water to separate the lighter material (typically plant remains), which floats to the top of the tank, from the heavier material (typically lithic material, faunal remains, and rocks), which sinks to the bottom. The heavy and light fraction material is then laid out separately to dry before being taken into the laboratory for further analysis. Separating the light and heavy fraction material was the initial step in the lab analysis of the heavy fraction lithic material.

Dry Screen Methodology

The dry screened lithic material was first sieved through a 3.2 mm mesh in the field and then further screened in the laboratory through 8 mm and 4 mm sieves, which resulted in three size categories: >8 mm, 4–8 mm, and 3.2–4 mm. Lithic artifacts within the >8 mm size class were individually weighed (grams) and measured (mm) for the maximum length, width, and thickness. Lithic artifacts within the 4–8 mm and 3.2–4 mm size classes that were identified as complete flakes and proximal flakes were also weighed and measured as described above. However, due to the prohibitive number of debitage pieces in some units, lithic artifacts identified as flake and angular shutter were counted and weighed in groups.

Lithic Classification

All cultural constituents were sorted after material was size-sorted with the help of a low-powered microscope. Lithics were then classified following Andrefsky (1998), Lightfoot and Gonzalez (2018), and Silliman (2000).

Our classification of chipped stone artifacts follows Andrefsky (1998), with slight modifications to classifications of debitage. Chipped stone artifacts were classified according to morphological characteristics and sorted into "tool" and "debitage" classes. Tools include cores (CO), bifaces (BI), projectile points (PP), and unifaces (UN) (see Figure 7.1). Cores can be defined as chunks of raw material, often derived from cobbles, from which flakes have been removed. Bifaces are formal tools that have been symmetrically shaped on both sides by flake removal, as opposed to unifaces, which have been systematically formed on one side. Projectile points are a specific kind of biface that may have been used as a dart or arrow point and usually exhibit notched corners.

Figure 7.1. Images of Core (1), Projectile Point (2), Biface (3), and Uniface (4) (Photos by Kathryn Field and Ariadna Gonzalez).

Figure 7.2. Images of Flake Shatter (1) and Angular Shatter (2) (Photos by Kathryn Field and Ariadna Gonzalez).

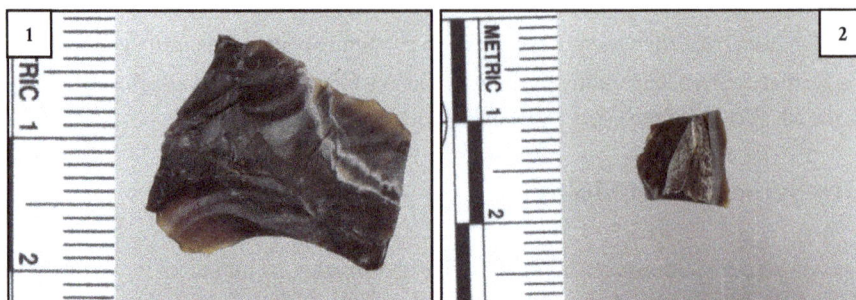

Figure 7.3. Images of Complete Flake (1) and Proximal Flake (2) (Photos by Kathryn Field and Ariadna Gonzalez).

Debitage was further classified into flake and shatter types and includes complete flakes (CP), proximal flakes (PF), flake shatters (FS), and angular shatter (AS) (see Figures 7.2 and 7.3) (Silliman 2000; Gonzalez 2011; Lightfoot and Gonzalez 2018). A complete flake exhibits a striking platform, a bulb of percussion, and either a feather or hinge termination. A proximal flake features a striking platform, a bulb of percussion, and a step fracture on the distal end. Flake shatter and angular shatter have neither a striking platform nor termination point. Unlike angular shatter, which has no diagnostic features and consists primarily of angular chunks, flake shatter exhibits an identifiable ventral and dorsal surface.

Our analysis of chipped stone artifacts included the identification of raw material, which includes chert (CH), obsidian (OB), basalt (BA), quartz/quartzite (QZ), quartz crystal (QC), chalcedony (CA), and sandstone (SA). We also made note of the amount of cortex surrounding flakes and categorized it as follows: primary decortication (PC), in which artifacts exhibit cortex on most of the dorsal or outer surface; secondary decortication (SC), including flakes partially covered with cortex; and tertiary decortication (TC), in which the flake exhibits no cortex.

Classifying Ground Stone

The classification of lithic basic groups includes a separate category for ground stone tools. In our analysis of material from the four Santa Cruz sites, we identified an assortment of ground stone artifacts. After soil floatation and dry-screen samples were processed, each ground stone artifact was weighed in grams and measured in millimeters for its maximum length, width, and thickness.

Lithic ground stone artifacts (LG) have smoothed sides or surfaces with intentional shaping by grinding, pecking, or polishing. Ground stone artifacts found within our Santa Cruz study include the following: basin milling stones or metates (BM), which are diagnostic through the evidence of grinding along the flat surface of stone; milling hand stones, or manos (MH), which are hand-sized spherical tools with one or more grinding surfaces; pestles (PE), which contain evidence of battering on the proximal and distal surfaces of a cylindrical shaped tool; mortars (MO), which are bowl-shaped and exhibit evidence of pecking and grinding; and hammerstones (HA), which are round-shaped stones with evidence of battering or hitting against other stone material (see Figure 7.4). Other ground stones that do not fit into these classifications are put into the lithic other (LO) category. This classification could contain fire-cracked rocks (FC), which are typically angular-shaped unmodified cobbles with evidence of burning or heating, as well as fire-cracked ground stone fragments (FG), which are fire-cracked stones that reveal evidence of smoothing or grinding on at least one side (see Figure 7.5)

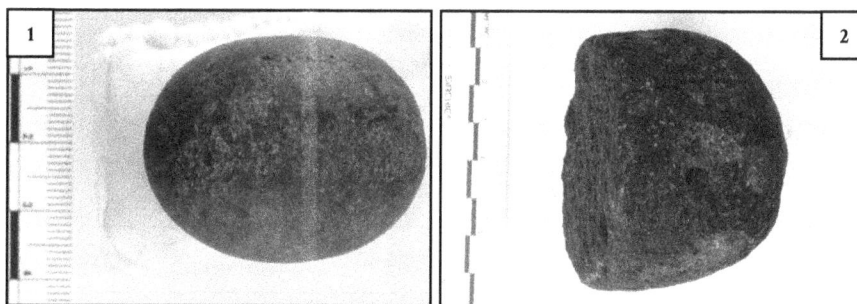

Figure 7.4. Image of Hammerstone (1) and Pestle Fragment (2) (Photos by Kathryn Field and Ariadna Gonzalez).

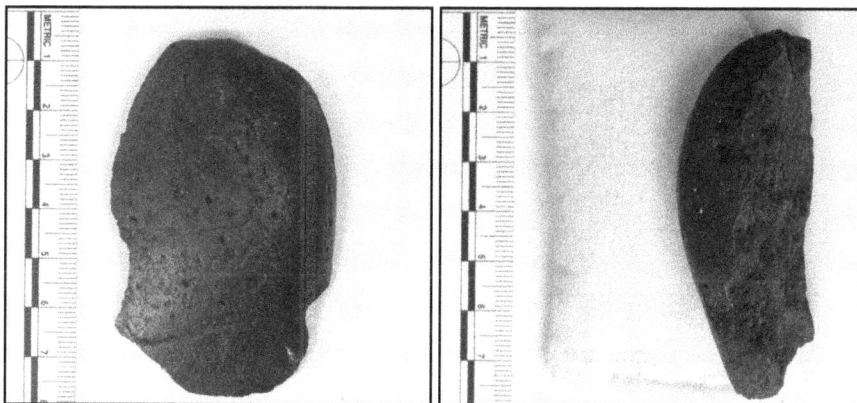

Figure 7.5. Images of Fire-Cracked Rock (Photos by Kathryn Field and Ariadna Gonzalez)

Results

CA-SCR-7

Excavation at CA-SCR-7 consisted of seven column units and three augers. Auger units were placed in contexts involving deep depositions or during the initial testing of GPR anomalies, as outlined in Chapter 4, while column samples were employed to record and expose profiles. A total of 357.5 liters of soil were recovered from CA-SCR-7. Samples from column units were floated, while samples collected from auger units were either dry screened or floated.

Table 7.1 CA-SCR-7 Raw Material.

CA-SCR-7	Auger #1	Auger #2	Auger #3	Column #1
Raw Material	Heavy Fraction	Heavy Fraction	Heavy Fraction	Heavy Fraction
Andesite (AN)	0	0	0	1
Basalt (BA)	0	0	0	1
Chert other (CO)	0	0	0	0
Franciscan chert (FC)	0	0	0	2
Granite (GN)	0	1	0	0
Jasper (JA)	0	0	0	0
Magnesite (MA)	0	0	0	0
Monterey chert (MC)	6	46	119	1068
Obsidian (OB)	0	0	0	13
Quartz crystal (QC)	0	0	0	1
Quartzite (QT)	0	0	0	0
Rhyolite (RH)	0	0	0	0
Sandstone (SA)	0	0	0	0
Schist (SC)	0	0	0	0
Steatite (ST)	0	0	0	0
Unidentified (UNID)	0	0	0	0
Total	6	47	119	1086
Liters	6.5	7	10	89.5
Density (lithics/liter)	0.92	6.71	11.9	12.13

Column #2	Column #3	Column #4	Column #5	Column #6	Column #7	Total
Heavy Fraction	Heavy Fraction	Heavy Fraction	Heavy Fraction	Heavy Fraction	Heavy Fraction	
2	0	0	0	0	0	**3**
1	0	0	0	1	0	**3**
0	0	0	0	0	0	**0**
0	0	0	0	1	0	**3**
0	0	0	0	0	0	**1**
0	0	0	0	0	0	**0**
0	0	0	0	0	0	**0**
1086	315	123	37	132	79	**3011**
8	1	0	0	3	0	**25**
1	0	0	0	0	0	**2**
1	0	0	0	0	0	**1**
0	0	0	0	0	0	**0**
1	0	0	0	0	0	**1**
0	0	0	0	0	0	**0**
0	0	0	0	0	0	**0**
2	0	0	0	0	0	**2**
1102	316	123	37	137	79	**3052**
103.5	65.5	21.5	7.5	34	12.5	**357.5**
10.64	4.82	5.72	4.93	4.02	6.32	**8.5**

Table 7.2 CA-SCR-7 Lithic Class.

CA-SCR-7	Auger #1	Auger #2	Auger #3	Column #1
Lithic Class	Heavy Fraction	Heavy Fraction	Heavy Fraction	Heavy Fraction
Core (CO)	0	1	1	5
Biface (BI)	1	0	0	0
Projectile Point (PP)	0	0	0	0
Uniface (UN)	0	0	0	0
Other formal tool (OF)	0	0	0	2
Complete flake (CP)	2	3	12	60
Proximal flake (PX)	1	2	4	68
Bipolar flake (BP)	0	0	0	0
Flake shatter (FS)	2	25	75	714
Angular shatter (AS)	0	16	27	237
Basin milling stone (BM)	0	0	0	0
Slab milling stone (SM)	0	0	0	0
Milling handstone (MH)	0	0	0	0
Mortar (MO)	0	0	0	0
Pestle (PE)	0	0	0	0
Anvil stone (AV)	0	0	0	0
Battered cobble (BC)	0	0	0	0
Hammerstone (HA)	0	0	0	0
Net weight (NW)	0	0	0	0
Fire-cracked rock (FC)	0	0	0	0
Fire-cracked groundstone (FG)	0	0	0	0
Other	0	0	0	0
Total	6	47	119	1086
Liters	6.5	7	10	89.5
Density (lithics/liter)	0.92	6.71	11.9	12.13

Column #2	Column #3	Column #4	Column #5	Column #6	Column #7	Total
Heavy Fraction	Heavy Fraction	Heavy Fraction	Heavy Fraction	Heavy Fraction	Heavy Fraction	
3	1	0	0	1	0	**12**
0	0	0	0	0	0	**1**
0	0	0	0	0	0	**0**
0	0	0	0	0	0	**0**
3	0	0	0	0	1	**6**
61	20	9	0	17	5	**189**
66	30	8	2	6	2	**189**
0	0	0	0	0	0	**0**
725	202	53	20	74	45	**1935**
244	63	53	15	38	26	**719**
0	0	0	0	0	0	**0**
0	0	0	0	0	0	**0**
0	0	0	0	0	0	**0**
0	0	0	0	0	0	**0**
0	0	0	0	0	0	**0**
0	0	0	0	0	0	**0**
0	0	0	0	0	0	**0**
0	0	0	0	1	0	**1**
0	0	0	0	0	0	**0**
0	0	0	0	0	0	**0**
0	0	0	0	0	0	**0**
0	0	0	0	0	0	**0**
1102	316	123	37	137	79	**3052**
103.5	65.5	21.5	7.5	34	12.5	**357.5**
10.64	4.82	5.72	4.93	4.02	6.32	**8.5**

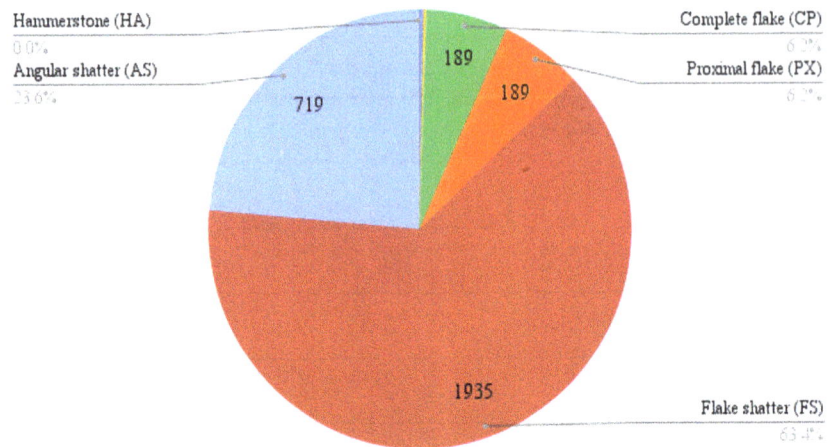

Figure 7.6. Pie Chart Showing Lithic Classification Percentages for CA-SCR-7.

The strata of CA-SCR-7 were occupied from 4800 to 2200 cal BCE based on 27 AMS dates (Chapter 6). The excavation of 3 augers and 7 column units produced a total of 3,052 artifacts from a total of 357.5 liters of soil and a total artifact density of 8.5 lithics/liter (Tables 7.1, 7.2). The lithic assemblage from CA-SCR-7 consists primarily of Monterey chert, which would have been locally sourced. Other materials, such as obsidian, were present in the assemblage in lower quantities and consisted primarily of small debitage flakes. Flake shatter and angular shatter dominate most of the CA-SCR-7 assemblage, with flake shatter making up 63% of our assemblage and angular shatter constituting 23.6%. Complete and proximal flakes each comprise 6.2% of our assemblage (see Figure 7.6). Notable to mention was the collection of 12 cores, 1 biface, and 1 hammerstone.

Our analysis and identification of 12 cores, and the large quantities of debitage at CA-SCR-7, suggest that lithic production was taking place at this site. All stages of debitage involved in the production of expedient flake tools, bifaces, and other formal tools were identified. The large amount of tertiary debitage supports the idea that lithic reduction involved the manufacture and retouching of bifaces and other formal chipped stone tools. Though we did not recover many bifaces, it is possible that many of these artifacts have been removed from the surface over the last century or more. The production of formal chipped stone tools at this site is plausible when considering the large number of projectile points and bifaces that artifact collectors have amassed from walking the plowed fields near the site (Hylkema 2021). As for ground stone artifacts such as mortars, pestles, and milling stones that may have been employed in the processing of plant foods, we recovered little evidence. Likewise, there was little evidence for fire-cracked rocks resulting from possible cooking activities. The low quantities of grinding implements at CA-SCR-7 may relate to the abundance of ocean plants available for harvest, such as kelp and seaweed, that would require minimal processing using ground stone tools (see Chapter 9).

CA-SCR-10

Table 7.3. CA-SCR-10 Raw Material Type.

CA-SCR-10	Auger #1	Auger #2	Auger #3	Exc. Unit #1	Exc. Unit #1	Total
Raw Material	Heavy Fraction	Heavy Fraction	Heavy Fraction	Heavy Fraction	Dry Screen	
Andesite (AN)	0	0	0	0	0	**0**
Basalt (BA)	0	0	0	0	11	**11**
Chert other (CO)	0	0	0	0	0	**0**
Franciscan chert (FC)	0	0	0	0	7	**7**
Granite (GN)	0	0	0	0	6	**6**
Jasper (JA)	0	0	0	0	0	**0**
Magnesite (MA)	0	0	0	0	0	**0**
Monterey chert (MC)	5	9	13	557	4123	**4707**
Obsidian (OB)	0	0	0	2	25	**27**
Quartz crystal (QC)	0	0	0	0	1	**1**
Quartzite (QT)	0	0	0	1	0	**1**
Rhyolite (RH)	0	0	0	0	0	**0**
Sandstone (SA)	0	0	0	0	0	**0**
Schist (SC)	0	0	0	0	0	**0**
Steatite (ST)	0	0	0	0	0	**0**
Unidentified (UNID)	0	0	0	0	4	**4**
Total	5	9	13	560	4177	**4764**
Liters Excavated	3	4.5	3.5	85.8	1033	**1129.8**
Density (lithic/liter)	1.66	2	3.71	6.52	4.04	**4.21**

Table 7.4. CA-SCR-10 Lithic Class.

CA-SCR-10	Auger #1	Auger #2	Auger #3	Exc. Unit #1	Exc. Unit #1	Total
Lithic Class	Heavy Fraction	Heavy Fraction	Heavy Fraction	Heavy Fraction	Dry Screen	
Core (CO)	0	0	0	1	20	**21**
Biface (BI)	0	0	0	0	5	**5**
Projectile point (PP)	0	0	0	0	0	**0**
Uniface (UN)	0	0	0	0	1	**1**
Other formal tool (OF)	0	0	0	0	0	**0**
Complete flake (CP)	1	0	2	45	77	**125**
Proximal flake (PX)	0	0	1	37	170	**208**
Bipolar flake (BP)	0	0	0	0	0	**0**
Flake shatter (FS)	4	1	7	285	2999	**3296**
Angular shatter (AS)	0	8	3	192	895	**1098**
Basin milling stone (BM)	0	0	0	0	0	**0**
Slab milling stone (SM)	0	0	0	0	0	**0**
Milling handstone (MH)	0	0	0	0	0	**0**
Mortar (MO)	0	0	0	0	0	**0**
Pestle (PE)	0	0	0	0	2	**2**
Anvil stone (AV)	0	0	0	0	0	**0**
Battered cobble (BC)	0	0	0	0	0	**0**
Hammerstone (HA)	0	0	0	0	6	**6**
Net weight (NW)	0	0	0	0	0	**0**
Fire-cracked rock (FC)	0	0	0	0	2	**2**
Fire-cracked groundstone (FG)	0	0	0	0	0	**0**
Other	0	0	0	0	0	**0**
Total	5	9	13	560	4177	**4764**
Liters Excavated	3	4.5	3.5	85.8	1033	**1129.8**
Density (lithic/liter)	1.66	2	3.71	6.52	4.04	**4.21**

At CA-SCR-10, 1,129.8 liters were excavated from three auger units and one 1x1 m excavation unit (EU 1). This formed a total of 4,764 artifacts, the densest of all four sites for both liters excavated and artifacts recovered (Tables 7.3, 7.4). The artifact density of CA-SCR-10 is 4.21 lithics per liter. The auger units from the central area of the site dated from 3800 to 3200 cal BCE based on three AMS dates, while intact deposits in the excavation unit date from 680 to 880 cal CE based on eight AMS dates (Chapter 6). EU 1 recovered the most artifacts, making up 99.4% of the artifact assemblage of this site.

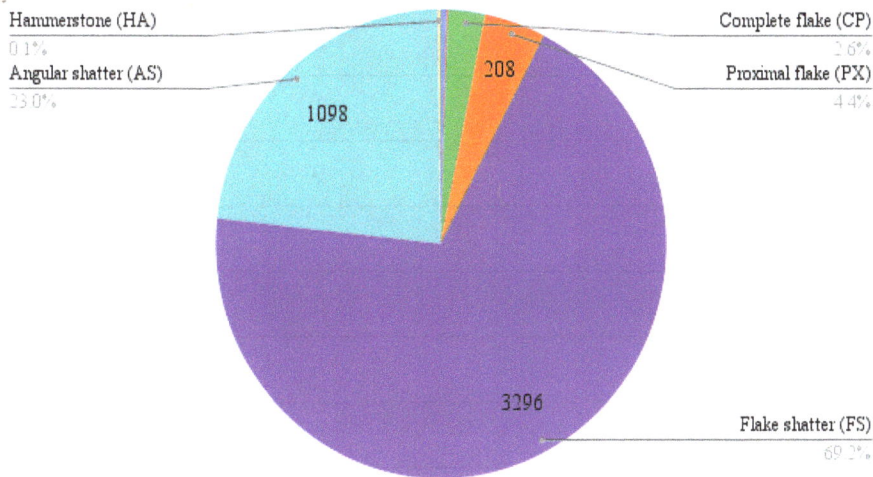

Figure 7.7. Pie Chart Showing Lithic Classification Percentages for CA-SCR-10.

Similar to CA-SCR-7, Monterey chert was the most commonly found lithic material type, totaling to 4,707 pieces overall, with obsidian trailing at 27 identified pieces. Flake shatter made up 69.3% of the assemblage with 3,296 pieces and angular shatter made up 23.1% with 1,098 pieces, making these the two most commonly found lithic classes (see Figure 7.7). A total of 21 cores (0.4%) were identified, as well as 208 proximal flakes (4.4%) and 125 complete flakes (2.6%). Two pestles, six hammerstones, and 2 fire-cracked were also identified.

Although Monterey chert dominated the assemblage, the presence of obsidian is significant. Obsidian is a non-local material, revealing evidence of trade with more distant communities. Additionally, the presence of pestles signifies the potential for food and plant processing occurring at this site. The mass amount of angular and flake shatter reveals that this space was also used as a tool manufacturing and tool retouch activity area.

It is important to note that this site has been heavily impacted over the years due to agricultural development. As discussed in Chapters 4 and 5, the site has evidence of plowing and bioturbation, which has impacted the spatial integrity of some archaeological contexts. One non-cultural manuport with evidence of plow scars was found during excavation, providing concrete evidence that the integrity of some of the archaeological deposits was disrupted through agricultural practices.

CA-SCR-14

Table 7.5 CA-SCR-14 Raw Material Type.

CA-SCR-14	Exc. Unit #1	Exc. Unit #2	Exc. Unit #1	Exc. Unit #2	Total
Raw Material	Heavy Fraction	Heavy Fraction	Dry Screen	Dry Screen	
Andesite (AN)	0	0	0	0	**0**
Basalt (BA)	0	0	0	1	**1**
Chert other (CO)	0	0	0	0	**0**
Franciscan chert (FC)	1	0	0	0	**1**
Granite (GN)	1	0	0	0	**1**
Jasper (JA)	0	0	0	0	**0**
Magnesite (MA)	0	0	0	0	**0**
Monterey chert (MC)	45	49	31	35	**160**
Obsidian (OB)	0	0	1	0	**1**
Quartz crystal (QC)	0	0	0	0	**0**
Quartzite (QT)	0	0	0	0	**0**
Rhyolite (RH)	0	0	0	0	**0**
Sandstone (SA)	0	0	1	0	**1**
Schist (SC)	0	0	0	0	**0**
Steatite (ST)	0	0	0	0	**0**
Unidentified (UNID)	0	0	0	2	**2**
Total	47	49	33	38	**167**
Liters	22	26.5	130	169	**347.5**
Density (lithic/liters)	2.13	1.84	0.25	0.22	**0.48**

Table 7.6 CA-SCR-14 Lithic Class.

CA-SCR-14	Exc. Unit #1	Exc. Unit #2	Exc. Unit #1	Exc. Unit #2	Total
Lithic Class	Heavy Fraction	Heavy Fraction	Dry Screen	Dry Screen	
Core (CO)	0	0	0	0	**0**
Biface (BI)	0	0	0	0	**0**
Projectile point (PP)	0	0	0	0	**0**
Uniface (UN)	0	0	0	0	**0**
Other formal tool (OF)	0	0	0	0	**0**
Complete flake (CP)	1	2	2	0	**5**
Proximal flake (PX)	1	2	1	3	**7**
Bipolar flake (BP)	0	0	0	0	**0**
Flake shatter (FS)	26	27	21	27	**101**
Angular shatter (AS)	18	18	8	8	**52**
Basin milling stone (BM)	0	0	0	0	**0**
Slab milling stone (SM)	0	0	0	0	**0**
Milling handstone (MH)	0	0	0	0	**0**
Mortar (MO)	0	0	0	0	**0**
Pestle (PE)	0	0	1	0	**1**
Anvil stone (AV)	0	0	0	0	**0**
Battered cobble (BC)	0	0	0	0	**0**
Hammerstone (HA)	1	0	0	0	**1**
Net weight (NW)	0	0	0	0	**0**
Fire-cracked rock (FC)	0	0	0	0	**0**
Fire-cracked groundstone (FG)	0	0	0	0	**0**
Other	0	0	0	0	**0**
Total	47	49	33	38	**167**
Liters	22	26.5	130	169	**347.5**
Density (lithic/liters)	2.13	1.84	0.25	0.22	**0.48**

Based on eight AMS dates, CA-SCR-14 dates from about 1000 to 1700 cal CE—and more conservatively, with the removal of two dates from questionable contexts, it dates from 1000 to 1510 cal CE (see Chapter 6). The excavation of two 50x50 cm units at this upland resting site produced a total of 167 artifacts from 347.5 liters of soil. A total of 96 of the artifacts were recovered from soil flotation, while 71 were recovered in the field through 3.2 mm dry screen. EU 1 and EU 2 yielded an almost equal amount of artifacts, with EU 1 totaling 80 artifacts recovered and EU 2 producing 87, forming a density of 0.48 lithics per liter of soil excavated (Tables 7.5 and 7.6). As with other sites, Monterey chert dominated the manufacturing material with 160 artifacts. Flake shatter and angular shatter were also the most abundant, making up about 61% and 31% of the assemblage, respectively (see Figure 7.8). One pestle fragment and one hammerstone were also recovered from the site.

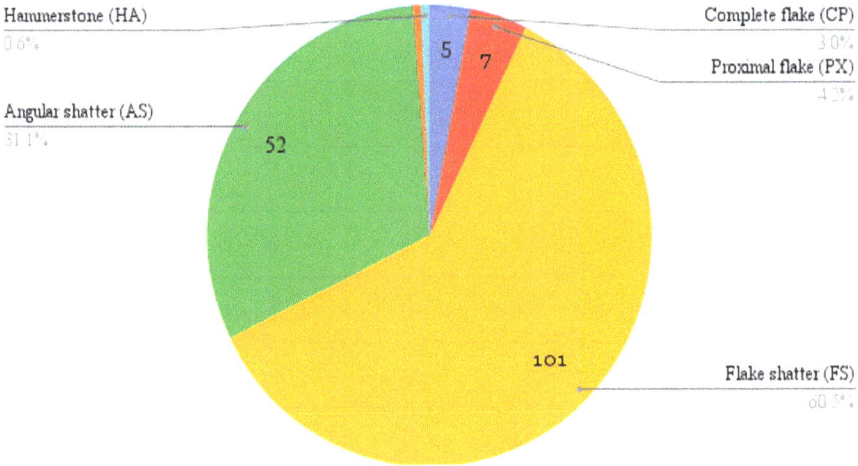

Figure 7.8. Pie Chart Showing Lithic Classification Percentages for CA-SCR-14.

Similar to CA-SCR-10, flake and angular shatter dominated the lithic assemblage, signifying the potential use of this space as a lithic tool production site. Additionally, the presence of ground stone, including one pestle fragment and one hammerstone, reveal the likelihood of plant processing.

CA-SCR-15

Table 7.7 CA-SCR-15 Raw Material Type.

CA-SCR-15	Auger #1	Exc. Unit #1	Exc. Unit #3	Exc. Unit #1	Exc. Unit #3	Total
Raw Material	Heavy Fraction	Heavy Fraction	Heavy Fraction	Dry Screen	Dry Screen	
Andesite (AN)	0	0	0	0	0	**0**
Basalt (BA)	0	0	0	0	0	**0**
Chert other (CO)	0	0	0	0	0	**0**
Franciscan chert (FC)	0	0	0	0	0	**0**
Granite (GN)	0	0	0	0	0	**0**
Jasper (JA)	0	0	0	0	0	**0**
Magnesite (MA)	0	0	0	0	0	**0**
Monterey chert (MC)	1	258	91	187	34	**571**
Obsidian (OB)	0	1	1	1	0	**3**
Quartz crystal (QC)	0	0	0	2	0	**2**
Quartzite (QT)	0	0	3	0	0	**3**
Rhyolite (RH)	0	0	0	0	0	**0**
Sandstone (SA)	0	0	0	0	0	**0**
Schist (SC)	0	0	0	0	0	**0**
Steatite (ST)	0	0	0	0	0	**0**
Unidentified (UNID)	0	0	0	0	0	**0**
Total	1	259	95	190	34	**579**
Liters Excavated	6	34	32.5	147	128	**347.5**
Density (lithic/liters)	0.16	7.61	2.92	1.29	0.26	**1.66**

Table 7.8 CA-SCR-15 Lithic Class.

CA-SCR-15	Auger #1	Exc. Unit #1	Exc. Unit #3	Exc. Unit #1	Exc. Unit #3	Total
Lithic Class	Heavy Fraction	Heavy Fraction	Heavy Fraction	Dry Screen	Dry Screen	
Core (CO)	0	0	0	0	2	**2**
Biface (BF)	0	0	0	0	0	**0**
Projectile point (PP)	0	0	0	0	0	**0**
Uniface (UN)	0	0	0	0	0	**0**
Other formal tool (OF)	0	0	0	0	0	**0**
Complete flake (CP)	0	15	4	7	0	**26**
Proximal flake (PX)	1	14	5	29	6	**55**
Bipolar flake (BP)	0	0	0	0	0	**0**
Flake shatter (FS)	0	138	44	102	18	**302**
Angular shatter (AS)	0	92	42	52	8	**194**
Basin milling stone (BM)	0	0	0	0	0	**0**
Slab milling stone (SM)	0	0	0	0	0	**0**
Milling handstone (MH)	0	0	0	0	0	**0**
Mortar (MO)	0	0	0	0	0	**0**
Pestle (PE)	0	0	0	0	0	**0**
Anvil stone (AV)	0	0	0	0	0	**0**
Battered cobble (BC)	0	0	0	0	0	**0**
Hammerstone (HA)	0	0	0	0	0	**0**
Net weight (NW)	0	0	0	0	0	**0**
Fire-cracked rock (FC)	0	0	0	0	0	**0**
Fire-cracked groundstone (FG)	0	0	0	0	0	**0**
Other	0	0	0	0	0	**0**
Total	1	259	95	190	34	**579**
Liters Excavated	6	34	32.5	147	128	**347.5**
Density (lithic/liters)	0.16	7.61	2.92	1.29	0.26	**1.66**

Excavation of CA-SCR-15 includes the placement of two 50x50 cm units and one auger unit, which produced a total of 347.5 liters of soil. A total of 72.5 liters of soil were collected for flotation, while 275 liters of soil were dry screened through 3.2 mm mesh (Tables 7.7, 7.8). Based on seven AMS dates from intact contexts, the strata we investigated at CA-SCR-15 dates to the late Holocene, from 1050 to 1400 cal CE (see Chapter 6). A total of 579 lithic artifacts were recovered from the 347.5 liters of soil and produced an artifact density of 1.66 lithics per liter. The two excavation units were primarily comprised of flakes and debitage, with most of the assemblage coming from the dry screen and flotation from EU 1. We identified two cores, with flake shatter making up 52% of the assemblage and angular shatter making up 33%. Complete and proximal flakes contributed 4% and 9%, respectively (see Figure 7.9). No formal tools were recovered at this site. CA-SCR-15 is very similar to both CA-SCR-10 and CA-SCR-14 in that flake and angular shatter were again the dominant lithic classes with the addition of two cores being found, all of which reveal stone tool production was occurring at this site.

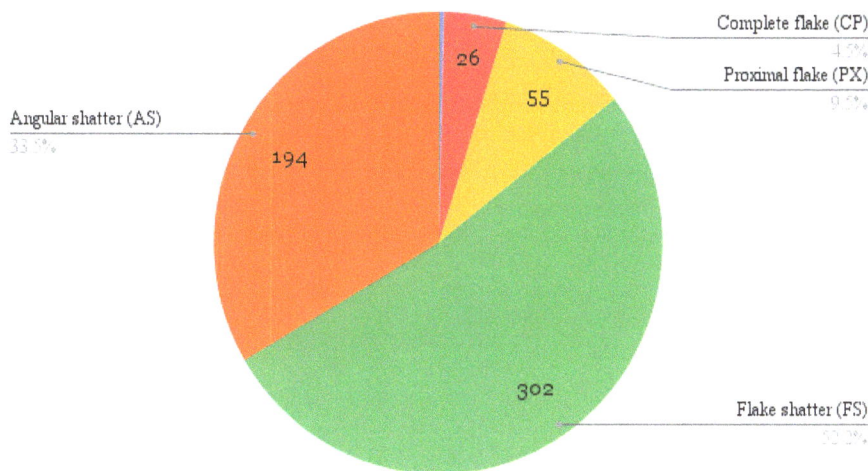

Figure 7.9. Pie Chart Showing Lithic Classification Percentages for CA-SCR-15.

Conclusion

The lithic artifact recovery from the four sites yielded a large amount of material that broadened our understanding of stone tool production on the Santa Cruz coastline. Monterey chert is the dominant lithic material found in all four sites. This would have been locally sourced and readily available for lithic production. Debitage consisting of flake shatter, angular shatter, complete and proximal flakes, as well as cores were recovered at all four sites. Obsidian was the second most common material type; however, unlike the Monterey chert cores and large flakes, only small obsidian flakes, possibly from pressure flaking and retouching of formal tools, were recovered. Little evidence of ground stone tools was identified at CA-SCR-7 in contrast to CA-SCR-10, CA-SCR-14, and CA-SCR-15, where ground stone and fire-cracked rock were recovered. These findings would support the idea that the processing of plant and food material was occurring at these three sites.

The lithic assemblage from the four Santa Cruz sites is largely made up of remnants of the early stages of tool production. The presence of readily available Monterey chert cores, along with the extensive amount of debitage recovered, suggest that lithic production of expedient flakes was taking place as was the retouching and manufacturing of formal chipped stone tools. The low density of formal tools found within our study area could be a result of historical developments at CA-SCR-7 and CA-SCR-10, where farming and aquaculture have caused site disturbance. It may also reflect the relatively low volume of sediments excavated at the four sites. Though not many formal tools are present in our assemblage, it is possible that many of these artifacts have been removed from the surface. Artifact collectors have amassed a large collection of projectile points and bifaces from walking the plowed fields nearby the sites (Hylkema 2021).

Acknowledgements

We would like to thank the Amah Mutsun Land Trust and Stewardship Core, Mark Hylkema and California State Parks, Dr. Kent Lightfoot for all of his guidance throughout the project, as well as Dr. Grone, Dr. Sanchez, and Alec Apodaca for their continual assistance. We would like to thank students from the UC Berkeley Undergraduate Research Apprentice Program (URAP) for their help in the initial sorting and lithic analysis, as well as the Hearst Museum of Anthropology for their assistance in photography.

References

Andrefsky, W.
   1998   *Lithics: Macroscopic Approaches to Analysis.* Cambridge University Press, Cambridge.

Gonzalez, Sara L.
   2011   Creating Trails from Tradition: The Kashaya Pomo Interpretive Trail at Fort Ross State Historic Park. PhD Dissertation, Department of Anthropology, University of California, Berkeley.

Hylkema, Mark G.
   1991   Prehistoric Native American Adaptations along the Central California Coast of San Mateo and Santa Cruz counties. MA Thesis, Department of Anthropology, San Jose State University.
   2021   Middle Holocene Projectile Points from the Santa Cruz Coast of Northern Monterey Bay. Paper Presented at the 2021 Annual Meeting for the Society for California Archaeology, A Virtual Event. Paper Archived by the Society for California Archaeology.

Lightfoot, Kent G., and Sara L. Gonzalez
   2018   *Metini Village: An Archaeological Study of Sustained Colonialism.* Volume 3, The Archaeology and Ethnohistory of Fort Ross, California. Berkeley, California: Archaeological Research Facility.

Lightfoot, Kent G., Rob Q. Cuthrell, Mark G. Hylkema, Valentin Lopez, Diane Gifford-Gonzalez, Roberta A. Jewett, Michael A. Grone, Gabriel M. Sanchez, Peter A. Nelson, Alec J. Apodaca, Ariadna Gonzalez, Kathryn Field, Jordan Brown, Alexii Sigona, and Paul V. A. Fine
   2021   The Eco-Archaeological Investigation of Indigenous Stewardship Practices on the Santa Cruz Coast. *Journal of California and Great Basin Anthropology* 41(2): 187–205.

Silliman, Stephen W.
   2000   Colonial Worlds, Indigenous Practices: The Archaeology of Labor on a 19th-Century California Rancho. PhD Dissertation, Anthropology Department, Department of Anthropology, University of California, Berkeley.

LITHIC ARTIFACTS RECOVERED FROM AUGER, COLUMN, AND EXCAVATION UNITS

KEY TO CODE

Material Category:	LF = Lithic Flake Stone LG = Lithic Ground Stone LO = Lithic Othe

Lithic Class:	CO = Core	MO = Mortar
	BF = Biface	PE = Pestle
	BI = Projectile Point	AV = Anvil stone
	UN = Uniface	BC = Battered Cobble
	OF = Other Formal tool	HA = Hammer Stone
	CP = Complete Flake	NW = Net weight
	PX = Proximal Flake	FC = Fire-cracked Rock
	AS = Angular Shatter	FG = Fire-cracked Ground Stone
	BM = Basin Milling Stone	OT = Other
	MH = Milling Hand Stone	

Raw Material:	AN = Andesite	OB = Obsidian
	BA = Basalt	QC = Quartz Crystal
	CO = Chert Other	QT = Quartzite
	FC = Franciscan Chert	RH = Rhyolite
	GN = Granite	SA = Sandstone
	JA = Jasper	SC = Schist
	MA = Magnesite	UNID = Unidentified
	MC = Monterey Chert	

Unit:	AUG = Auger COL = Column EXC = Excavation

CHAPTER 8

Archaeobotanical Research at Nine Archaeological Sites West of the Santa Cruz Mountains: Implications for Subsistence Practices and Anthropogenic Burning

ROB Q. CUTHRELL

Introduction

Through analysis of archaeological plant remains and silica phytoliths from soil, this chapter explores how Indigenous people living west of the Santa Cruz Mountains used and managed plant resources, with a particular focus on prescribed burning. Employing macrobotanical, anthracological (wood charcoal), and/or silica phytolith data from nine archaeological sites, this chapter investigates differences in plant food consumption practices by site location (coastal vs. inland) and through time. This study also considers implications of these data for landscape stewardship practices, specifically focusing on evidence pertaining to the development and use of cultural prescribed burning systems. Indigenous people across California used prescribed burning to maintain expansive coastal prairies, reduce understory fuels in forests, and revitalize stands of ethnobotanical plants among other purposes (Anderson 2005; Lightfoot and Parrish 2009).

Coastal prairies on foothills and marine terraces west of the Santa Cruz Mountains host a diverse suite of reliable herbaceous food, medicine, and crafting resources, with highest productivity from late winter through late summer. Prior eco-archaeological and historical ecological research on the Central Coast indicates Indigenous people used prescribed burning to maintain extensive coastal prairies over at least the last ca. 1,000–1,500 years at Quiroste Valley north of Año Nuevo Point (Cuthrell 2013a, 2013b; Lightfoot et al. 2013). By the onset of Spanish colonization in the 1770s, historical documentary evidence shows that Indigenous prescribed burning was used to maintain expansive swaths of prairie in near-coastal areas that would have been dominated by shrublands or forests in the absence of regular burning (Cuthrell 2013b).

In coastal portions of San Mateo and Santa Cruz Counties, as elsewhere on the coast of Central and Northern California, coastal prairies and other grasslands or forblands are generally disturbance-dependent vegetation types. In the absence of regular disturbance factors such as grazing, mowing, plowing, or fire, woody vegetation will typically recruit in prairies and convert the prairies to shrublands or forests within several decades (Cuthrell 2013b; Ford and Hayes 2007; McBride 1974; Williams et al. 1987). Exceptions to this expectation include areas where woody vegetation establishment is limited by soil characteristics or hydrological factors. The rate of type conversion from coastal prairie to woody vegetation depends on many factors, such as proximity to existing woody vegetation stands, plant community species composition, and local environmental conditions. At Quiroste Valley, historical aerial photography shows that in the absence of vegetation disturbance factors, prairie cover in a ca. 50-hectare area was reduced by 67% in 30 years (1982–2012; Cuthrell 2013a).

A study of vegetation change following the 1995 Vision Fire at Point Reyes provided high resolution data on post-fire vegetation succession in near-coastal environments in the broader region (Forrestel et al. 2011). The Vision Fire burned areas covered primarily by northern coastal scrub, Douglas fir forest (*Pseudotsuga menziesii*), and Bishop pine (*Pinus muricata*) forest. Ten years after the fire, ephemeral post-fire prairie vegetation had been succeeded by fire-compatible woody vegetation types, with substantial expansion of vegetation types dominated by fire-refractory species, such as California lilac (or blueblossom, *Ceanothus thyrsiflorus*) and Bishop pine. During the same period, there was about a 50% reduction in the cover of Douglas fir forest, as Douglas fir is relatively susceptible to fire mortality.

Northern coastal scrub, usually the dominant shrubland type in coastal areas of Central California, is dominated by species that are highly resilient to top-killing. Between 2009 and 2020, established stands of northern coastal scrub vegetation in Quiroste Valley recovered quickly from multiple episodes of top-killing. In areas where northern coastal scrub was top-killed, resprouting shrubs displaced the ephemeral post-disturbance prairie in two to five years, even after three to four repetitions of top-killing over 10–15 years. These observations suggest that relatively short fire return intervals of less than 5–10 years would probably be required to type convert and maintain this vegetation type as coastal prairie over the long term (Cuthrell 2013b). In contrast, natural fire regimes in California's Central Coast and North Coast regions would probably have been characterized by fire return intervals on the order of many decades to over a century (Greenlee and Langenheim 1990; Lightfoot et al. 2013).

This study synthesizes and interprets three data sets with respect to Indigenous subsistence practices, long-term landscape vegetation history, and, by proxy, fire regimes. These include macrobotanical data (charred seeds and other vegetative material) and anthracological data (wood charcoal) collected through flotation of materials from archaeological sites, as well as silica phytolith data collected from near-surface soils in proximity to sites. Data were collected from eight archaeological sites in Santa Cruz and San Mateo Counties, including five "coastal" sites located in close proximity to the Pacific shoreline and three "inland" sites located along riparian corridors farther from the coast (Figure 8.1). The data set includes one site from the Early Period (ca. 3500–1000 BCE), three from the Middle Period (ca. 1000 BCE–1050 CE), and four from the Late Period (ca. 1050–1770 CE). Data from the eight sites analyzed in this study are also compared to results previously reported from Late Period inland site CA-SMA-113 near Año Nuevo Point (Cuthrell 2013a, 2013b).

Macrobotanical materials from flotation samples were analyzed to record the variety of taxa and quantities of charred plant remains at each site, providing information about Indigenous plant food use and local vegetation. Our research team's expectation was that coastal prairie would have comprised a small proportion of landscape vegetation in the absence of Indigenous prescribed burning. Under this scenario, it is unlikely that prairies could have produced seed food resources in quantities large enough to support sustained use by Indigenous people over the long term. Accordingly, we would expect people to have relied more heavily on nut and fruit foods from forests and shrublands. Conversely, if Tribes regularly burned the landscape at short intervals of less than five to ten years over the long term, we would expect the landscape to host extensive coastal prairies that could provide abundant greens, geophyte, and seed foods from late winter to early fall (Cuthrell 2013b). Archaeologically, greens and geophytes are rarely preserved because they lack rigid anatomical structures amenable to preservation by charring; however, prairie seeds are commonly preserved archaeologically and recovered in high densities from Emergent Period sites in interior Central California, where Indigenous seed food use is well attested (see Wohlgemuth 2004).

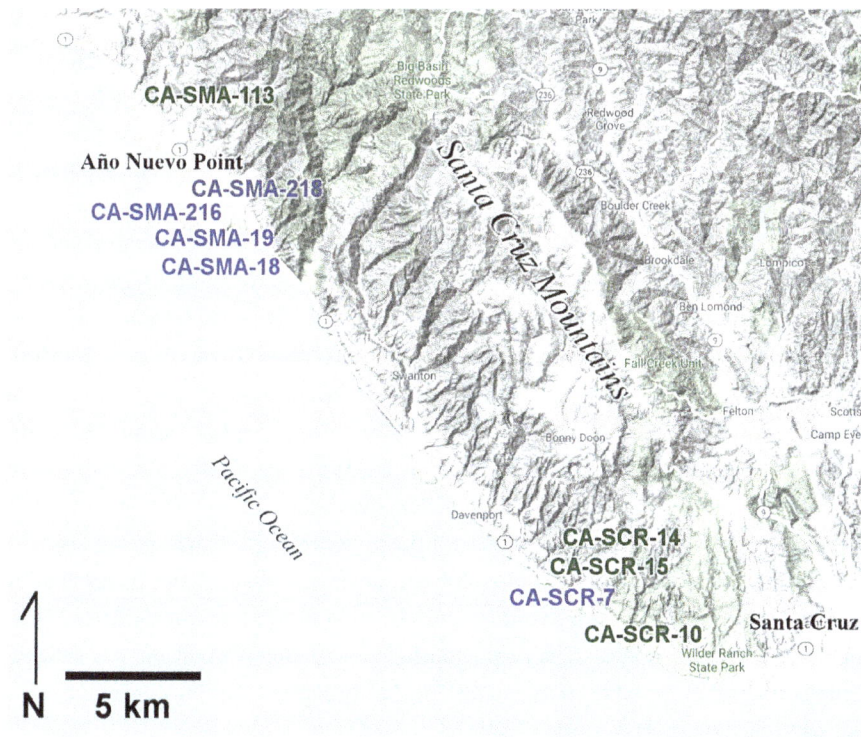

Figure 8.1. Map of Archaeological Sites Analyzed and Referenced in this Study. Site names are color-coded, with blue labels indicating near-coastal sites and green labels indicating inland sites. Google Maps imagery.

Anthracological specimens were sampled from the light fractions of flotation samples and analyzed to investigate the relative fire compatibility of wood fuels used by Indigenous people. Because anthracological analysis requires substantial labor, data was only collected from four of the eight sites analyzed in this study. In the absence of regular prescribed burning, we expect that northern coastal scrub shrub taxa—e.g., coyotebrush (*Baccharis pilularis*), poison oak (*Toxicodendron diversilobum*), and California coffeeberry (*Frangula californica*)—and Douglas fir would probably have comprised a large majority of woody landscape vegetation on the foothills and terraces west of the Santa Cruz Mountains. If Indigenous people did not have strong preferences for wood fuel selection, we would expect these taxa to represent a substantial proportion of anthracological specimens observed at archaeological sites in the absence of regular prescribed burning.

Although mature Douglas fir trees become resistant to fire damage through development of thick bark, younger Douglas fir stands can display over 90% mortality even from relatively low-intensity fires (Engber and Varner 2012; Ryan et al. 1988). Prescribed burning conducted at intervals less than ca. 20–30 years would be expected to severely curtail the abundance of Douglas fir on the landscape. Under an anthropogenic fire regime characterized by sub-decadal fire return intervals, it is expected that wood fuels would have been harvested primarily from trees and shrubs compatible with frequent burning, such as redwood (*Sequoia sempervirens*) and coast live oak (*Quercus agrifolia*). All of the dominant shrubs in local northern coastal scrub vegetation are also resilient to fire, resprouting vigorously when top-killed. Observation of these taxa among anthracological assemblages would be consistent with regular burning at intervals of around five to ten years or more. More frequent burning, such as at two-to-five-year intervals (and possibly longer, depending on local conditions), would be required to type convert northern coastal scrub shrublands to prairie. As a result, low proportions of northern coastal scrub shrub taxa in anthracological assemblages could indicate a lack of shrublands relative to prairies. Alternatively, such a pattern could indicate a preference for tree over shrub fuels.

The third data set employed in this study is the amount and composition of silica phytoliths in near-surface soils. Silica phytoliths are microscopic structures produced in and around the cells of certain plant taxa (Piperno 2006). On the Central Coast of California, grasslands and sedge-dominated wetlands are vegetation types that produce phytoliths abundantly, including morphological forms diagnostic to the grass family (Poaceae). In contrast, shrubland and forest vegetation types are typically dominated by taxa that produce few or no silica phytoliths (Evett and Bartolome 2013; Evett et al. 2012). Phytolith researchers in California have proposed that observation of ca. >0.3% silica phytolith content in soil, along with high densities of diagnostic grass taxa morphological types (ca. >200k n/g), provides a strong indicator of long-term grassland vegetation (Evett and Cuthrell 2013; Evett and Bartolome 2013; Evett et al. 2012). Previous silica phytolith research in Quiroste Valley associated with Late Period site CA-SMA-113 displayed strong and consistent evidence for the presence of grasslands over at least the last ca. 1000–1500 years, with soil phytolith content levels ranging from ca. 0.9 to 1.7% and grass short cell phytolith densities typically >1M n/g (Evett and Cuthrell 2013). Observation of similar levels of silica phytoliths in association with inland sites in Santa Cruz County would provide proxy support for the use of prescribed burning to maintain prairie vegetation in these areas.

Methods

Archaeobotanical Recovery and Analysis Methods

Archaeobotanical data collection followed a comprehensive flotation sampling strategy in which a sediment sample generally one to five liters in size was collected from each excavated context (except some auger units). Flotation samples were collected from excavation units, auger units, column sample units in exposed profiles, and directly from exposed profiles according to natural stratigraphy. Flotation samples were processed using SMAP-type tanks constructed by the author (see Pearsall 2000) and analyzed according to typical disciplinary conventions, which included sieving samples into various size fractions and recording counts and weights of macrobotanical specimens. A detailed description of archaeobotanical sample collection, processing, and analysis methods is presented in Appendix 8.1

Anthracological Sampling and Analysis Methods

Anthracological analysis was conducted on materials from sites CA-SCR-7, CA-SCR-10, CA-SCR-14, and CA-SCR-15. Flotation samples from each site were judgmentally selected for anthracological analysis based on the abundance of wood charcoal in the sample and the integrity of the context from which the sample was collected (i.e., highly disturbed contexts were not sampled). In each sample selected for anthracological analysis, wood charcoal >2 mm in size was sieved through a 4 mm USGS standard sieve. All charcoal specimens >4 mm in size were analyzed. If the sample contained fewer than 20 charcoal specimens >4 mm in size, additional charcoal specimens were selected from the 2–4 mm charcoal fraction (targeting the largest specimens) until at least 20 charcoal specimens were analyzed. Additional detailed information about anthracological sample processing and analysis is provided in Appendix 8.1

Silica Phytolith Sample Processing Methods

Phytolith sample collection, processing, and analysis methods employed in this study were based on techniques described in several prior studies and manuals (Coil et al. 2003; Cuthrell 2013b; Evett and Cuthrell 2013; Lentfer et al. 2003; Parr et al. 2001; Piperno 2006; Wu and Wong 2009). The phytolith sample processing method produced three data sets: 1) the overall soil phytolith content, expressed as a percentage of soil dry weight; 2) estimates of the number of diagnostic grass short cells (e.g., bilobate, rondel, crenate type phytoliths) per gram of soil; and 3) estimates of the number of diatoms (siliceous algae) per gram of soil. Phytoliths were sampled from near-surface soils in proximity to sites CA-SCR-7, CA-SCR-10, CA-SCR-14, and CA-SCR-15. Sample processing involved separation of phytoliths from other soil constituents (e.g., sand, clay, minerals, organic matter), mounting a measured proportion of material extracted from each sample on slides for quantitative analysis, and employing digital image analysis techniques and manual specimen counting to estimate soil phytolith percentage content and density of selected phytolith morphotypes. Detailed information about phytolith sample collection, processing, and analysis methods is presented in Appendix 8.1.

Results

General Summary of Analytical Results

Results of Macrobotanical Analysis. Macrobotanical analysis presented here includes a total of 99 flotation samples representing 646.0 liters of soil from the eight sites analyzed. Summaries of total macrobotanical specimen counts and weights by site are presented in Appendix 8.2: Table A1; macrobotanical densities (i.e., items or weight per liter of soil) are presented in Appendix 8.2: Table A2; and macrobotanical percentages are presented in Appendix 8.2: Table A3. Table 8.1 presents macrobotanical data from selected synthetic analytical categories by site. The preceding tables also include previously published comparative data from inland Late Period site CA-SMA-113 (only data from Loci 101–104 and 401–409, representing the central portion of the site and features, were included; adapted from Cuthrell 2013b). Macrobotanical data from CA-SMA-113 are not described in the Results section of this report but are incorporated into the Discussion section. Based on results of radiocarbon dating, samples from site CA-SCR-7 were divided into two temporally distinct components: "A," ca. 4800–3600 BCE and "B," ca. 2700–2200 BCE (see Chapter 6). In Appendix 8.2 (Tables A1 through A3), taxonomic categories represented by fewer than five specimens in the overall assemblage were omitted. These included Acmispon, Artemisia, Calandrinia, Cardamine, Chenopodium, Clinopodium, Daucus, Deinandra, Fragaria, Pinus, Ranunculus, Solanum, Caryophyllaceae, Lamiaceae, Papaveraceae, Rosaceae, Centaurea, Medicago, Erodium, and Domesticated Grain. Complete macrobotanical taxon count, weight, and density data for each flotation sample analyzed are presented in Appendix 8.2 (Tables A4 through A29).

Seventeen flotation samples were excluded from analysis for one or more of the following reasons: a) the sample represented a highly disturbed context (e.g., near-surface strata), a sterile or nearly sterile archaeological context in a site containing otherwise rich contexts (e.g., an ash lens with all contents completely combusted, representing a distinct set of formation processes from other contexts at the site), or a likely non-archaeological context (e.g., subsoil); b) the sample was collected from a chronologically different or ambiguous site component than other samples, and this component did not contain sufficient data to treat as a separate analytical unit; c) the sample had ambiguous contextual information; and/or d) the sample was collected using an auger, producing a small sample with poor stratigraphic control and little data. Data associated with samples excluded from analysis are included in Appendix 8.2 macrobotanical data tables, and a summary of excluded samples is presented in Appendix 8.2: Table A30.

Flotation samples analyzed per site or component ranged from 3 (CA-SMA-18; Appendix 8.2: Table A4) to 26 (CA-SCR-7 A; Appendix 8.2: Table A1). The total identified macrobotanical specimens per site/component were highly variable, ranging from

29 specimens at CA-SMA-218 to 2,778 specimens at CA-SCR-15 (Table 8.1). Density of identified macrobotanical remains by site/component varied by more than two orders of magnitude, ranging from 0.6 n/l at CA-SMA-218 to 86.6 n/l at CA-SCR-14. Identified macrobotanical specimen density among all coastal sites was <13 n/l, but >58 n/l at all inland sites (Table 8.1). Of the 9 sites, 4 analyzed contained assemblages with fewer than 200 identified specimens and densities lower than 20 n/l (CA-SCR-7 A, CA-SCR-7 B, CA-SMA-218, and CA-SMA-18). At these sites, interpretation of Indigenous plant use practices is uncertain because small assemblage sizes and low densities obscure whether these plant remains represent cultural use or incidental charring. All of these sites also displayed very low densities of wood charcoal (0.03–0.29 g/l), indicating either relatively less use of fire by site inhabitants, greater taphonomic impacts on botanical materials, or both, as compared to other sites. As these four sites are also the oldest ones sampled, it seems likely that taphonomic processes could have substantially modified the composition of these assemblages. Based on field observations, contexts from the near-surface deposits at sites CA-SMA-218 and CA-SMA-18 were more disturbed than those from CA-SCR-7, which displayed well-consolidated and apparently intact deposits in exposed profiles. Among all sites/components, the percentage of identified macrobotanical specimens among all potentially identifiable specimens was generally high, with values of >90% at all sites except CA-SMA-218 (63.0%) and CA-SMA-18 (84.2%; see "% Identified of Identifiable Specimens," Table 8.1). The unusually low percentage of identified specimens at CA-SMA-218 was attributable to the low overall number of specimens recovered and poor preservation conditions.

Results of Anthracological Analysis. Anthracological analysis was carried out on materials from four sites: CA-SCR-7, CA-SCR-10, CA-SCR-14, and CA-SCR-15. Results of anthracological analysis are presented in Table 8.2 (with previously published results from CA-SMA-113). Between 76 and 113 anthracological specimens were analyzed at each site. Due to the substantial time required to identify individual anthracological specimens, sample sizes at each site were relatively small, and results should be interpreted with appropriate caution. Overall, 76.2% of specimens were identified to family or to a lower taxonomic level, while the remainder were identifiable only as angiosperms, gymnosperms, or indeterminate wood type. Factors preventing specimen identification included degradation, decomposition prior to charring, small size, and irregular wood anatomy. In total, 21 taxa were identified to genus level, and 4 taxa were identified to family level. The category "Sequoia (cf.)" contains specimens consistent with redwood but lacking diagnostic features that would differentiate them from other genera in the family Cupressaceae. Since redwood is the only naturally occurring local species in Cupressaceae—that is, excluding Monterey cypress (*Hesperocyparis macrocarpa*), which was planted historically in these areas and observed in proximity to sampled sites—these specimens are likely redwood. With the exception of a single specimen tentatively identified as poplar (from CA-SMA-113, and possibly a misidentification), all taxa detected currently occur in the local region within ca. 10 km of sites.

Results of Phytolith Analysis. Phytolith sampling was carried out near four sites: CA-SCR-7, CA-SCR-10, CA-SCR-14, and CA-SCR-15. In the vicinity of each site, three to four samples of near-surface soil were processed. As described above, phytolith researchers in California have proposed that soils displaying overall phytolith content

Table 8.1. Summary Botanical Data Categories by Site. Category details: "Edible Nut" includes Corylus, Notholithocarpus, Quercus, and Umbellularia; "Edible Seed" includes the species Scirpus microcarpus, the genera Calandrinia, Chenopodium, Claytonia, Deinandra, Lupinus, Madia, Rumex, and Trifolium, and the families Chenopodiaceae, Montiaceae, and Poaceae; "Identified Specimens" includes all specimens identified to family or lower taxonomic level.

Site	CA-SCR-7 A	CA-SCR-7 B	CA-SMA-218	CA-SMA-18
Location	Coastal	Coastal	Coastal	Coastal
Period	Early	Early	Middle	Middle
Calibrated c14 Date Range	4800–3600 BCE	2700–2200 BCE	400 BCE– 50 CE	550–850 CE
Flot Samples (n)	26	18	4	3
Flot Sample Volume (l)	195.5	157.0	48.5	39.5
SUMMARY DATA—COUNTS and WEIGHTS				
Wood wt (g)	35.4157	19.9284	1.5112	11.5136
Edible Nut ct (n)	26	55	4	10
Total Edible Nut wt	0.0843	0.2442	0.0069	0.0164
Edible Seed ct (n)	84	86	18	50
Edible Nut & Seed ct (n)	110	141	22	60
Identified Specimen ct (n)	128	151	29	80
Identifiable Seed ct (n)	13	12	17	15
Unidentified Seed ct (n)	72	88	40	71
SUMMARY DATA—DENSITIES				
Wood wt (g/l)	0.1812	0.1269	0.0312	0.2915
Edible Nut ct (n/l)	0.13	0.35	0.08	0.25
Edible Nut wt (g/l)	0.0004	0.0016	0.0001	0.0004
Edible Seed ct (n/l)	0.43	0.55	0.37	1.27
Identified Specimen ct (n/l)	0.65	0.96	0.60	2.03
SUMMARY DATA—OTHER				
% Edible Nut of Identified Specimens	20.3	36.4	13.8	12.5
% Edible Seed of Identified Specimens	65.6	57.0	62.1	62.5
% Edible Nut & Seed of Identified Specimens	85.9	93.4	75.9	75.0
% Identified of All Specimens	64.0	63.2	42.0	53.0
% Identified of Identifiable Specimens	90.8	92.6	63.0	84.2
Edible Seed: Nut Ratio	3.2	1.6	4.5	5.0

CA-SMA-216	CA-SMA-19	CA-SCR-10	CA-SMA-113	CA-SCR-15	CA-SCR-14
Coastal	Coastal	Inland	Inland	Inland	Inland
Late	Late	Middle	Late	Late	Late
1300–1650 CE	1450–1650 CE	700–1000 CE	1000–1350 CE	1050–1400 CE	1000–1700 CE
13	6	12	77	10	7
32.7	67.0	30.3	494.1	46.0	29.5
101.6708	48.6273	13.7057	191.6503	16.6394	11.9263
19	47	176	1611	502	539
0.0466	0.0561	0.9241	8.2471	2.2838	3.2559
184	93	1461	34399	1998	1533
203	140	1637	36010	2500	2072
417	337	1782	42178	2778	2555
33	26	102	2159	120	178
273	273	932	19185	748	951
3.1092	0.7258	0.4523	0.3879	0.3617	0.4043
0.58	0.70	5.81	3.26	10.91	18.27
0.0014	0.0008	0.0305	0.0167	0.0496	0.1104
5.63	1.39	48.22	69.62	43.43	51.97
12.75	5.03	58.81	85.36	60.39	86.61
4.6	13.9	9.9	3.8	18.1	21.1
44.1	27.6	82.0	81.6	71.9	60.0
48.7	41.5	91.9	85.4	90.0	81.1
60.4	55.2	65.7	68.7	78.8	72.9
92.7	92.8	94.6	95.1	95.9	93.5
9.7	2.0	8.3	21.4	4.0	2.8

Table 8.2. Results of Anthracological Analysis—Counts and Percentages of Taxa. Notes: "Sequoia (cf.)" = specimens in the family Cupressaceae and matching characteristics of Sequoia but lacking unequivocal diagnostic characteristics of Sequoia (see text); "UnID Angio." = unidentified angiosperm; "UnID Gymno." = unidentified gymnosperm; "UnID Other" = irregular or degraded specimens and specimens whose anatomical characteristics did not match a taxon in the identification key. The category "Total Trees" includes Aesculus, Alnus, Notholithocarpus, Populus, Quercus, Salix, Umbellularia, Pinus, Pseudotsuga, and Sequoia. All other taxa are classified as "Shrubs and Small Trees."

Site		CA-SCR-7		CA-SCR-10	
Location		Coastal		Inland	
Period		Early		Middle	
Calibrated c14 Date Range		4800–2200 BCE		700–1000 CE	
ANGIOSPERMS	Common Name	n	%	n	%
Adenostoma	chamise			2	2.0
Aesculus	Cal. buckeye				
Alnus	alder	7	6.2	9	9.0
Arctostaphylos	manzanita			1	1.0
Artemisia	sagebrush	1	0.9	2	2.0
Baccharis	coyotebrush				
Ceanothus	Cal. lilac			4	4.0
Cornus	dogwood			4	4.0
Corylus	hazelnut				
Frangula	coffeeberry			1	1.0
Heteromeles	toyon			1	1.0
Morella	Cal. wax myrtle			2	2.0
Notholithocarpus	tanoak			1	1.0
Populus	poplar				
Quercus	oak			2	2.0
Salix	willow			15	15.0
Sambucus	elderberry			2	2.0
Umbellularia	Cal. bay			8	8.0
Asteraceae	sunflower fam.			1	1.0
Caprifoliaceae	honeysuckle fam.				
Ericaceae	heath fam.				
Rhamnaceae	buckthorn fam.			1	1.0

CA-SMA-113 Inland Late 1000–350 CE		CA-SCR-15 Inland Late 1050–1400 CE		CA-SCR-14 Inland Late 1000–1700 CE		TOTAL	
n	%	n	%	n	%	n	%
						2	0.4
1	0.9					1	0.2
9	8.2	6	7.9			31	6.5
						1	0.2
						3	0.6
		1	1.3			1	0.2
19	17.3	7	9.2	6	7.5	36	7.5
						4	0.8
				1	1.3	1	0.2
		2	2.6	3	3.8	6	1.3
1	0.9					2	0.4
2	1.8	2	2.6	1	1.3	7	1.5
		2	2.6	3	3.8	6	1.3
1	0.9					1	0.2
		3	3.9	6	7.5	11	2.3
1	0.9	1	1.3			17	3.5
						2	0.4
						8	1.7
						1	0.2
		4	5.3			4	0.8
1	0.9					1	0.2
						1	0.2

Table 8.2. Continued.

Site		CA-SCR-7		CA-SCR-10	
Location		Coastal		Inland	
Period		Early		Middle	
Calibrated c14 Date Range		4800–2200 BCE		700–1000 CE	
GYMNOSPERMS	**Common Name**	**n**	**%**	**n**	**%**
Pinus	pine				
Pseudotsuga	Douglas fir			1	1.0
Sequoia	redwood	29	25.7	7	7.0
Sequoia (cf.)	redwood	55	48.7	9	9.0
UNIDENTIFIED					
UnID Angio.		9	8.0	17	17.0
UnID Gymno.		9	8.0	7	7.0
UnID Indet.		3	2.7	3	3.0
Total Shrubs and Small Trees		1	0.9	21	21.0
Total Trees		91	80.5	52	52.0
TOTAL		**113**	**100.0**	**100**	**100.0**

CA-SMA-113		CA-SCR-15		CA-SCR-14		TOTAL	
Inland		Inland		Inland			
Late		Late		Late			
1000–350 CE		1050–1400 CE		1000–1700 CE			
n	%	n	%	n	%	n	%
1	0.9			5	6.3	**6**	**1.3**
7	6.4	2	2.6	12	15.0	**22**	**4.6**
36	32.7	16	21.1	7	8.8	**95**	**19.8**
		13	17.1	18	22.5	**95**	**19.8**
		4	5.3	9	11.3	**39**	**8.1**
17	15.5	7	9.2	3	3.8	**43**	**9.0**
14	12.7	6	7.9	6	7.5	**32**	**6.7**
23	20.9	16	21.1	11	13.8	**72**	**15.0**
56	50.9	43	56.6	51	63.8	**293**	**61.2**
110	**100.0**	**76**	**100.0**	**80**	**100.0**	**479**	**100.0**

ca. >0.3% by weight and high densities of diagnostic grass short cells (ca. >200,000 n/g) indicate the presence of grasslands over the long term (Evett and Bartolome 2013; Evett and Cuthrell 2013; Evett et al. 2012). Soil phytolith percentage content data are presented in Table 8.3 and Figure 8.2. All soil samples except one (Sample PHY14 from a riparian forest area east of CA-SCR-14) displayed phytolith content above 0.3%, with values generally ranging from ca. 0.5 to 1.2%. These levels of phytolith density indicate the presence of phytolith-producing vegetation in proximity to each site over the long term. Estimates of the density of grass short cell phytoliths and diatoms are presented in Table 8.4 and Figure 8.3. Among all sites, samples were highly variable in density of grass short cell phytolith morphotypes, with values ranging from ca. 25k to 200k n/g. No samples displayed values well above the 200k n/g threshold that would, in concert with high phytolith content values, strongly indicate the presence of grasslands over the long term.

Phytolith evidence for long-term grasslands was variable among the locations sampled. In soils near Early Period site CA-SCR-7, phytolith content values ranged from ca. 0.4 to 0.8%, while grass short cell density values ranged from 27 to 181k n/g, with three of the four samples analyzed containing density values ca. ≤100k n/g. The relatively weak evidence for long term grasslands observed near this site is unsurprising, as the area was once an active dune field, which would be expected to support more ephemeral grasslands than locations with stable soils. Phytoliths deposited in the dune field also could have been transported out of the area as dunes shifted. At Middle Period site CA-SCR-10, phytolith content ranged from 0.5 to 1.0%, with grass short cell densities of 86–208k n/g. Phytolith evidence for long-term grasslands was only slightly stronger at CA-SCR-10 than at CA-SCR-7. Samples collected in proximity to CA-SCR-14 and CA-SCR-15 presented the strongest evidence for long-term grasslands among sites analyzed in this study. One of the five samples analyzed in this area was low in phytolith content (0.3%) and grass short cell density (34k n/g). This could result from small-scale variability in geomorphological processes, such as erosion or colluvial events. The other four samples collected from this area were relatively consistent in phytolith data, with phytolith content between 0.8% and 1.2% and grass short cell density ranging from 150 to 193k n/g.

Table 8.3. Estimates of Percentage Phytolith Content in Soils Near Archaeological Sites. Low and high estimates of soil phytolith percent are calculated based on the average phytolith percentage of phytolith extract plus or minus one standard error of the mean.

Sample #	Site	Sample Weight (g)	Phyto Extract Weight (g)	Phyto % of Extract Spot 1	Phyto % of Extract Spot 2	Avg. Phyto % of Extract	Std. Err. Phyto % of Extract	Avg. Phyto % of Soil	Low Phyto % of Soil	High Phyto % of Soil
PHY13	CA-SCR-14, -15	5.21	0.0985	42.4	44.6	43.5	1.1	**0.82**	0.80	0.84
PHY14	CA-SCR-14, -15	5.11	0.0741	14.9	22.1	18.5	3.6	**0.27**	0.22	0.32
PHY15	CA-SCR-14, -15	5.21	0.0821	75.8	71.0	73.4	2.4	**1.16**	1.12	1.19
PHY16	CA-SCR-14, -15	4.85	0.0902	43.3	43.1	43.2	0.1	**0.80**	0.80	0.81
PHY17	CA-SCR-14, -15	5.24	0.0685	74.7	71.1	72.9	1.8	**0.95**	0.93	0.98
PHY18	CA-SCR-7	4.87	0.0295	57.9	69.2	63.6	5.7	**0.39**	0.35	0.42
PHY19	CA-SCR-7	4.88	0.0549	69.9	75.0	72.5	2.6	**0.82**	0.79	0.84
PHY20	CA-SCR-7	5.24	0.0422	64.1	51.4	57.7	6.4	**0.46**	0.41	0.52
PHY21	CA-SCR-7	5.21	0.0601	70.3	61.5	65.9	4.4	**0.76**	0.71	0.81
PHY22	CA-SCR-10	5.08	0.1042	41.9	40.2	41.0	0.8	**0.84**	0.82	0.86
PHY23	CA-SCR-10	5.1	0.0664	79.1	69.4	74.3	4.8	**0.97**	0.90	1.03
PHY24	CA-SCR-10	5.18	0.0451	65.0	49.7	57.3	7.6	**0.50**	0.43	0.57

Table 8.4. Estimates of Grass Short Cell Phytolith and Diatom Densities in Soils Near Archaeological Sites. Density values are expressed in thousands of specimens per gram of soil.

Sample #	Site	Bilobate Density (n/g)	Rondel Density (n/g)	Crenate Density (n/g)	Grass Short Cell Density (n/g)	Diatom Density (n/g)	Location Note
PHY13	CA-SCR-14/15	70	14	84	**167**	2	Riparian forest, between CA-SCR-14 and CA-SCR-15
PHY14	CA-SCR-14/15	9	0	25	**34**	3	Riparian forest, east of CA-SCR-14
PHY15	CA-SCR-14/15	67	9	101	**176**	14	Grassland, east of CA-SCR-15
PHY16	CA-SCR-14/15	81	17	52	**150**	0	Riparian forest, between CA-SCR-14 and CA-SCR-15
PHY17	CA-SCR-14/15	99	18	77	**193**	13	Grassland, east of CA-SCR-15
PHY18	CA-SCR-7	11	5	12	**27**	5	Fallow agricultural field, northeast of CA-SCR-7
PHY19	CA-SCR-7	39	18	43	**101**	39	Fallow agricultural field, southeast of CA-SCR-7
PHY20	CA-SCR-7	20	7	20	**47**	23	Fallow agricultural field, east of CA-SCR-7
PHY21	CA-SCR-7	72	23	87	**181**	17	Fallow agricultural field, east of CA-SCR-7
PHY22	CA-SCR-10	26	9	59	**94**	32	Fluvial terrace, north of CA-SCR-10
PHY23	CA-SCR-10	103	25	79	**208**	8	Agricultural field, west of CA-SCR-10
PHY24	CA-SCR-10	32	7	46	**86**	89	Agricultural field, northwest of CA-SCR-10

Analytical Results by Site

The specific results of macrobotanical, anthracological, and phytolith analysis at each site in this study are too lengthy to include in this text. Accordingly, contextual information about each site, as well as presentation and interpretation of analytical results for each data set, is presented in Appendix 8.3.

Comparative Analytical Results

This section integrates data from each of the sites studied in this project, as well as previously published data from site CA-SMA-113, interpreting results with respect to a) subsistence practices, b) fuel use and vegetation types, and c) phytolith evidence for

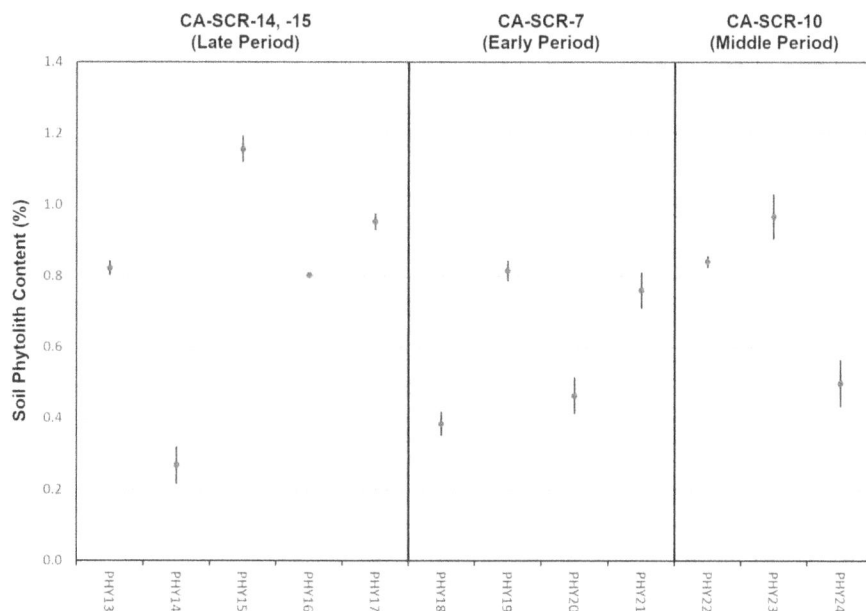

Figure 8.2. Soil phytolith content expressed as a percentage of soil dry weight. Vertical bars indicate error range of plus or minus one standard error of the mean.

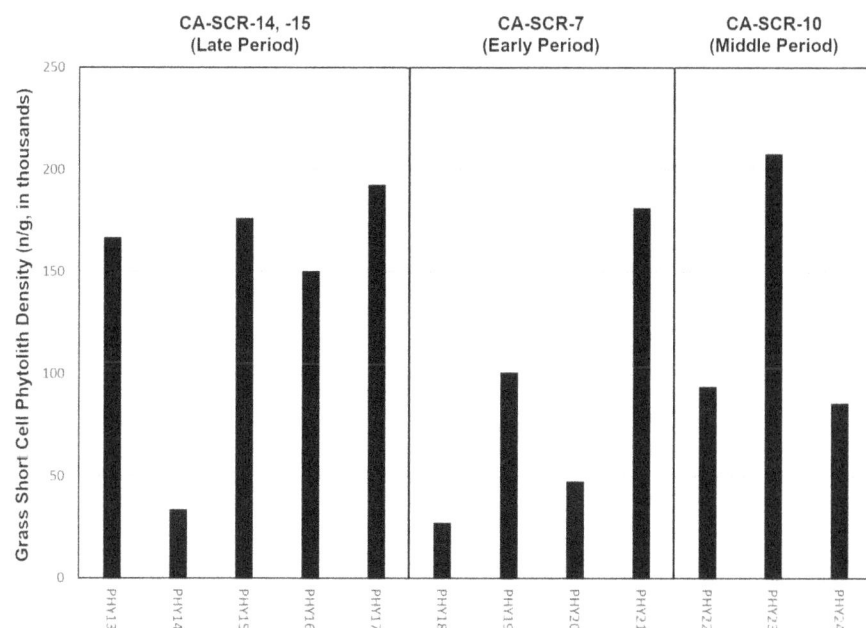

Figure 8.3. Density of Diagnostic Grass Short Cell Phytoliths (Rondel, Bilobate, and Crenate Morphotypes) Among Soil Samples Analyzed in this Project. Values are number of specimens per gram, in thousands.

long-term grasslands. Distinctions between sites are considered primarily on the basis of location (coastal vs. inland) and chronology (temporal period).

Comparative Results of Macrobotanical Analysis and Implications for Subsistence Practices. There were clear and consistent distinctions between the archaeobotanical assemblages of coastal versus inland sites; however, differences between temporal periods were comparatively minor (Figure 8.4). Densities of edible nuts differed profoundly between coastal and inland sites, with all coastal sites containing median edible nut density <0.8 mg/l (<0.6 n/l) and all inland sites displaying median edible nut density

>8.2 mg/l (>2.0 n/l). Applying t-tests on natural log-transformed values of edible nut density by count and by weight between coastal and inland sites indicated the difference in edible nut density by site location was highly statistically significant (p<0.0001). Due to the high non-normality of data distributions and a non-trivial number of samples containing no edible nut remains in the overall assemblage (preventing log-transformation without data loss), it was not possible to use the Tukey HSD test to evaluate statistical significance of multiple comparisons in edible nut density by site. Box plots in Figure 8.4 indicate that among inland sites, edible nut density by weight was consistently higher in the three Santa Cruz County sites than at CA-SMA-113. Pooling data from the three Santa Cruz County inland sites and comparing natural log-transformed values of edible nut density by weight and by count to values from CA-SMA-113 using t-tests showed that edible nut density was statistically significantly higher (p<0.0001) at the Santa Cruz County sites in both cases. The same results are obtained if Middle Period site CA-SCR-10 is excluded from analysis, thus comparing only Late Period sites. CA-SMA-113 also displayed a lower percentage of edible nut remains than other inland sites, with edible nut remains comprising only 3.8% of the identified macrobotanical assemblage, compared to values ranging from 9.9 to 21.1% at other inland sites (Appendix 8.2: Table A3). Chi-square comparisons of edible nut remains vs. other identified macrobotanical specimens between CA-SMA-113 and each other inland site showed these proportional differences were highly statistically significant in all cases (p<0.0001).

Of all sites analyzed, the highest percentage of edible nuts among identified macrobotanical specimens was observed in the assemblage from CA-SCR-7 Component B, with 36.4% edible nut remains (quantified by count; Figure 8.5). CA-SCR-7 Component A also contained a relatively high percentage of edible nut remains, at 20.3%. Other coastal sites displayed edible nut percentages from 4.6 to 13.9%, and inland sites had values ranging from 3.8 to 21.1%. Chi square comparisons of edible nut count vs. other identified macrobotanical specimen count between CA-SCR-7 Component B and all other coastal sites/components returned statistically significant differences in all cases (p<0.05). A similar comparison of CA-SCR-7 Component A and all other coastal sites/components returned statistically significant results only in relation to CA-SCR-7 Component B (p=0.003) and CA-SMA-216 (p<0.0001).

Taphonomic effects are anticipated to be greater in older sites, and CA-SCR-7 deposits are ca. 2,000–4,000 years older than those from any other site analyzed in this project. Although edible nut densities were low at the site, the relatively higher percentages of edible nuts observed, particularly in CA-SCR-7 Component B, could indicate relatively more nut processing at this site than at other coastal sites. This possibility does not apply to nuts of Monterey pine (*Pinus radiata*), for which there was indirect evidence of processing at sites CA-SMA-18, CA-SMA-19, and CA-SMA-218 in the form of charred cones, but little direct evidence (with one nut identified at CA-SMA-216) because Monterey pine nut shells are very thin and thus unlikely to preserve archaeologically.

Comparative analysis of nut assemblages between sites shows marked differences in the types of nuts processed at coastal vs. inland sites (Figure 8.6 and Table 8.5). Among coastal sites, tanoak (*Notholithocarpus densiflorus*) comprised 65–90% of edible nut remains at three sites, while at inland sites, tanoak was only 4–32% of edible nuts (quantified by count; sites with 10 or less total edible nut specimens were excluded due to small sample size). Hazelnut (*Corylus cornuta* ssp. *californica*) was consistently the dominant edible nut recovered from inland sites (60–94% of edible nut remains), while it was almost entirely absent from coastal sites. The edible nut assemblage from coastal site CA-SMA-19 was distinct from that of all other sites in the study because it was almost entirely comprised of California bay (*Umbellularia californica*, 98% of edible nut remains). Chi-square comparisons of counts of each nut taxon vs. counts of all other edible nuts observed in coastal. vs. inland sites showed these proportional differences were highly statistically significant (p<0.0001) in all cases except for oak acorn (*Quercus* sp.; p=0.86).

The near absence of hazelnut at coastal sites is particularly notable. As the only edible nut expected to occur in close proximity to coastal sites is Monterey pine nut (and only on Año Nuevo Point), the observed differences in the types of edible nuts consumed at coastal vs. inland sites cannot be easily explained by proximity to nut resources. While California bay nuts could have been harvested from low-lying riparian corridors closer to the coast, both tanoak and hazelnut grow in locations higher in elevation and farther inland. Inhabitants of inland sites, positioned alongside riparian corridors at the base of marine terraces and foothills, would probably have had ready access to every type of nut resource. If coastal site inhabitants based nut-harvesting practices on proximity, we would expect to see greater proportional representation of oak and bay, which grow in riparian corridors close to the coast. We would also expect hazelnuts and tanoak acorn to occur together, since both tend to grow in similar inland locations.

Since hazelnut is expected to be the hardiest of all local edible nut remains and coastal sites are predicted to generally have worse preservation conditions than inland sites due to having looser, less stable, and more porous sandy substrate, differences in species composition of nut assemblages cannot be attributed to taphonomy. The observed pattern could be related to seasonality, as tanoak and California bay nuts both ripen in the fall (for California bay nuts, possibly into early winter), while hazelnuts usually ripen from late summer to early fall.

Higher densities of edible seeds were consistently observed among inland sites in comparison to coastal sites. Median edible seed density at inland sites ranged from 14.1 to 52.5 n/l, but only from 0.3 to 1.4 n/l at coastal sites (Figure 8.4). As described above, seed food processing was clearly indicated at all inland sites by high densities of edible seeds in a subset of contexts analyzed, while similar practices were not apparent at any of the coastal sites. A Tukey HSD test on natural log-transformed edible seed density values between inland sites showed only one statistically significant difference in edible seed density, between sites CA-SMA-113 and CA-SCR-15 (p<0.05). However, this difference resulted from inclusion of marginal Excavation Unit 1 (EU 1) at CA-SCR-15 in the overall data set (which contained little archaeobotanical material; see Appendix 8.3, "CA-SCR-15" section). When data from this unit are excluded, the test indicates no statistically significant differences in edible seed density between inland sites.

Among inland sites, the edible seed assemblages of the three sites in Santa Cruz County were dominated by seeds of grasses, which comprised 66.0–87.5% of edible seeds and 39.6–71.8% of identified macrobotanical specimens (Appendix 8.2: Table A3 and Figure 8.6). At site CA-SMA-113 in Quiroste Valley, the proportion of grass seeds was considerably lower, accounting for 44.1% of edible seeds and 36.0% of identified specimens. The lower proportional representation of grasses at CA-SMA-113 was due

Table 8.5. Percentages of Edible Nuts at Each Site/Component, Quantified by Count.

Site		CA-SCR-7 A	CA-SCR-7 B	CA-SMA-218
Location		Coastal	Coastal	Coastal
Period		Early	Early	Middle
Calibrated ^{14}C Date Range		4800–3600 BCE	2700–2200 BCE	400 BCE–50 CE
Flot Samples (n)		26	18	4
Flot Volume (l)		195.5	157.0	48.5
NUTSHELL CT (n)	**Common Name**			
Corylus	hazelnut		1	1
Notholithocarpus	tanoak	17	50	3
Quercus	oak	3	4	
Umbellularia	California bay	6		
TOTAL		**26**	**55**	**4**
NUTSHELL %	**Common Name**			
Corylus	hazelnut		1.8	25.0
Notholithocarpus	tanoak	65.4	90.9	75.0
Quercus	oak	11.5	7.3	
Umbellularia	California bay	23.1	0.0	0.0

CA-SMA-18	CA-SMA-216	CA-SMA-19	CA-SCR-10	CA-SMA-113	CA-SCR-15	CA-SCR-14
Coastal	Coastal	Coastal	Inland	Inland	Inland	Inland
Middle	Late	Late	Middle	Late	Late	Late
550–850 CE	1300–1650 CE	1450–1650 CE	700–1000 CE	1000–1350 CE	1050–1400 CE	1000–1700 CE
3	13	6	12	77	10	7
39.5	32.7	67.0	30.3	494.1	46.0	29.5
			160	964	432	504
3	17	1	7	516	64	26
1			8	115	2	7
6	2	46	1	16	4	2
10	**19**	**47**	**176**	**1611**	**502**	**539**
			90.9	59.8	86.1	93.5
30.0	89.5	2.1	4.0	32.0	12.7	4.8
10.0			4.5	7.1	0.4	1.3
60.0	10.5	97.9	0.6	1.0	0.8	0.4

primarily to the abundance of panicled bulrush (*Scirpus microcarpus*) seeds at the site, which comprised 34.6% of edible seeds and 28.2% of identified specimens in the assemblage. The unusually high proportional representation of panicled bulrush seeds in the CA-SMA-113 assemblage was the clearest difference in taxonomic composition of edible seed assemblages among inland sites.

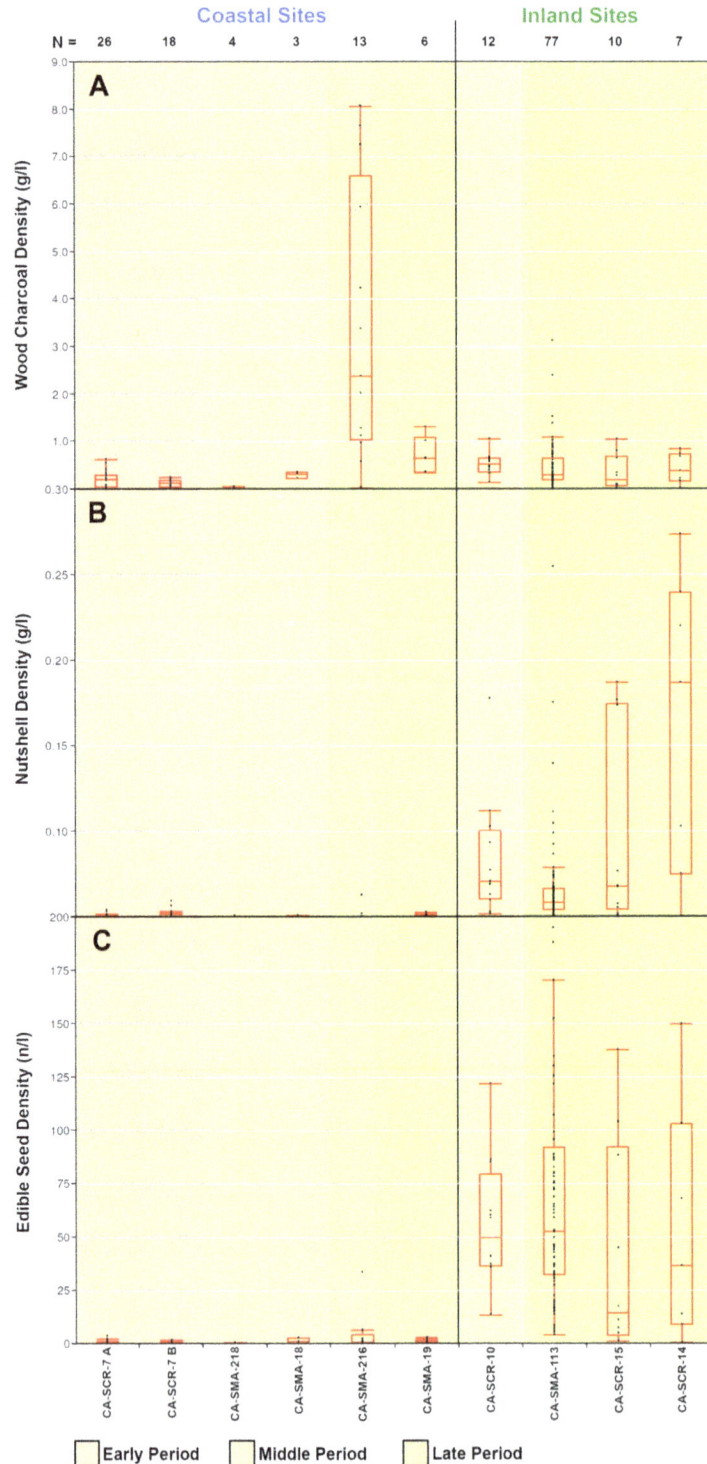

Figure 8.4. Box Plot Inter-Site Comparison of: **A** – Wood Charcoal Density (g/l); **B** – Edible Nutshell Density (g/l); And **C** – Edible Seed Density (n/l). Each data point represents a flotation sample. Note: In graph **C**, for site CA-SMA-113, seven outlier samples with edible seed density >200 n/l are not displayed.

Indigenous Landscape and Seascape Stewardship

164

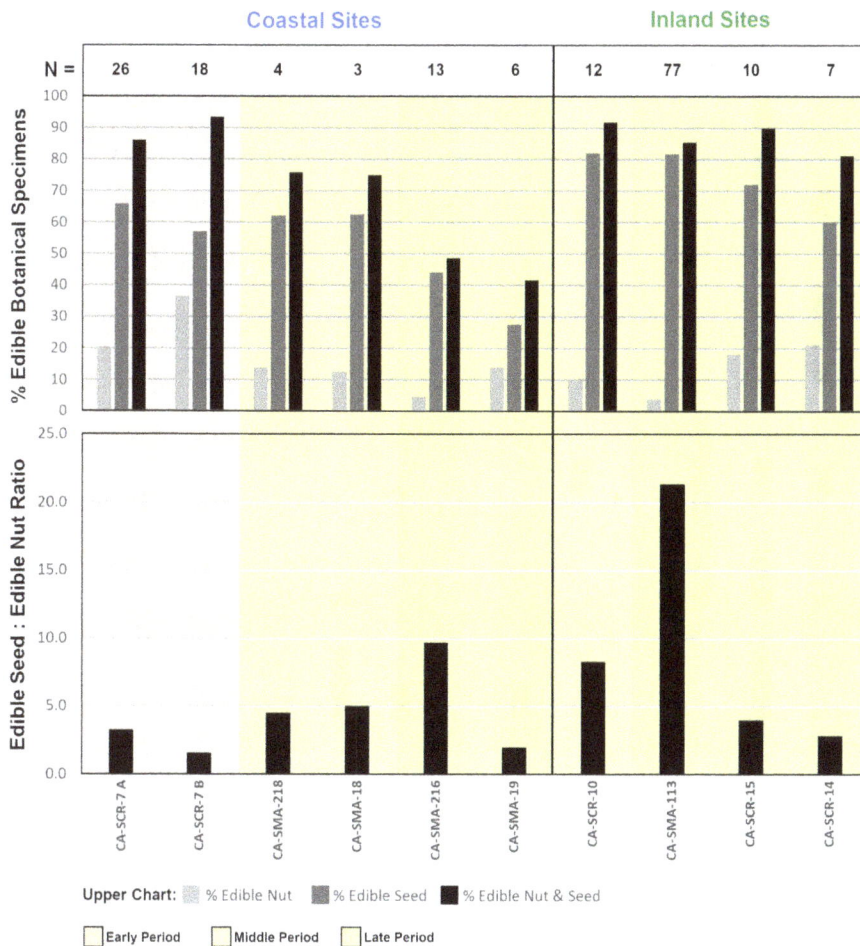

Figure 8.5. Top: Inter-site Comparison of Percentages of Edible Nuts (Using Number of Nutshell Fragments), Edible Seeds, and Edible Nut Plus Seeds. Values represent each category as a percentage of all macrobotanical specimens identified to family, genus, or species level. **Bottom**:Inter-site Comparison of Ratio of Edible Seeds to Edible Nuts (Using Number of Nutshell Fragments).

The proportion of the archaeobotanical assemblage comprised of edible plant foods was consistently >80% among inland sites, while Middle and Late Period coastal sites had lower edible plant food proportions ranging from ca. 40 to 75% (Figure 8.5). Interestingly, both temporal components of Early Period site CA-SCR-7 displayed edible plant food proportions similar to those of inland sites, ranging from 85.9 to 93.4% of identified specimens. The higher percentage observed at CA-SCR-7 results from greater proportional representation of grass seeds (Figure 8.6). If the low densities of macrobotanical remains observed at CA-SCR-7 are attributable primarily to taphonomic effects, the data could indicate use of grass seed foods at the site. However, this possibility cannot be distinguished from the alternative interpretation that grass seed representation at the site results from incidental charring, and that the difference in proportional representation of edible plant foods at this site compared to other coastal sites reflects differences in taxonomic composition between the local habitat at Año Nuevo Point (where all other coastal sites are located) and the habitat in the vicinity of Sand Hill Bluff, located ca. 20 km to the southeast.

Among coastal sites, the ratio of edible seeds to fragments of edible nutshell (the "seed-to-nut ratio") ranged from 1.6 to 9.7 (Figure 8.5), displaying a high degree of

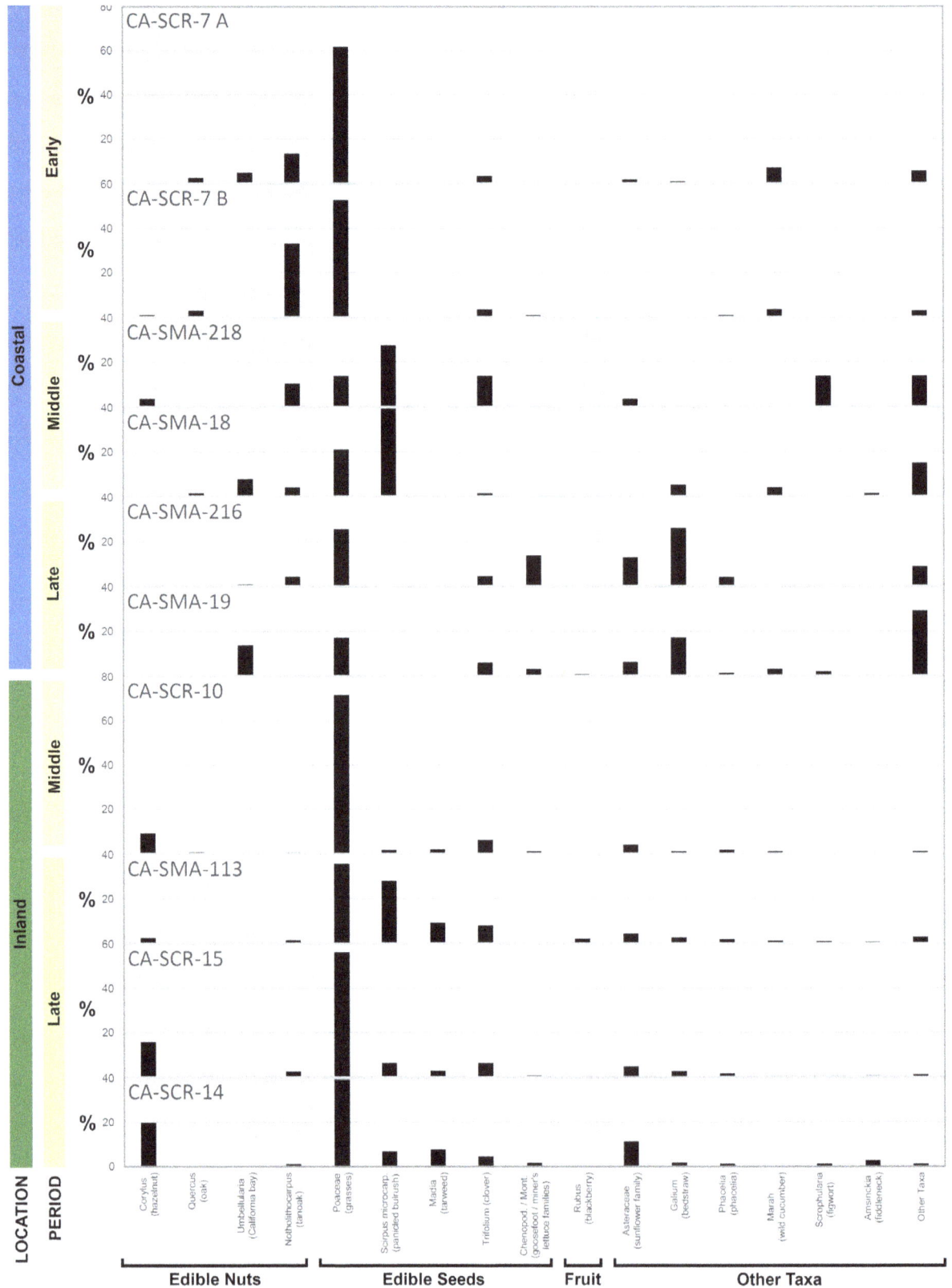

Figure 8.6. Inter-site Comparison of Selected Taxon Percentages. This graph displays taxa representing >0.5% of the overall project assemblage, as well as "Chenopodiaceae / Montiaceae" (goosefoot and miner's lettuce families; inclusive of genera within these families, 0.31% of assemblage), *Quercus* (oak; 0.28% of assemblage), and *Umbellularia* (California bay; 0.16% of assemblage).

variability. However, because it seems most likely that the assemblage of edible seeds at these sites represents incidental charring rather than cultural use, the seed-to-nut ratio does not appear to be interpretable with respect to subsistence practices, particularly in light of the general sparsity of macrobotanical remains among coastal sites. Among inland sites, where consumption of seed foods was consistently indicated, the seed-to-nut ratio indicates substantial differences in the use of these resource types. Pairwise comparisons of seed and nut counts between inland sites using the chi-square test showed that the overall seed-to-nut ratio at each inland site was highly statistically significantly different (p<0.0001) from that of every other inland site (Table 8.6). A Tukey HSD test on seed-to-nut ratios between sites did not indicate statistically significant differences between any pair of sites; however, this result appears to be attributable to the small number of flotation samples analyzed at sites other than CA-SMA-113 (n=6–12 among Santa Cruz County sites vs. n=73 at CA-SMA-113; note that flotation samples with no edible nut remains could not be included in analysis due to division by zero errors).

Sites CA-SCR-14 and CA-SCR-15, located in close proximity to one another, had the lowest and most similar seed-to-nut ratios, at 2.8 and 4.0, respectively. The seed-to-nut ratio at CA-SCR-10 was about twice that observed at the two Late Period Santa Cruz County sites, at 8.3, and the ratio at CA-SMA-113 was more than double that of CA-SCR-10, at 21.4. The much higher seed-to-nut ratio observed at CA-SMA-113 was mainly attributable to the abundance of panicled bulrush seeds at the site (n=11904, 24.1 n/l, 28.2% of identified specimens), which was a relatively minor constituent of the edible seed assemblage at other inland sites (see Appendix 8.2: Tables A1 through A3). If panicled bulrush seeds are excluded from analysis, the seed-to-nut ratio at CA-SMA-113 would be 14.0, a value that is still much higher than the values observed among other inland sites. These striking differences in relative proportional representation of seed and nut foods among inland sites likely indicate meaningful distinctions in subsistence practices between temporal periods and/or historic Tribal territories.

Proportional taxonomic composition of seed assemblages varied between coastal sites with respect to temporal period. Both temporal components of Early Period site CA-SCR-7 contained relatively high percentages of grass seeds (61.7% in Component A and 53.0% in Component B), while all other later coastal sites displayed grass seed percentages from 13.8 to 25.7% (Appendix 8.2: Table A3, Figure 8.6). Each other seed taxon observed at CA-SCR-7 (in both temporal components) comprised <4% of the identified assemblage, further illustrating the proportional dominance of grasses at the site. As described above, this distinction between CA-SCR-7 and other coastal sites could reflect differences in local habitats between sites on Año Nuevo Point and Sand Hill Bluff at the time of site occupation or differences in plant use by site inhabitants.

Middle Period coastal sites CA-SMA-218 and CA-SMA-18 contained much higher proportions of panicled bulrush than other coastal sites (27.6% and 40.0%, respectively, vs. 0.0–0.3% at other sites), with levels similar to those observed at Late Period inland site CA-SMA-113 (Appendix 8.2: Table A3, Figure 8.6). However, it should be noted that these high proportional values are relatively uncertain due to the very small sample size of identified macrobotanical specimens at both sites (n=29 at CA-SMA-218 and n=80 at CA-SMA-18). At both sites, densities of this taxon were very low (0.2 n/l at CA-SMA-218 and 0.8 n/l at CA-SMA-18; Tables 8.2 and 8.3). As described above, the

Table 8.6. Pairwise Statistical Significance Results of Chi-Square Test (p values) on Edible Nutshell and Edible Seed Counts between Archaeological Sites Analyzed. Cells are shaded as follows: red: p>0.0500 (not statistically significant at 95% confidence); yellow: 0.0500>p> 0.0100 (statistically significant at 95–99% confidence); green: p<0.0100 (statistically significant at >99% confidence). The value "0.0000" indicates p<0.00005.

Site	CA-SCR-7 A	CA-SCR-7 B	CA-SMA-218	CA-SMA-18
Location	Coastal	Coastal	Coastal	Coastal
Period	Early	Early	Middle	Middle
Calibrated c14 Date Range	4800–3600 BCE	2700–2200 BCE	400 BCE– 50 CE	550–850 CE
CA-SCR-7 A	-	0.0097	0.5773	0.2878
CA-SCR-7 B	0.0097	-	0.0587	0.0019
CA-SMA-218	0.5773	0.0587	-	0.8717
CA-SMA-18	0.2878	0.0019	0.8717	-
CA-SMA-216	0.0006	0.0000	0.1945	0.1124
CA-SMA-19	0.0864	0.3434	0.1485	0.0152
CA-SCR-10	0.0000	0.0000	0.2656	0.1497
CA-SMA-113	0.0000	0.0000	0.0019	0.0000
CA-SCR-15	0.3635	0.0000	0.8248	0.5136
CA-SCR-14	0.5792	0.0008	0.4044	0.1026

CA-SMA-216	CA-SMA-19	CA-SCR-10	CA-SMA-113	CA-SCR-15	CA-SCR-14
Coastal	Coastal	Inland	Inland	Inland	Inland
Late	Late	Middle	Late	Late	Late
1350–1650 CE	1450–1650 CE	700–1000 CE	1000–1350 CE	1050–1400 CE	1000–1700 CE
0.0006	0.0864	0.0000	0.0000	0.3635	0.5792
0.0000	0.3434	0.0000	0.0000	0.0000	0.0008
0.1945	0.1485	0.2656	0.0019	0.8248	0.4044
0.1124	0.0152	0.1497	0.0000	0.5136	0.1026
-	0.0000	0.5434	0.0008	0.0002	0.0000
0.0000	-	0.0000	0.0000	0.0001	0.0498
0.5434	0.0000	-	0.0000	0.0000	0.0000
0.0008	0.0000	0.0000	-	0.0000	0.0000
0.0002	0.0001	0.0000	0.0000	-	0.0000
0.0000	0.0498	0.0000	0.0000	0.0000	-

immediate proximity of each site would have been characterized by dune fields, where panicled bulrush would probably not naturally grow; however, it is possible that local wetland areas or seeps may have supported populations of this plant. As the densities of panicled bulrush and of all other potentially edible seeds at these sites were very low, the presence of this taxon may indicate use for crafting or construction purposes during the Middle Period rather than consumption.

Late Period coastal sites CA-SMA-216 and CA-SMA-19 contained relatively high proportions of bedstraw (*Galium* sp.; 26.1% and 17.2%, respectively) compared to all other coastal sites (0.0–5.0%; Appendix 8.2: Table A3 and Figure 8.6). This could indicate the plant was used culturally for stuffing or bedding (or other purposes), or it could reflect incidental charring of naturally occurring seeds. At CA-SMA-19, all flotation samples contained bedstraw seeds and densities were uniformly low (0.3–2.3 n/l; Appendix 8.2: Table A5), tending to support the latter interpretation. However, at site CA-SMA-216, 98.2% of bedstraw specimens were present in a single flotation sample (n=107 of 109 total specimens; Sample #02-0003-000; n=12 flotation samples analyzed; Appendix 8.2: Tables A6 and A8), and this sample contained a relatively high density of bedstraw seeds, at 23.3. n/l (Appendix 8.2: Table A7). This illustrates a high level of variability in plant use between contexts at the site and indicates cultural use of the plant; however, its mode of use is unclear. Alternatively, this observation could represent intentional burning of herbaceous vegetation for disposal.

Site CA-SMA-19 contained a relatively high proportion of poison oak (*Toxicodendron diversilobum*) seeds (n=48, 14.2% of identified botanical remains, included in "Other Taxa" in Figure 8.6), a taxon that was not observed in any other coastal site, and which was observed at only one inland site (CA-SMA-113; n=3, 0.01%; Appendix 8.2: Table A3). Although poison oak has ethnographically reported cultural uses in the broader region (Bocek 1984), none involve seeds of the plant. Poison oak seeds were present in all flotation samples from the site at low densities ranging from 0.1 to 1.5 n/l. Poison oak is a common constituent of northern coastal scrub shrublands and coastal strand vegetation, which dominates the area in proximity to CA-SMA-19 today. Incidental charring of seeds from locally occurring plants appears to be the most likely explanation for the unusually high incidence of poison oak at this site.

Among inland sites, grasses were consistently the largest category of edible seeds, representing 36.0–71.8% of identified macrobotanical assemblages (Appendix 8.2: Table A3, Figure 8.6). The relatively low proportion of grass seeds at CA-SMA-113 (36.0%) is attributable to the unusually high proportion of panicled bulrush seeds in its assemblage (n=11904, 28.2%; Tables 8.1 and 8.3). If panicled bulrush was excluded, the percentage of grasses at the site would be 50.2%. "Overall density" of a taxon at a site indicates the total number of specimens observed divided by the total volume of flotation samples analyzed, while "median density" of a taxon is the density value observed in the flotation sample, which represents the median of density values observed among all flotation samples analyzed. At inland sites, overall grass seed densities were consistently high, ranging from 30.7 to 42.2 n/l (Appendix 8.2: Table A2), with median grass seed densities ranging from 10.1 to 43.2 n/l (Figure 8.7). Overall and median grass seed density values at CA-SCR-15 were lowered substantially by the inclusion of samples from marginal EU 1. Excluding samples from this unit raises the overall grass seed density to 64.3 n/l and the median grass

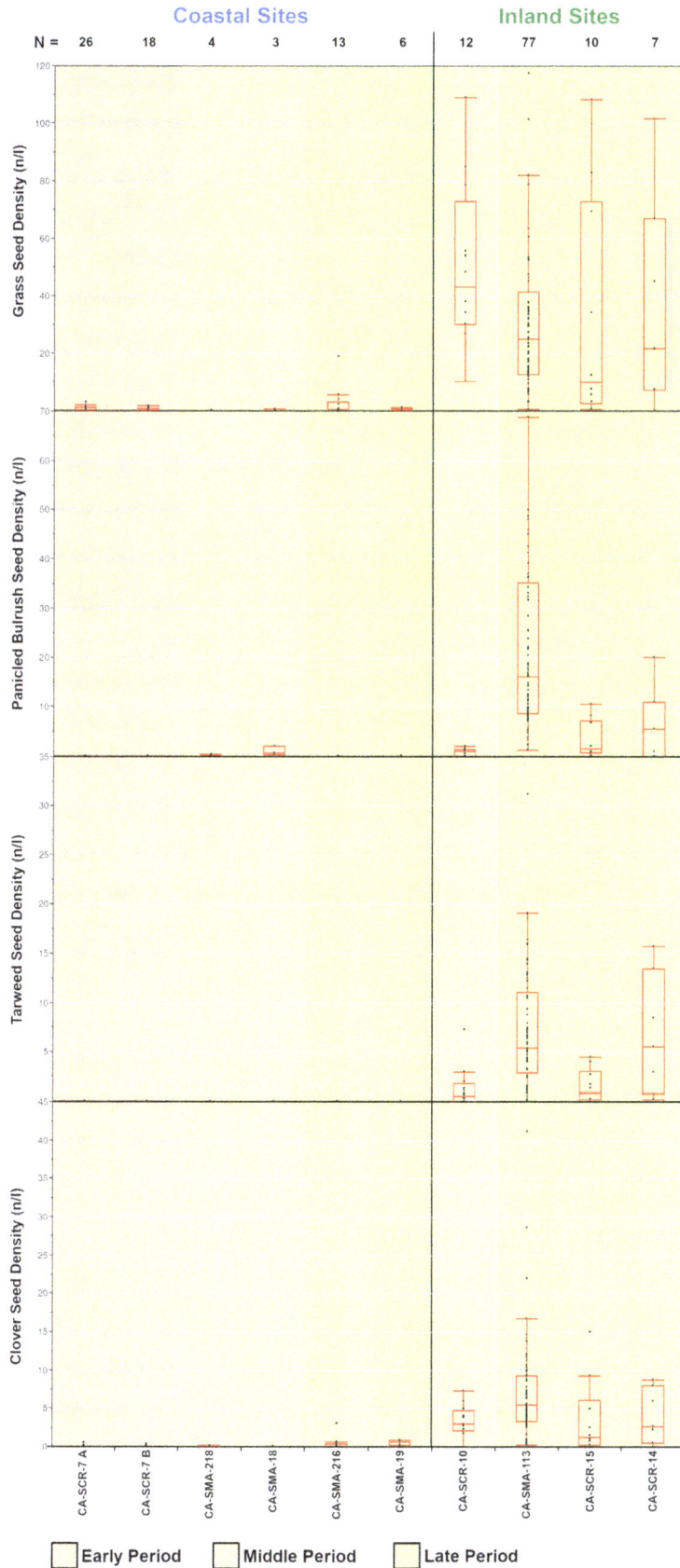

Figure 8.7. Inter-site Comparison of Edible Seed Densities by Taxon for Grasses (Poaceae), Panicled bulrush (*Scirpus microcarpus*), Tarweed (*Madia* sp.), and Clover (*Trifolium* sp.). Data points not shown include six samples from CA-SMA-113 with grass seed density >120 n/l; seven samples from CA-SMA-113 with panicled bulrush density >70 n/l; and one sample from CA-SMA-113 with tarweed density >35 n/l.

seed density to 69.5 n/l. These values represent the highest median grass seed densities observed among all sites in the project and are clear evidence of grass seed food processing (see Appendix 8.2: Tables A18 and A19). Interestingly, the second highest overall grass seed density (42.2 n/l) and median grass seed density (43.2 n/l) was observed at Middle Period site CA-SCR-10. A Tukey HSD test on per-sample grass seed density values between inland sites showed no statistically significant differences at the 95% confidence level.

The unusually high overall density (24.1 n/l), median density (16.1 n/l), and percentage (28.2%) of panicled bulrush seeds at Late Period inland site CA-SMA-113 led Cuthrell (2013b) to propose that seeds of this plant were consumed by site inhabitants (Appendix 8.2: Tables A2 and A3; Figure 8.7). At inland sites analyzed in this project, overall panicled bulrush densities were much lower (0.9–5.9 n/l), as were median densities (1.0–5.5 n/l) and percentages (1.5–6.9%; Appendix 8.2: Tables A2 and A3; Figures 8.6 and 8.7). Since this wetland plant would not be expected to occur naturally in the immediate vicinity of any of these sites, consistent representation of its seeds strongly indicates intentional cultural use; however, it is not clear whether the plant was used for crafting/construction, as a food, or for both purposes.

Tarweed (*Madia* sp.) seeds were not observed at any coastal site; and among inland sites, overall densities were variable, ranging from 1.7 to 9.2 n/l (Appendix 8.2: Table A3). At sites CA-SCR-10 and CA-SCR-15, overall tarweed seed density was <3 n/l, median tarweed densities were <1 n/l, and all but one flotation sample displayed densities <5 n/l. At these levels, tarweed seed consumption could be indicated, but it is also possible that the low densities represent incidental charring. At site CA-SMA-113, median tarweed density was 5.4 n/l, but the upper quartile of samples had density values >11.1 n/l, with a maximum of 63.1 n/l, strongly indicating seed food processing (Figure 8.7; Cuthrell 2013b: Appendix N). The median density of tarweed at site CA-SCR-14 was similar to that of CA-SMA-113 at 5.6 n/l; however, the maximum density observed was much lower at 15.8 n/l (Figure 8.7 and Appendix 8.2: Table A15). Processing of tarweed as a seed food appears likely at CA-SCR-14, though the evidence is not as strong as at CA-SMA-113.

Clover (*Trifolium* sp.) occurred sporadically and at low densities among all coastal sites, with a maximum density of 3.0 n/l observed among all samples (CA-SMA-216; Figure 8.7; Appendix 8.2: Tables A7 and A9). These values likely reflect incidental charring, as native clovers could occur naturally among all vegetation types in proximity to these sites. Among inland sites, the strongest evidence for intentional cultural use of clover seeds was at CA-SMA-113, with an overall clover density of 6.7 n/l, a median clover density of 5.5 n/l, densities of >9.3 n/l among the highest quartile of samples, and a maximum density of 41.2 n/l among samples analyzed (Appendix 8.2: Table A2 and Figure 8.7; Cuthrell 2013b: Appendix N). Among other inland sites, overall clover densities ranged from 3.5 to 3.9 n/l, with median densities of 1.3–3.0 n/l, and only one flotation sample had a density >10 n/l (CA-SCR-15, Sample #015-0006-04, 15.0 n/l; Figure 8.7 and Appendix 8.2: Table A19). While each of the Santa Cruz County inland sites contained flotation samples with clover seed densities higher than would be expected as a byproduct of incidental charring, none present compelling evidence of intentional cultural use of clover as a seed food. As clover was used primarily as a greens food among Native peoples of California (Anderson 2005), it is possible that non-incidental archaeobotanical representation of clover seeds is a byproduct of its use as a greens food rather than indicative of its use as a seed food.

The Chenopodiaceae and Montiaceae families—"Cheno. / Mont."; including the genus *Claytonia*—contain taxa, such as miner's lettuce (*Claytonia* sp.), that were primarily used as a greens food, as well as red maids (*Calandrinia* sp.) and goosefoot (*Chenopodium* sp.), which were mainly used as seed foods. Seeds of plants in these Cheno. / Mont. families were sporadically observed at coastal sites and regularly observed at inland sites, with low densities at inland sites ranging from 0.04 to 1.32 n/l. At these density levels, intentional cultural use is not indicated. As described in supplemental materials (Appendix 8.3, "CA-SMA-216" section), the occurrence of 51 charred Cheno. / Mont. seeds (11.1 n/l; Appendix 8.2: Tables A6 and A7; specifically, *Claytonia* sp.) in a single flotation sample from CA-SMA-216 could indicate seed food processing; yet, due to its association with an unusually high density of inedible bedstraw seeds, this occurrence more likely represents burning of herbaceous material as tinder or for vegetation disposal.

Typically, archaeological sites with examples of charred Cheno. / Mont. seeds also contained specimens of uncharred Cheno. / Mont. seeds—or seeds for which the status of charring could not be determined because seed testas of some taxa in these families are naturally jet black, resembling charred seeds. The data presented in this analysis include only Cheno. / Mont. seeds that were unambiguously charred. Most of the time, only charred specimens of seeds are enumerated in archaeobotanical assemblages because it is assumed that uncharred specimens would decay. However, this may not be the case for some species of Cheno. / Mont. At CA-SMA-113, a radiocarbon date on a pooled sample of 12 uncharred Cheno. / Mont. testas returned a calibrated date range of 1470–1650 CE, indicating these seeds may be preserved in an uncharred condition for hundreds of years (Cuthrell 2013b). Additionally, several flotation samples from CA-SMA-113 contained relatively high densities of uncharred Cheno. / Mont. seeds (>10 n/l), indicating probable use as a seed food. Among sites analyzed in this project, none contained relatively high densities of uncharred Cheno. / Mont. seeds, which would be indicative of intentional use. The maximum density of uncharred seeds observed was 3.3 n/l in a flotation sample from CA-SCR-10 (Sample #010-0023-04, Appendix 8.2: Table A11; see Appendix 8.2 data tables for numbers and densities of uncharred Cheno. / Mont. seeds in all flotation samples).

Hierarchical Cluster Analysis of Edible Nuts and Seeds

Hierarchical cluster analysis is a multivariate exploratory data analysis technique that efficiently illustrates relationships between members of a class (e.g., archaeological sites) given the distribution of data observed among multiple variables, as well as between variables given the distribution of data observed among members of a class (referred to as a two-way cluster analysis). Figure 8.8 presents a two-way cluster analysis of overall nutshell density and edible seed density among sites/components analyzed in this project. Along the horizontal axis, cluster analysis dendrogram branch spacing is proportional to the overall difference between classes, given the spread of data among variables. For example, two sites connected by dendrogram branches that join relatively close to the left side of the dendrogram (i.e., at a "lower" level on the dendrogram), indicate sites with relatively similar distributions of data between variables, while two sites connected by dendrogram branches that join close to the right side of the dendrogram (i.e., at a "higher" level on the dendrogram) indicate sites with highly dissimilar distributions of data.

In the cluster analysis of nutshell density (Figure 8.8 Part A), the primary division between clusters of sites is between coastal and inland sites. Coastal sites were relatively similar to one another in nutshell density because they typically displayed very low nutshell densities compared to inland sites. All coastal sites/components except CA-SMA-19 are connected by a relatively low-level dendrogram branch for this reason. CA-SMA-19 splits from other coastal sites at a high level because the unusually high density of California bay nutshell at the site distinguishes it from all other sites analyzed. All inland sites are connected by dendrogram branches that split apart at relatively high levels, indicating a higher degree of dissimilarity between the distributions of nutshell types among inland sites as compared to coastal sites. The dendrogram associated with variables (below the color map) indicates that among all sites, hazelnut (*Corylus cornuta* ssp. *californica*) and oak acorn (*Quercus* sp.) had the most similar distributions, while the distribution of California bay was most dissimilar from that of other nuts (as it was generally rare or absent and appeared in higher densities at only one site, CA-SMA-19). Summarily, the cluster analysis indicates a greater uniformity in nut use among most coastal sites (where nutshell density was typically low) than among inland sites (where nutshell density was higher and nutshell densities were more variable). Because sites CA-SCR-14 and CA-SCR-15 are located in such close proximity that residents of both sites would have had the same edible nut catchment area, the results suggest nut use differences between these sites cannot be attributed to differences in the physical environment alone. However, if these two sites were not occupied contemporaneously, the differences could be linked to dynamics between vegetation composition and fire regimes on decadal time scales. Another alternative is that these differences reflect distinctions in the foodways practices of site inhabitants.

Two-way cluster analysis of edible seed density among sites is presented in Figure 8.8 Part B. Again, the main division in the site dendrogram is between coastal and inland sites, with inland sites consistently displaying higher, and more highly variable, densities of edible seeds. Among coastal sites, CA-SMA-216 is distinguished from other sites because it contained a much higher density of Cheno. / Mont. seeds, while all other coastal sites displayed very low densities of all categories of edible seeds. Interestingly, among inland sites, the edible seed assemblage of CA-SCR-15 was most similar to that of CA-SCR-10, an older site located closer to the coast, rather than to CA-SCR-14, a contemporaneous site located just a few hundred meters away. As described above, the distinctions in edible seed use between the two sites in close proximity could be linked to fire regime / vegetation dynamics (if sites were not occupied contemporaneously) or to cultural practices. The edible seed assemblage of CA-SMA-113 was relatively distinct from all other inland sites, primarily due to the high density of paniceled bulrush seeds observed there. Among edible seed taxa, the distribution of Cheno. / Mont. seeds among sites was highly distinct from that of all other taxa primarily due to its unusually high density at site CA-SMA-216. Tarweed and paniceled bulrush were distributed similarly in relation to one another across all sites except CA-SCR-14 and CA-SMA-113. Although both had relatively high tarweed densities, only the latter had high paniceled bulrush density. Clover and grass seeds were likewise distributed similarly in relation to one another among all sites, with all coastal sites displaying low densities and all inland sites containing moderate to high densities of these taxa.

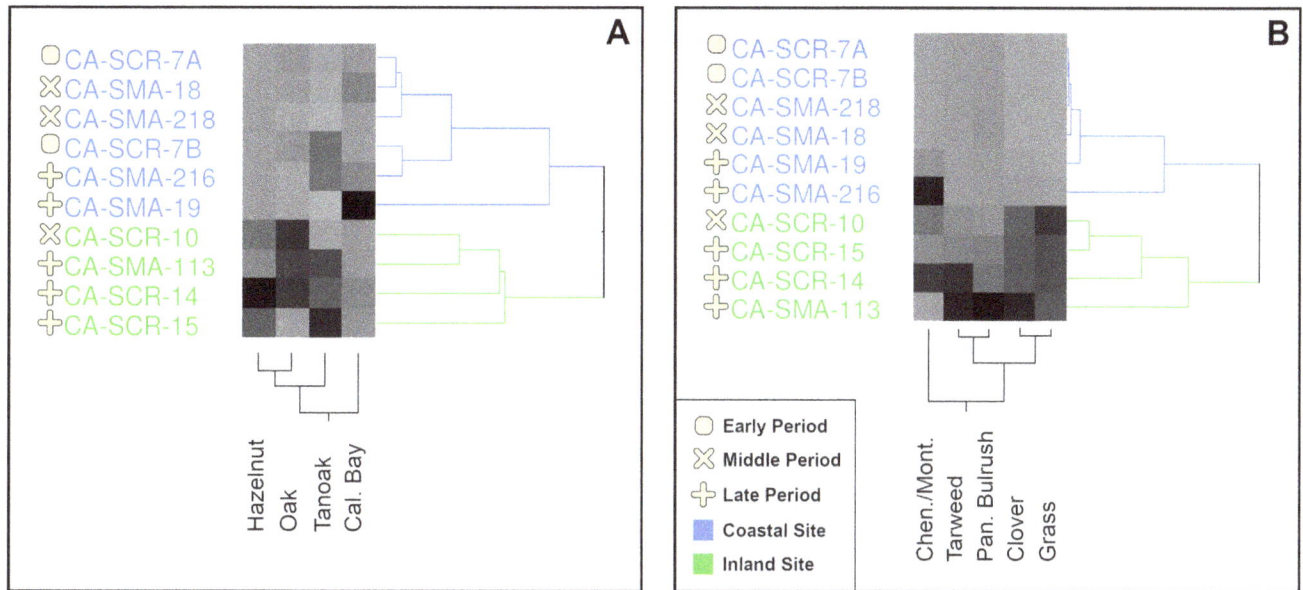

Figure 8.8. A: Two-way Hierarchical Cluster Analysis of Edible Seed Density by Site/Component. (*Corylus* = hazelnut; *Quercus* = oak; *Notholithocarpus* = tanoak; *Umbellularia* = California bay). **B**: Two-way Hierarchical Cluster Analysis of Edible Seed Density by Site/Component (*Chenopodiaceae / Montiaceae* = goosefoot / miner's lettuce / red maids; *Madia* = tarweed; *Scirpus microcarpus* = paniced bulrush; *Trifolium* = clover; Poaceae = grasses). Symbols beside site labels indicate temporal period, while site labor color indicates coastal vs. inland site location.

Diversity of Edible Nuts and Seeds Among Inland Sites

As coastal sites displayed little evidence for use of plant foods, the macrobotanical assemblages from these sites were not suitable for diversity analysis. However, the macrobotanical assemblages of all inland sites were sufficiently robust to explore differences between sites in the diversity of potential plant foods observed. Qualitatively, Middle Period contexts at site CA-SCR-10 appeared to contain a less diverse assemblage of plant foods than other inland sites. Seeds of grasses comprised a larger proportion of the macrobotanical assemblage at CA-SCR-10 (71.8%) than at other Late Period sites (36.0–56.5%; Appendix 8.2: Table A3), while displaying grass seed densities similar to other Late Period sites (ca. 30–70 n/l interquartile range; Figure 8.7). Among other potential seed foods, clover comprised 6.0% of the CA-SCR-10 assemblage, while paniced bulrush, tarweed, and Cheno. / Mont. each made up less than 2.0% of the assemblage. Together, these four taxa comprised 10.2% of the CA-SCR-10 assemblage, compared to 15.5–45.3% of the assemblages at other Late Period inland sites. Additionally, among individual flotation samples at CA-SCR-10, these taxa were typically observed in densities too low to clearly indicate intentional processing, with density values generally at <5 n/l (Figure 8.7). Among nuts, hazelnut dominated the CA-SCR-10 assemblage, comprising 9.0% of identified macrobotanical specimens, while all other edible nuts were <0.5% of specimens (Appendix 8.2: Table A3). Summarily, CA-SCR-10 appears to indicate a stronger focus on grass seeds and hazelnuts than was observed at Late Period inland sites, where additional food taxa made up larger proportions of macrobotanical assemblages. In some cases, this difference was minor—for example, between CA-SCR-10 and CA-SCR-15 where grasses plus hazelnut comprised 80.8% and 71.9%, respectively, of the assemblage—while in other cases this distinction was pronounced, such as in comparison with CA-SMA-113 (38.3% of the assemblage).

To quantitatively assess differences in plant food diversity between inland sites, Simpson's diversity index ("Simpson's D") was calculated for each flotation sample using the following formula: Formula 1: $D = 1 - \Sigma(n/N)^2$.

In Formula 1, "D" is Simpson's diversity index, "n" is the number of specimens of each individual taxon included in analysis, and "N" is the sum of specimens of all taxa included in analysis. The resulting index value ranges from 0.0 to 1.0, with zero indicating no diversity (i.e., observation of only a single taxon) and one indicating maximum diversity (i.e., equal proportions of every taxon included in the analysis). The following 9 food plants and potential food plants were used to calculate the index: hazelnut, tanoak acorn, oak acorn, California bay, tarweed, panicled bulrush, clover, grasses, and "Cheno. / Mont" (including specimens identified as *Calandrinia* sp., *Claytonia* sp., and *Chenopodium* sp.). Flotation samples with less than 20 total identified specimens in the preceding taxonomic categories were excluded from analysis. Simpson's D data and box plots are presented in Figure 8.9. A Tukey HSD test on Simpson's D values showed plant food diversity at Middle Period site CA-SCR-10 was statistically significantly lower than all Late Period sites ($p<0.01$; Table 8.7). Diversity at site CA-SCR-15 was also lower than at site CA-SMA-113 ($p<0.05$). Results of this test support the interpretation that inhabitants of CA-SCR-10 focused on using a narrower suite of plant food resources than inhabitants of Late Period sites.

Comparative Results of Macrobotanical Data and Implications for Land-scape Vegetation. All sites contained taxa that cannot be clearly associated with ethnobotanical use and likely represent incidental charring of seeds originating from local vegetation, either through natural dispersal of seeds into fires, charring of seeds in soil seed banks, or charring of seeds collected by site inhabitants unintentionally along with edible seeds. Examples include members of the sunflower family (Asteraceae), bedstraw (*Galium* sp.), phacelia (*Phacelia* sp.), and others (see "Other Taxa" group in Figure 8.6). These taxa provide information about the taxonomic composition of local vegetation at the time of site occupation.

Among coastal sites, bedstraw, figwort (*Scrophularia californica*), and poison oak (*Toxicodendron diversilobum*) each comprised >10% of the assemblage from at least one coastal site (Figure 8.6 and Appendix 8.2: Table A3). These taxa are typical members of northern coastal scrub, coastal prairie, and coastal strand vegetation types that characterize the vicinities of these archaeological sites today. Poison oak and figwort most commonly occur in northern coastal scrub vegetation, suggesting there were shrublands in the vicinity of sites CA-SMA-218 and CA-SMA-19, both on Año Nuevo Point. These taxa also occur in forests, particularly along forest margins; however, the sand dune substrate in the vicinity of these sites probably could not have supported forest vegetation. Among inland sites, evidence for shrubland vegetation in the seed assemblage was sparse. Three of the four inland sites contained figwort in low proportions (<1.1%) and at low densities (<1.0 n/l), which could indicate the presence of some shrublands or could represent figwort plants growing among prairies or forests (Appendix 8.2: Tables A2 and A3). Among inland sites, poison oak seeds were only observed at site CA-SMA-113, where they comprised 0.01% of the assemblage. The data suggest poison oak was much less common in the vicinity of these inland sites than at coastal site CA-SMA-19, where it comprised 14.2% of the macrobotanical assemblage. Today, poison oak is abundant in the vicinity of all inland sites.

All inland sites contained low proportions (<3.0%) and densities (<2.5 n/l) of fiddleneck (*Amsinckia* sp.) and phacelia (*Phacelia* sp.). Fiddleneck is associated with prairies, while phacelia can occur in prairies or woody vegetation types. These taxa are either absent or uncommon in the vicinity of all inland sites today, and both genera contain fire-following species (Keeley 1991; Pavlik et al. 1993). It is possible that the paucity of these taxa on the contemporary landscape is connected to a lack of landscape fire since the mid-1900s or before. At the time of site occupation, the abundance of these taxa may have been increased by fires within Native settlements (if these seeds were charred by hearth fires while lying in-situ in soil seed banks) or by more extensive landscape burning (if these seeds were collected incidentally along with seed foods). As both genera produce plants that are typically shorter than and morphologically distinct from common seed food plants, the former explanation seems more likely. A final possibility is that the presence of these taxa could be associated with cultural use of the plants for medicinal or other purposes. However, although species from both of these genera were reportedly used medicinally, the uses documented did not involve seeds (Bocek 1984).

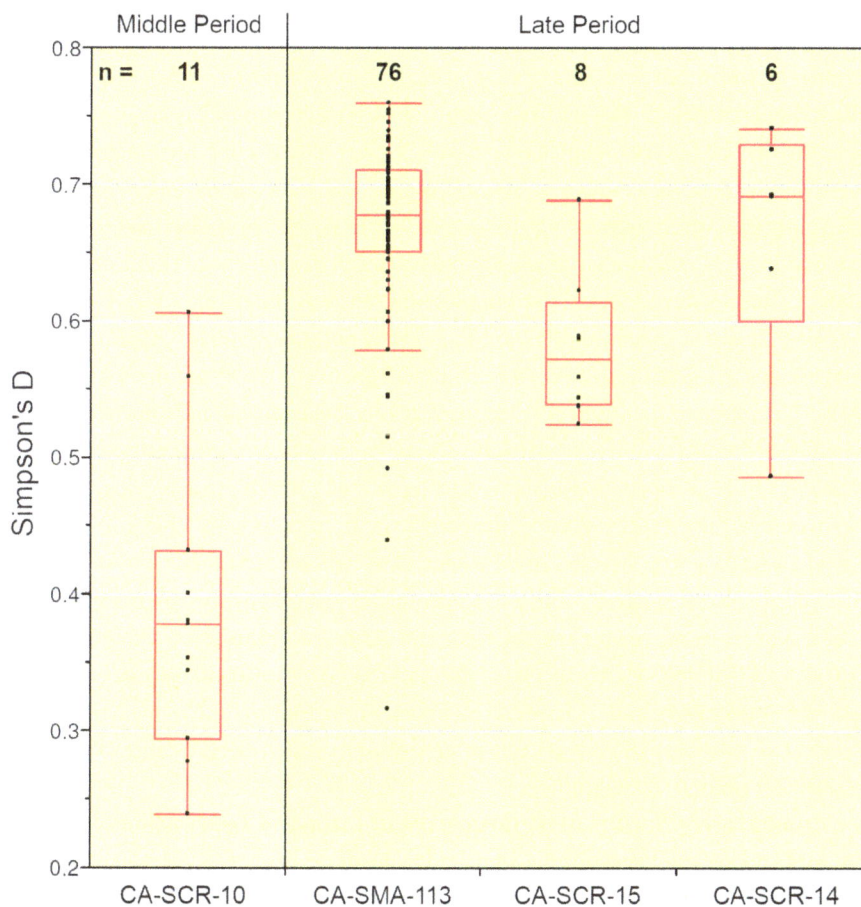

Figure 8.9. Simpson's Diversity Index ("Simpson's D") of Food Plants and Potential Food Plants at Inland Sites. Lower Simpson's D values indicate lower diversity. The following taxa were included in analysis: hazelnut (*Corylus* sp.), tanoak acorn (*Notholithocarpus densiflorus*), oak acorn (*Quercus* sp.), California bay (*Umbellularia californica*), tarweed (*Madia* sp.), panicled bulrush (*Scirpus microcarpus*), clover (*Trifolium* sp.), grasses (Poaceae), and "Cheno. / Mont." (Chenopodiaceae and Montiaceae, including specimens identified as *Calandrinia* sp., *Claytonia* sp., and *Chenopodium* sp.). Each data point represents a flotation sample. Flotation samples with fewer than 20 total identified specimens from potential food taxa were excluded from analysis.

Table 8.7. Results of Tukey HSD Test on Simpson's Diversity Index of Food Plants and Potential Food Plants from Inland Sites. At each statistical significance level (p<0.05 and p<0.01), sites not connected by the same letter are statistically significantly different.

Site	Period	p<0.05		p<0.01		Mean
CA-SMA-113	Late	A		A		0.67
CA-SCR-14	Late	A	B	A		0.66
CA-SCR-15	Late		B	A		0.58
CA-SCR-10	Middle		C		B	0.39

Fragments of wild cucumber (*Marah* sp.) seeds were recovered from all sites except CA-SMA-218, which was also the site with the smallest identified archaeobotanical assemblage (n=29; note that in data tables specimens of *Marah* are listed under "Nutshell" due to the large seed size; Appendix 8.2: Tables A1 and Table 8.1). Wild cucumber, a long-lived geophyte vine, is also known as "man-root" because it develops a large tuber that can reach over 90 kg in weight (Martin 2009). Wild cucumber grows among prairies, shrublands, and forests, and it is a relatively common plant in the research area. Although wild cucumber does not require fire to resprout seasonally or to propagate, it resprouts vigorously in the first two years after a burn (Borchert 2016) and is often considered a fire-follower. Because a medicinal use of this plant common among Tribes in Central and Southern California involved roasting its seeds (Bocek 1984; Martin 2009), some proportion of wild cucumber in the archaeobotanical assemblage of these sites could reflect cultural use, particularly in samples where it was observed in relatively high densities. However, densities of >2 n/l were rarely observed among flotation samples from all sites analyzed (with only two examples from CA-SCR-10 and three from CA-SMA-113; Appendix 8.2: Table A13; Cuthrell 2013b: Appendix N), and only one sample from CA-SMA-113 appeared to strongly indicate cultural use, with a density of >10 n/l. Wild cucumber seeds develop inside a distinctive spiny round capsule; consequently, they would not have been collected incidentally in association with other edible seeds. If wild cucumber in archaeobotanical assemblages reflects incidental charring rather than cultural use, it suggests hearth fire burning within Native settlements may have created a micro-niche, increasing the vigor and incidence of this plant. Of course, it is also possible that the presence of a favorable anthropogenic micro-niche was associated with the development and widespread occurrence of its cultural use as medicine, such that the two possible pathways leading to its archaeological preservation (medicinal vs. incidental) cannot be distinguished.

Comparative Results of Anthracological Data and Implications for Fuel Use and Landscape Vegetation. Compared to all other sites, CA-SMA-216 contained extremely high and highly variable densities of wood charcoal (Figure 8.4). At 2.37 g/l, median charcoal density at this site was 3.7 times higher than at any other site, and the interquartile range of charcoal density (5.56 g/l) was 7.6 times higher than that observed at any other site. A Tukey HSD test on natural log-transformed values of per-sample charcoal density by site showed CA-SMA-216 had statistically significantly higher (p<0.05) charcoal density than all other sites in the assemblage except CA-SMA-19, CA-SMA-18, and CA-SCR-10. The anomalously high wood charcoal density at CA-SMA-216 suggests intensive burning at the site, likely associated with processing of marine food resources. High charcoal density at the site is probably also due to particularly favorable preservation conditions. This site consisted of relatively thin lenses of cultural deposits separated by nearly sterile layers of sand. Presumably, each lens of cultural material represents a short-term (possibly seasonal) episode of occupation, with cultural deposits from each occupation episode quickly covered by a protective layer of sand and remaining in-situ until the time of sample collection.

Among coastal sites, Late Period sites CA-SMA-216 and CA-SMA-19 both displayed higher densities of wood charcoal than Early and Middle Period sites. To some degree, the higher charcoal density observed in the more recent sites is related to taphonomic effects. We expect charcoal preservation over the long term would be relatively poorer at coastal sites than at inland ones due to the abrasive and poorly consolidated sand dune substrate. The relatively greater frequency of wetting and drying cycles associated with higher substrate porosity would also be expected to cause macrobotanical remains to fracture. However, archaeobotanical materials recovered from the well-consolidated and apparently intact deposits from CA-SCR-7 Component A demonstrate that charcoal in coastal contexts that is several thousand years old can be preserved at densities similar to those observed in relatively recent inland sites when conditions are favorable.

Among inland sites, median charcoal density values differed little, ranging from ca. 0.2 to 0.5 g/l. Although median charcoal density at CA-SCR-15 appears quite low compared to other sites in Figure 8.4, this is primarily due to inclusion of data from marginal EU 1, which contained little archaeobotanical or other cultural material and substantially reduced the site's median charcoal density value. When data from EU 1 is excluded, median charcoal density at the site rises to 0.63 g/l, the highest value among inland sites, and the range of median charcoal density values among sites becomes 0.3–0.6 g/l. The low variability in charcoal densities among inland sites indicates general similarities in modes of wood fuel use and in the effects of taphonomy.

Comparative results of anthracological analysis are presented in Figure 8.10 (note that the categories "Sequoia" and "Sequoia cf." from Table 8.2 are consolidated in this figure). CA-SCR-7, the only coastal site selected for anthracological analysis, displayed a highly distinctive charcoal assemblage, with redwood (*Sequoia sempervirens*) comprising the overwhelming majority of specimens (74.7%), along with a small proportion (6.2%) of alder (*Alnus* sp.). As described in supplemental materials (Appendix 8.3, "CA-SCR-7" section), driftwood appears to have been the main source of wood fuel at this site, as it is unlikely that redwood or alder could have grown in proximity to the site, even in the mid-Holocene. A Late Pleistocene to Late Holocene pollen record from Laguna de las Trancas, located ca. 15 km NW of the site, recorded redwood as the dominant source of tree pollen in the local region since the terminal Pleistocene and also confirmed that alder has persisted in the area throughout this time (Cowart 2014). However, it should be noted that the pollen record contained a gap between ca. 6 and 13 kya.

Anthracological results from inland sites are well-suited for comparison, as all sites were occupied within a ca. 1,000-year span between ca. 700 and 1700 CE, and all were located in relatively similar environmental settings along the western margins of the Santa Cruz Mountain foothills and marine terraces. Redwood was proportionally the most abundant taxon at each inland site, and it was much more abundant (by >15 percentage points) than any other taxon at all sites except CA-SCR-10. As noted above, redwood is highly resistant and resilient to fire, and in the local region it would be expected to comprise a major proportion of forest cover in any location with suitable habitat under highly variable fire regimes (Cuthrell 2013b: 156).

Other notable fire-adapted and fire-resilient taxa common among anthracological assemblages were California lilac (*Ceanothus* sp., most likely *C. thyrsiflorus*) and oak (*Quercus* sp., most likely *Q. agrifolia*). California lilac was observed at all sites, representing 4.0–17.3% of anthracological assemblages. This shrub or small tree rarely comprises more than a minor proportion (i.e., more than 1%) of woody vegetation cover on the landscape today, and it is absent in many locations. However, this plant produces fire-refractory seeds that germinate en masse post-fire, and it can become a dominant component of woody vegetation cover in the two to three decades after a fire (Forrestel et al. 2011). Oak was observed at all inland sites except CA-SMA-113, and in all cases it comprised small proportions of the wood fuel assemblages (2.0–7.5%). Mature coast live oak develops thick bark that makes it relatively fire-resistant, and it is able to survive through relatively high-intensity fires that cause total canopy mortality and substantial branch charring (Dagit 1995, 2002; Holmes et al. 2008).

Among other taxa observed in anthracological assemblages, alder (*Alnus* sp.), willow (*Salix* sp.), and California bay (*Umbellularia californica*) represent genera that typically occur within riparian forest corridors, although California bay also occurs outside of riparian zones. These taxa were most abundant at site CA-SCR-10, with each comprising 8.0–15.0% of the anthracological assemblage (Figure 8.10 and Table 8.2). From among taxa at CA-SMA-113 and CA-SCR-15, only alder comprised more than 5% of anthracological assemblages (8.2% and 7.9%, respectively), and none of these taxa were observed at site CA-SCR-14. CA-SCR-10 also displayed the richest anthracological assemblage among sites analyzed, with 16 taxa identified to genus or species level, while all other sites contained 10 or fewer taxa.

Differences in the proportional representation of riparian taxa and in assemblage richness between sites may be associated with the amount of redwood in proximity to sites. Based on contemporary observations, at lower elevations and in proximity to the coast, redwood is generally restricted to riparian corridors. In these areas, it is observed occasionally as a minor component of riparian forest cover. Moving inland and up into higher elevations, redwood becomes the dominant riparian corridor tree species and sometimes the dominant taxon on the broader landscape. Historic logging in concert with fire suppression has likely substantially reduced the overall cover of redwood on the contemporary landscape (and specifically the proportional cover of redwood in relation to Douglas fir); and it is probable that redwood was a more widespread and dominant component of forests in the local region prior to Euro/American colonization than it is today.

All inland sites analyzed in this study were located along the margins of riparian corridors. At all sites except CA-SCR-10, redwood would likely have grown either in the immediate vicinity of the site or within a few hundred meters (see supplemental Appendix 8.3, "CA-SCR-10" section). This could explain the substantial differences observed between the proportional representation of redwood at CA-SCR-10 (16%) and other sites (>30%). Pairwise chi-square comparisons of redwood specimens vs. other charcoal specimens between CA-SCR-10 and all other inland sites returned statistically significant results in each case (p<0.02). Residents of CA-SCR-10 probably would have had to travel >500 m inland to acquire redwood, and they may have alternatively or supplementally obtained redwood in the form of driftwood from the coastline, located ca. 500 m to the south.

Douglas fir was a minor component of anthracological assemblages (1.0–6.4%) at all inland sites except CA-SCR-14, where it comprised 15.0% of charcoal specimens and was the second most abundant taxon observed after redwood. Pairwise chi-square comparisons of Douglas fir specimens vs. other charcoal specimens between CA-SCR-14 and all other inland sites returned highly statistically significant results in comparison to CA-SCR-10 and CA-SCR-15 (p<0.01) and marginally significant results in comparison to CA-SMA-113 (p=0.05). Proportional representation of Douglas fir in comparison to redwood is likely related to fire regime. Douglas fir trees are much more susceptible to fire mortality than redwoods, particularly when young. Consequently, as fire return intervals in forests become shorter, we would expect to observe fewer Douglas firs relative to redwood trees.

Today, due largely to historic redwood logging and fire suppression, Douglas fir is relatively common among forests covering foothills west of the Santa Cruz Mountains, even forming multi-hectare monocultures near Año Nuevo Point (Cuthrell 2013b). The low representation of Douglas fir at most inland sites, along with strong representation of the fire-adapted species redwood and California lilac, strongly indicates a change in fire regime between the time of site occupation and today. Prior to Euro/American colonization, relatively frequent prescribed burning by Native peoples probably kept proportions of Douglas fir on the landscape low while expanding the environmental niche for California lilac and reducing competition for fire-adapted redwood. The relatively high proportion of Douglas fir charcoal at CA-SCR-14 could indicate that there were some locations on the broader landscape that burned infrequently enough to support higher proportional cover of Douglas fir, or that site inhabitants targeted young Douglas fir trees killed by landscape fires as a fuel source. Potential explanations for the differences in charcoal assemblages observed between CA-SCR-14 and CA-SCR-15, located in close proximity to one another, are considered below.

Discussion

Subsistence Practices

Archaeobotanical Seed Representation and Seed Consumption. Given the Mediterranean climate of California's Central Coast region, nearly all types of seeds produced by the regional flora must be charred if they are to be preserved in the archaeological record. Any time site inhabitants built a fire, it is expected that some seeds would have become charred incidentally through mechanisms unrelated to the cultural use of plants, or specifically to their use as food. Examples include charring of seeds in soil seed banks in contact with fires, seeds carried into fires by wind, and seeds incidentally contained in fine fuels. Consequently, occasional representation of taxa and/or regular representation in very low densities does not necessarily indicate intentional cultural use. In evaluating whether any particular taxon observed in an archaeobotanical assemblage results from incidental charring or indicates cultural use, it is necessary to consider the ecology, seed dispersal mechanism, modes of processing, and ethnographically reported uses of the plant. With respect to seed consumption, in some cases it is relatively clear whether a taxon's representation can be attributed to its use as a food, while in other cases it is equivocal.

Seeds of grasses (Poaceae; particularly relatively large-seeded grasses, such as *Bromus* spp., *Elymus* spp., *Stipa* spp., *Phalaris* spp.) and coast tarweed (*Madia sativa*) are well-documented ethnographically as food sources, and these taxa were consistently observed in densities that provide clear evidence for seed food processing at all inland sites. Historic and ethnographic accounts from the region describe how Native peoples roasted grass and forb seeds by tossing them in shallow baskets containing hot coals (Margolin 1989: 88). After roasting, seeds could be used in a variety of ways, such as in making atole (a soup or gruel), pinole (seed meal, sometimes held together by oils in seeds), or possibly other types of foods such as acorn bread. Based on the author's observations of Amah Mutsun Tribe members roasting seeds today, each time a basket of seeds is roasted in this way a small proportion of seeds will become charred. Any subsequent loss of roasted seeds would be expected to include some charred seeds that could become preserved archaeologically. It seems likely that this was the predominant mode through which seed foods entered the archaeological record. Another potential mode of seed food charring is intentional burning of small quantities of seed foods as an offering prior to consumption, a tradition practiced by contemporary Amah Mutsun Tribe members. However, it is unclear whether Quiroste and Cotoni Tribes, for whom we have very few historic accounts and no ethnographic accounts regarding food use, carried out similar practices.

Among inland sites examined in this study, it is equivocal whether seeds of clover (*Trifolium* spp.) and "Cheno. / Mont." (Chenopodiaceae and Montiaceae, the goosefoot and miner's lettuce families) should be interpreted as evidence for seed consumption. In California, clover seed consumption has been reported among Cahuilla, Mendocino, Paiute, and Luiseño groups (Moerman 2020), but this practice was not reported among Ohlone groups (use of clover as a greens food was reported; Bocek 1984). It should be noted that Ohlone ethnographic informants were interviewed in the 1920s–1930s, after nearly 150 years of Euro-American colonization efforts in the region, which sought to destroy Indigenous ways of life and traditional practices (see Milliken 1995; Rizzo 2016), and many pre-Contact food practices had been lost or become dormant by that time. In addition, informants interviewed by John P. Harrington and other ethnographers in the early 1900s were members of Tribes who spoke Mutsun, Rumsen, and Chochenyo languages. None descended from Quiroste or Cotoni Tribes, who spoke Awaswas and possibly other southern San Francisco Bay languages, and who may have had distinct foodways practices from Tribes who are better represented ethnographically. Clover

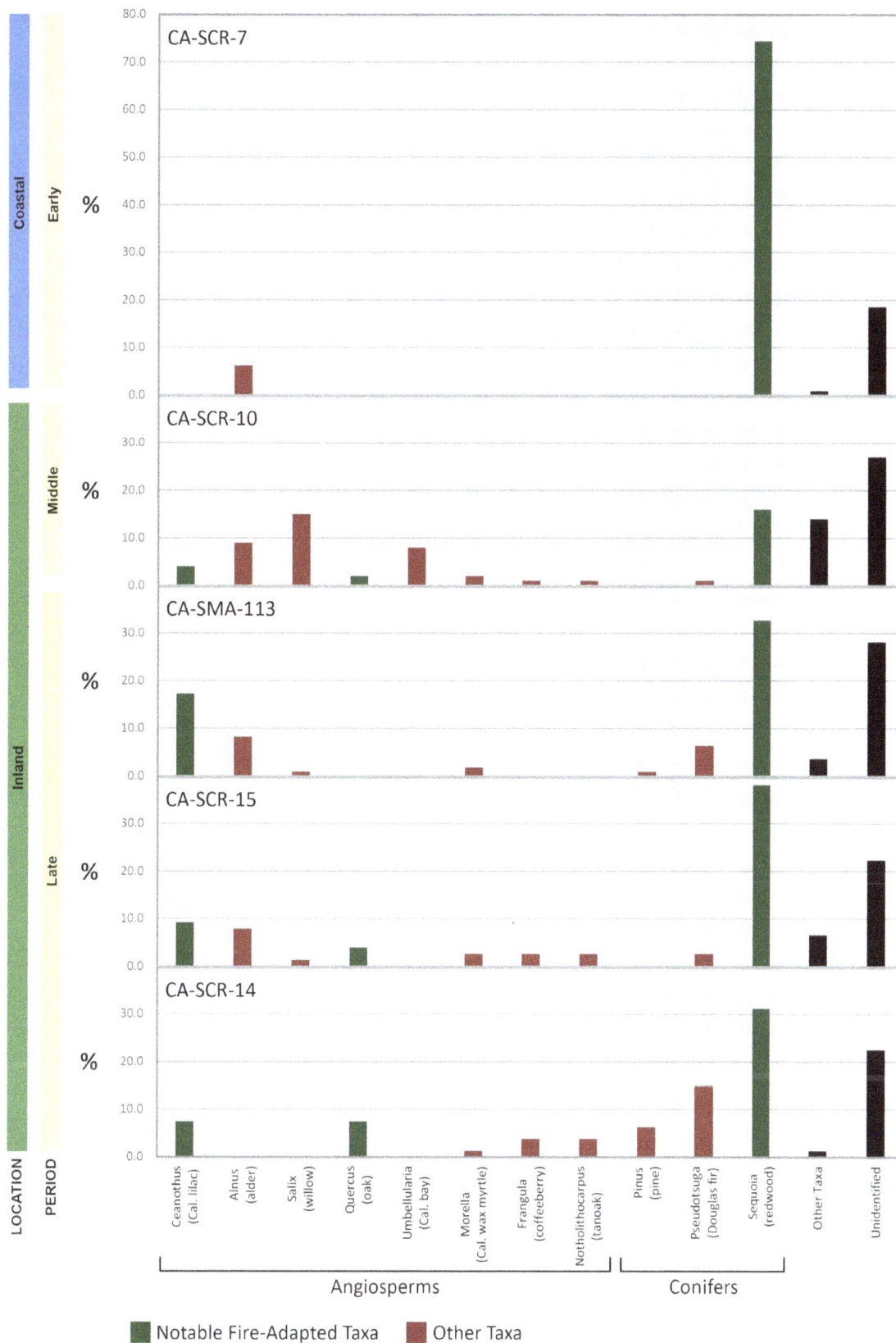

Figure 8.10. Inter-site Comparison of Percentages of Wood Charcoal Taxa. This graph displays taxa representing >1.0% of the overall project assemblage (5+ specimens). "Notable Fire-Adapted Taxa" are highly resilient to fire or display fire-refractory germination.

seeds were consistently observed in moderate densities among inland sites, with overall density values at each site ranging from 3.5 to 6.7 n/l. At CA-SMA-113, 17 of 77 flotation samples (22.1%) contained clover densities >10 n/l, with a maximum density of 41.2 n/l (Cuthrell 2013b: Appendix N). Among other inland sites, CA-SCR-15 contained a single flotation sample with clover seed density >10 n/l (#015-0006-04; 15.0 n/l; Appendix 8.2: Table A19). The data from CA-SMA-113 presents relatively compelling evidence for use of clover seeds as food, but the data from the three inland sites in Santa Cruz County is much weaker. While it is reasonable to conclude that clover seeds may have been consumed by inhabitants of all inland sites, the data from sites other than CA-SMA-113 could potentially indicate incidental charring of clover seeds in seed banks or charring of seeds in association with other food uses, rather than intentional clover seed processing and consumption.

The Cheno. / Mont. group of seeds includes specimens from taxa such as miner's lettuce (*Claytonia* spp.), red maids (*Calandrinia* spp.), goosefoot (*Chenopodium* spp.), and possibly others (e.g., *Atriplex* spp.). Charred examples of these seeds are often fragmentary and/or warped, making them difficult to identify to genus level. Among California Tribes, consumption of red maids and goosefoot seeds was ethnographically attested by multiple Tribes, while miner's lettuce was exclusively reported as a greens food (Moerman 2020). Red maids, also a potential greens food, was a favorite traditional seed food of local Amah Mutsun ethnographic informants who worked with John P. Harrington in the 1920s–1930s (Harrington 1921–1939: Reel 58). As described above, the Cheno. / Mont. group contains seeds that have the potential to preserve archaeologically without charring, but because it is not possible to differentiate uncharred archaeological seeds from seeds deposited naturally post-occupation, uncharred seeds were not included in the analysis presented here (but see data tables in Appendix 8.2). Cheno. / Mont. seeds were consistently observed in low densities (<2 n/l) at all inland sites and in Late Period coastal sites CA-SMA-216 and CA-SMA-19. Among all sites, only one flotation sample from CA-SMA-216 appeared to clearly indicate seed processing, with a density of 11.1 n/l (all *Claytonia* sp.; Sample #02-0004-000; Appendix 8.2: Table A7). Cheno. / Mont. seeds in this sample were identified as miner's lettuce, which produces edible seeds but has not been reported ethnographically as a seed food in California. This sample could indicate miner's lettuce seed consumption, or these seeds could be associated with use of the plant as a greens food, if site inhabitants harvested entire plants and later plucked and discarded seed-bearing inflorescences into fires. Interestingly, this flotation sample also contained relatively high densities of grass seeds and bedstraw (*Galium* sp.) seeds (18.9 n/l and 23.3 n/l, respectively; Appendix 8.2: Table A7). Grasses are a potential seed food, but bedstraw is not. The high density of bedstraw along with potentially edible seeds raises the possibility that this context could represent intentional burning of a large amount of fine fuels, possibly as tinder or for vegetation disposal. Generally, representation of Cheno. / Mont. seeds among sites analyzed in this project cannot be attributed with certainty to either intentional cultural uses or to incidental charring.

Paniced bulrush seeds were recovered in unusually high densities at site CA-SMA-113 (24.1 n/l overall; Appendix 8.2: Table A2), which Cuthrell (2013b) interpreted as a probable indicator of seed consumption. Although seeds of paniced bulrush are very small (<1.0 mm), they occur in large seed heads that could be collected relatively easily. Sedges/bulrushes, and particularly tule (*Schoenoplectus acutus* var. *occidentalis*), were commonly used by Tribes in Central California for construction and crafting. Sedge/bulrush seed consumption was not ethnographically or historically recorded in the territory of Ohlone/Costanoan language speakers; however, consumption of rhizomes was recorded in this area (Bocek 1984), and sedge seed consumption has been recorded among Klamath, Northern Paiute, and Cahuilla groups (Moerman 2020). The presence of panicled bulrush seeds in three coastal sites that lacked evidence for seed food processing (CA-SMA-18, CA-SMA-19, and CA-SMA-218) suggests seeds of this taxon could become incorporated into archaeobotanical assemblages as a result of crafting, construction, or other uses. However, inflorescences (seed heads) probably would not have been intentionally harvested or used for crafting or construction purposes, and so we would not expect such uses to produce high densities of charred seeds. Densities of panicled bulrush seeds were uniformly low at coastal sites (<1.0 n/l), as well as at inland site CA-SCR-10. However, inland sites CA-SCR-14 and CA-SCR-15 displayed moderate densities of panicled bulrush seeds (5.9 n/l and 3.7 n/l, respectively). It is unclear whether these density levels could result from non-food uses or whether the data indicate seed consumption. As an exclusively wetland plant, panicled bulrush probably could not have occurred naturally at these sites.

Plant Food Use at Coastal and Inland Sites. Archaeobotanical data show clear differences in the use of plant foods between coastal and inland sites. Coastal sites consistently lacked evidence for seed food processing and for regular nut food processing. The presence of nutshell from edible taxa that could not have grown in the vicinity of coastal sites demonstrates that site inhabitants occasionally processed and consumed nuts, but nut representation among coastal sites was sporadic and sparse. All coastal sites on Año Nuevo Point contained indirect evidence (i.e., charred cone fragments) or direct evidence (i.e., charred nuts) for consumption of pine (*Pinus* sp., likely *P. radiata*; or less likely knobcone pine, *P. attenuata*) nuts. Monterey pine nuts are unlikely to preserve archaeologically because the nutshells are very thin and nuts are high in oil content, leading to a high degree of deformation during charring and a greater likelihood of total combustion to ash. Evidence for Monterey pine nut processing was strong at Middle Period sites CA-SMA-218 and CA-SMA-18, where pine cone scales were noted in every flotation sample. At CA-SMA-18, pine cone fragments even comprised the majority of "wood" charcoal in every flotation sample. At Late Period coastal sites CA-SMA-216 and CA-SMA-19, pine nuts or cone fragments were only observed in a single flotation sample from each site.

All inland sites displayed moderate to high densities of prairie seed foods, with samples from intact portions of each site often containing ca. 25–150 n/l edible seeds, clearly indicating seed food processing. Inland sites also had low to moderate densities of edible nuts, with most samples from each site containing ca. 2–40 n/l nutshell specimens (or ca. 30–200 mg/l). Because it is not possible to control for potential cultural differences in nut and seed processing practices between temporal periods or between contemporaneous Tribes, nor for differences in preservation between sites, inferences about the relative dietary importance of seed vs. nut foods based on comparison of abundances of each category of remains between sites should be made with caution. With this caveat in mind, the difference in the seed-to-nut ratio at CA-SMA-113 (21.4) and the three inland sites analyzed in this project (2.8–8.3; Figure 8.5) seems too large to result from differences in preservation alone—given that all inland sites were occupied within a ca. 1,000-year span—and thus more likely indicates a distinction in cultural practices. While the data could indicate that CA-SMA-113 inhabitants consumed fewer nuts relative to seeds, it could also indicate a difference in processing location of nuts (e.g., in the frequency of nut processing at logistical sites vs. residential sites) or a difference in nutshell disposal practices (e.g., in the frequency of disposal of nutshell into fires vs. other modes of discard). Exploring these possibilities further would require additional research at contemporaneous sites located at higher elevations and farther inland, where most people would have likely resided in the fall and winter months during and after the nut harvesting season.

Among the three inland sites analyzed in this project, Middle Period site CA-SCR-10 displayed the clearest and most consistent evidence for seed food processing, with 11 of 12 flotation samples analyzed containing >35 n/l (and up to 121.8 n/l) edible seeds. CA-SCR-15 and CA-SCR-14 also produced a subset of samples with approximately the same range of edible seed density, but the data from these sites was more variable because the two excavation units collected at each site differed substantially in their representation of cultural activities. At each site, one excavation unit was located in a relatively marginal cultural context, with distinctly lower densities of most or all categories of ecofacts and artifacts than the other unit (at both sites, EU 1 was the more marginal unit). This reduced the number of flotation samples from data-rich contexts representing food processing or food waste discard areas to 5 at CA-SCR-15 and 4 at CA-SCR-14. At all three inland sites analyzed in this project, sample size was relatively low compared to CA-SMA-113, where 77 flotation samples from data-rich contexts were analyzed, providing clear and consistent evidence of seed food processing over a ca. 350-year period of the site's occupation (Cuthrell 2013b). Although the archaeobotanical assemblages at all three inland sites analyzed in this project presented strong evidence for seed food processing, the small number of excavation units and flotation samples collected prevented evaluating how well the data characterizes each site as a whole. The small sample size also constrained interpretive potential because the data sets produced could not be evaluated through some parametric statistical tests—or if they could, large error ranges resulting from small sample sizes reduced the statistical significance of results.

Despite the relatively small sample sizes, the archaeobotanical assemblages of all three inland sites examined in this study indicated consumption of seed foods, particularly seeds of grasses. Tarweed consumption appears probable at CA-SCR-14, and consumption of tarweed, clover, and panicled bulrush seeds may have occurred at all sites. Seed food use is a proxy for the presence of prairies (or wetlands in the case of panicled bulrush), which must have existed within the catchment area of these sites at the time of occupation. Nuts were apparently processed more frequently at inland sites relative to coastal sites, and the three inland sites analyzed in this study suggest greater use of nuts than was observed at site CA-SMA-113. At coastal sites, inhabitants primarily focused on harvesting marine resources, but in some cases, they also processed Monterey pine nuts from the historic local pine forest. At inland sites, inhabitants harvested a more diverse suite of plant foods from prairies, forests, and possibly wetlands.

Temporal Changes in Plant Food Use. Among coastal sites, densities of nut foods and potential seed foods were consistently very low during all time periods. The only potential temporal difference in plant food use was related to indirect evidence for the processing of Monterey pine nuts among sites on Año Nuevo Point. At Middle Period sites CA-SMA-18 and CA-SMA-218, pine cone fragments were observed in all flotation samples, and at the former site, these comprised the majority of charcoal specimens. However, at Late Period sites CA-SMA-19 and CA-SMA-216, evidence for pine nut processing was sparse, with pine cone fragments or pine nuts observed in only one flotation sample from each site. As all four sites are located within ca. 1 km of one another, site catchment areas would have largely overlapped, and so this difference in pine nut use cannot be attributed entirely to site location.

Temporal differences in pine nut use could be explained by a change in the extent of the historic Monterey pine forest between occupation of Middle Period site CA-SMA-18 (ca. 550–850 CE) and Late Period site CA-SMA-216 (ca. 1300–1650 CE), or the difference could reflect changes in cultural practices related to resource use at coastal sites between the two periods. The earliest documentary account of pine forests in this area was recorded by Fr. Juan Crespi during the first Spanish expedition through the area, led by Gaspar de Portolà in 1769. Crespi observed "a small, very dense grove of pine-nut pinewoods dropping down through some knolls from the mountains" (Brown 2001: 577) on November 23, 1769, the day on which the party traveled from Waddell Creek to Quiroste Valley and passed through the area containing the contemporary Monterey pine forest on Año Nuevo Point. Although this passage appears most likely to refer to the Monterey pine forest on Año Nuevo Point, due to ambiguity in the organization of Crespi's text, it could alternatively indicate some other pine grove located in closer proximity to Quiroste Valley. However, this seems unlikely, as pines are rare to absent in proximity to Quiroste Valley today, and there are no indications that pines were historically more abundant there. Additionally, Late Period site CA-SMA-113 in Quiroste Valley displayed no evidence for pine nut processing or use and very little evidence for the use of pine wood as fuel (pine comprised 0.9% of anthracological specimens, see Table 2; Cuthrell 2013b).

A historic vegetation map from 1935 shows Monterey pine forest located just to the east of Highway 1, with forested areas surrounded by cultivated land (Figure 8.11). Today, this forest is part of Año Nuevo State Park, and areas that were cultivated in the early 1900s have been undisturbed for several decades, permitting natural vegetation succession to occur. As a result, the Monterey pine forest now extends ca. 200 m west of Highway 1, suggesting it could be expanding back into areas where pine forests were cleared for cultivation in the nineteenth and twentieth centuries. The contemporary forest is located ca. 1.5–2.8 km from coastal sites analyzed in this study, well within their probable catchment area. Given the considerations presented above, it seems most probable that this Monterey pine forest was present within the catchment areas of Late Period sites CA-SMA-19 and CA-SMA-216 when they were occupied. Consequently, the relative lack of evidence for pine nut processing at these sites could relate to increased specialization in resource acquisition activities at coastal sites in the Late Period, with site inhabitants focusing almost exclusively on collection and processing of marine resources during this time.

The three Late Period inland sites considered here had overlapping occupations during the period from ca. 1000 to 1400 CE. Relatively small sample sizes in concert with limited chronological control prevent evaluation of temporal changes within and

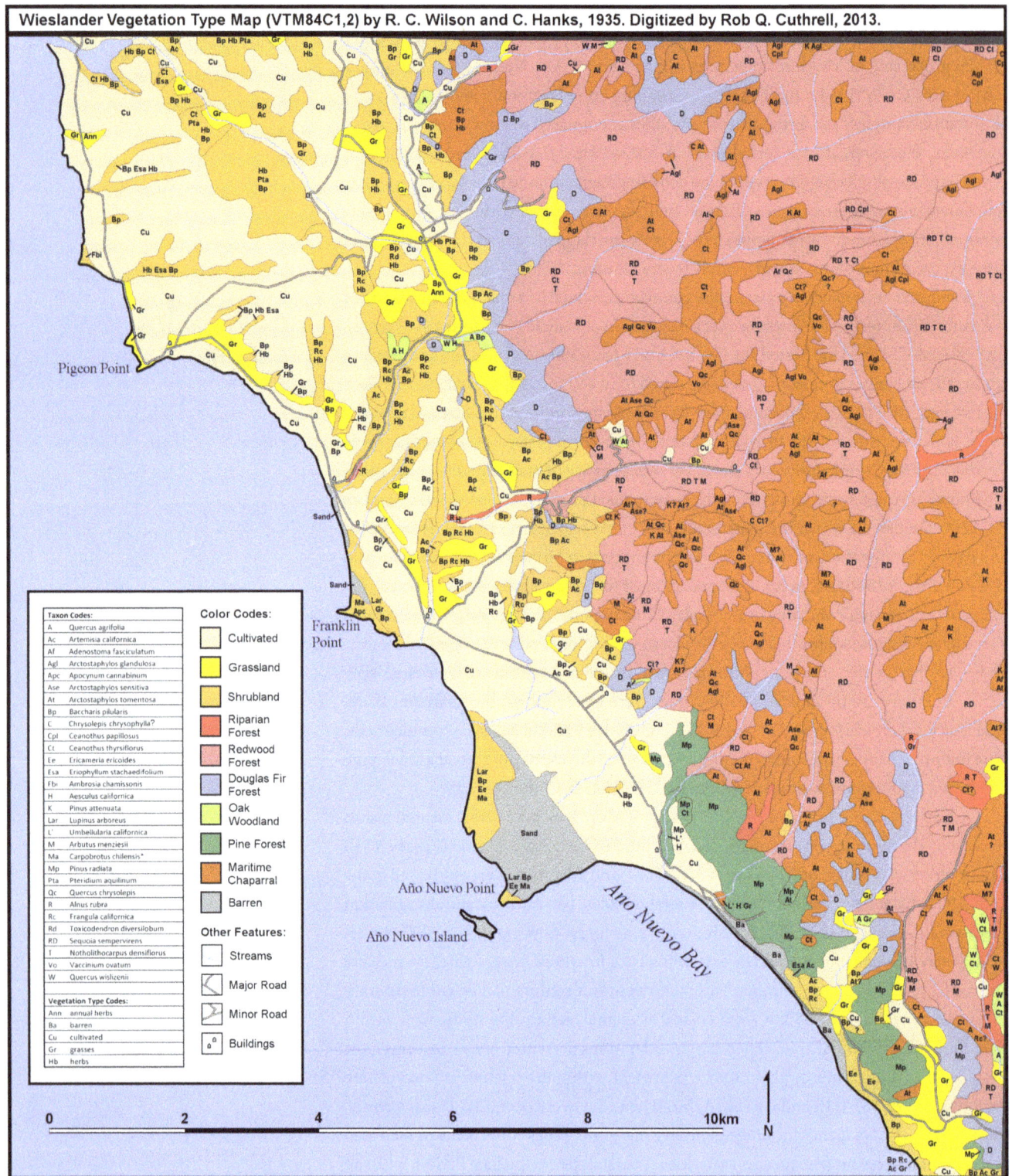

Figure 8.11. Map of 1935 Vegetation in the Vicinity of Año Nuevo Point. This map was originally produced by R. C. Wilson and C. Hanks (Wilson and Hanks 1935) in association with the Wieslander Vegetation Type Mapping Project (WVTM 2013). This digitized version of the map was reproduced from Cuthrell (2013b). Note: Some color codes were modified from the original map. The original map contained elevation data and contours, which are not reproduced here. Ambiguous taxon codes are indicated with a question mark. Areas with coast live oak (*Quercus agrifolia*) or interior live oak (*Quercus wislizeni*) listed as the most common taxon are categorized as oak woodlands, while areas with canyon live oak (*Quercus chrysolepis*) are categorized as maritime chaparral.

among these Late Period sites. However, Middle Period contexts sampled at CA-SCR-10 consistently returned dates within a prior span between ca. 700 and 1000 CE, presenting a favorable data set for comparison to the three Late Period sites. As described above, CA-SCR-10 displayed a statistically significant lower diversity of plant foods than Late Period sites. Compared to inhabitants of Late Period sites, people at CA-SCR-10 seem to have placed relatively greater emphasis on harvesting grass seeds and hazelnuts. At Late Period sites, additional taxa were apparently consumed more frequently. Although this distinction in diversity was highly statistically significant, it was not strongly pronounced qualitatively. The archaeobotanical assemblage at CA-SCR-10 contained examples of every type of edible nut observed at Late Period sites, as well as densities of clover, tarweed, and Cheno. / Mont. comparable to Late Period sites.

Vegetation, Fire Ecology, and Prescribed Burning

In California's Central Coast region, natural lightning fires are rare, with natural fire return intervals expected to range from ca. 75 to 100 years or more (Cuthrell 2013b). On foothills and marine terraces west of the Santa Cruz Mountains, vegetation succession would result in landscapes dominated by shrublands and forests under a natural fire regime. Prior research at Quiroste Valley produced strong and consistent multi-proxy evidence (including archaeobotanical, anthracalogical, and phytolith data) for the presence of prairies over the last ca. 1,000–1,500 years (Cuthrell 2013b; Lightfoot et al. 2013).

Archaeobotanical assemblages from coastal sites analyzed in this study primarily contained seeds of taxa associated with prairies, northern coastal scrub, and coastal strand vegetation types. Because no coastal sites presented strong evidence for seed processing and consumption, it is most likely that the majority of archaeobotanical remains in these assemblages were charred incidentally, and therefore represent taxa occurring in the immediate vicinity of sites. Near-coastal landscapes are subjected to ecological factors absent from inland landscapes, such as stronger prevailing winds, salt spray, shifting dune fields, and less fertile, sandier soils. As a result, near-coastal areas would be expected to display higher variability in natural vegetation succession outcomes (i.e., in relative proportions of major vegetation types such as prairie, shrubland, and coastal strand) than inland sites. Due to these considerations, archaeobotanical data from coastal sites has little potential to inform inferences about the fire regime at the time of site occupation.

All inland sites examined in this project presented clear evidence for processing of prairie seed foods. Due to the small number of excavation units and flotation samples collected at each site, it is uncertain how well the data generally characterizes each site either spatially or temporally. However, the data suggests that inhabitants of these sites accessed prairies that were large enough to produce seed foods in quantities suitable for consumption, which would have required much more extensive prairies than would be produced by anthropogenic disturbances other than prescribed burning (e.g., such as small-scale disturbances within and in the immediate proximity of habitation sites). The prairie seed food data is consequently consistent with the presence of an anthropogenic fire regime characterized by short fire return intervals, which would have expanded and maintained more extensive prairies than would have occurred in these locales under a natural fire regime. Significantly, Middle Period contexts at CA-SCR-10 dating from ca.

700 to 1000 CE suggest prescribed burning may have been employed prior to the Late Period (which began ca. 1000 CE). This finding is consistent with phytolith data from Quiroste Valley (Cuthrell 2013b: 237–240).

Anthracological data on wood fuel use at inland sites was also consistent with an anthropogenic fire regime characterized by short fire return intervals. At all Late Period sites (including CA-SMA-113), the fire-compatible redwood tree was proportionally the most common wood fuel, comprising 30–40% of anthracological assemblages. Fire-compatible California lilac and oak together contributed another ca. 15–20% of charcoal specimens in each assemblage, while fire-susceptible Douglas fir was generally uncommon, except at site CA-SCR-14, where it was 15% of the assemblage. The presence of California lilac at each site, and the observation that it was the second most common wood fuel observed at CA-SMA-113 and CA-SCR-15, strongly indicates prescribed burning. As described above, in the local region, this plant tends to be uncommon or absent in landscapes that have not burned for many decades, but its fire-refractory seeds can produce a population explosion in the first two decades after a fire (Forrestel et al. 2011). The relatively higher proportion of Douglas fir observed at CA-SCR-14 could indicate that some forested areas burned infrequently enough to permit stands of Douglas fir to reach maturity and develop thick fire-resistant bark that would then allow them to persist through subsequent low-intensity burns; or it could reflect harvesting of wood fuel from cohorts of Douglas fir saplings and immature trees killed by prescribed fires. Considered together, the data from all Late Period sites suggests lower proportional cover of Douglas fir on the landscape than is present today, after over 150 years of redwood timber harvesting and nearly a century of fire suppression. At all Late Period sites, northern coastal scrub shrub taxa were uncommon or absent, suggesting that either site inhabitants avoided harvesting shrub fuels or that this vegetation type was less extensive at the time of site occupation than it is today.

In Middle Period contexts at CA-SCR-10, redwood was also the most proportionally abundant wood fuel observed, but it was closely followed by willow and other riparian taxa such as alder and California bay. Redwood probably would not have grown in proximity to CA-SCR-10, and site inhabitants apparently gathered firewood from taxa occurring in the riparian corridor adjacent to the site. Notably, Douglas fir grows in close proximity to the site today (within ca. 200 m), but it comprised only 1% of the anthracological assemblage. Wood fuels at CA-SCR-10 also included a more diverse suite and a larger proportion of shrubs than Late Period sites. This could indicate that trees were relatively less abundant near CA-SCR-10 than at Late Period sites, leading inhabitants to harvest finer shrub fuels instead.

Anthracological assemblages of Middle and Late Period sites display a diversity of woody taxa that would be expected to occur in proximity to sites. The data is compatible with a model in which regular prescribed burning kept the proportional cover of fire-susceptible Douglas fir and shrubs relatively low, making riparian taxa and fire-compatible trees the most abundant fuel sources on the landscape. The data do not indicate strong preferences for wood fuel selection among site inhabitants. Contemporary AMTB members report that hardwoods such as oak, manzanita, and madrone are preferred for fuel because their wood burns hotter and longer than that of conifers (Lopez 2020, personal communication). Wood fuel analysis indicates that oak produces about 40–80% more heat than redwood when quantified by volume (Corder 1973; World Forest Industries 2020). Accordingly, redwood was probably used by archaeological site inhabitants because it was relatively abundant—rather than because it was a preferred fuel type.

Phytolith data from soils in proximity to sites analyzed in this project displayed weaker evidence for the presence of long-term grasslands than soils in proximity to Late Period site CA-SMA-113 in Quiroste Valley (Cuthrell 2013b; Evett and Cuthrell 2013). In Quiroste Valley, near-surface soils (ca. 5 cm depth) from a transect of seven sampling columns displayed phytolith content values ranging from 0.9 to 1.7% and grass short cell densities ranging from 101 to 1967k n/g (Cuthrell 2013b: Tables D1 and D.3). Five of the seven sampling columns from Quiroste Valley displayed very high densities of grass short cell morphotypes, with values of >1000k n/g. Among sites where phytolith analysis was carried out in this study (CA-SCR-7, -10, -14, and -15), at least one soil sample collected in proximity to each site contained phytolith content of >0.8%, or about the same range as near-surface soils in Quiroste Valley (Figure 8.2). However, grass short cell phytolith density at all sites analyzed in this study was uniformly much lower than the values observed in Quiroste Valley, with values typically ranging from ca. 80 to 210k n/g (Figure 8.3), or about 10–20% of the values typically observed among Quiroste Valley soil samples. The data indicate that phytolith-producing vegetation other than grasses, such as sedges (Cyperaceae), horsetails (Equisetaceae), or other unknown plants, may have comprised a relatively larger proportion of landscape vegetation in these areas over the long term than at Quiroste Valley.

The relatively lower phytolith content and grass short cell density values observed at CA-SCR-7 are likely attributable to the long-term presence of an active dune field in the area. Other distinctions in geomorphological history could also account for differences observed between soils in Quiroste Valley and soils associated with sites analyzed in this project. At Quiroste Valley, geomorphological research established that the valley floor (from which all phytolith samples were collected) was formed ca. 3,000–4,000 years ago and has remained relatively stable since that time. However, in the vicinity of sites CA-SCR-10, CA-SCR-14, and CA-SCR-15, similar information about geomorphological history is not available. Consequently, phytolith pools in soils around these sites cannot be temporally bounded. While soils in proximity to these three sites displayed modest evidence for the presence of grasslands over the long term, the data is less conclusive than the data from Quiroste Valley. Although the phytolith data is more consistent with the presence of Indigenous prescribed burning over the long term than with its absence, the data does not strongly support or refute either interpretation; this is due to relatively low observed densities of grass short cell phytoliths, lack of geomorphological context, and consequent temporal uncertainty.

Conclusion

This comparative diachronic study of plant use at coastal and inland sites west of the Santa Cruz Mountains has produced new information about site use, subsistence practices, landscape vegetation, and prescribed burning. During all time periods, coastal sites were used primarily for harvesting and processing marine resources, with limited evidence for plant food processing. Middle Period coastal sites on Año Nuevo Point consistently displayed indirect evidence for pine nut processing in the form of abundant charred cone fragments, while Late Period coastal sites lacked such evidence, possibly indicating increasing specialization in site activities. In contrast to coastal sites, Middle and Late Period inland sites were places where people acquired and processed a broad suite of marine and terrestrial resources, including a variety of seed and nut foods. All inland sites contained relatively low densities of archaeobotanical nut remains in comparison to seed food remains. Possibly, nuts were primarily processed at seasonal habitation sites located farther inland and at higher elevations in the Santa Cruz Mountains.

While proportional representation of seed and nut foods varied somewhat between inland sites, grass seeds were consistently the most abundant type of plant food observed. The high density and proportional representation of the potential seed food panicled bulrush observed at site CA-SMA-113 in comparison to sites in Santa Cruz County could indicate a difference in local access to wetland resources, or a distinction in foodways practices between historic Quiroste and Cotoni Tribes. Diversity of plant foods was highly statistically significantly lower at Middle Period site CA-SCR-10 than at all Late Period sites, with inhabitants of the earlier site placing greater emphasis on consumption of grass seeds and hazelnuts. However, because all samples at CA-SCR-10 were collected from a single excavation unit, the data may not characterize the site generally.

Inland sites in the terminal Middle Period and Late Period consistently displayed evidence for prairie seed food processing, and fire-compatible tree and shrub species such as redwood and California lilac comprised the largest proportion of wood fuel taxa observed at all sites. Together, these data sets indicate that Indigenous people in the local region probably employed prescribed burning during the Late Period, and that inhabitants of northern Santa Cruz County probably used prescribed burning in the terminal Middle Period. In the absence of anthropogenic burning, it is unlikely that prairies would have been extensive enough in areas west of the Santa Cruz Mountains to provide a regular source of food for site inhabitants. Additionally, fire-susceptible Douglas fir would be expected to comprise much larger proportions of wood fuel assemblages than was observed, and fire-adapted California lilac would have been much rarer. Due to the limited sample sizes collected from sites excavated in Santa Cruz County, these interpretations are not supported as robustly as at site CA-SMA-113 in Quiroste Valley, where excavations were more extensive and sample sizes were much larger.

Soils in proximity to sites in Santa Cruz County contained proportional phytolith content within the range that previous researchers have proposed indicates the presence of long-term grasslands, but direct evidence for grasses in the form of diagnostic grass phytoliths was relatively weak. At Early Period coastal site CA-SCR-7, the lack of phytoliths in soils around the site is probably attributable to geomorphological history, as the dune field that once covered the area may have been incapable of supporting long-term grasslands, and any phytoliths produced would have been subject to wind transport away from the area along with sand dune matrix. At all other inland sites in Santa Cruz County, densities of diagnostic grass phytoliths represented only a small fraction of the values observed in proximity to site CA-SMA-113 in Quiroste Valley. These differences could be attributable to distinctions in paleovegetation composition, geomorphological history, or both. The lack of temporal control over phytolith samples associated with sites in Santa Cruz County substantially limited the interpretive potential of this data set.

Acknowledging that archaeological excavation is a destructive process, and that impacts to sites are of great concern to many Tribes, archaeologists engaging in discretionary research have long grappled with the question of how to balance the detriments of adverse impacts to archaeological sites against the benefits of answering research questions. This study has illustrated some of the limitations of a very low-impact approach to research design. First, because only one or two excavation units were employed at most sites, the number of flotation samples collected was small, reducing the potential to employ parametric statistical tests and to establish the statistical significance of differences observed in plant use between sites. Second, due also to the small number of excavation units placed at each site, it was unclear how well the data recovered from units characterized each site as a whole. Finally, excavation units that were discovered to represent marginal areas of sites (and thus produced little data) had a proportionally greater impact on the site assemblage than would have been the case at a site where larger numbers of excavation units were collected. In future studies with similar goals, it would be beneficial to place at least six to eight excavation units at each site analyzed, perhaps employing smaller 1.0 x 0.5 m units instead of typical 1.0 x 1.0 m units to reduce the total area impacted.

This project, along with previous research, has documented consistent differences in foodways practices between coastal and inland sites through time, as well as a consistent pattern of prairie seed food use and use of fire-compatible wood fuels in Late Period inland sites in San Mateo and Santa Cruz Counties. The results of this study indicate two paths for future archaeobotanical research on subsistence practices and anthropogenic prescribed burning. First, all inland sites studied so far have been located in relatively low elevation foothills and marine terraces west of the Santa Cruz Mountains. Under the current regional model of Late Period settlement in the Santa Cruz Mountains region, it is expected that Tribes would have had seasonal settlements in lowland areas that would have been inhabited during the spring and summer, as well as higher elevation settlements located in proximity to nut food resources, particularly acorns, that would have been inhabited in the fall and winter (Cuthrell 2013b: 116, 396; Milliken et al. 2007). A clear next step in regional archaeobotanical research is to investigate Late Period habitation sites in locations farther inland and at higher elevations to explore how plant resource use differed between upland and lowland areas and to produce new data that will contribute to a better understanding of regional settlement systems. Second, detection of grass seed food consumption at terminal Middle Period contexts from site CA-SCR-10 intriguingly demonstrates development of small seed food processing practices and possible initiation of anthropogenic burning systems before the Late Period, begging the question of when and how these practices developed. In future research, a focus on inland archaeological sites from the latter half of the Middle Period (ca. 0–1000 CE) could explore this topic further.

References

Anderson, M. Kat
    2005   *Tending the Wild: Native American Knowledge and the Management of California's Natural Resources.* University of California Press, Berkeley.

Bocek, Barbara R
    1984   Ethnobotany of Costanoan Indians, California, Based on Collections by John P. Harrington. *Economic Botany* 38: 240–255.

Borchert, Mark
    2016   Vertebrate Seed Dispersal of Marah Macrocarpus (Cucurbitaceae) after Fire in the Western Transverse Ranches of California. *Écoscience* 11(4): 463–471.

Brown, Alan K.
    2011   *With Anza to California, 1775–1776: The Journal of Pedro Font, O.F.M.* The Arthur H. Clark Company, Norman, Oklahoma.

Calflora
    2019   Calflora: Information on California Plants for Education, Research, and Conservation. Electronic database, https://www.calflora.org, accessed October 2019.

Coil, James, M. Alejandra Korstanje, Steven Archer, and Christine Hastorf
    2003   Laboratory Goals and Considerations for Multiple Microfossil Extraction in Archaeology. *Journal of Archaeological Science* 30(8): 991–1008.

Corder, Stanley
    1973   *Wood and Bark as Fuel.* Research Bulletin 14. Forest Research Laboratory, School of Forestry, Oregon State University. Corvallis, Oregon.

Cowart, Alicia D.
    2014   Paleoenvironmental Change in Central California in the Late Pleistocene and Holocene: Impacts of Climate Change and Human Land Use on Vegetation and Fire Regimes. PhD Dissertation, Department of Geography, University of California, Berkeley.

Cuthrell, Rob Q.
    2013a   Archaeobotanical Evidence for Indigenous Burning Practices and Foodways at CA-SMA-113. *California Archaeology* 5(2): 265–290.
    2013b   An Eco-Archaeological Study of Late Holocene Indigenous Foodways and Landscape Management Practices at Quiroste Valley Cultural Preserve, San Mateo County, California. PhD Dissertation, Department of Anthropology, University of California, Berkeley.

Dagit, Rosi

1995   Recovery of Oaks (*Quercus agrifolia*) Following the Old Topanga Fire, November 1993. In *Brushfires in California Wildlands: Ecology and Resource Management*, edited by Jon E. Keeley and T. Scott, pp. 189–191. International Association of Wildland Fire, Fairfield, Washington.

2002   Post-fire Monitoring of Coast Live Oaks (Quercus agrifolia) Burned in the 1993 Old Topanga Fire. In *USDA Forest Service General Technical Report PSW-GTR-184*, pp. 243–249. US Department of Agriculture, Forest Service, Pacific Southwest Research Station, Albany, California.

Engber, Eamon A., and J. Morgan Varner

2012   Predicting Douglas-fir Sapling Mortality Following Prescribed Fire in an Encroached Grassland. *Restoration Ecology* 20(6): 665–668.

Evett, Rand R., and James W. Bartolome

2013   Phytolith Evidence for the Extent and Nature of Prehistoric Californian Grasslands. *The Holocene* 23(11): 1642–1647.

Evett, Rand R., and Rob Q. Cuthrell

2013   Phytolith Evidence for a Grass-dominated Prairie Landscape at Quiroste Valley on the Central Coast of California. *California Archaeology* 5(2): 319–335.

Evett, Rand R., Arthur Dawson, and James W. Bartolome

2012   Estimating Vegetation Reference Conditions by Combining Historical Source Analysis and Soil Phytolith Analysis at Pepperwood Preserve, Northern California Coast Ranges, U.S.A. *Restoration Ecology* 21: 464–473.

Ford, Lawrence D., and Grey F. Hayes

2007   Northern Coastal Scrub and Coastal Prairie. In *California Grasslands: Ecology and Management*, edited by Mark R. Stromberg, Jeffrey D. Corbin, and Carla M. D'Antonio, pp. 180–207. University of California Press, Berkeley.

Forrestel, Alison B., Max A. Moritz, and Scott L. Stephens

2011   Landscape-scale Vegetation Change Following Fire in Point Reyes, California, USA. *Fire Ecology* 7: 114–128.

Greenlee, Jason M., and Jean H. Langenheim

1990   Historic Fire Regimes and Their Relation to Vegetation Patterns in the Monterey Bay Area of California. *American Midland Naturalist* 124(2): 239–253.

Hammett, Julia E.

2000   Out of California: Cultural Geography of Western North American Tobacco. In *Tobacco Use by Native North Americans: Sacred Smoke and Silent Killer*, edited by J. C. Winter, pp. 128–140. University of Oklahoma Press, Norman.

Harrington, John P.
   1921–39   Northern and Central California: Costanoan. John Peabody
   Harrington Papers. Unpublished Archival Materials. National Anthropological
   Archives, Smithsonian Institution, Washington, DC.

Holmes, Katherine A., Kari E. Veblen, Truman P. Young, and Alison M. Berry
   2008   California Oaks and Fire: A Review and Case Study. In *USDA Forest Service
   General Technical Report PNW PSW-GTR-217*, pp. 551–565. U.S. Department
   of Agriculture, Forest Service, Pacific Southwest Research Station, Albany,
   California.

InsideWood
   2018   The InsideWood Working Group (IWG). NC State University. https://
   insidewood.lib.ncsu.edu/search?3, accessed 2017–2018.

Keeley, Jon E.
   1991   Seed Germination and Life History Syndromes in the California
   Chaparral. *Botanical Review* 57: 81–116.

Lentfer, Carol, Maria Cotter, and William Boyd
   2003   Particle Settling Times for Gravity Sedimentation and Centrifugation: A
   Practical Guide for Palynologists. *Journal of Archaeological Science* 30(2): 149–168.

Lightfoot, Kent G., and Otis Parrish
   2009   *California Indians and Their Environment: An Introduction.* California Natural
   History Guide Series, no. 96. University of California Press, Berkeley.

Lightfoot, Kent G., Rob Q. Cuthrell, Cristie M. Boone, Roger Byrne, Andreas
S. Chavez, Laurel Collins, Alicia Cowart, Rand R. Evett, Paul V. A. Fine, Diane
Gifford-Gonzalez, Mark G. Hylkema, Valentin Lopez, Tracy M. Misiewicz, and
Rachel E. B. Reid
   2013   Anthropogenic Burning on the Central California Coast in Late Holocene
   and Early Historical Times: Findings, Implications, and Future Directions.
   California Archaeology 5(2): 371–390.

Lopez, Valentin
   2020   Chair, Amah Mutsun Tribal Band. Personal communication, July 2020.

Margolin, Malcolm (editor)
   1989   *Monterey in 1786: The Journals of Jean Francois de La Perouse.* Heydey Books,
   Berkeley, California.

Martin, Steve L.
   2009   The Use of *Marah Macrocarpus* by the Prehistoric Indians of Coastal
   Southern California. *Journal of Ethnobiology* 29(1): 77–93.

McBride, Joe R.

1974   Plant Succession in the Berkeley Hills, California. *Madroño* 22: 317–329.

Milliken, Randall T.

1995   *A Time of Little Choice: Disintegration of Tribal Culture in the San Francisco Bay Area, 1769–1810.* Ballena Press, Menlo Park, California.

Milliken, Randall T., Richard T. Fitzgerald, Mark G. Hylkema, Randy Groza, Tom Origer, David Bieling, Alan Levanthal, Randy Wiberg, Andrew Gottsfield, Donna Gillette, Viviana Bellifemine, Eric Strother, Robert Cartier, and David A. Fredrickson

2007   Punctuated Culture Change in the San Francisco Bay Area. In *California Prehistory: Colonization, Culture, and Complexity*, edited by Terry L. Jones and Katherine Klar, pp. 99–124. AltaMira Press, Lanham, Maryland.

Moerman, Dan

2020   Native American Ethnobotany: A Database of Foods, Drugs, Dyes and Fibers of Native American Peoples, Derived from Plants. Electronic database, http://naeb.brit.org, accessed June, 2020.

Parr, Jeffrey, Vesna Dolic, Graham Lancaster, and William Boyd

2001   A Microwave Digestion Method for the Extraction of Phytoliths from Herbarium Specimens. *Review of Palaeobotany and Palynology* 116: 203–212.

Pavlik, Bruce M., Daniel L. Nickrent, and Ann M. Howald

1993   The Recovery of an Endangered Plant. I. Creating a New Population of *Amsinckia grandiflora. Conservation Biology* 7: 510–526.

Pearsall, Deborah M.

2000   *Paleoethnobotany: A Handbook of Procedures.* 2nd ed. Academic Press, San Diego, California.

Piperno, Dolores R.

2006   *Phytoliths: A Comprehensive Guide for Archaeologists and Paleoecologists.* Rowman and Littlefield Publishers, Inc., Lanham, Maryland.

Priestley, Herbert I. (editor)

1937   *A Historical, Political, and Natural Description of California by Pedro Fages, Soldier of Spain.* Translated by Herbert I. Priestley. University of California Press, Berkeley.

Rizzo, Martin

2016   No Somos Animales: Indigenous Survival and Perseverance in 19th Century Santa Cruz, California. PhD Dissertation, Department of History, University of California, Santa Cruz.

Ryan, Kevin C., David L. Peterson, and Elizabeth D. Reinhardt
1988   Modeling Long-term Fire-caused Mortality of Douglas-fir. *Forest Science* 34: 190–199.

Wheeler, Elisabeth A.
2011   Inside Wood—A Web Resource for Hardwood Anatomy. *IAWA Journal* 32(2): 199–211.

Williams, K., R. J. Hobbs, and S. P. Hamburg
1987   Invasion of an Annual Grassland in Northern California by Baccharis pilularis ssp. consanguinea. *Oecologia* 72(3): 461–465.

Wilson, R. C., and C. Hanks
1935   Map #VTM84C1,2: Santa Cruz Quadrangle. Vegetation Type Mapping Project, A. E. Wieslander, Director, US Forest Service.

Wohlgemuth, Eric
2004   The Course of Plant Food Intensification in Native Central California. PhD Dissertation, Department of Anthropology, University of California, Davis.

World Forest Industries
2020   *Firewood BTU Ratings.* http://worldforestindustries.com/forest-biofuel/firewood/firewood-btu-ratings/, Accessed July 2020.

Wu, Yan, and Changsui Wong
2009   Extended Depth of Focus Image for Phytolith Analysis. *Journal of Archaeological Science* 36(10): 2253–2257.

WVTM
2013   Wieslander Vegetation Type Mapping Project. http://vtm.berkeley.edu, accessed August 2013.

CHAPTER 9

Ancient Mussel Bed Harvesting and Indigenous Stewardship Practices on the Central California Coast

MICHAEL A. GRONE

Introduction

Marine resources have been critical components of Native foodways on the Central California coast for millennia (Braje et al. 2006; Classen 1998; Erlandson 1988; Rick and Erlandson 2008; Jones et al. 2008; Lightfoot and Parrish 2009). The abundance and diversity of shellfish, fish, and marine mammals, combined with a carbohydrate-rich diet from plants, enabled human population densities and social stratification often associated with agriculture in some areas (Erlandson 1988; Moss and Erlandson 2001). Indeed, the presence and antiquity of shell middens worldwide demonstrates an early focus on lower-ranked resources such as shellfish, suggesting shellfish-harvesting practices that persisted for millennia and contributed to an otherwise carbohydrate-rich diet, which made coastal settlement a priority for some groups (Classen 1998; Erlandson 1988; McGuire and Hildebrandt 1994; Jones et al. 2008). Shellfish have been viewed in the archaeological literature as a low-ranked resource based on the amount of caloric energy they provide for the amount of time and energy required to access and harvest them, with most shellfish offering lower net gains than larger-bodied prey (Osborn 1977, Raab 1992). However, shellfish are high in critical micro and macronutrients such as iron and protein while also being abundant, stable, and seasonally reliable in coastal environments (Erlandson 1988).

The archaeological investigation of shellfish resources in this region has demonstrated changes in their populations that have been interpreted as resource depression and intensification in the Late Holocene (Campbell and Braje 2015; Basgall 1987; Broughton 1994; Erlandson 1988; Jones 1994)—a trend that is well-documented in Central California (Beaton 1991; Bettinger 2001; Broughton 1994). Resource intensification is here defined as decreased foraging efficiency due to increased pressure from human predation, leading to more reliance on low-ranked resources, such as shellfish and plant foods, due to decreases in the presence of more highly ranked resources (Belovsky 1988). These foundational studies chronicle increases in the diversity of low-ranked invertebrate remains, depression of high-ranked large-bodied prey species, and a shift toward storage-based economies, all indicating increased pressures on the environment through time resulting from human predation and population growth. Most of these studies acknowledge non-anthropogenic variables—such as environmental variation, habitat, sea surface temperature, tidal elevation, and predation—which also influence shellfish populations and potentially confound archaeological interpretations (Blanchette et al. 2009; Classen 1998; Thakar et al. 2017). It is reported that, in some cases, resource intensification led to the reduction of key species and the degradation of food webs (Broughton 1994) as population growth and increased sedentism required people to augment the productivity of resources within their local territories.

However, people are not inherently inclined to degrade or improve their environments, though both scenarios have played out through time around the globe (Balée 2006; Crumley 2003; Fitzhugh and Habu 2002). There is a growing body of literature suggesting that Native peoples enacted various practices that increased the productivity of their local landscapes through stewardship and landscape management (Anderson 2005; Cuthrell 2013; Cuthrell et al. 2012; Lightfoot et al. 2013a). In some areas, people developed sophisticated management strategies to increase the productivity and extent of vital resources for food, medicines, and technology (Anderson 2005; Lightfoot and Parrish 2009).

For example, a rich body of ethnographic literature, Native oral traditions, and archaeological data documents the importance of marine resources to Indigenous people on the Pacific Northwest Coast of North America. Research in this region has demonstrated that Indigenous people managed the intertidal zone through the construction and maintenance of clam gardens and fish weirs (Ames 1994; Byram and Witter 2000; Moss and Erlandson 2001; Lepofsky and Caldwell 2013; Groesbeck et al. 2014; Smith 2001). These management practices also serve as cornerstones of socio-ecological systems, with four aspects of traditional management systems highlighted by Lepofsky and Caldwell (2013): harvesting methods, enhancement strategies, tenure systems, and worldview and social relations. The last two of these require ethnographic or oral tradition, while the first two may be detectable in archaeological deposits.

While there has been considerable archaeological research regarding Indigenous stewardship and management of marine resources in the Pacific Northwest, especially the construction and maintenance of clam gardens, comparatively less emphasis has been placed on the study of similar practices on the California coast (Anderson 2005; Erlandson and Rick 2010; Lightfoot et al. 2011, 2021; Whitaker 2008). It is recognized from the outset that assessing the possibility of ancient Indigenous shoreline management practices in the archaeological record on the Central Coast of California is complicated by the paucity of known built environments, such as rock-walled clam gardens and fish weirs. While these features may be hard to detect in this region, systematic efforts to detect and record such features have not yet been undertaken. Fortunately, there are numerous examples of archaeometric approaches for assessing shifts in populations through time resulting from human predation and harvesting practices (Campbell and Braje 2015; Sanchez 2020; Singh and McKechnie 2015). Such methods are particularly pertinent for the study of California mussel (*Mytilus californianus*), as they can be employed to examine archaeological faunal assemblages to evaluate if they were exploited so intensively that their populations became depleted—or harvested in a sustainable manner that maintained their populations for many decades or centuries.

Assessing Indigenous stewardship practices in this region is further complicated by coastal development, agriculture, sea-level rise, and successive waves of Euro-American settlers who removed Native peoples from their homelands. A significant legacy of colonialism is that many coastal Indigenous people have been denied access to their resources through land ownership and laws restricting gathering(see Chapter 2). This has resulted in the suppression of Traditional Ecological Knowledge (TEK) regarding many marine resources in Central California. The effects of removal from traditional territories are especially true for the Amah Mutsun Tribal Band (AMTB), whose traditional homelands stretch from the coast of San Mateo County down to Monterey Bay and eastward to the Central Valley (see Figure 1.1). Historically composed of more than 20 politically distinct Tribelets, the modern Tribe represents the surviving descendants of these groups who were taken to Missions Santa Cruz and San Juan Bautista. Research in the ancestral homelands of the Amah Mutsun demonstrates that the use of traditional resource and environmental management (TREM) practices (Fowler and Lepofsky 2011) increased the extent and productivity of coastal grasslands in the Late Holocene (3000–500 BP) on the northern coast of Santa Cruz County and the southern coast of San Mateo County (Cuthrell 2013; Cuthrell et al. 2012; Lightfoot and Lopez 2013). It is possible that there are comparable scenarios in which alternative harvesting methods and stewardship practices were employed to maintain stability in targeted shellfish populations over time, making this region especially interesting for exploring the possibility of shoreline management practices evidenced in the archaeological record. Researchers have suggested that practices that enhanced the productivity and extent of terrestrial plants may have been mirrored by management of shellfish populations, such as clams and mussels (Baker 1992; Blackburn and Anderson 1993; Mirschitzka 1992). This chapter evaluates whether Native peoples may have employed similar kinds of resource management practices used for enhancing the productivity of terrestrial resources for mussel beds in this region.

Santa Cruz Coast Study Area

For the past decade, a team of scholars from the AMTB, California State Parks, and the University of California campuses at Berkeley and Santa Cruz have been investigating coastal sites along the northern Santa Cruz and southern San Mateo coastlines spanning the Middle Holocene (7000–3000 BP) and Late Holocene (3000–500 BP) to historical times (Cuthrell 2013; Lightfoot and Lopez 2013; Lightfoot et al. 2021). Changes in Indigenous resource management practices in this region can be examined in the context of long-term coastal occupation, climatic and environmental variability, and the development of Indigenous, Spanish, Mexican, and American relationships with the environment. According to evidence summarized by Lightfoot and colleagues (2013; 2021) Indigenous people used fire in this region during the last 1,300 years to enhance the extent and productivity of grassland seed foods through the maintenance of coastal prairies. These resource stewardship practices are argued to be part of sophisticated stewardship practices, which may have been necessary to sustain increased anthropogenic pressures on the environment during the Late Holocene (Cuthrell 2013). They also reflect local Indigenous worldviews, resource management practices, and long-term stewardship of the environment (Lepofsky and Caldwell 2013).

This research program incorporates multiple lines of evidence to understand the historical ecology of landscapes in this area, providing crucial information to the Amah Mutsun for restoring TEK and resource management practices, as well as to California State Parks' restoration ecology efforts. Much of this work has focused on the study of TREM practices to enhance terrestrial resources, such as grasslands in the Late Holocene on the coast of Santa Cruz and San Mateo counties. To address questions of Tribal interests, AMTB has extended the scope of this research to focus on marine resources, such as small schooling fish, shellfish, and seaweed. To assess this, a historical ecological approach was taken, using multiple lines of evidence to examine broader regional trends in shellfish-harvesting practices through time, incorporating stable isotope analysis and experimental morphometrics to assess changes in harvesting practices and resource stewardship through a comparative, diachronic framework.

Previous work in the study area by Cuthrell (2013) outlines approaches including experimental morphometrics, stable isotope analysis, mussel integrity indices, and the relative abundance of mussel bed associates such as the gooseneck barnacle (*Pollicipes polymerus*) for assessing harvesting practices of mussels that may have led to resource depression, expansion, or stability. This study builds upon this body of work by comparing invertebrate assemblages from Middle and Late Holocene coastal and inland sites to assess whether Native peoples were enacting harvesting practices for shellfish populations that may be analogous to those used for terrestrial resources in the region.

Amah Mutsun Coastal Stewardship

Shellfish and other marine resources, such as seaweed and small schooling forage fishes, were an important component of Amah Mutsun foodways for millennia, as evidenced by ethnographic and archaeological data (Amah Mutsun 2020; Cuthrell 2013). However, of over 1,000 words in the constantly growing Amah Mutsun Ethnobiology Database—a database of Tribal ecological knowledge based on Tribal, ethnographic, and archaeological sources compiled by the Amah Mutsun Land Trust (AMLT; see Chapter 3, this volume)—only 34 relate to marine and intertidal resources (Amah Mutsun 2020). The suppression of TEK of marine resources is due in part to the forcible removal of the Amah Mutsun from the coast and their displacement to Missions Santa Cruz and San Juan Bautista during the Spanish Mission Period. Many others fled the region and headed further inland to escape the brutal conditions experienced in the missions (Valentin Lopez, personal communication; and Rizzo 2016). This historical flight—combined with contemporary restrictions to accessing the coast via private land and the restricted harvesting of traditional resources due to policies of public agencies in the past—creates a scenario with very limited opportunity for the Amah Mutsun to access and steward their traditional shorelines. Today, members of the Tribe are widely dispersed throughout California, citing both historical factors of removal and high rent and real estate prices in their traditional homelands as challenges to returning to the Santa Cruz coast (Valentin Lopez, personal communication).

Despite this history of missionization and removal from ancestral lands, the AMTB are mobilizing ethnographic and archaeological data to aid in awakening dormant knowledge, which they recognize as critical to restoring and reclaiming their traditional practices and lifeways. This concept of "Dormant Knowledge" was first introduced to our research team by Tribal Chair Valentin Lopez (personal communication), who recounted the shame and historical trauma felt by many Amah Mutsun members who have lost touch with traditional forms of knowledge due to removal from traditional homelands and the lasting effects of colonialism. Lopez maintains that this knowledge is not lost but lying dormant, awaiting rediscovery and revitalization. According to the Amah Mutsun creation story, Creator bestowed upon them the duty to steward their lands and the plants and animals that relied on them. That obligation was never rescinded, regardless of missionization, removal, and the struggles they face in returning to their traditional territories (Lightfoot and Lopez 2013; and Chapter 3, this volume). To that end, AMLT is now making efforts to revitalize dormant TEK, partnering with researchers at UC Berkeley and Santa Cruz, who are using eco-archaeological data to help inform and restore ancient traditional resource management practices in the Tribe's traditional territories.

Understanding ancient methods of mussel harvesting can be directly applied to current efforts to restore connections to coastal resources and food sovereignty for the Amah Mutsun. Access to and control of the production of cultural foods is a critical factor in food sovereignty and community wellness efforts. Within this narrative, archaeology can provide essential information for restoring traditional foodways and revitalizing dormant TEK of the Amah Mutsun with further insights, gleaned from research, for reinstating modified TREM practices. Thus, ancient data can inform new strategies among the AMLT and their Native Stewardship Corps for managing the lands and coasts in their traditional homelands, in close collaboration with archaeologists and ecologists. Evidence of terrestrial foodways practiced by the Amah Mutsun has been enhanced by information from ethnography, oral traditions, and the archaeological record. Efforts are underway to establish Tribal-led prescribed burns as a management tool for cultural foods and terrestrial resources. Building upon this applied work, my study hopes to extend the scope of traditional management practices to marine resources, providing the Tribe with critical information for their developing Coastal Stewardship Program. This study thus demonstrates how tightly interwoven marine resource management was with Indigenous foodways and efforts of resistance and persistence. Indeed, one of the sites included in this study extends well into the historic period and displays continuity of Indigenous foodways and mussel-harvesting practices despite the pressures of colonialism. Such interpretations of hinterland sites are important for increasing our understanding of varied Native responses to colonial pressures, helping to dispel notions of colonial takeover and passive Indigenous societal collapse (Schneider et al. 2012; Panich 2013; Lightfoot et al. 2013).

California Mussels as a Resource

California mussels played a critical role in Native diets for millennia, and have been the subject of much archaeological inquiry, as they are often the most abundant and ubiquitous shellfish taxa in coastal Californian middens (Braje and Erlandson 2009; Erlandson 1988; Jones and Richman 1995). Rich in protein and many vitamins and minerals, California mussels provided key nutrients for diets based primarily on grasslands seed foods otherwise rich in fats and carbohydrates (Bettinger et al. 1997; Erlandson 1988; Cuthrell 2013). These sessile, low trophic level filter-feeders are also a very sustainable and easily accessible resource. Though filter-feeders can be affected by harmful algal blooms—as displayed by the annually observed mussel quarantine in California from May 1 to October 31 due to an increase in domoic acid (which causes paralytic shellfish poisoning) in their tissues during this period—they can provide a temporally and spatially predictable resource base throughout much of the year (Jones and Richman 1995; Jones et al. 2008). Though understanding of ancient red tide extent and severity is still poor (Waselkov 1987), it is likely that harmful algal blooms or "red tides" were less frequent and intense in the past; this is due to contemporary warming sea surface temperatures and the increased presence of nitrogen-rich fertilizers in California's coastal waters, resulting from agricultural runoff. Archaeological research by Jones and colleagues (2008) suggests year-round harvesting and seasonal stability of mussel on the Central Coast in the late Holocene, with most harvesting taking place during the summer months.

Habitat Variability

There has been considerable research and debate on the issue of differentiating the cause of changes in mussel assemblages through time (Braje 2010; Flores Fernandez 2014; Thakar et al. 2017; Glassow and Wilcoxon 1988), with archaeologists assessing the relative influence and impacts of natural and environmental variability versus human predation on mussel populations. As demonstrated by Thakar and colleagues (2017), size differences in *Mytilus californianus* populations vary along tidal gradients, with increased growth rates and larger terminal sizes observed in lower tidal elevations and reduced growth rates and smaller terminal sizes observed in higher tidal elevations. This variation within habitats may contribute significantly to variation in mussel size represented in the archaeological record, along with many other zoological factors (Campbell 2015). In their study of mass harvesting of molluscs in coastal environments, Braje and Erlandson (2009) suggest that prey size is not necessarily an accurate measure of prey rank in coastal environments due to habitat variability. They have argued that these interpretations must be backed by solid archaeological data. Ultimately, interpretations regarding diachronic changes in mussel size as an indication of resource depression or habitat variation must be offered cautiously and in the context of broader patterns of regional prehistory and local ecology (Braje et al. 2017).

Seasonality

The issue of seasonality of shellfish harvest on the California coast throughout the Holocene has been addressed by many researchers, using isotopic ratios of o18 and o13 as a proxy for seasonal harvesting patterns (Eerkens et al. 2013; Jones et al. 2008; Jew et al. 2014). On the Central Coast, previous work by Jones and collaborators (2008) suggests that coastal Tribes were harvesting mussels nearly year-round during the Middle Holocene. Based on expectations laid out by Thakar and colleagues (2017), a harvest profile of mussels gathered year-round rather than seasonally would likely be comprised of more individuals from higher tidal margins, as year-round harvesters would be subject to more neap tides than seasonal harvesters, which were likely focused on more advantageous spring tides. This study uses isotopic analyses of mussels from Middle Holocene and Late Holocene sites on the coasts of Santa Cruz and San Mateo Counties to assess variation in seasonal harvesting practices through time.

Harvesting Strategies

It has been argued that sessile bivalves can be easily overharvested. If bivalves were intensively exploited over some length of time, then archaeologists would expect to see a decrease in mean shell size and an increase in the diversity of lower ranked invertebrate resources exploited through time as a signal of resource depression and intensification (Beaton 1991; Braje et al. 2006; Broughton 1994; Classen 1998; Erlandson 1988; Rick and Erlandson 2008). However, like plants, which are also sessile, filter-feeding shellfish can be harvested for either short-term efficiency, by taking all of the largest members of a population and gradually depressing the resource, or for long-term productivity, by employing methods of sustainable harvest that allow for resource stability (Whitaker 2008). For example, research in the Pacific Northwest documents traditional shellfish-harvesting practices that enhance substrate for larval clams, which has been shown to improve the development, continuity, and sustainability of clam populations through time (Lepofsky and Caldwell 2013; Groesbeck et al. 2013).

Researchers have devoted much time to developing methods for assessing the archaeological signatures of different harvesting strategies of California mussel (Bettinger et al. 1997; Basgall 1987; Cuthrell 2013; Jones and Richman 1995; Whitaker 2008). Two primary methods of harvesting have been proposed and modeled: plucking individual mussels and stripping entire beds. According to a study conducted by Bettinger and collaborators (1997), plucking is always a superior method of harvesting based on energy expenditure return rates. However, it has been demonstrated (Bouey and Basgall 1991) that return rates for California mussel beds are higher when mussels beds have been periodically disturbed by human predation, similar to some plants (e.g., native grasses) that become more productive when subject to disturbances, such as fires. This could suggest that contexts with high numbers of smaller mussels could have resulted from harvesting and disturbing the same patch, in effect increasing the rate of return of that patch relative to unharvested patches over the long term.

The exploitation of mussel populations for long-term productivity and sustainability via stripping by Native peoples in California has been proposed by Whittaker (2008). He suggests that stripping entire mussel beds at two-year intervals would result in an assemblage of small to medium (<40 mm) mussels. I expect that a two-year interval of mussel harvesting would be evidenced in the archaeological record by an assemblage of small to medium mussels with a specific seasonal harvest profile, suggesting consistent harvesting practices, which allow mussels to grow back to a size of around 40 mm before being harvested again. According to Whittaker's study, three small to medium mussels can occupy the same space as one large mussel. A patch that can yield six small to medium mussels every four years is therefore more productive than a patch with one large mussel. Which is to say, just because smaller mussels show up in the archaeological record does not necessarily mean they are the consequence of resource depression; this may simply reflect harvesting strategies focusing on higher tidal margins where mussels tend to be smaller, as noted by Thakar and collaborators (2017), and/or a stripping method of harvesting focusing on smaller to medium-sized samples. While Bouey and Basgall (1991) suggest that a stripping method of harvest will always be less optimal than plucking larger individuals, stripping could be a component of harvesting strategies aimed to enhance the extent or productivity of locally owned mussel beds over the longer term, where continued disturbance of patches leads to greater net gains over time (Whitaker 2008). This can be compared to terrestrial resource management practices such as burning different patches on rotational cycles for grassland production, allowing for fallow periods for regrowth and renewal (Anderson 2005; Cuthrell 2013; Lightfoot et al. 2013). Just as burning a grassland might temporarily reduce the amount of harvestable food available but eventually lead to greater productivity and returns, so too could a method of strip harvesting of mussels at regular intervals produce greater net gains in the long run, as Whitaker (2008) suggests.

While stripping leaves areas once populated by mussels and their associate species barren until recolonization, which can take up to two years, Classen (1998) argues that human predation on mussels is unlikely to lead to extirpation of these ubiquitous invertebrates. There are additional reasons that people would strip beds even if it was not the most optimal strategy, such as cultural and individual taste, preferred methods of cooking, ease of transport, or for critical micronutrients such as iron (Bettinger et al. 1997). In order to detect these practices in the archaeological record, one must consider several variables including the size of mussels harvested, the ability to detect individual harvesting events, the season of harvest, as well as the presence of species associated with mussel beds that suggest stripping events (Cuthrell 2013; Jones and Richman 1995; Whittaker 2008).

Modeling archaeological shellfish remains for nuanced harvesting strategies, which may not always operate at maximum efficiency but rather employ strategies aimed toward long-term productivity, can be an elusive task (Lepofsky et al. 2013). Through morphometric analyses of modern and archaeological *Mytilus californianus* specimens, archaeologists have developed multiple regression formulae for estimating the length of individual mussels from anatomical landmarks (Campbell and Braje 2015; Singh and McKechnie 2015). By making the most of fragmented archaeological remains, the methods can help us reconstruct individual size and harvest profiles through time, which may suggest methods of harvesting for either short-term efficiency or long-term productivity, or both (Whittaker 2008).

To further evaluate evidence for a stripping method of harvesting, I call upon the presence of taxa often considered to be "ridealongs," which share substrate with mussels and frequently clump together in their beds. I propose that the presence of these species in archaeological deposits is indicative of a stripping method of harvest, where entire patches of mussels are removed from their substrate and the animals clumped with them come along for the ride. A plucking method of harvest may also result in a few ridealongs, though likely significantly less than a stripping method. My analysis indicates that a relative abundance of these associated species suggests stripping of beds rather than plucking of individual mussels.

The gooseneck barnacle (*Pollicipes polymerus*) is one of these associates that turns up in assemblages from the sites included in the analysis with increased relative abundance in <8 mm size classes, bolstering evidence of a stripping harvest method, as proposed by Cuthrell (2013). The file dogwinkle (*Nucella lima*) is a predatory whelk that preys upon mussels and is also present in most of these assemblages. As previously stated, an increase in lower-ranked prey in the archaeological record, such as these smaller-bodied barnacles and snails, is viewed in the archaeological literature as an indication of intensification and resource depression due to increased human harvesting pressures. The presence of the ridealongs, however, may be an exception to that rule, as they would not have been targeted directly for food but ended up in the archaeological record as a proxy for other targeted resources (Ainis et al. 2014). As demonstrated by Braje and Erlandson (2009), sites with an abundance of smaller-bodied shellfish cannot automatically be considered as evidence of resource intensification but must be considered in light of broader regional trends and local habitat, such as mussel beds.

Expectations and Approach

It is likely that there are examples of resource depression of mussel populations as well as sustainable management at local and regional scales, and that these practices may change in an area over time (Jones and Richman 1995; Broughton 1994; Whitaker 2008). Regardless, California mussel was an integral part of ancient foodways, and evidence of resource depression or sustainable management practices can help inform current efforts to integrate mussels into Native foodways. This study aims to assess harvesting practices along the Central California coast spanning over 6,000 years to test these hypotheses employing expectations for different harvesting strategies, as outlined in Table 9.1.

Table 9.1. Table Outlining Measurable Expectations for Detecting Harvesting Practices.

Harvesting practices	Expectations	Archaeological Evidence
Stripping	• Small to medium size (40 mm) • Narrow range in size • Seasonally specific • Mussel beds associates would also be removed	• Morphometric reconstructions • Isotopic Seasonality data displaying seasonal trend • Presence/Absence of ridealongs
Plucking	• Larger size earlier, decrease in size through time • Annual, opportunistic	• Morphometric reconstructions • Isotopic Seasonality data displaying no seasonal trend • Presence/Absence of ridealongs

This study uses a diachronic, comparative approach to assess changes in *Mytilus californianus* population size collected from several sites along the coast of Santa Cruz and San Mateo counties to detect harvesting practices and resource exploitation through time. I expect that the presence of a wide range of size classes, with small to medium average size, is likely indicative of a stripping method of harvesting. I also expect that harvesting profiles occurring at two-year intervals, as evidenced by morphometric and stable isotopic analysis, are also indicative of a stripping method of harvesting. Furthermore, assemblages with a significant presence of non-dietary, ridealong species associated with mussels are indicative of stripping practices and/or the harvesting of other marine resources, such as kelp and seaweed. For detecting plucking harvesting practices, I expect to see a decrease in mean shell size and an increase in diversity of resources exploited through time, with seasonal stability in harvesting practices. I also expect to see a greater average size and greater minimum size in the earliest deposits at shell midden sites.

Considering these expectations and prior research, I investigate the following questions:

Are there detectable discreet harvesting practices of *Mytilus californianus* in these archaeological deposits? Do these data suggest resource depression or stability and sustainability of mussel populations over time? How can this data inform contemporary Indigenous stewardship practices and restore TEK for the AMTB?

Methods and Materials

This study analyzes data from three sites (CA-SCR-7, CA-SCR-14, and CA-SMA-216) within Amah Mutsun territory with dates ranging from 4750 BCE to 1700 CE, as outlined in Figure 9.1. The work was undertaken as part of a broader project that analyzed the invertebrate remains from CA-SCR-7, CA-SCR-10, CA-SCR-14, and CA-SCR-15 on the Santa Cruz coast. The full results of this investigation are reported in Grone (2021). As part of a collaborative approach, and in keeping with the Tribe's request to minimize impacts to ancestral sites, we employed low-impact, surgically precise field methods to avoid disturbing sensitive cultural materials (see Chapter 4). We began with surface pedestrian survey, followed by the systematic "catch-and-release" surface survey technique refined by this project (see Chapters 1 and 4). The field methods also included the use of minimally invasive geophysical survey techniques, including ground-penetrating radar, electrical resistivity, and magnetometry. These approaches provide high resolution of subsurface features—such as house floors, burials, and middens lenses—which guides placement of subsurface sampling. In some cases, augers were used to assess site depth, integrity of deposits, and terminal dates. Opportunistic column sampling was employed when midden deposits were eroding from exposed faces. Finally, limited excavation units were used to reduce impacts to the sites and the possibility of disturbing sensitive materials. Members of the Amah Mutsun Stewardship Corps participated in each phase of this process, learning these methods and providing non-Tribal members with critical perspective and cultural knowledge.

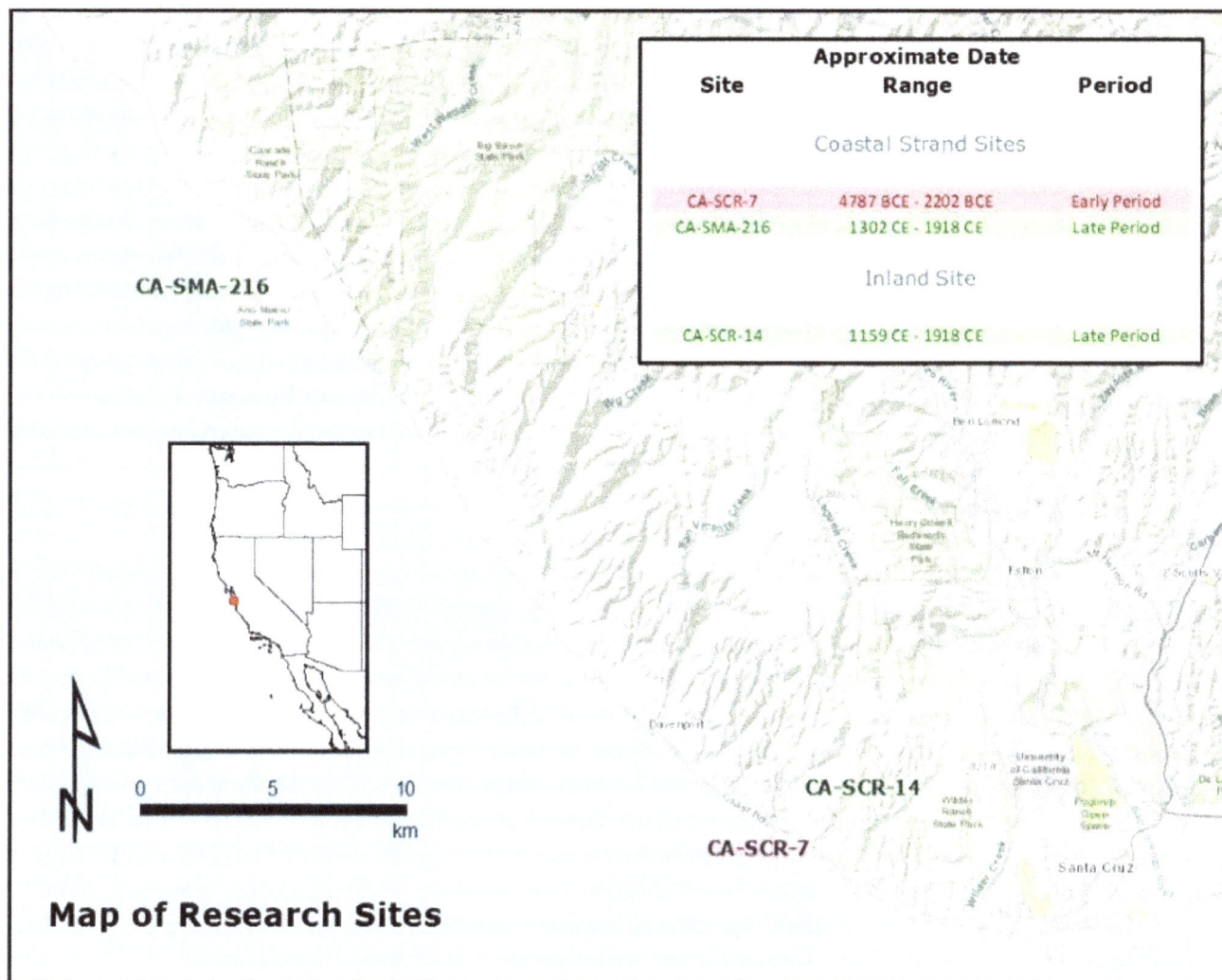

Figure 9.1. Map of Study Area and Sites Sampled.

Table 9.2. Table Outlining Most Abundant Shellfish Taxa by Site.

Sites	Site Type	Age Range	Most Abundant	Second Most Abundant	Third Most Abundant
CA-SCR-7	Coastal Midden	4787BCE 2202BCE	*Mytilus ca* (70.4%)	*Balanus* spp. (28.7%)	*Pollicipes polymerus* (.9%)
CA-SMA-216	Coastal Midden	1302CE–1640CE	*Mytilus ca* (40.9%)	*Tegula funebralis* (37.4%)	Chitons (8.3%)
CA-SCR-14	Upland Village	1159CE–1918 CE	*Mytilus ca* (93.2%)	*Balanus* spp. (2.9%)	*Pollicipes polymerus* (1%)
CA-SCR-14	Upland Village	1159CE–1918 CE	*Mytilus ca* (93.2%)	*Balanus* spp. (2.9%)	*Pollicipes polymerus* (1%)

Bulk sediment samples from column samples and excavation units were processed via flotation methods developed by Rob Cuthrell to separate light and heavy fraction materials (see Chapter 8). Sorting of 1–2 mm, 2–4 mm, 4–8 mm, and >8 mm heavy fraction materials separated with Tyler sieves was conducted in the California Archaeology Lab at UC Berkeley. Invertebrate remains were then separated from the >4 mm-size fraction, as smaller fractions were generally composed of highly fragmented shell that would be too time-intensive to sample, based on the quality of data it might yield. However, the 2–4 mm materials were sorted exclusively for non-dietary gastropods, which provide proxy data for ancient seagrass and kelp-harvesting practices (Ainis et al. 2014; Grone 2020). Shell remains were identified with the aid of comparative collections housed in the California Archaeology Lab. Quantitative measures included weight, relative abundance, density (grams of shell / liter of soil, listed as g/l) and MNI of each taxon (see Grone 2021 for full results).

Site Comparisons

Taxa from all three sites display a heavy reliance on marine resources in the Middle Holocene through Late Holocene, evidenced by a broad range of species that span the entirety of the intertidal zone. CA-SCR-7 invertebrate assemblages are characterized by a wide range of intertidal resources, though primarily dominated by mussel and barnacle. Of the seven columns and two augers analyzed from CA-SCR-7, *Mytilus ca.* and *Balanus* spp. were ubiquitous and high in relative abundance throughout all samples and levels. The average density of shell in these samples was 87.4 grams of shell per liter of soil, with a range of 35.06 g/l to 135.26 g/l. *Mytilus ca.* remains in the assemblages comprise 70.4% of the 12,197.6 grams of invertebrate remains sorted from all samples. In smaller size fractions (4–8 mm, 2–4 mm), a much greater diversity of intertidal taxa was represented, especially limpets, leaf barnacles, chitons, urchins, and turban snails, suggesting broad-spectrum and intensive harvesting of shoreline resources. Mid Holocene site CA-SCR-7 is a multi-component site with midden and hearth features, and invertebrate assemblages from this site can be interpreted as the result of processing, cooking, and discarding. The column samples analyzed in the study (CU 1, CU 2, CU 7) were taken from the northwest side of the site (from an eroding midden with radiocarbon dates spanning nearly 3,000 years across the Mid Holocene; see Chapter 4).

CA-SMA-216 evidences a diverse assemblage of intertidal invertebrate remains with much greater density and relative abundance of *Tegula funebralis* than other sites, consistent with expectations of resource intensification and focus on lower-ranked marine species in the late Holocene, though mussels still dominated at 40.9%.

Two excavation units (EU 1, EU 2) from CA-SCR-14, another Late Holocene site upland from CA-SCR-7, yielded a great diversity of intertidal species. EU 2 had a high density of shell at 68.4 grams per liter and the greatest diversity of invertebrate taxa represented of all samples taken from all sites. While *Mytilus ca.* (93.2%) and *Balanus* spp. (2.9%) were again the most abundant taxa represented, other taxa such as limpets, chitons, urchins, whelks, and turban snails were present in low quantities in nearly all levels. Purple sea urchin (*Strongylocentrotus purpuratus*) was ubiquitous in the unit as well. CA-SCR-14 assemblages were largely dominated by *Mytilus ca.*, though the assemblage at this upland site could be influenced by processing and transport. The mussel bed associate *Pollicipes polymerus* was present in all the assemblages, with an increased presence in fractions <8 mm.

Differences between Late Holocene site assemblages from CA-SCR-14 and CA-SMA-216 likely reflect differences in settlement patterns and the processing and transport of shellfish, as CA-SMA-216 is located on the coast and CA-SCR-14 is nearly a mile inland. The abundance and diversity of intertidal invertebrates in assemblages from coastal CA-SMA-216 suggests that this was a specialized processing site for shellfish and other marine resources. CA-SCR-14 is a village site with a diverse artifact assemblage and relatively homogenous invertebrate assemblage comprised mostly of California mussels. Other studies have shown that bivalves, such as mussels, transport well because of their ability to seal themselves, thereby retaining moisture longer than most other marine invertebrates (Jazwa et al. 2013), which suggests that differences in these assemblages may be more a factor of processing and transport than of harvesting and subsistence practices. Mobility on the landscape, well-defined Tribal territories, and resource processing/ transport strategies were surely important components of Indigenous lifeways in the Late Holocene, and likely account for great variability between coastal and upland assemblages and subsequent interpretations regarding subsistence practices and localized resource management (Hylkema 1991).

Mussel Umbo Study

Recent research on mussel assemblages has resulted in multiple regression formulae to make the most of fragmented *Mytilus ca.* remains, providing one approach for measuring mussel umbos to reconstruct individual size (Campbell and Braje 2015; Singh and McKechnie 2015). While there is some debate about the use and interpretation of these formulae (Campbell 2015, Singh et al. 2015), they appear to be an effective way to estimate mussel size using archaeological materials. These formulae were developed using modern comparative specimens collected in Southern California. Due to the observed biogeographical morphological variation of *Mytilus ca.* north and south of Point Conception (Glassow and Wilcoxon 1988), I developed an experimental morphometric formula from n=151 modern specimens collected from Pescadero State Beach by Rob Cuthrell. After measuring the same elements used by Campbell and Braje (2015) and Singh and McKechnie (2015), the author decided upon umbo thickness, as it tends to be the most well-preserved in archaeological specimens and had a strong correlation with mussel length, as outlined below in Figure 9.2.

To further understand the archaeological signatures of mussel-harvesting practices, stable isotope analysis of mussel shells was conducted from the contexts sampled in the morphometric study. For brevity and simplicity, the results are summarized below and are outlined further in Chapter 10 of this volume.

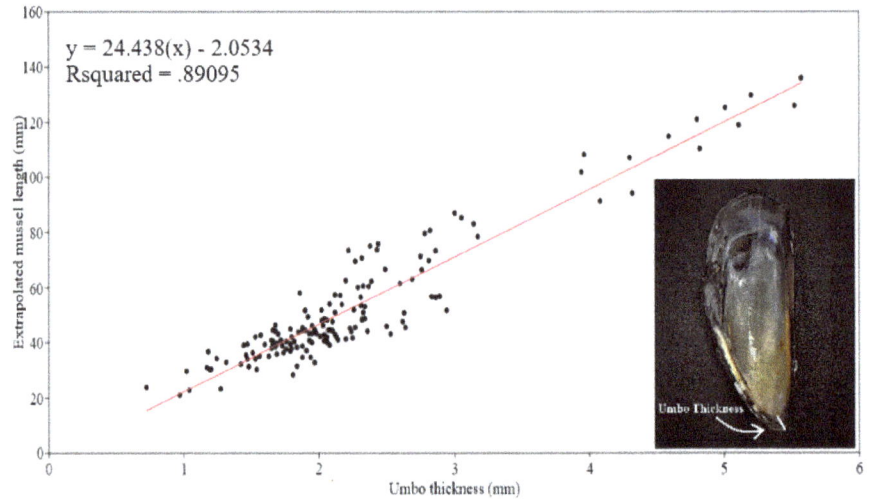

$y = 24.438(x) - 2.0534$
$Rsquared = .89095$

Figure 9.2. Graph Depicting Regression Formula for Predicting Total Mussel Length from Umbo Thickness.

Using δ¹⁸O Isotopes in to Estimate Season of Harvest

Stable oxygen isotopic ratios in biogenic carbonate have been used for several decades to reconstruct paleoclimate records, ancient land use, settlement patterns, and subsistence strategies. Archaeological studies of shellfish harvesting practices often incorporate the isotopic composition of shell carbonate to reconstruct seasonal harvesting patterns (Eerkens et al. 2013; Jew et al. 2013; Jones et al. 2008; Leng and Lewis 2016). These studies typically determine the season of death of individual specimens by comparing the temperature conditions (estimated from an oxygen isotope ratio-temperature regression relationship), under which the terminal growth band (TGB) formed, and the apparent temperature trend, based on samples from preceding growth bands. These data are then compared to changes in sea-surface temperature (SST) over the course of the year in the location where the specimens were likely gathered. Historical SST data was used to calibrate an estimator for season of death as a function of mean temperature and temperature trend, as recorded in distal growth band oxygen isotope values.

This method was employed to examine the seasonality of mussel harvesting for the shellfish assemblages analyzed for CA-SCR-7, CA-SMA-216, and CA-SCR-14, as outlined in Chapter 10. The goal of the study was to compare and predict season of death from the mean temperature values of 200 stable oxygen isotope (δ¹⁸O) microsamples taken from the distal growth bands of archaeological (n=40) mussel (*Mytilus californianus*) samples from coastal and upland sites from the Middle Holocene to Late Holocene. First, we normalized 100 years of instrumental sea surface temperature data to remove the linear increasing scale caused by modern warming trends. Second, we simulated temperature trends in a *M. californianus* population to statistically evaluate seasonality hypotheses. Third, we used a standard nonlinear least squares regression as a predictor for the mean temperatures recorded in archaeological isotopic oxygen microsample transects.

The stable oxygen isotope geochemistry of (n=40) archaeological *Mytilus californianus* specimens was analyzed to estimate the seasonality of mussel harvesting at CA-SCR-7, CA-SMA-216, and CA-SCR-14. The selection of specimens for analysis was done systematically and based on meeting three parameters: 1) the specimen has a visible terminal growth band (TGB), 2) the specimen is from an intact deposit with established AMS dates, and 3) the specimen is free of any observable pre- and post-life trauma, pathology, or modification (e.g., burned, predation scars). Once prepared, samples were submitted to the Stable Isotope Geochemistry Lab at UC Berkeley for analysis.

Samples from CA-SCR-7 show nearly year-round harvesting with a summer lull and a winter peak. CA-SMA-216 shows a similar pattern, but perhaps with a stronger seasonality signal, with most harvesting taking place between winter and spring, followed by a summer to fall hiatus. CA-SCR-14 shows a strong seasonality signal, with no evidence for shells harvested outside of winter. This analysis suggests a more year-round harvest of mussels during the Middle Holocene and a more seasonally contracted harvest during the Late Holocene. Since the sample was by no means exhaustive, we cannot discount the possibility that further sampling of mussel remains from these sites might reveal specimens harvested during other times of the year. However, given that none of the samples analyzed appear to have been harvested outside of this range, and that this reconstruction broadly agrees with the metrical and ecological data presented above, I feel that confidence in these preliminary conclusions is warranted. Further statistical analysis will be pursued in a subsequent publication, including corrections for potentially confounding biological and paleo-oceanographic factors.

Results

To assess mussel-harvesting practices, I consider morphometric data, stable isotope analysis, and ridealong presence to model mussel-harvesting profiles through time. First, I applied our morphometric formula to n=2901 umbos from CA-SCR-7, CA-SCR-14, and CA-SMA-216 (spanning 4787 BCE to 1918 CE). The results are outlined below in Figures 9.3–9.8 and Tables 9.3–9.8. Figures 9.3–9.8 present the morphometric data in boxplots with Tukey's HSD (Honestly Significant Difference) tests to assess statistical significance of variation and trends in the data. This was followed with an analysis of n=40 stable isotope samples taken from mussel shells within the same contexts as those used in the morphometric study. Finally, I integrate these data with ridealong presence in assemblages from each site and compare these three lines of evidence to the expectations laid out in Table 9.1 to integrate and interpret harvesting profiles in Table 9.9.

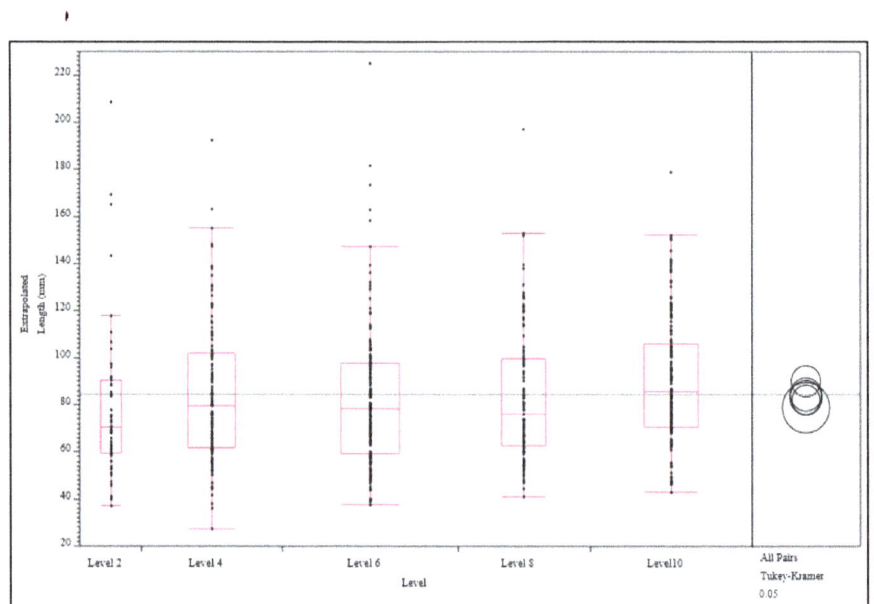

Figure 9.3. CA-SCR-7 Column 1; 3775–2200 BCE; n total = 603.

Table 9.3. Average Sizes of Mussel from Column 1 at CA-SCR-7.

Level	Age	n=	Average size	Std. dev.
2	2300–2202 BCE (0.99)	55	7.9 cm	3.0 cm
4	2286–2246 BCE (0.43); 2235–2195 BCE (0.36); 2173–2145 BCE (0.19)	126	8.5 cm	2.9 cm
6	2579–2488 BCE	158	8.2 cm	2.9 cm
8	3635–3621 BCE (0.13); 3606–3522 BCE (0.87)	119	8.3 cm	2.8 cm
10	3776–3694 BCE (0.97)	145	9.0 cm	2.7 cm

The results of the analysis of Column 1 from CA-SCR-7, which was taken from the northwest side of the midden from an eroding face, are outlined in Figure 9.3 and Table 9.3. The mean size of mussels in all levels averaged over 7 cm, with a slight decrease in average size over time, though the Tukey's HSD does not suggest that this is a significant decrease. These results are consistent with expectation of a plucking style of harvest, which would target larger mussels in a population and gradually depress the resource, leading to decreased size.

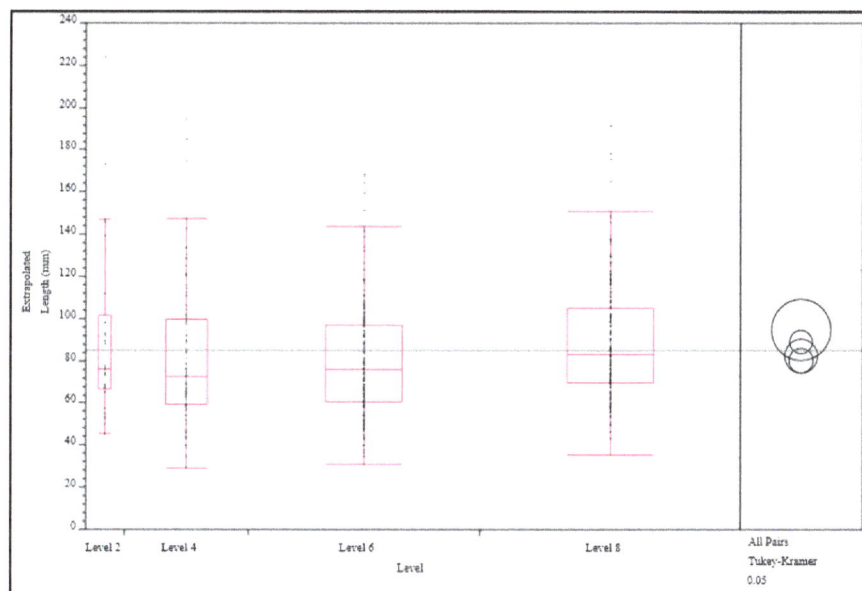

Figure 9.4. CA-SCR-7 Column 2; AMS Date Range: 3940–2420 BCE; n total = 520.

Table 9.4. Average Sizes of Mussel from Column 2 at CA-SCR-7.

Level	Age	n=	Average size	Std. dev.
2	NA	30	9.5 cm	4.7 cm
4	2457–2418 BCE (0.09)	99	8.0 cm	3.7 cm
6	2463–2335 BCE (0.93); 2324–2303 BCE (0.07)	184	8.2 cm	2.7 cm
8	3939–3869 BCE (0.61); 3813–3765 BCE (0.38)	207	8.9 cm	2.7 cm
10	3776–3694 BCE (0.97)	145	9.0 cm	2.7 cm

The results of the analysis of Column 2 from CA-SCR-7, which was also taken from the northwest side of the midden, from an eroding face, are outlined in Figure 9.4 and Table 9.4. The mean size of mussels in all levels averages 8 cm or higher, with no significant decrease or increase in size through time. This profile is consistent with expectation of a plucking style of harvest, as we see large average mussel size. However, the data does not indicate any reduction in size through time.

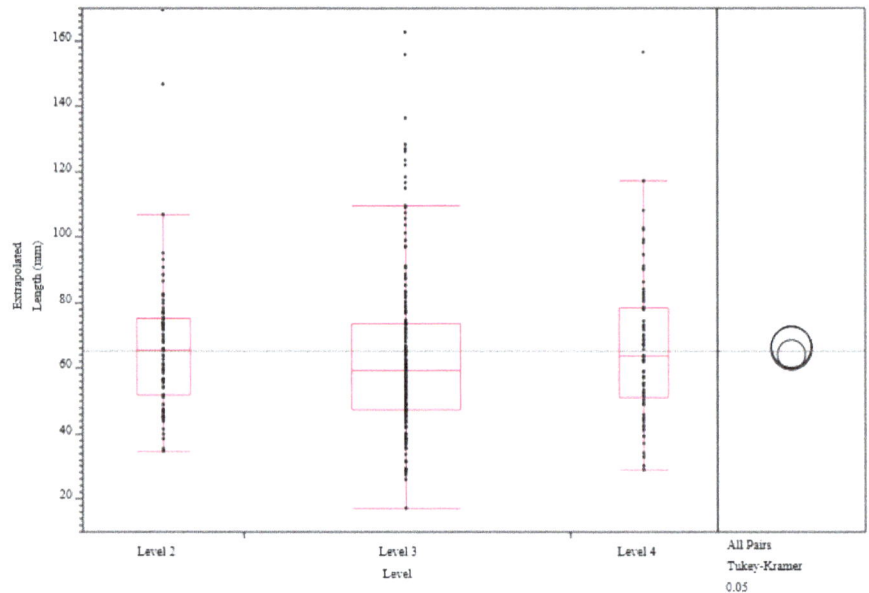

Figure 9.5. CA-SCR-7 Column 7; AMS Date Range: 4785–4555 BCE, n total = 316.

Table 9.5. Average Sizes of Mussel from Column 7 at CA-SCR-7.

Level	Age	n=	Average size	Std. dev.
2	4686–4577 BCE (0.90); 4575–4555 BCE (0.10)	162	6.7 cm	2.1 cm
3	4780–4694 BCE	80	6.4 cm	2.6 cm
4	4787–4713 BCE	74	6.6 cm	2.2 cm

The results of the analysis of Column 7 from CA-SCR-7, which was taken from the northwest side of the midden, are presented in Figure 9.5 and Table 9.5. The average size of mussels stays between 6.4 and 6.7 cm in all the contexts sampled, suggesting resource stability through time that may have been the result of both stripping and plucking harvesting. The average size is smaller than other contexts sampled from CA-SCR-7, which is interesting because this column is the oldest context sampled, suggesting mussel size did not significantly decrease through time at CA-SCR-7. In sum, in examining the three plots for CA-SCR-7, there appears to be a wide range of mussel sizes in earlier assemblages from CA-SCR-7, yet no statistically significant difference in mean size of individuals harvested from 6500 BP to 4000 BP, suggesting stability in mussel-harvesting practices in this region during the Middle Holocene. However, the average mussel size from CA-SCR-7 is considerably larger than the assemblages from CA-SCR-14 and CA-SMA-216, which may be indicative of resource depression over time and a combination of plucking and stripping harvesting practices. The standard deviation of mussel sizes is also greater for CA-SCR-7, displaying a wider range in sizes, which may also indicate a combination of stripping and plucking harvesting practices.

The results of the analysis of EU 2 from CA-SCR-14 are presented in Figure 9.6 and Table 9.6 below. The data display an increase in the average size of mussels through time, with most mussels falling in the small to medium size range, consistent with expectation of a stripping method of harvest and resource stability. As previously mentioned, the average size of mussels from CA-SCR-14 is much smaller than CA-SCR-7, suggesting that people in the Late Holocene may have been interacting with a coastal environment that had been subject to resource depression of mussel by previous inhabitants. The same could be said for the mussel assemblage at CA-SMA-216.

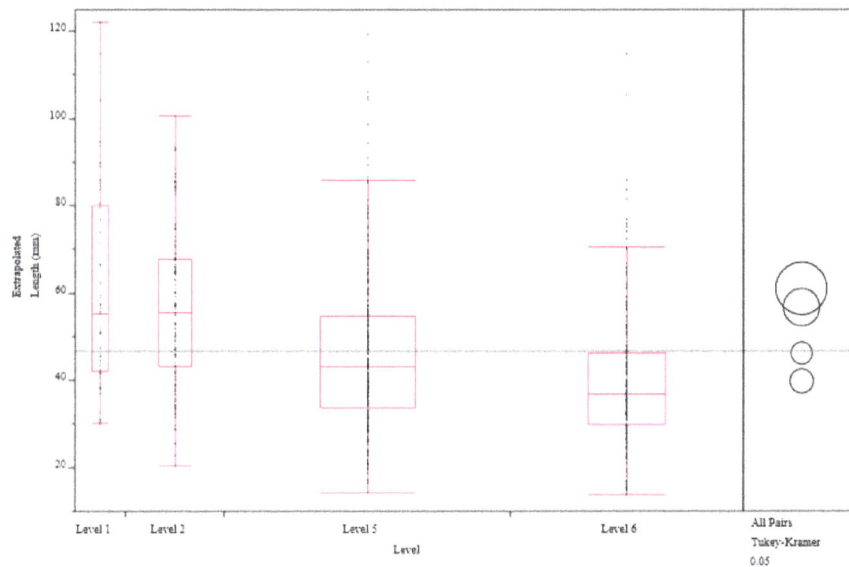

Figure 9.6. CA-SCR-14 Excavation Unit 2; AMS Range 1160–1920 CE; n total = 696.

Table 9.6. Average Size of Mussel from Excavation Unit 2 at CA-SCR-14.

Level	Age	n=	Average size	Std. dev.
1	NA	55	5.9 cm	2.4 cm
2	1695–1726 CE (0.27); 1813–1838 CE (0.20); 1868–1918 CE (0.49)	113	5.7 cm	1.8 cm
5	1267–1411 CE	321	4.6 cm	1.8 cm
6	1159–1212 CE	262	4.0 cm	1.5 cm

The results of the analysis of Area 1 and 2 at CA-SMA-216 are outlined in Figure 9.7 and Table 9.7, which display an increase in the average size of mussels through time, with most mussels falling in the small to medium size range, consistent with expectation of a stripping method of harvest and resource stability during the Late Holocene.

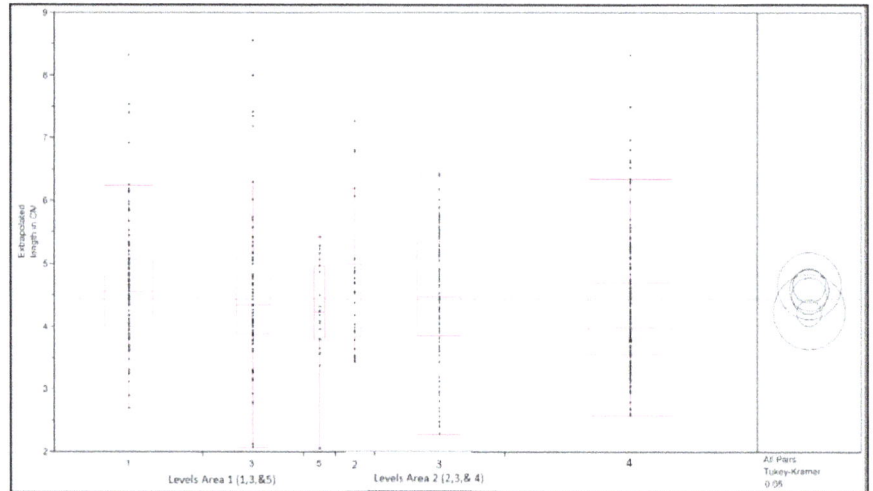

Figure 9.7. CA-SMA-216 Area 1 and 2; AMS Range 1300–1640 CE; n total = 796.

Table 9.7. Average Size of Mussel from Area 1 and 2 at CA-SMA-216.

Level	Age	n=	Average size	Std. dev.
1/1	NA	186	5.0 cm	1.5 cm
1/3	1302 CE–1420 CE	105	5.2 cm	1.8 cm
1/5	NA	27	4.8 cm	1.1 cm
2/2	NA	37	4.7 cm	2.1 cm
2/3	1460 CE–1640 CE	129	4.9 cm	1.8 cm
2/4	NA	312	4.5 cm	1.4

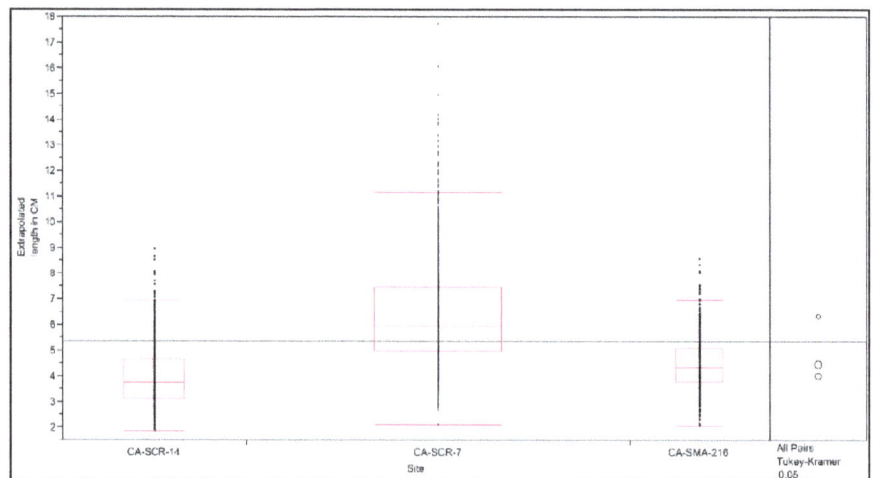

Figure 9.8. Comparison of Sites.

Table 9.8. Average Sizes of Mussels from all three sites.

Site	Age	n=	Average Size	Std. Deviation
CA-SCR-7	4787 BCE–2202 BCE	1409	8.0 cm	2.9 cm
CA-SMA-216	1302 CE–1640 CE	796	5.0 cm	1.9 cm
CA-SCR-14	1159 CE–1918 CE	696	4.7 cm	1. 8 cm

Comparison of Harvesting Patterns at the Three Sites

As displayed in Figure 9.8 and Table 9.8, the average size of mussels is significantly greater at CA-SCR-7 than at CA-SCR-14 or CA-SMA-218. This trend suggests that the mean size of mussels decreased over time from Middle Holocene to Late Holocene, which is consistent with expectations of resource depression through time. However, while the average size is smaller in the two Late Holocene sites, the average size increases slightly through time at these sites. These data suggest that people may have employed harvesting methods that maintained the stability of the mussel populations over several centuries in the Late Holocene. There is also a greater density of umbos per sample in the two Late Holocene sites, which may suggest mass harvesting events of greater numbers of individual mussels than at CA-SCR-7. These harvesting profiles may be indicative of a stripping method of harvesting, as laid out by Whitaker (2008), which could account for both the reduced size and greater number of individuals per context. The standard deviation of mussel size is also less for CA-SMA-216 and CA-SCR-14, consistent with expectation of a stripping method of harvest. When compared to CA-SCR-7, there is a higher standard deviation of mussel size, which may be indicative of plucking larger members as well as stripping entire beds. When considered alongside the results from the following study (Chapter 10, this volume) on seasonality of these contexts, we can begin to draw broader interpretations about harvesting practices through time.

Discussion of Results

The observed reduction in mussel size over time is consistent with expectations of resource depression. However, the average size of umbos from Late Holocene sites CA-SMA-216 and CA-SCR-14 fit with Whitaker's expected two-year harvest profile, with the average extrapolated length of mussels around 40 mm and displaying a seasonally contracted harvesting trend for both sites, as demonstrated in chapter 10 of this volume. This would follow my expectation of a stripping method of harvesting focused on small- to medium-sized mussels. Additionally, the slight but gradual increase in size of mussels through time in Late Holocene sites suggests stability and sustainable harvest of this resource rather than depression.

Table 9.9. Interpretations of Harvesting Practices from All Sites Based on Size, Seasonality, and Ridealong presence.

Site	Size	Seasonality	Ridealong Presence	Interpretation
CA-SCR-7	Large	Fall-Spring	Present	Plucking AND Stripping
CA-SCR-14	Medium	Winter	Present	Stripping
CA-SMA-216	Medium	Winter-Spring	Present	Stripping

When considered alongside seasonality assessments, these data would suggest that during the Middle Holocene, people were harvesting medium to large mussels with a combination of plucking and stripping practices throughout most of the year, while in the Late Holocene, people were harvesting small to medium mussels in the winter and spring using a stripping method of harvesting, as outlined in Table 9.9. These interpretations are consistent with Indigenous TEK of mussel harvesting, which throughout the year during months with an "R" (September-April) (Anderson 2005). Rather than depleting their resources, perhaps the people who lived in these coastal environments were refining their harvesting methods to account for increases in populations, enacting practices that may have coincided with their preferences and long-term plans. They also may have been dealing with an environment that had been subject to resource depression from increased anthropogenic pressures in the past, and learning from these lessons to develop and implement more sustainable methods of harvesting mussels. These practices may not have always operated at maximum efficiency but could have employed long-term strategies as a component of seasonally flexible foodways, which demonstrate a deep understanding of dynamic seascapes. Of course, interpretations of harvesting practices signaling management must be made cautiously and take into consideration broader regional trends, which point toward intensification and resource depression of intertidal resources in the Late Holocene (Beaton 1991).

Conclusion

Developing archaeological research programs for assessing the management of shellfish populations in the absence of physical features, such as rock-terraced clam gardens, is still incipient, though some have laid out expectations for what these signatures may look like (Whitaker 2008). Interpreting evidence of resource intensification, depression, expansion, stability, or management may also be largely influenced by theoretical underpinnings and data modeling. Future directions in this line of inquiry should include a larger number of coastal sites dating to Middle and Late Holocene times and the integration of more seasonality assessments using stable isotope geochemistry to enhance the resolution of invertebrate harvesting profiles through time. Additionally, building a better comparative database of sites assemblages along the coast into a similar research program will provide more robust interpretations.

This work also addresses goals of applied collaborative archaeology and community-engaged research with the AMTB. The data recovered during this project has implications and applications for the AMLT and their developing Coastal Stewardship Program (see Chapter 14, this volume). Such an intensive focus on marine resources, coupled with terrestrial resource management practices, reflects a deep level of engagement with their environment from sea to summit prior to European contact. Instances of resource depression and overexploitation can serve as a cautionary tale, while evidence of sustained management can provide a model for addressing contemporary issues in fisheries management and developing stewardship practices based on ancient TEK and resource management for the present and future.

Acknowledgements

I want to thank the Amah Mutsun Tribal Band for their collaboration, insight, and help with this project, as well as California State Parks, especially Mark Hylkema for his extensive knowledge of the study area and for facilitating all phases of field research. I would like to also thank the Undergraduate Research Apprenticeship Program at UC Berkeley and all the undergraduate researchers who aided in the sorting and analysis of this project, especially Paul Rigby and Kathryn Field. Funding for the project has been generously provided by the National Science Foundation (BCS-1523648).

References

Ainis, Amira F., René L. Vellanoweth, Queeny G. Lapeña, and Carol S. Thornber.
2014   Using non-dietary gastropods in coastal shell middens to infer kelp and seagrass harvesting and paleoenvironmental conditions. *Journal of Archaeological Science* 49: 343-360.

Amah Mutsun Tribal Band
2020   Amah Mutsun Ethnobiology Database. Confidential Electronic Resource Maintained by Amah Mutsun Tribal Band, Galt, California. Accessed June 2020.
2020   Map of Amah Mutsun Territory. http://amahmutsun.org/history/history-sub-page

Ames, Ken. M.
1994   The Northwest Coast: Complex Hunter-Gatherers, Ecology, and Social Evolution. *Annual Review of Anthropology* 23: 209–229

Anderson, M. Kat
2005   *Tending the Wild: Native American Knowledge and the Management of California's Natural Resources.* University of California Press, Berkeley.

Baker, Rob
1992   The Clam "Gardens" of Tomales Bay. *News from Native California* 6(2): 28–29.

Balée, William
2006   The Research Program of Historical Ecology. *Annual Review of Anthropology.* 35: 75–98.

Basgall, Mark E.
1987   Resource Intensification among Hunter-Gatherers: Acorn Economies in Prehistoric California. *Research in Economic Anthropology* 9: 21–52.

Beaton, John M.
1991   Extensification and Intensification in Central California Prehistory. *Antiquity* 65, (249): 946–952.

Belovsky, Gary E.
1988   An Optimal Foraging-Based Model of Hunter-Gatherer Population Dynamics. *Journal of Anthropological Archaeology* 7(4): 329–372.

Bettinger, Robert L.
2001   Holocene Hunter-Gatherers. In *Archaeology at the Millennium,* edited by Gary M. Feinman, T. Douglas Price, pp. 137–195. Academic/Plenum Publishers, New York.

Bettinger, Robert L., Ripan Malhi, and Helen McCarthy
    1997   Central Place Models of Acorn and Mussel Processing. *Journal of Archaeological Science* 24(10): 887–899.

Blackburn, Thomas C. and Kat Anderson
    1993   *Before the Wilderness: Environmental Management by Native Californians* (No. 40). Ballena Press, Menlo Park, California.

Blanchette, Carol A., C. Melissa Milner, Peter T. Raimondi, David Lohse, Kristen E. K. Heady, and Bernardo R. Broitman
    2008   Biogeographical Patterns of Rocky Intertidal Communities along the Pacific Coast of North America. *Journal of Biogeography* 35(9): 1593–1607.

Bouey, Paul Douglas, and Mark Basgall
    1991   Archaeological Patterns along the South Central Coast, Point Piedras Blancas, San Luis Obispo County, California: Archaeological Test Evaluation of Sites CA-SLO-264, SLO-266, SLO-267, SLO-268, SLO-1226, and SLO-1227. Far Western Anthropological Research Group, Davis, California.

Braje, Todd J.
    2010   *Modern Oceans, Ancient Sites: Archaeology and Marine Conservation on San Miguel Island, California.* University of Utah Press, Salt Lake.

Braje, Todd, and Jon Erlandson
    2009   Mollusks and Mass Harvesting in the Middle Holocene. *California Archaeology* 1(2): 269–289.

Braje, Todd J., Jon M. Erlandson, Douglas J. Kennett, and Torben C Rick
    2006   Archaeology and Marine Conservation. *The SAA Archaeological Record* 6(1): 14–19.

Braje, Todd J., Torben C. Rick, and Jon M. Erlandson
    2017   The Forest or the Trees: Interpreting Temporal Changes in California Mussel Shell Size. *Quaternary International* 427: 243–245.

Broughton, Jack. M.
    1994   Declines in Mammalian Foraging Efficiency During the Late Holocene, San Francisco Bay, California. *Journal of Anthropological Archaeology* 13(4): 371–401.

Byram, Scott., and Robert Whitter
    2000   Changing Landscapes: *Proceedings of the Coquille Indian Tribe Cultural Preservation Conference*, vol. 1, edited by Robert Losey. Coquille Indian Tribe, North Bend, Oregon.

Campbell, Greg
2015   "We want to go, where everyone knows, mussels are all the same...": a comment on some recent zooarchaeological research on Mytilus californianus size prediction. *Journal of Archaeological Science* 63: 156-159.

Campbell, Breana, and Todd J. Braje
2015   Estimating California Mussel (*Mytilus californianus*) Size From Hinge Fragments: A Methodological Application in Historical Ecology. *Journal of Archaeological Science* 58: 167–174.

Claassen, Cheryl
1998   *Shells.* Cambridge University Press, Cambridge.

Crumley, Carole
2003   *Historical Ecology: Integrated Thinking at Multiple Temporal and Spatial Scales. Conference: World System History and Global Environmental Change.* Lund University, Sweden.

Cuthrell, Rob Q.
2013   An Eco-Archaeological Study Of Late Holocene Indigenous Foodways And Landscape Management Practices At Quiroste Valley Cultural Preserve, San Mateo County, California. PhD Dissertation, Department of Anthropology, University of California, Berkeley.

Cuthrell, Rob Q., Chuck Striplen, Mark G. Hylkema, and Kent G. Lightfoot
2012   A Land of Fire: Anthropogenic Burning on the Central Coast of California. *Contemporary Issues in California Archaeology*, edited by Terry Jones, pp. 153–172. Left Coast Press, Walnut Creek.

Eerkens, Jelmer W., Brian F. Byrd, Howard J. Spero, and AnnaMarie K. Fritschi
2013   Stable Isotope Reconstructions of Shellfish Harvesting Seasonality in an Estuarine Environment: Implications for Late Holocene San Francisco Bay Settlement Patterns. *Journal of Archaeological Science* 40(4): 2014–2024.

Erlandson, Jon
1988   The Role of Shellfish in Prehistoric Economies: A Protein Perspective. *American Antiquity*: 102–109.
2013   *Early Hunter-Gatherers of the California Coast.* Springer Science & Business Media, New York.

Erlandson, Jon M., and Torben C. Rick
2010   Archaeology Meets Marine Ecology: The Antiquity of Maritime Cultures and Human Impacts on Marine Fisheries and Ecosystems. *Annual Review of Marine Science* 2: 231–251.

Fitzhugh, Ben, and Junko Habu (editors)

2002   *Beyond Foraging and Collecting: Evolutionary Change in Hunter-Gatherer Settlement Systems.* Springer Science & Business Media, New York.

Flores Fernandez, Carla F.

2014   Past Small-Scale Ecological and Oceanographic Variability around Santa Cruz Island, California. Implications for Human Foraging on *M. californianus* Beds during the Late Holocene (2200–500 cal BP). PhD Dissertation, Department of Anthropology, University of California, Santa Barbara.

Fowler, Catherine S., and Dana Lepofsky

2011   Traditional Resource and Environmental Management. In *Ethnobiology*, edited by E. N. Anderson, Deborah Pearsall, Eugene Hunn, and Nancy Turner, pp. 285–304. Wiley-Blackwell, New York.

Glassow, Michael A., and Larry R. Wilcoxon

1988   Coastal Adaptations Near Point Conception, California, With Particular Regard To Shellfish Exploitation. *American Antiquity* 53(1): 36–51.

Groesbeck, Amy S., Kirsten Rowell, Dana Lepofsky, and Anne K. Salomon

2014   Ancient Clam Gardens Increased Shellfish Production: Adaptive Strategies from the Past Can Inform Food Security Today. *PloS one* 9(3): e91235.

Grone, Michael A.

2020   Of Molluscs and Middens: Historical Ecology of Indigenous Shoreline Stewardship along the Central Coast of California. PhD Dissertation, Department of Anthropology, University of California, Berkeley.

2021   Zooarchaeological Analysis of Invertebrate Remains In *The Study of Indigenous Landscape and Seascape Stewardship Practices on the Santa Cruz Coast: A Collaborative Eco-Archaeological Approach*, edited by Kent G. Lightfoot, Rob Q. Cuthrell, Mark G. Hylkema, Valentin Lopez, Diane Gifford-Gonzalez, Roberta A. Jewett, Michael A. Grone, Gabriel M. Sanchez, Peter A. Nelson, Alex J. Apodaca, Ariadna Gonzalez, Kathryn Field, and Alexii Sigona. Report Prepared for California Department of Parks and Recreation, Santa Cruz District, Archaeological Research Facility, University of California, Berkeley.

Hylkema, Mark G.

1991   Prehistoric Native American Adaptations along the Central California Coast of San Mateo and Santa Cruz Counties. MA Thesis, Department of Anthropology, San Jose State University.

Jew, Nicholas P., Jon M. Erlandson, Torben C. Rick, and Leslie Reeder-Myers

2014   Oxygen Isotope Analysis Of California Mussel Shells: Seasonality And Human Sedentism at an 8,200-year-old Shell Midden on Santa Rosa Island, California. *Archaeological and Anthropological Sciences* 6(3): 293–303.

Jones, Terry L., and Jennifer R. Richman

    1995   On Mussels: *Mytilus californianus* as a Prehistoric Resource. *North American Archaeologist* 16(1): 33–58.

Jones, Terry L., Douglas J. Kennett, James P. Kennett, and Brian F. Codding

    2008   Seasonal Stability in Late Holocene Shellfish Harvesting on the Central California Coast. *Journal of Archaeological Science* 35(8): 2286–2294.

Leng, Melanie J., and Jonathan P. Lewis

    2016   Oxygen Isotopes in Molluscan Shell: Applications in Environmental Archaeology. *Environmental Archaeology* 21(3): 295–306.

Lepofsky, Dana, and M. Caldwell

    2013   Indigenous Marine Resource Management on the Northwest Coast of North America. *Ecological Processes* 2: 12.

Lightfoot, Kent G.

    2008   Collaborative Research Programs Implications for the Practice of North American Archaeology Kent G. Lightfoot. *Collaborating at the Trowel's Edge: Teaching and Learning in Indigenous Archaeology*, edited by Stephen Silliman, pp. 211–227. University of Arizona Press, Tucson.

Lightfoot, Kent G., and Otis Parrish

    2009   *California Indians and Their Environment: An Introduction*. University of California Press, Berkeley.

Lightfoot, Kent G., and Valentin Lopez

    2013   The Study of Indigenous Management Practices in California: An Introduction. *California Archaeology* 5 (2): 209–219.

Lightfoot, Kent G., Edward Luby, and L. Pesnichack

    2011   Evolutionary Typologies and Hunter-Gatherer Research: Rethinking the Mounded Landscapes of Central California. In *Hunter-Gatherer Archaeology as Historical Process*, edited by Ken Sassaman and Donald Holly, Jr., pp. 55–78. University of Arizona Press, Tucson.

Lightfoot, Kent G., Rob Cuthrell, Chuck Striplen, and Mark Hylkema

    2013   Rethinking the Study of Landscape Management Practices Among Hunter-Gatherers in North America. *American Antiquity* 78(2): 285–301.

Lightfoot, Kent G., Rob Q. Cuthrell, Mark G. Hylkema, Valentin Lopez, Diane Gifford-Gonzalez, Roberta A. Jewett, Michael A. Grone, Gabriel M. Sanchez, Peter A. Nelson, Alec J. Apodaca, Ariadna Gonzalez, Kathryn Field, Jordan F. Brown, Alexii Sigona, and Paul V. A. Fine

2021   The Eco-Archaeological Investigation of Indigenous Stewardship Practices on the Santa Cruz Coast. *Journal of California and Great Basin Anthropology* 41(2): 185–204.

Mirschitzka, Sussanne

1992   Usage and Management of Marine Resources on the Northwest Coast. *European Review of Native American Studies* 6(1): 9–16.

Moss, Madonna, and Jon M. Erlandson

1995   Reflections on North American Pacific Coast Prehistory. *Journal of World Prehistory* 9(1): 1–45.

1998   Early Holocene Adaptations on the Southern Northwest Coast. *Journal of California and Great Basin Anthropology* 20(1): 13–25.

Osborn, Alan J.

1977   Strandloopers, Mermaids, and Other Fairy Tales: Ecological Determinants of Marine Resource Utilization, the Peruvian case. In *For Theory Building in Archaeology: Essays on Faunal Remains, Aquatic Resources, Spatial Analysis, and Systemic Modeling*, edited by Lewis R. Binford, pp.157–206. Academic Press, New York.

Panich, Lee M.

2013   Archaeologies of Persistence: Reconsidering the Legacies of Colonialism in Native North America. *American Antiquity* 78(1): 105–122.

Raab, L. Mark.

1992   An Optimal Foraging Analysis of Prehistoric Shellfish Collecting on San Clemente Island, California. *Journal of Ethnobiology* 12(1): 63–80.

Rick, Torben C.

2007   *The Archaeology and Historical Ecology of Late Holocene San Miguel Island.* Cotsen Institute of Archaeology, UCLA, Los Angeles.

Rick, Torben C., and Jon M. Erlandson (editors)

2008   *Human Impacts on Ancient Marine Ecosystems: A Global Perspective.* University of California Press, Berkeley.

Rizzo, Martin

2016   No Somos Animales: Indigenous Survival and Perseverance in 19th Century Santa Cruz, California. PhD Dissertation, Department of History, University of California, Santa Cruz.

Sanchez, Gabriel M.

2020 Indigenous Stewardship of Marine and Estuarine Fisheries?: Reconstructing the Ancient Size of Pacific Herring through Linear Regression Models. *Journal of Archaeological Science: Reports* 29: 102061. DOI:10.1016/j.jasrep.2019.102061.

Schneider, Tsim D., Sara L. Gonzalez, Kent G. Lightfoot, Lee M. Panich, and Mathew A. Russell

2012 A Land of Cultural Pluralism: Case Studies from California's Colonial Frontier. In *Contemporary Issues in California Archaeology*, edited by Terry Jones and Jennifer Perry, pp. 319–338. Left Coast Press, Walnut Creek, New York.

Singh, Gerald G., and Iain McKechnie

2015 Making the Most of Fragments: A Method for Estimating Shell Length from Fragmentary Mussels (*Mytilus californianus* and *Mytilus trossulus*) on the Pacific Coast of North America. *Journal of Archaeological Science* 58: 175–183.

Smith, Bruce D.

2001 Low-Level Food Production. *Journal of Archaeological Research* 9(1): 1–43.

Thakar, Heather B., Michael A. Glassow, and Carol A. Blanchette

2017 The Forest and the Trees: Small-Scale Ecological Variability and Archaeological Interpretations of Temporal Changes in California Mussel Shell Size. *Quaternary International* 427: 246–249.

Waselkov, Gregory A.

1987 Shellfish Gathering and Shell Midden Archaeology. *Advances in Archaeological Method and Theory* 10: 93–210.

Whitaker, Adrian

2008 Incipient Aquaculture in Prehistoric California? Long-term Productivity and Sustainability vs. Immediate Returns for the Harvest of Marine Invertebrates. *Journal of Archaeological Science* 35(4): 1114–1123.

CHAPTER 10

Seasonality of Mussel Harvesting at Three Holocene Sites on the Santa Cruz Coast: Insights from Isotopic Variation in Marine Mollusks

---

ALEC J. APODACA, JORDAN F. BROWN, AND MICHAEL A. GRONE

The isotopic content in shell produced by marine invertebrates can be examined to provide a better understanding of shellfish harvesting patterns in the past (Bailey et al. 1983; Claassen 1998). This study describes the results from oxygen isotope analysis of archaeological California mussel (*Mytilus californianus*) specimens from three Holocene sites on the Santa Cruz coast. We aim to use these data as an auxiliary line of evidence to better understand the legacy of shellfishing practices and the harvesting regimes that have been part of the coastal ecosystems and foodways of Indigenous people inhabiting the Santa Cruz coast throughout the Holocene (see Chapter 9, this volume). This chapter also discusses how archaeological isotope research may be relevant to the contemporary stewardship goals of California Tribes engaging with marine resources, as well as discussing the challenges inherent in isotopic seasonality studies that rely on archaeological shell.

The Amah Mutsun Tribal Band (AMTB) has made significant progress in the revitalization of marine resource stewardship over recent years. Several environmental issues are adversely impacting ancestral Amah Mutsun waters, such as the declining health of kelp forests, instability of native shellfish populations, poor conditions of anadromous fish spawning habitat, sea level rise (Reeder et al. 2012), and increased rates of coastal erosion (Newland et al. 2019). The AMTB are participating and leading restoration efforts to help address these environmental issues, while engaging researchers to further understand them. As outlined in this volume, AMTB is also incorporating archaeological research to restore and affirm ancient knowledge of traditional resource management practices, which were suppressed during colonization and are lying dormant. Since stable oxygen isotopes in archaeological shell can inform us about the environment in which shellfish were living, as well as harvesting seasonality, such data can be relevant when planning the reintroduction of traditional harvesting practices. This practical contribution is a major expansion of the traditional role of isotope-based seasonality and also enhances the diachronic picture of harvesting patterns, site occupation, and residential mobility on the California coast (e.g., Glassow et al. 1994; Hylkema 1991, 2002; Jones et al. 2002; Kennett and Voorhies 1996).

Archaeologists working along the Central California coast have used oxygen isotope analysis of a diverse number of molluscan shellfish taxa to interpret site seasonality. Several beach and bay species, such as Washington clam (*Saxidomus nuttalli*) (Jones et al. 2002), bent-nose clam (*Macoma nasuta*) (Culleton et al. 2009; Eerkens et al. 2014), bay mussel (*Mytilus trossulus*) (Finstad et al. 2013; Schweikhardt al. 2011), and Pacific oyster (*Ostrea lurida*) (Harold et al. 2019) have been examined, while species from rocky open coastlines, such as abalone (*Haliotis* sp.) (Rick et al. 2006) and California mussel (Cuthrell 2013; Glassow et al. 2012; Jew et al. 2013), have also been studied.

Research Design and Questions

The mussel specimens studied here were excavated from CA-SCR-7, CA-SCR-14, and CA-SMA-216. CA-SCR-7 is an Early Period coastal midden site with Middle and Late Holocene components. CA-SCR-14 is a Late Period inland village site directly upland from CA-SCR-7. CA-SMA-216 is a Late Period coastal midden to the north of both CA-SCR-7 and CA-SCR-14. These sites were selected because they provide spatial representation of both coastal and inland habitation areas and temporal representation from Middle Holocene to Late Holocene. These sites also contained intact stratigraphic components rich with mussel shells and have contexts that were studied for archaeobotanical, zooarchaeological, and other archaeological evidence. (See Chapters 4 and 9, this volume, for detailed descriptions of each site.)

The research presented here aims to address a variety of questions relating to the harvest and processing or consumption of shellfish at these sites. First, at what time of year were the shellfish recovered from these three sites originally harvested? Does each site exhibit a clear seasonal signature, or was harvest activity variable throughout the year? Second, has the seasonal timing of harvesting changed over the long term? This study compares the seasonality trends of the three sites to examine if seasonality of harvesting changes from the Middle to the Late Holocene. Mussel shell is one of the most common ecofacts in coastal sites in California, yet its abundance, diversity, and ubiquity requires more scrutiny. The people who lived at these sites forged important relationships with marine ecosystem resources and were part of the ecological processes of shorelines. Understanding how shellfish harvesting regimes and the stewardship practices related to shellfishing developed over time as a subsistence strategy is of particular interest to the AMTB and archaeologists.

Third, how can archeological isotopes be useful for the contemporary stewardship of coastal Indigenous resources today? Isotope analysis can offer several benefits toward this goal. It is a less-destructive technique that has become relatively affordable for running a larger number of samples. Multiple inferences can be drawn from isotopic readings in mussel, such as paleoclimatic information on ancient sea temperatures, seasonality of harvest, and site occupancy patterns. Isotope data can be another "tool in the shed" when the AMTB plans for restoring and revitalizing coastal Indigenous stewardship and foodways.

Approaches to Shellfish Seasonality Reconstructions

Oxygen isotope values of molluscan shell carbonate are commonly used to study the seasonality of past human activity at archaeological sites where mollusks were eaten or processed (Culleton et al. 2009; Kennett and Voorhies 1996; Rick et al. 2006; Shackleton 1973). For rocky intertidal species, sea-surface temperature (SST) is the main driver of intra-annual $\delta^{18}O$ variation in shell carbonate geochemistry.

There are a few common ways oxygen isotope measurements from archaeological marine mollusk shell are used to determine season of death as a proxy for harvesting activity. All approaches make use of the sinusoidal shape of SST variation over the course of the year, as well as the well-established relationship between SST and $\delta^{18}O$ of shell carbonate (calcite, in the case of *M. californianus*), which allows $\delta^{18}O$ to be used as a "paleothermometer" to infer the SST under which the calcite was formed.

The approach advocated by Shackleton (1973)—which we will refer to as the "seasonal classification approach" for the purposes of this chapter—focuses on placing each shell within one of four seasonal brackets, using two basic pieces of information: (1) the SST value recorded by the terminal growth band (TGB); and (2) the direction of change in SST between that growth band and the preceding growth band (TGB+1). For example, if the TGB indicates a middling SST value and the TGB+1 indicates that this was preceded by a lower temperature, the shell is considered to have been harvested in spring (i.e., when temperatures are in the middle of the annual range and rising). Note that there is more than one way to construct seasonal classes that are characterized by identifiable features of the SST trend (see Table 10.1).

The other approach—which we refer to as "curve-fitting" for the purposes of this chapter—is exemplified by Killingley (1981) and uses a larger number of densely spaced carbonate samples in an attempt to reproduce the shape of the annual SST cycle. The position of the TGB is then associated with a specific month based on the researcher's judgment of where it falls in that cycle. In practice, this can be challenging for various reasons, including variations in mollusk growth rates and seasonal temperature cycles (Bailey et al. 1983), and thus demands a high density of sampling, which is not always feasible.

While there are relatively few examples of the latter approach in recent years (but see Cuthrell 2013), a popular approach has been to increase sampling density or longitudinal time span to tighten seasonal resolution in a hybrid approach that incorporates aspects of curve-fitting and seasonal classification (e.g., Burchell et al. 2018; Jew et al. 2014; Jones et al. 2008; Mannino et al. 2011). Yet there are a few commonly known issues with isotope seasonality reconstructions regardless of technique. For example, the need for subjective analysis of each individual specimen becomes increasingly onerous as sample numbers increase. Additionally, the boundaries of the four traditional (winter, fall, spring, and

Table 10.1. Two Possible Approaches to Seasonal Classification by Temperature Value and Temperature Trend, Generalized from the Specific Methods Described in Jew et al. (2013) and Jones et al. (2008).

Seasonal Classification Method I (after Jew et al. 2013)

Winter	Low	Increasing
Spring	High	Increasing
Summer	High	Decreasing
Fall	Low	Decreasing

OR

Seasonal Classification Method II (after Jones et al. 2008)

Winter	Very Low	—
Spring	Medium	Increasing
Summer	Very High	—
Fall	Medium	Decreasing

summer) seasonal classes are determined based on the ability to recognize a characteristic shape in the SST curve. Yet there is no necessary reason that these "seasons" (defined based on the morphology of the annual SST curve) should correspond to seasonal divisions of ecological or cultural significance, i.e., of a sort that would dictate shellfish harvesting patterns. This presents a considerable challenge, as that patterning in the seasonality record could be obscured by the imposition of arbitrary coarse-grained classification schemes. Also, the use of *absolute* temperature values for seasonal classification assumes that paleo-SST and species-specific isotope fractionation patterns can be reliably estimated, which is not always the case.

Paleoclimatologists have developed quantitative methods for dealing with seasonal cycles in mollusk isotope records (e.g., de Brauwere et al. 2009; Wang et al. 2015), but these often depend on very high-resolution sampling procedures for relatively small numbers of specimens (see Burchell et al. 2018). It would be difficult to extend such analyses to the large numbers of specimens required to assess changes in mollusk-harvesting dynamics over millennia and across large regions. Note that all the methods mentioned here are enhanced by traditional sclerochronological work, i.e., growth-band counting and analysis (Claassen 1998; Deith 1985; West et al. 2018).

The approach advanced in this study builds upon the oxygen isotope sampling and interpretive designs described above, while presenting certain innovations designed to support the specific research goals and stewardship partnerships that are characteristic of collaborative eco-archaeology. A main driver behind the design of the study was to explore the ways in which stable oxygen isotopes from archaeological shells may be integrated into the broader coastal stewardship plans of the AMTB. Therefore, keeping track of the several lines of uncertainty involved in the interpretation of isotopic data is crucial when comparing isotopic results about harvesting with Tribal knowledge, ethnographic data, and other evidentiary lines.

An Alternative Approach to Isotopic Seasonality Reconstructions

All the aforementioned techniques for season-of-death assignment rely on some specification of an annual SST curve. The curve-fitting approach attempts to match an entire shell profile to the curve, while the seasonal classification approach divides the annual curve into seasonal bins with identifiable markers (as shown in Table 10.1). These SST curves usually are constructed from historical instrumental datasets recording daily SST over decades, establishing a mean ocean climate for the region of interest. Isotope readings are used to infer the temperature at which the carbonate was formed (via carbonate paleothermometer relationships), and these reconstructed temperatures are matched to some time of year by comparison with a historically derived annual SST curve.

In an ideal world, we would not have to take this circuitous route to establish the time of year at which a shell was harvested. One could imagine a large-scale experiment in which large numbers of mussels are harvested all throughout the year (in our region of archaeological interest); then, their shells would be sampled according to whatever sampling method we prefer, and subsequently measured for oxygen isotope content. We would thereby build up an extensive database of mussels with known harvest dates and known isotope profiles, which we could use as a comparative collection for archaeological mussel specimens collected from sites in this region: if an archaeological mussel showed a similar isotopic profile to the "experimental" mussels harvested during—for example, May—then we could infer a May harvest date for that archaeological specimen. This is of course a simplistic illustration—our experiment would benefit from controlling such factors as mussel age, intertidal location, exposure to climatic oscillations, anomalous events, and so on, and our inferences from experimental to archaeological data would benefit from statistical control.

However, this thought experiment makes clear that the historical SST data that we have used to inform our picture of annual SST variation, in fact, has much more to offer us. By inverting the paleothermometer relationship, we may estimate the expected isotopic content of mussel calcite deposited on a given day in our historical record. If we make certain assumptions about mussel shell growth-rate, we may string together several days at intervals, in order to mimic our sampling method (in our case, TGB+4 at 2 mm intervals). By this method, the historical SST record becomes the basis for generating a myriad of "simulated" mussel shell isotope profiles, which have a known "harvest" date (i.e., the date which we have chosen for the TGB-equivalent estimated isotope value) and can thus be used in the same manner as the comparative database of "experimental" mussels envisioned above.

This is an essentially empirical approach, which relies on the following assumptions: (1) the historical SST dataset in use is comparable to the annual SST conditions experienced by the archaeological mussels of ultimate interest, (2) the paleothermometer relationship is reliable for the species in question, and (3) the shell growth-rate variation for this species is approximately known. All three of these assumptions are also required by the traditional approaches described above, and our approach in fact weakens (i.e., makes less stringent) the first two assumptions, since a correlation-based approach (described below) means that (1) the historical SST dataset need only exhibit similar *relative* temperature variation as would have occurred during the archaeological period in question, and (2) the paleothermometer need only reliably record *relative* temperature variation within a given specimen; that is, not all mussels need share exactly the same paleothermometer relationship, so long as each organism is *internally* consistent. Regarding growth-rate, our quantitative approach at least requires that we make this assumption explicit, so that it may be criticized or varied in a principled manner.

Methodology

Sample Selection and Preparation

The stable oxygen isotope geochemistry of (n=40) archaeological *Mytilus californianus* specimens were analyzed from three Indigenous archaeological sites. Analysis specimens were selected systematically and based on meeting three parameters: 1) a visible terminal growth band (TGB) on the specimen, 2) specimen is from intact deposit with established AMS dates, and 3) the specimen is free of any observable pre- and post-life trauma, pathology, or modification (e.g., burned, predation scars). The first two specimens selected randomly from a context that met the criteria above were chosen for isotopic sampling, reflecting a total of 20 archaeological contexts. The samples were first cleaned with a gentle brush and rinsed with deionized water to remove loose sediment before being placed in an aqueous hydrochloride (50% HCL diluted with water) solution for 30 seconds to abrade the surface of any organic contaminants. The samples were then placed in an ultrasonic bath for 60 seconds and rinsed before drying in an oven at 30 degrees Celsius overnight.

Procedures for Sampling Biogenic Carbonate

The goal of sampling is to obtain readings for locations on the shell that are assumed to correlate with specific chronological markers (Claasen 1998; Jew et al. 2014). A tungsten carbide 0.5 mm engraving bit attached to a Dremel rotary tool at medium speed was used to obtain a spot sample transect along the lengthwise growth axis of *M. californianus* samples. Five spot samples of approximately 20–100 µg of powdered upper shell layer were collected per sample to generate isotopic profiles for each specimen. In most cases, we elected to collect additional carbonate material adjacent to the spot sample to ensure adequate weights of carbonate material were collected for Mass Spectroscopy.

The first sample taken is of the TGB, or the final growth band that was precipitated by the specimen immediately preceding its harvest, and subsequent death. Then, four additional samples are taken along the growth axis at 2 mm increments (a total distance of 8 mm from TGB) per sample (Figure 10.1). Each spot sample was assigned a letter, so the first spot sample from the fifth specimen is labeled as #5A, and sample #5E is 8 mm apart from the TGB on the same specimen. This sampling method resulted in a total of 200 carbonate microsamples from 40 archaeological mussel specimens. This method was guided by other oxygen isotope studies on *M. californianus* in coastal California that apply similar spot-sampling methods (Culleton et al. 2009; Jew et al. 2013; Cuthrell 2013; Schneider 2015; Harold et al. 2019).

The microsamples were submitted to the Center for Stable Isotope Geochemistry at UC Berkeley for analysis in a GV IsoPrime mass spectrometer with a long-term external precision of ± 0.07‰. The $\delta^{18}O$ results are expressed as the difference between carbonate VPDB standard and the absolute carbonate value as parts per mil.

Figure 10.1. Schematic of Spot Sampling Strategy.

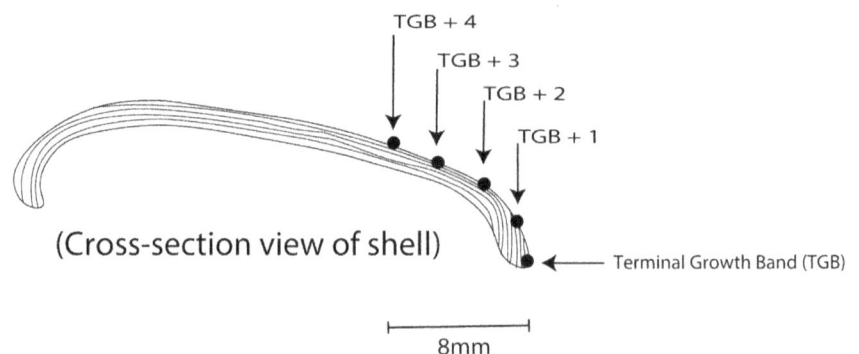

Modeling Isotopic Data from Archaeological Samples

A century-long instrumental SST record from the southern end of Monterey Bay at the Pacific Grove Shore Station (36° 37.3′ N, 121° 54.2′ W) maintained by Hopkins Marine Station was used. SST was recorded each morning between 0600 and 0900 hours, and conditions are considered representative of the Pacific swell (Barry et al. 1995). It is plausible that this record also represents SST conditions as they would have been experienced by *M. californianus* living in the intertidal zone slightly north of Monterey Bay as well, particularly when allowing for differences in *absolute* temperature, as mentioned above. The 40-year period (1920–1959) was chosen to encompass one cool and one warm period in global SST—as outlined by Jones and colleagues (1991; see also Sagarin et al. 1999)—without skewing trends toward more recent increases in summer temperatures (Barry et al. 1995: Figure 2).

We approximated $\delta^{18}O_{seawater}$ by using the 1919–1975 average salinity value (33.52 per mil) reported in Fox and colleagues (1991) and geolocated salinity-$\delta^{18}O_{seawater}$ data from the NASA Global Seawater Oxygen-18 Database (Schmidt et al. 1999) to estimate a $\delta^{18}O_{seawater}$ of -30.37 per mil PDB. No correction for differing paleosalinity or paleo-$\delta^{18}O_{seawater}$ was made, since such corrections are negligible (~0.01 per mil) during the period we are interested in, i.e., the last 6,000 years (Legrande and Schmidt 2009, Figure 4).

We used the best-estimate oxygen isotope paleothermometer equation developed by Ford and colleagues (2010: Eq. 4). We have also tested various paleothermometer relations, including those of Kim and O'Neill (1997) and Ford and colleagues (2010), with upper and lower 95% confidence limits. Since these relations are all relatively linear over a small range of temperatures and we are interested in relative—rather than absolute—temperature variation, the choice is insignificant.

As noted, the growth-rate assumption is the most challenging. For lack of empirical study of *M. californianus* growth patterns on the Central Coast, we estimate plausible values based on existing studies elsewhere. Behrens Yamada and Dunham (1989) note a range of possible growth-rates from Santa Barbara north toward British Columbia: 2.3 to 0.4 mm/month. Smith and colleagues (2009) record growth rates on the northern coast (i.e., at the Golden Gate and northward) of about 1 mm/month and below, and growth rates south of the Bight are around 2 mm/month and above. Both studies (Behrens Yamada and Dunham 1989; Smith et al. 2009) describe relatively continuous shell growth throughout the year, as do Ford and colleagues (2010), though the latter records a growth-rate of ~3 mm/month (as expected in the relatively warmer waters of San Diego). Based on this north-south variation in growth-rate, we assume a 1.5 mm/month constant growth-rate for our specimens. We also test 1.0 and 2.0 mm/month growth-rates as to their effect on seasonality determinations. Since isotope samples are not taken with daily precision, they generally incorporate calcite from throughout an entire tidal growth band cycle. We estimate, based on growth-line counts reported in Ford and colleagues (2010), that a typical *M. californianus* growth band forms over ca. 13 days. These estimates then allow us to simulate mussel isotope profiles similar to those of mussels that would have been collected in archaeological sites.

Statistical Procedures

Since our historical SST data have daily resolution, in principle, we could seek daily resolution in our harvest estimates as well. However, this level of resolution seems unnecessary for our present interpretive aims and would greatly increase computational burden, so we seek monthly resolution for the current analysis.

We must choose which metric we will use to assess the similarity between two isotope profiles. As alluded to above, a straightforward approach is to use correlation, which essentially measures the (scale-independent) similarity in shape between two profiles. That is, if each profile shows two peaks, the two will be highly correlated, no matter if the peaks of one profile are quite high, while the others are relatively low. This is helpful in the event that different organisms tend to precipitate calcite within slightly different $\delta^{18}O$ ranges. In a similar vein, the average annual SST may vary considerably between two years separated by several decades, while the shape of the SST curve does not; a correlation-based approach would not be stymied by this change in *absolute* temperature values. This latter point is especially important for comparing archaeological shells that grew under ancient oceanic conditions against a modern SST record. Some authors have addressed this issue by trying to use a densely sampled archaeological specimen, for example, to reconstruct absolute paleo-SST (e.g., Jew et al. 2013). However, this records only a short snapshot of variation—perhaps the year sampled was an anomalously cold one—and assumes that the chosen shell records in an isotopic range are representative of all other shells in the assemblage, which is far from guaranteed.

We then compared the simulated mussel profiles against all others. We group the simulated profiles by harvest month and compare each profile to the others in turn. For each profile, we compare it first against a large sample of January-harvested profiles, then against February-harvested profiles, and so on, each time recording the average (median) correlation value, so that each simulated profile generates a list of twelve median correlation values, one for each month. Then, we group the profiles together by harvest month and plot twelve distributions of correlation values for each (monthly) group, for example: (1.1) the distribution of average correlation values between January profiles and all (other) January profiles; (1.2) the distribution of average correlation values between January profiles and all February profiles; (1.3) the distribution of average correlation values between January profiles and all March profiles; (1.12) the distribution of average correlation values between January profiles and all December profiles; (2.1) the distribution of average correlation values between February profiles and all January profiles; (2.2) the distribution of average correlation values between February profiles and all (other) February profiles; and so on.

Each of these distributions may be used as a sort of reference table to answer the question of how likely it would be for a profile from a given month to show a given correlation value with profiles from some reference month. This latter value, the correlation with a chosen reference month, is what we can in fact compute for our archaeological specimens. So, for any given archaeological specimen, we may compute its average correlation with simulated profiles from each of the twelve months, resulting in a list of twelve correlation values. We can then ask the question: "Assuming this archaeological mussel was harvested in February, how likely is it that it would have this set of twelve correlation values with the twelve months?" We can then compare each of the twelve possibilities (e.g., "Assuming this mussel was harvested in March...") and determine which is most likely to have produced the pattern of correlations that we do, in fact, see.

We describe this process in qualitative terms here, for ease of understanding only. As elsewhere in our approach, these steps have been automated and made quantitative: the average correlations are computed empirically from the simulated historical profiles; the monthly distributions are interpolated by kernel density estimation; the relative probability of harvest in any particular month is calculated using Bayes' Rule; and the ensemble of monthly harvest probabilities for a given specimen is presented as a probability mass function.

Results

The most probable month of harvest for each archaeological mussel specimen is presented in Table 10.2. The reconstructions presented in the table rely on the Pacific Grove 1920–1960 SST dataset, and an assumed mussel shell growth-rate of 1.5 mm/month, as discussed above. Each archaeological specimen showed an isolated peak in its probability distribution over months of the year, so that these maximum probability estimates accord well with the probability mass functions shown in Figure 10.2. These latter represent the probability that the specimen in question was harvested during a given month, under the assumptions outlined above as regards growth-rates, harvesting locations, and so on.

Table 10.2. Estimated (Most Likely) Month of Harvest by Specimen, Listed with Site and Associated Radiometric Dating Information.

Specimen ID	Month of Harvest (maximum probability estimate)	Site ID	2-Sigma Date Range (BP) (upper and lower bounds only)	Calibration Curve
1	May	SMA-216	650–310	INTCAL09
2	Apr	SMA-216	650–310	INTCAL09
3	Jan	SMA-216	650–310	INTCAL09
4	Jul	SMA-216	650–310	INTCAL09
5	Apr	SMA-216	650–310	INTCAL09
6	Dec	SMA-216	650–310	INTCAL09
7	Nov	SMA-216	650–310	INTCAL09
8	Apr	SMA-216	650–310	INTCAL09
9	May	SMA-216	650–310	INTCAL09
10	Nov	SMA-216	650–310	INTCAL09
11	Dec	SMA-216	650–310	INTCAL09

Table 10.2. Continued.

Specimen ID	Month of Harvest (maximum probability estimate)	Site ID	2-Sigma Date Range (BP) (upper and lower bounds only)	Calibration Curve
12	Feb	SMA-216	650–310	INTCAL09
13	Nov	SCR-14	255–30	INTCAL13
14	Feb	SCR-14	255–30	INTCAL13
15	Dec	SCR-14	685–540	INTCAL13
16	Nov	SCR-14	685–540	INTCAL13
17	Nov	SCR-14	790–740	INTCAL13
18	Jan	SCR-14	790–740	INTCAL13
19	Nov	SCR-7	6635–6505	INTCAL13
20	Jan	SCR-7	6635–6505	INTCAL13
21	Nov	SCR-7	6735–6665	INTCAL13
22	Mar	SCR-7	6735–6665	INTCAL13
23	May	SCR-7	5890–5715	INTCAL13
24	Nov	SCR-7	5890–5715	INTCAL13
25	Dec	SCR-7	6730–6645	INTCAL13
26	Mar	SCR-7	6730–6645	INTCAL13
27	Sep	SCR-7	4405–4180	INTCAL13
28	Apr	SCR-7	4405–4180	INTCAL13
29	Jan	SCR-7	4415–4255	INTCAL13
30	Nov	SCR-7	4415–4255	INTCAL13
31	Feb	SCR-7	5725–5645	INTCAL13
32	Oct	SCR-7	5725–5645	INTCAL13
33	May	SCR-7	4530–4440	INTCAL13
34	Sep	SCR-7	4530–4440	INTCAL13
35	Mar	SCR-7	5585–5470	INTCAL13
36	Jan	SCR-7	5585–5470	INTCAL13
37	Jan	SCR-7	4235–4095	INTCAL13
38	Feb	SCR-7	4235–4095	INTCAL13
39	Sep	SCR-7	4250–4150	INTCAL13
40	Nov	SCR-7	4250–4150	INTCAL13

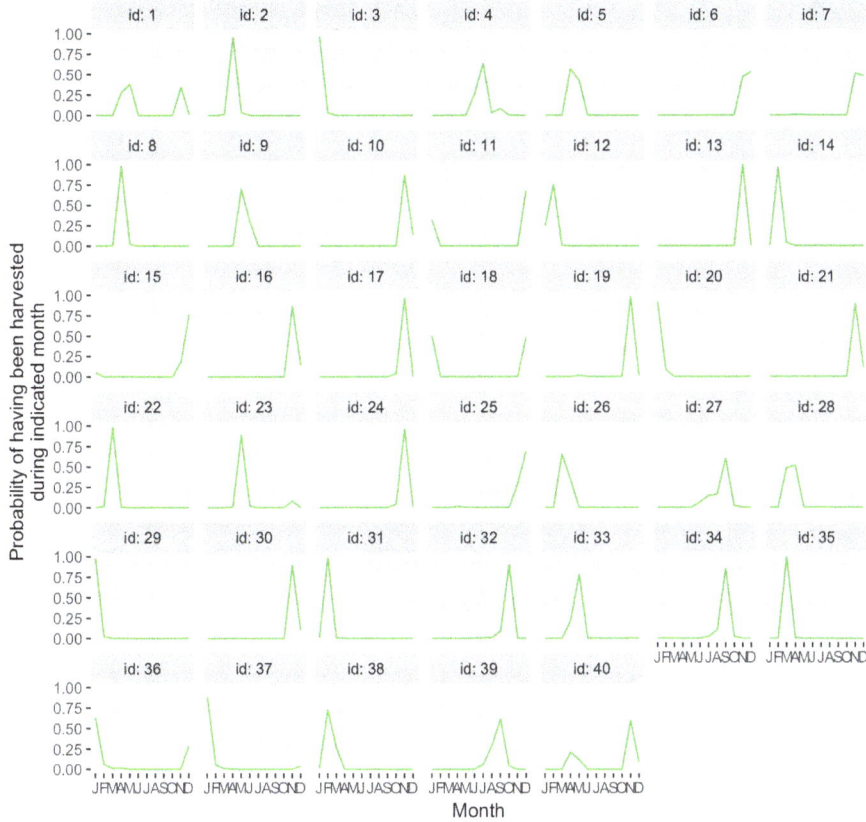

Figure 10.2. Probability of Harvest by Month per Archaeological Specimen.

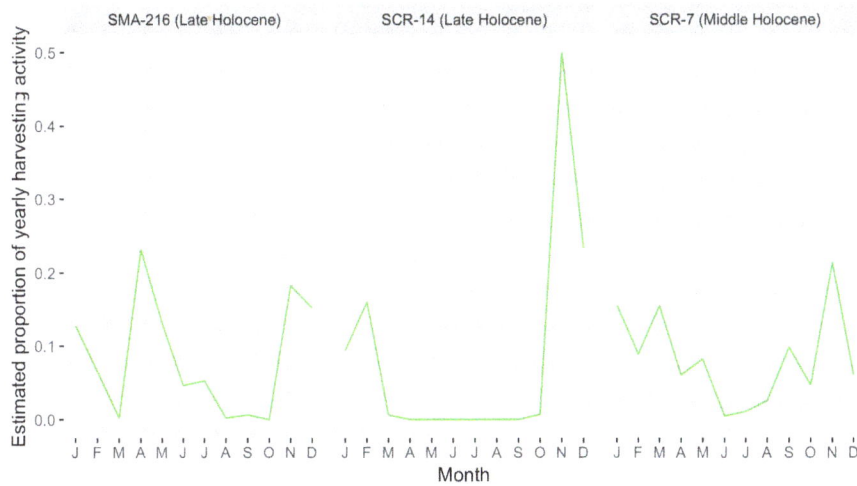

Figure 10.3. Harvesting Activity Averages per Month.

The specimen-by-specimen estimates are summed over sites in Figure 10.3. These whole-site summaries may be viewed as estimates of the proportion of shellfish-harvesting activity taking place during each month of the year. Note, however, that the distinction between shellfish harvesting and on-site shell deposition must be closely considered, especially in the case of the inland site, CA-SCR-14.

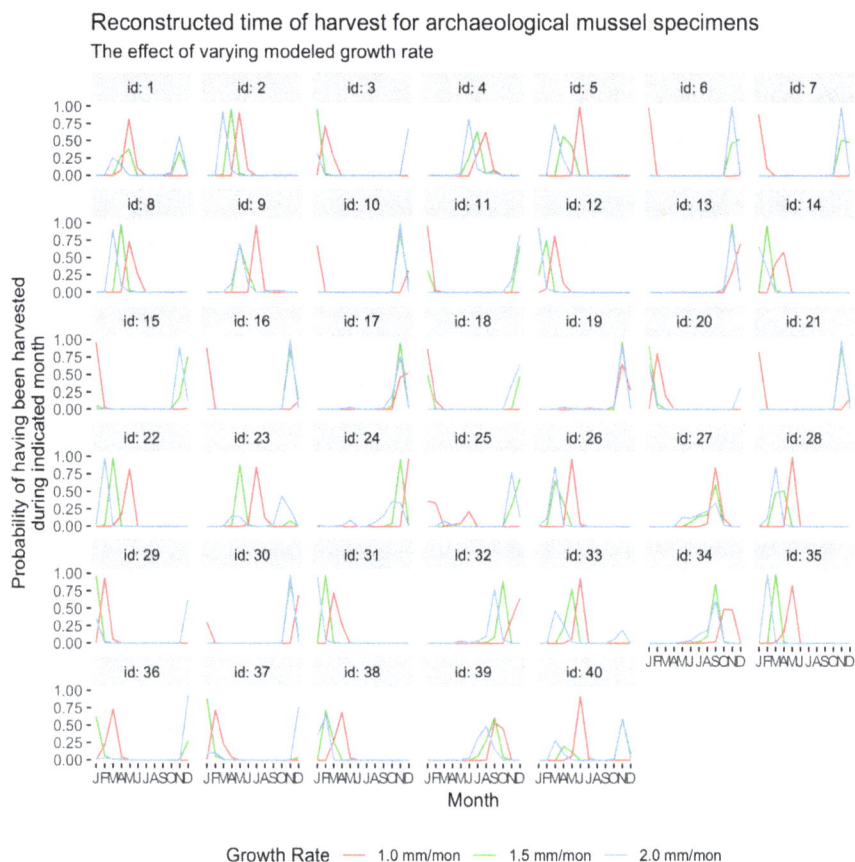

Figure 10.4. Probability of Harvesting Month per Archaeological Specimen with Effect of Differential Growth Rate.

Comparing Sources of Variation in Archaeological Isotope Modeling

Known growth rate functions for mussels at different latitudes along the California coast were applied to our archaeological dataset as an attempt to understand how this may affect our results and seasonality interpretations. Figures 10.4 and 10.5 show results when 2.0 mm and 1.0 mm per-month growth-rates are applied to the archaeological materials. Based on the graphs, it is reasonable to assume that isotopic seasonality estimates can be affected by changes in the growth rate, even with distances as small as half a millimeter. Despite these small-scale effects, most of the reconstructed trends stay generally consistent with a fair degree of overlap. If changes in growth rates remain in reasonably small magnitudes, on the order of plus or minus 0.5 mm, then growth rate appears not to be a major issue. This finding is consistent with the results from a study on oxygen isotope fractionation and mussel growth rates in Southern California (Ford et al. 2010).

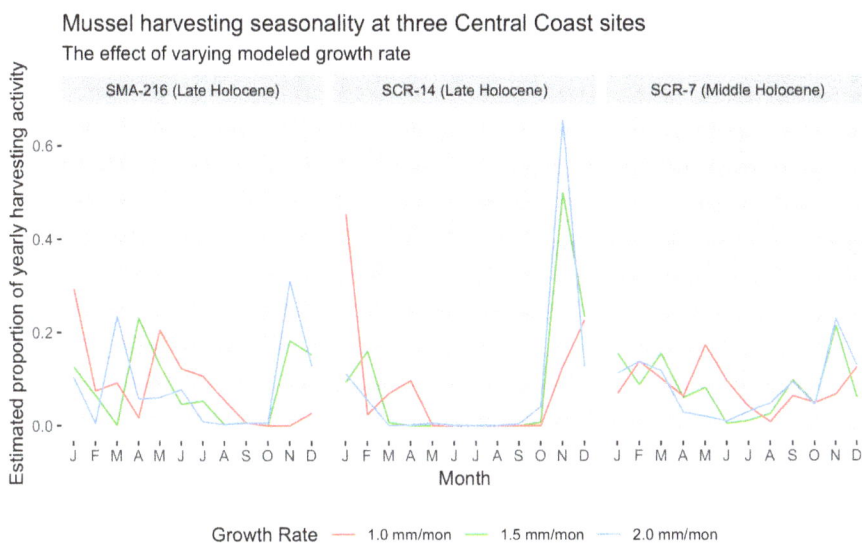

Figure 10.5. Harvesting Activity per Month with Effect of Differential Growth Rate.

Another internal test that we conducted was exploring the effect that spatial and temporal variation in the historical SST record may have on how isotopic values are reconstructed for seasonality. Figure 10.6 displays three SST records applied to the archaeological dataset: 1) Pacific Grove 1920–1960 (in red), 2) Pacific Grove 1975–1995 (in green), and 3) Granite Canyon 1975–1995 (in blue). The latter two 1975–1995 records are used so that a direct comparison may be drawn between Pacific Grove and Granite Canyon; the latter is a younger SST recording station and does not have instrumental records from before 1975. Only very slight differences can be observed in the reconstructed season of harvest between the two Pacific Grove stations—unimportant for our interpretations. However, the Granite Canyon-based reconstructions differ dramatically in estimated peak harvesting season. We believe that Pacific Grove better represents the SST conditions under which the archaeological mussels from the three sites studied here were collected (primarily because Granite Canyon is significantly further south and thus more distant from these sites). However, ca. 70–90 km of shoreline distance separate Pacific Grove on the Monterey Peninsula from the sampled archaeological sites on the northern Santa Cruz coast, and we must consider the possibility that the local SST conditions at shellfish-harvesting locales differed from those assumed based on the Pacific Grove record. As seen in Figure 10.6, such discrepancies may have a significant effect on interpretations. Therefore, seasonality interpretations drawn in this study must be considered tentative until more detailed oceanographic justifications for comparability of SST conditions between instrumental records and presumed shellfishing locales can be presented.

Figure 10.6. Harvesting Activity Averages per Month with Effect Differential SST Location and Dates.

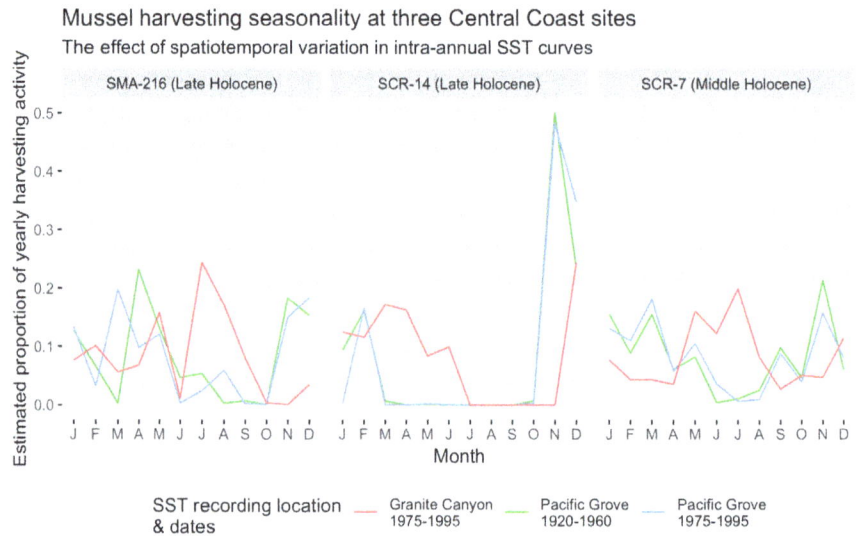

Figure 10.7. Seasonality Reconstruction by Individual Specimen.

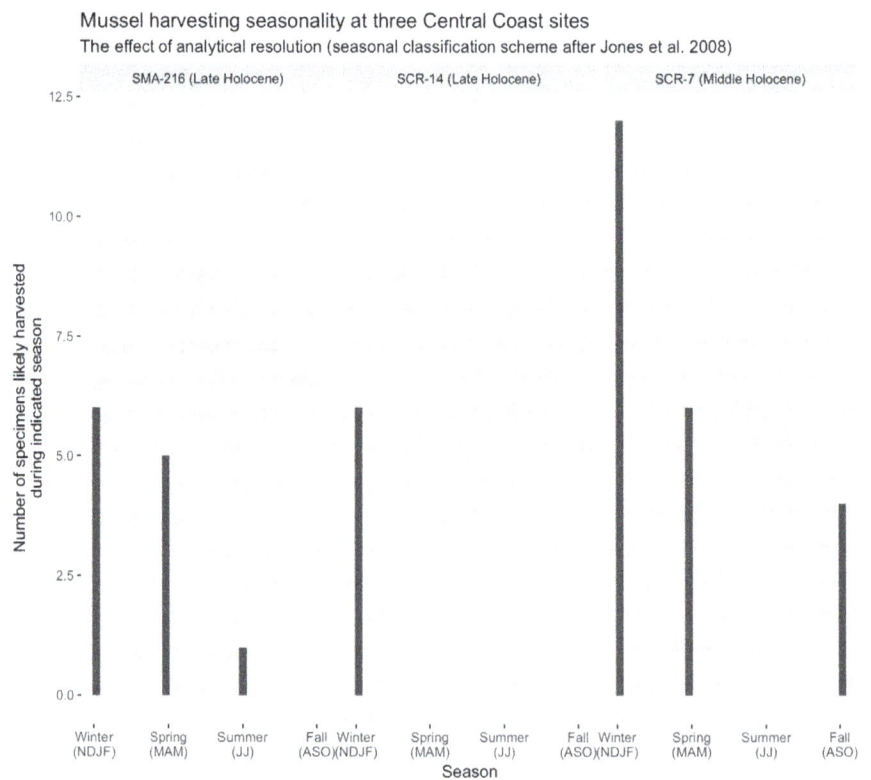

We also found that the way we displayed our data impacted the way we interpreted the results of the isotopic reconstructions. Using discrete seasonal categories—such as spring, summer, fall, and winter in bar charts—is helpful to determine the general proportionality of harvest estimates, but some values can occur in the middle between two categories, which forces the analyst to subjectively decide which seasonal category to place an ambiguous value in. To balance out this perceived issue, displaying the seasonality estimates at higher resolution throughout the year can paint a different picture. Using both approaches to display the seasonality reconstructions can provide a more flexible and realistic interpretation of the results. Figure 10.7 indicates that peaks in the high-resolution data are in general accordance with the trends in the bar charts; however, the high-resolution data reveal that the potential for year-round harvesting cannot be ruled out as a possibility.

Seasonality Reconstructions for CA-SCR-7, CA- SCR-14, and CA-SMA-216

Now that we have demonstrated the relative robustness of site-level seasonal shellfish-harvesting trends, with respect to varying growth rates (between 1.0 and 2.0 mm/month) and dates of historical SST record (using 1920–1960 and 1975–1995 data), let us focus on the features that appear in Figure 10.3. We will take this graph as our major point of departure in isolating trends and making interpretations from here forward. First, we observe the following:

1. CA-SCR-7 (n=22) shows harvesting picking up in fall (Sept–Oct), peaking in winter (Nov–Mar), decreasing through spring (Apr–May), and reaching its lowest levels of activity in summer (Jun–Aug);
2. CA-SCR-14 (n=6) presents a distinct harvest period during winter (Nov–Feb) and little indication of activity at other times of the year;
3. CA-SMA-216 (n=12) shows a longer harvesting period during winter and spring (Nov–May), with little activity during summer and fall.

To generalize, we may say that CA-SCR-7 shows nearly year-round harvesting with a lull in summer and a peak in winter. CA-SMA-216 shows a similar pattern, but perhaps stronger seasonality: winter-spring activity and summer-fall hiatus. Meanwhile, CA-SCR-14 shows a strong seasonality signal, with no evidence for shells harvested during any season other than winter. CA-SMA-216 seems to follow the same basic seasonal rhythms of shellfishing as CA- SCR-7, but is perhaps more focused during peak season, i.e., winter-spring. (Note, however, that this apparent contrast may be a consequence of a smaller sample size.) Meanwhile, CA-SCR-14 shows a considerably restricted period of mussel consumption, perhaps associated with the onset of peak shellfishing season.

Discussion

It appears that there was continuity in shellfishing practices from Middle to Late Holocene, at least between coastal sites. Both CA-SCR-7 and CA-SMA-216 show the same broad annual rhythm: peak in winter, hiatus in summer. However, there is suggestive but inconclusive evidence for a longer summer-fall hiatus at CA-SMA-216, possibly indicating a focusing of shellfishing activity during the peak winter-spring harvesting period. If true, this would have left time during summer and fall for the harvesting of grassland and parkland resources. The strongest departure from Middle Holocene (or perhaps simply coastal) patterns of on-site shellfish use occurs at CA-SCR-14. The season-of-harvest of shells recovered from CA-SCR-14 overlaps only a small part of the longer harvest season shown in the broadly contemporaneous Late Holocene site CA-SMA-216. Perhaps this is because the two sites are ~20 km apart and not directly associated (but CA-SCR-7 does show a similar harvest pattern to CA-SMA-216, even if at a much earlier date). Perhaps, instead, the discrepancy is due to a disjuncture between shellfish harvesting and shellfish transport to the inland village site that CA-SCR-14 represents. Mussels may have been brought back to the village only during "peak mussel season" in wintertime, despite the fact that mussels were still being gathered and eaten at oceanfront sites throughout the spring.

Further research is needed at more and different sites (including sites further inland and during different time periods, especially during the millennia that intervene between CA-SCR-7 and the other sites studied here), in order to elucidate the suggestive contrasts exhibited here. As an aside, we note that the diachronic spread of the contexts sampled within CA-SCR-7 does not seem to significantly skew our seasonality estimates. As shown in Figure 10.8, the three major temporal divisions sampled from this site pattern rather similarly with respect to season of harvest.

Figure 10.8. Diachronic Harvesting Activity by Chronological Period.

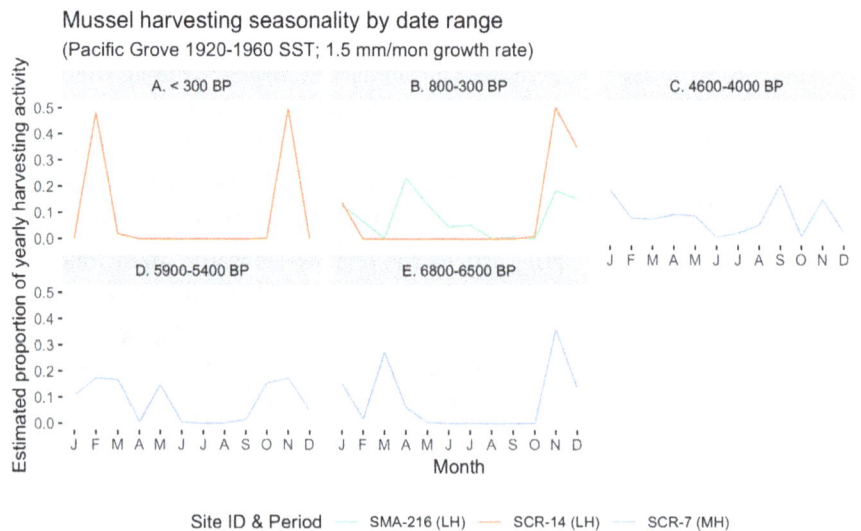

244

Both coastal middens CA-SCR-7 and CA-SMA-216 appear to follow a similar seasonal rhythm of shellfishing, but CA-SMA-216, which reflects Late Holocene deposits, has a more focused winter harvesting activity and hiatus in summer. This interpretation is generally consistent with other evidentiary lines, such as ethnographic testaments of shellfishing described by Grone (Chapter 9, this volume). The widely repeated folk wisdom (not apparently California-specific) that counsels to only consume shellfish during "months with an 'R'" marks September through April, coinciding with our harvesting reconstructions. A pair of ecological and logistical reasons for this might be (1) the danger of summertime harmful algal blooms in causing shellfish toxicity (Anderson 2005: 53) and (2) the easy availability of shellfish at low tide during the winter (Anderson 2005: 52). Anderson (2005: 53) also notes that Coastal Pomo people would cease harvesting shellfish when elderberry (*Sambucus* sp.) began to flower and resume when the elderberries ripened. However, the species of elderberry is not specified in this Coastal Pomo example, and there are differing flowering times between the two common taxa of elderberry along the Santa Cruz coast: blue elderberry (*Sambucus nigra* ssp. *Caerulea*) and red elderberry (*Sambucus racemosa*). Red elderberry blooms during the summer, which would closely match our current interpretations of mussel harvesting seasonality, while blue elderberry has a much longer flowering period from spring into summer.

Incorporating other lines of seasonality evidence, such as archaeobotanical and faunal data, has also been challenging. Although CA-SCR-14 yielded plant remains such as nutshells and edible grassland seeds, these foods may have been stored and eaten at a later time after harvest. There were also limited amounts of avifaunal and faunal remains recovered, possibly due to preservation issues. Measuring heavier isotopes, such as strontium of bird bone and deer bone, and measuring isotopic ratios in other ecofacts, such as terrestrial gastropods, small rodents, and others, may reveal additional information regarding the seasonality of site use and the overall environmental conditions. In the future, we intend to expand oxygen isotope sampling of other taxa previously studied in California such as clams, abalone, and urchin.

Conclusion

Given the results from the oxygen isotope seasonality analysis presented above, there is a proportionally greater likelihood that mussel harvesting occurred along a broader seasonal window for the coastal midden sites (CA-SCR-7 and CA-SMA-216). This differs from the readings obtained from isotope samples from the Late Holocene inland village CA-SCR-14, where the shellfish recovered appear to have been harvested during a much tighter seasonal window consistent with winter months. The studied sites show that mussel harvesting patterns have a stronger relationship with site type and spatial proximity to shoreline rather than as a function of chronological time. Mussel harvesting activity was more concentrated during the winter at inland sites, and a much more common activity throughout the year for the sites adjacent to the coast. Considering the limited sample size, the interpretations regarding harvesting trends at the three sites should be considered tentative.

Another contribution of this study is describing the effect that spatial variation, differences in growth rate, and statistical approach may have on how oxygen isotope data from archaeological shell are modeled and interpreted. Based on these findings, we caution that the location where historical instrumental data on sea surface temperature is recorded should be as similar as possible to the marine conditions in which the archaeological mussel specimens may have been harvested (e.g., open coast, protected bay). The variability in the growth rates throughout a mussel's life cycle is a factor that should be closely considered as well. The rate of shell growth is influenced by more complex factors such as shifts in salinity, water temperature, nutrient availability, pathogenic histories, predation, and perhaps cultural harvesting practices themselves (e.g., stripping and plucking strategies; see Whitaker 2008; and Grone, Chapter 9, this volume). Also, the method in which quantitative results are displayed may also affect how the data is interpreted. Displaying data continuously rather than categorically may lead to more realistic interpretations of derived sea surface temperature values from ancient biogenic carbonate.

The efficacy of archaeological oxygen isotopes for supporting the contemporary goals of Indigenous marine stewardship and revitalizing foodways is still being understood. However, this study has occurred at an opportunistic time considering the AMTB is currently developing a Marine Stewardship Program that carries out surveys of traditional coastal cultural resources and seeks to rebuild connections with marine resources, such as mussels, through monitoring, researching, harvesting, eating, and stewarding. Isotopic reconstructions of ancestral materials provide an additional line of evidence regarding mussel harvesting and can be weighed against other archaeological, ethnographic, or Tribal knowledge. A more extensive isotope sampling project of archaeological materials can provide guidance to the traditional season of harvest of mussels and be implemented into the broader planning of Indigenous marine stewardship. A crucial next step would be to have Amah Mutsun Tribe members establish experimental intertidal research plots where mussel growth rates and environmental conditions can be closely monitored. This localized information about mussel growth and ecology could help track the effects that environmental variables have on isotopic reconstructions of seasonality.

Acknowledgments

This oxygen isotope study was funded by the National Science Foundation (BCS-1523648). The work would not have been possible without the support from members of the Amah Mutsun Tribal Band. We also thank Dr. Wenbo Yang for assisting with the analysis of carbonate samples at the Center for Stable Isotope Biogeochemistry on the University of California, Berkeley campus.

References

Anderson, M. Kat
    2005   *Tending the Wild: Native American Knowledge and the Management of California's Natural Resources.* University of California Press, Berkeley.

Bailey, Geoffrey N., Margaret R. Deith, and Nicholas J. Shackleton
    1983   Oxygen Isotope Analysis and Seasonality Determinations: Limits and Potential of a New Technique. *American Antiquity*: 39 0–398.

Barry, James P., Chuck H. Baxter, Rafe D. Sagarin, and Sarah E. Gilman
    1995   Climate-Related, Long-Term Faunal Changes in a California Rocky Intertidal Community. *Science* 267(5198): 672–675.

Behrens Yamada, Sylvia, and Jason B. Dunham
    1989   Mytilus Californianus, a New Aquaculture Species? *Aquaculture* 81(3–4): 275–284.

Burchell, Meghan, Marianne P. Stopp, Aubrey Cannon, Nadine Hallmann, and Bernd R. Schöne
    2018   Determining Seasonality of Mussel Collection from an Early Historic Inuit Site, Labrador, Canada: Comparing Thin-Sections with High-Resolution Stable Oxygen Isotope Analysis. *Journal of Archaeological Science: Reports* 21: 1215–1224.

Claassen, Cheryl
    1998   *Shells.* Cambridge University Press, Cambridge.

Culleton, Brendan J., Douglas J. Kennett, and Terry L. Jones
    2009   Oxygen Isotope Seasonality in a Temperate Estuarine Shell Midden: a Case Study from CA-ALA-17 on the San Francisco Bay, California. *Journal of Archaeological Science* 36(7): 1354–1363.

Cuthrell, Rob Q.
    2013   An Eco-Archaeological Study of Late Holocene Indigenous Foodways and Landscape Management Practices at Quiroste Valley Cultural Preserve, San Mateo County, California. PhD Dissertation, Department of Anthropology, University of California, Berkeley.

de Brauwere, Anouk, Fjo De Ridder, Rik Pintelon, Johan Schoukens, and Frank Dehairs
    2009   A Comparative Study of Methods to Reconstruct a Periodic Time Series from an Environmental Proxy Record. *Earth-Science Reviews* 95(3–4): 97–118.

Deith, Margaret R.

1985   Seasonality from Shells: an Evaluation of Two Techniques for Seasonal Dating of Marine Molluscs. In *Palaeoenvironmental Investigations: Research Design, Methods and Data Analysis,* edited by N. R. J. Feller, D. D. Gilbertson, and N. G. A. Ralph, pp. 119–130. Symposia of the Association for Environmental Archaeology, no. 5B, BAR International Series 258, Oxford.

Eerkens, Jelmer W., Alex DeGeorgey, Howard J. Spero, and Christophe Descantes

2014   Seasonality of Late Prehistoric Clamming on San Francisco Bay: Oxygen Isotope Analyses of Macoma Nasuta Shells from a Stege Mound, CA-CCO-297. *California Archaeology* 6(1): 23–46.

Finstad, Kari M., B. Lynn Ingram, Peter Schweikhardt, Kent G. Lightfoot, Edward M. Luby, and George Coles

2013   New Insights about the Construction and Use of Shell Mounds from the Geochemical Analysis of Mollusks: An Example from the Greater San Francisco Bay. *Journal of Archaeological Science* 40: 2648–2658.

Ford, Heather L., Stephen A. Schellenberg, Bonnie J. Becker, Douglas L. Deutschman, Kelsey A. Dyck, and Paul L. Koch

2010   Evaluating the Skeletal Chemistry of Mytilus californianus as a Temperature Proxy: Effects of Microenvironment and Ontogeny. *Paleoceanography* 25(1). doi:10.1029/2008PA001677.

Fox, J. P., T. R. Mongan, and William J. Miller

1991   Long-term Annual and Seasonal Trends in Surface Salinity of San Francisco Bay. *Journal of Hydrology* 122(1–4): 93–117.

Glassow, Michael A., Douglas J. Kennett, James P. Kennett, and Larry R. Wilcoxon.

1994   Confirmation of Middle Holocene Ocean Cooling Inferred from Stable Isotopic Analysis of Prehistoric Shells from Santa Cruz Island, California. In *The Fourth California Islands Symposium: Update on the Status of Resources*, edited by W. L. Halvorson and G. L. Maender, pp. 222–232. Santa Barbara Museum of Natural History, Santa Barbara, California.

Glassow, Michael A., Heather B. Thakar, and Douglas J. Kennett

2012   Red Abalone Collecting and Marine Water Temperature During the Middle Holocene occupation of Santa Cruz Island, California. *Journal of Archaeological Science* 39(7): 2574–2582.

Harold, Laura B., Brian F. Byrd, and Jill Eubanks

2019   Clam, Mussel, and Oyster Harvest at CA-SFR-114: Estimating Seasonality of Shellfish Harvest Using Oxygen Isotopes. *Proceedings of the Society for California Archaeology* 33: 97–115.

Hylkema, Mark Gerald

1991   Prehistoric Native American Adaptations Along the Central California
Coast of San Mateo and Santa Cruz counties. MA Thesis, Department of
Anthropology, San Jose State University.

2002   Tidal Marsh, Oak Woodlands, and Cultural Florescence in the Southern
San Francisco Bay Region. In *Catalysts to Complexity: Late Holocene Societies of the
California Coast*, edited by Jon M. Erlandson and Terry L. Jones pp. 233–262.
Cotsen Institute of Archaeology, University of California, Los Angeles.

Jew, Nicholas P., Jon M. Erlandson, Jack Watts, and Frances J. White

2013   Shellfish, Seasonality, and Stable Isotope Sampling: $\delta^{18}O$ Analysis of
Mussel Shells from an 8,800-year-old Shell Midden on California's Channel
Islands. *The Journal of Island and Coastal Archaeology* 8(2): 170–189.

Jew, Nicholas P., Jon M. Erlandson, Torben C. Rick, and Leslie Reeder-Myers

2014   Oxygen Isotope Analysis of California Mussel Shells: Seasonality and
Human Sedentism at an 8,200-year-old Shell Midden on Santa Rosa Island,
California. *Archaeological and Anthropological Sciences* 6(3): 293–303.

Jones, P. D., T. M. L. Wigley, and G. Farmer

1991   Marine and Land Temperature Data Sets: A Comparison and a
Look at Recent Trends. In *Developments in Atmospheric Science*, vol. 19, pp.
153–172. Elsevier.

Jones, Terry L., Douglas J. Kennett, James P. Kennett, and Brian F. Codding

2008   Seasonal Stability in Late Holocene Shellfish Harvesting on the Central
California Coast. *Journal of Archaeological Science* 35(8): 2286–2294.

Jones, Terry L., Richard T. Fitzgerald, Douglas J. Kennett, Charles H. Miksicek, John
L. Fagan, John Sharp, and Jon M. Erlandson

2002   The Cross Creek site (CA-SLO-1797) and its Implications for New World
Colonization. *American Antiquity* 67(2): 213–230.

Kennett, Douglas J., and Barbara Voorhies

1996   Oxygen Isotopic Analysis of Archaeological Shells to Detect Seasonal Use
of Wetlands on the Southern Pacific Coast of Mexico. *Journal of Archaeological
Science* 23(5): 689–704.

Killingley, John S.

1981   Seasonality of Mollusk Collecting Determined from O-18 Profiles of
Midden Shells. *American Antiquity* 46(1): 152–158.

Killingley, John S., and W. H. Berger

1979   Stable Isotopes in a Mollusk Shell: Detection of Upwelling Events. *Science*
205: 186–188.

Kim, Sang-Tae, and James R. O'Neil
   1997   Equilibrium and Nonequilibrium Oxygen Isotope Effects in Synthetic
   Carbonates. *Geochimica et cosmochimica acta* 61(16): 3461–3475.

LeGrande, A. N., and G. A Schmidt
   2009   Sources of Holocene Variability of Oxygen Isotopes in Paleoclimate
   Archives. *Climate of the Past* 5(3).

Mannino, Marcello A., K. D. Thomas, M. J. Leng, R. Di Salvo, and Michael P. Richards
   2011   Stuck to the Shore? Investigating Prehistoric Hunter-gatherer Subsistence,
   Mobility and Territoriality in a Mediterranean Coastal Landscape Through
   Isotope Analyses on Marine Mollusc Shell Carbonates and Human Bone Collagen.
   Quaternary International 244(1): 88–104.

Newland, Michael, Sandra Pentney, Reno Keoni Franklin, Nick Tipon, Suntayea
Steinruck, Jeannine Pedersen-Guzman, and Jere H. Lipps
   2019   Racing Against Time: Preparing for the Impacts of Climate Change on
   California's Archaeological Resources. In *Public Archaeology and Climate Change*,
   edited by Tom Dawson, Courtney Nimura, Elias Lopez-Romero, and Marie-
   Yvane Daire, pp. 115–125. Oxbow Books, Oxford, England.

Reeder, Leslie A., Torben C. Rick, and Jon Erlandson
   2012   Our Disappearing Past: AS GIS Analysis of the Vulnerability of Coastal
   Archaeological Resources in California's Santa Barbara Channel region. *Journal of
   Coastal Conservation* 16(2): 187–197.

Rick, Torben C., John A. Robbins, and Kurt M. Ferguson
   2006   Stable Isotopes from Marine Shells, Ancient Environments, and Human
   Subsistence on Middle Holocene Santa Rosa Island, California, USA. *Journal of
   Island & Coastal Archaeology* 1(2): 233–254.

Sagarin, Raphael D., James P. Barry, Sarah E. Gilman, and Charles H. Baxter
   1999   Climate-related Change in an Intertidal Community Over Short and Long
   Time Scales. *Ecological monographs* 69(4): 465–490.

Schneider, Tsim D.
   2015   Envisioning Colonial Landscapes Using Mission Registers, Radiocarbon,
   and Stable Isotopes: An Experimental Approach from San Francisco Bay. *American
   Antiquity* 80(3): 511–529.

Schweikhardt, Peter, B. Lynn Ingram, Kent G. Lightfoot, and Edward M. Luby
   2011   Geochemical Methods for Inferring Seasonal Occupation of an Estuarine
   Shellmound: A Case Study from San Francisco Bay. *Journal of Archaeological Science*
   38: 2301–2312.

Scripps Institution of Oceanography

    2021   Surface Salinity and Water Temperatures at Granite Canyon Marine
    Pollution Studies Labs, 1975–1995. University of California, San Diego.

    2021   Surface Salinity and Water Temperatures at Pacific Grove Hopkins
    Marine Station, 1920–1960, 1975–1995. Stanford University, California.

Shackleton, Nicholas J.

    1973   Oxygen Isotope Analysis as a Means of Determining Season of Occupation
    of Prehistoric Midden Sites. *Archaeometry* 15(1): 133–141.

Schmidt, G. A., G. R. Bigg, and E. J. Rohling

    1999   Global Seawater Oxygen-18 Database. https://data.giss.nasa.gov/o18data/

Smith, Jayson R., Peggy Fong, and Richard F. Ambrose

    2009   Spatial Patterns in Recruitment and Growth of the Mussel Mytilus
    Californianus (Conrad) in Southern and Northern California, USA, Two Regions
    with Differing Oceanographic Conditions. *Journal of Sea Research* 61(3): 165–173.

Wang, Ting, Donna Surge, and Jonathan M. Lees

    2015   ClamR: A Statistical Evaluation of Isotopic and Temperature Records
    in Sclerochronologic Studies. *Palaeogeography, Palaeoclimatology, Palaeoecology*
    437: 26–32.

West, Catherine F., Meghan Burchell, and C. Fred T. Andrus

    2018   Molluscs and Paleoenvironmental Reconstruction in Island and Coastal
    Settings: Variability, Seasonality, and Sampling. In *Zooarchaeology in Practice*,
    edited by C. Giovas and M. LeFebvre, pp. 191–208. Springer, New York.

Whitaker, Adrian

    2008   Incipient Aquaculture in Prehistoric California? Long-term Productivity
    and Sustainability vs. Immediate Returns for the Harvest of Marine Invertebrates.
    Journal of Archaeological Science 35(4): 1114–1123.

CHAPTER 11

Middle and Late Holocene Fisheries of the Santa Cruz County Coast

GABRIEL M. SANCHEZ

Archaeological investigations of the Santa Cruz County coast offer an exceptional opportunity to understand human-environmental relationships from the Terminal Pleistocene through the Holocene (Cartier 1989; Hylkema 1991, 2002; Jones et al. 2007; Sanchez et al. 2017). This chapter presents the results of a study directed toward understanding ancient and historic fishing practices along the Santa Cruz coast. I sampled four archaeological sites from the Middle Holocene to the Contact Period to investigate long-term human-fish relationships through low-impact excavations and fine-grained analyses (i.e., ≥2 mm).

Previous studies of the ancient and historic fisheries of Santa Cruz County have been limited, and based on materials recovered using coarse-grained recovery methods (Hylkema 1991; Jones and Hildebrandt 1990, 1994; Nims et al. 2016). The first part of the chapter synthesizes fisheries data from previous investigations using ≥3.2 mm or larger mesh. The second part of the chapter describes the findings from our current project employing small-scale excavations with fine-grained (≥2 mm) and coarse-grained (≥3.2 mm) recovery methods. Conducted at the landscape level through the theoretical framework of historical ecology, the study offers an exceptional opportunity to trace long-term fishing practices (Balée 1992, 1998, 2006, 2010, 2018). Finally, I situate the findings of this case study in relation to previous studies reporting fish remains from Santa Cruz County. The ultimate goal of the research is to determine the fishes used by Native peoples and employ knowledge of past fisheries in modern conservation and Tribal cultural revitalization efforts.

Background

Previous Research and Fisheries Data of the Santa Cruz Coast

Previous studies reporting fish remains from archaeological sites on the Santa Cruz County coast have been limited. The lack of fisheries data from this region is likely driven by a history of coarse-grained excavations and the use of large mesh sieves. As described below, even when large volumes of sediments are sampled and analyzed, most fish remains are recovered from 3.2 mm mesh sieves rather than 6.2 mm screens. This section summarizes the findings from these previous Santa Cruz coast excavations, including field methods, recovery techniques, excavation volume, and recovered fish remains (Appendix 11.1).

CA-SCR-7. CA-SCR-7, known as the Sandhill Bluff site, is an archaeological locality that contains shell midden deposits dispersed within remnant sand dunes (Figure 11.1). Previous investigations indicate the site may have been inhabited from 5,880–6,410 to 3,400–2,830 cal BP (Jones and Hildebrandt 1990). The site contains three loci with shell midden deposits that represent discrete, short-term, or multiple occupation episodes dispersed across a large site area. Early excavations occurred in the late nineteenth century by A. W. Saxe from the California Academy of Sciences (Moratto 1984). In 1950, the site was formally documented by surveyors from the University of California (Jones and Hildebrandt 1990). In 1989, fieldwork was conducted in loci 1 and 2, with most materials recovered using 6.4 mm mesh sieves. However, within locus 1, a 100x50 cm control unit and a 20x20 cm column sample for shellfish analysis taken from unit 9 were sifted using 3.2 mm sieves. Locus 2 was sampled using three 3x3 m units excavated with 6.4 mm mesh sieves. The fish remains recovered from the excavations by Jones and Hildebrandt (1990) totaled 3 number of identified specimens (NISP) and were recovered from a control unit in locus 1 (unit 12). The fish remains from the locus 1 control unit were recovered with 3.2 mm mesh (2 NISP). In addition, one specimen was recovered from the 20x20 cm shell column sample (taken from unit 9) and recovered with 3.2 mm mesh (1 NISP).

CA-SCR-10. CA-SCR-10 is located adjacent to Baldwin Creek and was recorded by researchers from the Department of Anthropology, University of California, Berkeley (UC Berkeley), in 1950. The field crew recovered projectile points, scrapers, and ground stone artifacts during that initial fieldwork. Human remains were also encountered and buried in an undisclosed location. The site was excavated in the 1990s by Jones and Hildebrandt (1994), and while the site excavations sifted 8,000 liters of sediment through 6.4 mm (6,700 l) and 3.2 mm (1,300 l) mesh sieves, only 25 bone fragments were recovered with no fish remains reported.

CA-SCR-123/38. Site CA-SCR-123/38 is located within Wilder Ranch State Park and adjacent to Wilder Creek. The site includes CA-SCR-123, the primary shell midden representing the basal deposits. Shell midden CA-SCR-38 is upslope of CA-SCR-123, but due to erosion, these deposits now overlay CA-SCR-123. A Mexican/American period (the 1830s–1850s) adobe is built upon and lies within these two shell midden deposits. CA-SCR-123/38 was excavated by Jones and Hildebrandt (1994) in the 1990s, and over 15,250 liters of sediment were recovered using 6.4 mm (12,250 l) and 3.2 mm (3,000 l) mesh sieves. However, only 3 NISP fish remains were recovered from these excavations.

Figure 11.1. An Overview Map of the California Coast with an Inset Map Showing the Sites Discussed. ▲ represent sites sampled in this study while sites previously sampled are designated by .

CA-SCR-9. CA-SCR-9 lies in the Santa Cruz mountains, ~6 km from the Pacific Ocean. The site dates from 3,080–2,750 cal BP to 1,190–905 cal BP. However, caution should be used in interpreting these dates as they are based on composite shell and charcoal samples (Nims et al. 2016) cultural, and environmental change. However, the precision of the archaeological analysis and interpretation is dependent on a firm understanding of the site chronology. Fish remains are reported by Nims and colleagues (2016) and include 18 NISP. In addition, Gobalet and colleagues (2004) reported on the fish remains from the site, as did Cristie Boone in Nims and colleagues (2016). These remains are from excavations that employed 6.4 mm and 3.2 mm mesh sieves. The bony and cartilaginous fishes represented at the site are medium- and large-bodied organisms, likely due to the large mesh sieves used during excavation. Nonetheless, these data suggest that people harvested fish from the rocky intertidal coastline and possibly nearby streams to capture surfperches, including pile perch, salmon and trout, monkeyface pricklebacks, lingcod, and cabezon.

CA-SCR-35. CA-SCR-35 is a shell midden site adjacent to Majors Creek that lies between Santa Cruz and the town of Davenport. It was recorded in 1950 as part of the California survey by archaeologists from UC Berkeley. Formal excavations at the site have occurred in two phases. First, in 1967, Gordon O'Bannon of Cabrillo Community College excavated the site. However, the materials from this excavation and field notes have never been relocated (Gifford and Marshall 1984). Second, in 1971, John Fritz of the University of California, Santa Cruz (UC Santa Cruz), conducted salvage excavations after the property owner encountered human remains during construction. The work by Fritz was a large-scale excavation with at least 69 1x1 m units excavated with sediments screened using 6.4 mm mesh sieves (Gifford and Marshall 1984).

Gifford and Marshall (1984) analyzed less than a quarter of the site assemblage, with no fish remains reported. However, subsequent analyses by Sweeney (1986) report on the fish remains from the site. W. I. Follett and Stuart Poss of the California Academy of Sciences identified the fish specimens in that report. These remains suggest that the fishes are comprised of New World silversides, skates, herrings, trout and salmon, and surfperches.

CA-SCR-44. Research at CA-SCR-44 near Watsonville was conducted from the 1950s through the 1990s; the latter work was part of a cultural resource management mitigation project. Radiocarbon dating of deer (*Odocoileus* sp.) remains and an abalone artifact from the site suggests occupation from 2,950–2,750 cal BP to 380–130 cal BP (Breschini and Haversat 2000). Faunal remains from the site were recovered with 1.59 mm mesh sieves (Gobalet and Jones 1995; Langenwalter 2000). These data suggest that a variety of freshwater and marine organisms are present, with the majority of the assemblage comprised of night smelt, suckers, and cyprinids, including hitch and Sacramento sucker. Sacramento perch and tule perch also make up a portion of the freshwater assemblage.

CA-SCR-60/130. Site CA-SCR-60/130 is located in dune remnants in the Pajaro River floodplain near the confluence of Watsonville and Harkins sloughs near Watsonville (Culleton et al. 2005; Gobalet et al. 2004). The site contains an Early Holocene component dated between 7,650 and 6,400 cal BP and a Middle Holocene component that dates between 4,790 and 4,150 cal BP (Culleton et al. 2005). The site was excavated using 3.2 mm mesh sieves as part of a mitigation project due to development. Four control units (CU18A-D) were analyzed to establish which fishes were present (Culleton et al. 2005). A total of 196 NISP, including cartilaginous fishes and marine and freshwater bony fishes, were recovered.

Summary. Five sites along the Santa Cruz coast have fish remains reported; most of these sites have high excavation volumes with a low density of fish remains recovered per liter (Table 11.1). The overall pattern from these previous excavations is a trend toward low fish densities, and low specimen counts. These issues affect archaeological interpretations regarding human-environmental relationships, and the use of archaeological fish remains to inform fisheries management and cultural revitalization efforts among local Indigenous groups.

Table 11.1. Summary of Fish Remains from Previous Excavations from the Santa Cruz Coast.

Site	Liters	Sieve (mm)	Fish NISP	Density (Per Liter)	Source
CA-SCR-7 (locus 1)	11600.0	6.4			Jones and Hildebrandt (1990)
CA-SCR-7 (locus 1)	500.0	3.2			Jones and Hildebrandt (1990)
CA-SCR-7 (locus 2)	8100.0	6.4			Jones and Hildebrandt (1990)
CA-SCR-7 (locus 2)	0.0	3.2			Jones and Hildebrandt (1990)
Total	20200.0		3.0	0.0001	
CA-SCR-9	8550.0	6.4			Nims et al. (2016)
CA-SCR-9	5300.0	3.2			Nims et al. (2016)
Total	13850.0		18.0	0.0013	
CA-SCR-10	6700.0	6.4			Jones and Hildebrandt (1994)
CA-SCR-10	1300.0	3.2			Jones and Hildebrandt (1994)
Total	8000.0		0.0	0.0	
CA-SCR-60/130					Culleton et al. (2005)
Total	5900.0	3.2	196.0	0.0332	
CA-SCR-123/38	12250.0	6.4			Jones and Hildebrandt (1994)
CA-SCR-123/38	3000.0	3.2			Jones and Hildebrandt (1994)
Total	15250.0		3.0	0.0002	

Sampling Biases and Fisheries Studies

Sampling biases in the field of archaeology, especially those derived by the use of coarse-grained recovery methods (e.g., ≥6.4 mm and 3.2 mm sieves), and their effects on the representation of archaeological materials and interpretations derived from these assemblages, have been a concern for decades (Barker 1975; Casteel 1970, 1972, 1976; Colley 1990; Fitch 1969; Gobalet 1989; Meighan 1950; Payne 1972; Thomas 1969; Wheeler and Jones 1989). Comparisons were made between fish faunas recovered from entire excavation units and sorted in the field by the excavators and fish faunas from column samples recovered in the same units and examined in the laboratory with the aid of a microscope. The results indicated that the smaller column samples provided adequate control for defining the number of species utilized and often surpassed the whole unit analyses by showing the presence of smaller individuals of species not recovered from the larger units. Microscopic examination of small samples appears to provide a much needed addition to the commonly utilized methods of sampling fish faunas. In general, zooarchaeological studies demonstrate that fisheries-based research requires fine-grained recovery techniques with at least >2 mm or >1.59 mm mesh sieves (Cannon 2000; Gobalet 1989; Fitch 1969; Moss et al. 2017; Sanchez et al. 2018). Archaeological research conducted with coarse-grained methods will typically recover a reduced range of organisms, overrepresenting large-bodied fishes and underrepresenting small- and medium-bodied fishes (Casteel 1972, 1976; Colley 1990; Moss et al. 2017; Ross and Duffy 2000; Sanchez et al. 2018; Shaffer 1992; Thomas 1969; Tushingham and Bencze 2013; Wheeler and Jones 1989). Research conducted with fine-grained recovery approaches is essential for identifying the full suite of fish species harvested and understanding capture techniques and technologies, such as mass-harvesting and net fishing. Past research in the study area has resulted in the limited recovery of fish remains (Culleton et al. 2005; Jones and Hildebrandt 1990, 1994; Langenwalter 2000; Nims et al. 2016). Therefore, consideration of recovery methods and their effects on fish representation in the region is highly relevant.

Introduction to Current Research

During the summers of 2016–2017, fieldwork was conducted at four sites—CA-SCR-7, CA-SCR-10, CA-SCR-14, and CA-SCR-15. The research is part of a collaborative eco-archaeological project involving students and faculty from UC Berkeley and UC Santa Cruz, the Amah Mutsun Tribal Band (AMTB), including members of the Amah Mutsun Native Stewardship Corps (NSC), and staff from California State Parks. The project was designed to assess the temporal and material records of sites threatened by sea level rise, ongoing disturbance, coastal erosion, and destruction from agricultural activities, and to evaluate two sites largely protected from significant impacts on private property. In addition, these sites were studied to contribute to the ongoing landscape and seascape management research. As part of this project, crew members surveyed, recorded, and tested the four archaeological sites, sampling from major habitats along the coast (i.e., open coast, reef sites, inland localities, and sites in the coast range), with all sites located adjacent to Laguna Creek.

Given the history of previous coarse-grained excavations at two of the four sites sampled in this study—CA-SCR-7 and CA-SCR-10—the most recent iteration of field research at these sites offers an opportunity to examine the kinds of sampling biases that exist in the coarse-grained assemblages when compared to the fine-grained data derived from our research. While variation between earlier excavations and assemblages and those recovered from this project may be driven by differences in excavation unit placement and context, a general illustration of sampling biases may be revealed through a broad comparison of fish NISP and density data.

Methods and Materials

The sites in this study were surveyed and sampled using low-impact and minimally invasive archaeological methodologies summarized in Lightfoot (2008). Through low-impact diagnosis of surface and near-surface materials supplemented by focused geophysical survey—including ground penetrating radar (GPR), resistivity, and magnetometry—the research design sought to avoid disturbing human burials while guiding the placement of excavation units. All field research was conducted in close collaboration between UC Berkeley faculty, post-doctoral fellows, graduate and undergraduate students, the AMTB, and the Amah Mutsun NSC. We evaluated sensitive contexts and findings with the AMTB through in-field discussion and consultation before initiating or continuing archaeological fieldwork or excavation (see Chapter 4, this volume).

To conduct surface and near-surface sampling of materials, we applied the "dog-leash" method (Binford 1964) along with "catch-and-release" survey, or in-field analysis of archaeological materials. The "dog-leash" technique serves as a practical and expedient survey tool. When combined with in-field documentation of artifacts returned to near their original context, this approach has been called the "catch-and-release" method (Gonzalez 2016; Gonzalez et al. 2006). The results of the catch-and-release and geophysical survey guided the placement of excavation units, column samples, and auger units (see Chapters 4 and 5).

Field and Laboratory Processing of Bulk Sediment Samples

Previous research by Casteel (1970, 1976), Cannon (2000), and others suggests that flotation samples derived from column and auger sampling serve as an accurate and efficient method for accessing the focus and intensity of site-specific and regional fisheries. Subsequently, we collected column samples, auger samples (10 cm diameter taken at 20 cm intervals), and bulk sediment samples from each excavation unit and individual levels, features, or contexts, and these samples were bagged *en toto* (Table 11.2).

Bulk sediment samples and associated artifactual materials were separated from matrix through water flotation at Wilder Ranch State Park, dividing materials into light and heavy fraction samples (see Chapter 8, this volume). In total, 538.1 liters of heavy fraction materials were analyzed in this study (Table 11.2). Samples were processed using a modified SMAP-type tank (Pearsall 2000) with 1 mm heavy-fraction mesh and ca. 0.2 mm light-fraction mesh.

Table 11.2. Summary of Fish Remains from Previous Excavations from the Santa Cruz Coast.

Site	Flotation Volume Analyzed (Liters)	Dry-Screen (3.2 mm) Volume (Liters)
CA-SCR-7	370.5	60
CA-SCR-10	41.6	1300
CA-SCR-14	48.5	250
CA-SCR-15	77.5	275
Total	**538.1**	**1885**

After drying the heavy fraction materials, samples were sieved at the California Archaeology Laboratory at UC Berkeley into the following size fractions through nested geologic sieves: >4 mm, 2–4 mm, and 1–2 mm. Heavy fraction materials were separated into artifact classes, and all archaeofaunal remains were sorted based on size classes. Archaeofaunas were identified from the >4 mm and 2–4 mm size fractions.

The recovered fish remains were identified using comparative collections from the California Academy of Sciences, San Francisco. Laboratory protocols and faunal identifications were conservative in identifying elements (Driver 2011; Gobalet 2001). Sanchez completed the faunal analysis, and Gobalet confirmed or revised Sanchez's identifications.

A dissecting stereomicroscope was used to discern diagnostic features that allowed the most exclusive taxon designation, usually a family. I follow Page and colleagues (2013) in using scientific and common names. Osteological and provenience data were recorded for each specimen, with the results cataloged in Microsoft Excel and quantified using the measure of NISP (Grayson 1984; Lyman 2008). With the minor exception of some elasmobranch remains, non-diagnostic specimens were identified as Actinopterygii.

Five measures are applied in this study. First, I calculated the relative abundance of identified skeletal specimens of a particular taxon in relation to the total (%NISP) NISP (Grayson 1984; Lyman 2008). Second, I calculated the NISP per liter (NISP/l) to measure density. Third, I measured the diversity of the samples by calculating the taxonomic richness, or the number of genera identified within the samples (Grayson 1984; Harper 1999; Lyman 2008), through the Shannon-Weiner Index in Paleontological Statistics software (PAST) version 3.22. The Shannon-Weiner Index typically ranges from 1.5 to 3.5 and rarely exceeds 4.5, with larger values signifying greater diversity or heterogeneity (Harper 1999; Lyman 2008; Magurran 1988). Fourth, I calculated the equitability of organism relative abundance—or evenness in PAST (Faith and Du 2017; Grayson 1984; Harper 1999; Lyman 2008; Magurran 1988). In measuring equitability through the Shannon-Weiner Index, values are between 0 and 1, with 1 indicating that all genera are equally abundant or even (Lyman 2008; Magurran 1988). Lastly, I compared the recovery rates of fish remains, and the diversity of organisms recovered from the sample excavation levels between paired coarse-grained (>3.2 mm) and fine-grained (>2 mm) mesh recovery methods.

Results

Zooarchaeological Analyses

The results of the faunal analyses suggest that the >3.2 mm and >2 mm Santa Cruz County assemblage includes 2,041 NISP. Much of the assemblage, ~63%, comprises NISP that could only be classified as Actinopterygii or ray-finned fish (Table 11.3).

CA-SCR-7. The most abundant organisms at CA-SCR-7 in the >2 mm samples are surfperches, which comprise ~28% of the assemblage (Table 11.4). Surfperches include pile perch, shiner perch, and barred, calico, or redtail surfperch. Greenlings comprise 21% of the assemblage by relative abundance and include lingcod, kelp, rock, or masked greenling. Skates make up 15% of the assemblage. Rockfishes make up another 10%. Together, surfperches, greenlings, rockfishes, and skates make up 74% of the assemblage, with 18 genera making up the remaining 26%. The density of fish remains recovered and identified within the assemblage to at least the taxon Actinopterygii is 3.5 NISP/l. The Shannon-Weiner Index for the assemblage, calculated at the genera level, is 2.4, suggesting greater heterogeneity, with evenness or equitability measured at 0.85, signifying an assemblage closer to equal distribution.

The dry-screened materials from CA-SCR-7 were from highly disturbed contexts within the plow zone of agricultural fields. Given that the contexts of the >3.2 mm samples are not paired with the >2 mm samples, as is the case at the other sites, I exclude the fish remains from CA-SCR-7 dry screening from further consideration.

CA-SCR-10. For CA-SCR-10, the quantification and analyses were separated by screen size and recovery method. Therefore, I divide those samples recovered through dry screening with >3.2 mm mesh sieves and those recovered through flotation methods with >2 mm sieves. At CA-SCR-10, the most abundant organisms recovered through flotation with >2 mm sieves were herrings at 34%, possibly including Pacific herring and Pacific sardine (Table 11.5). Surfperches comprise 23% of the assemblage and include rubberlip seaperch and barred, calico, or redtail surfperch. Northern anchovy comprises 11%, and New World silversides 8% of the site assemblage. Herrings, surfperches, and New World silversides totaled 75% of the site assemblage, with 11 genera making up the remaining 25%. The density of fish remains recovered and identified within the assemblage to at least Actinopterygii is 8.2 NISP/l. The Shannon-Weiner Index for the assemblage, calculated at the genera level, is 2.3, suggesting greater heterogeneity, with evenness or equitability measured at 0.85, signifying an assemblage closer to equal distribution.

For the dry-screened materials at SCR-10, the most abundant organisms recovered with >3.2 mm sieves are surfperches. They comprise 33% of the assemblage and include rubberlip seaperches and rockfishes at 25% (Table 11.6). Together, surfperches and rockfish make up 58% of the assemblage. Herrings, plainfin midshipman, greenlings, cabezon, and monkeyface pricklebacks make up the remaining 42%, and their relative abundance is evenly distributed among these organisms. The density of fish remains recovered and identified within the >3.2 mm assemblage to at least the class level is 0.2 NISP/l. The Shannon-Weiner Index for the assemblage, calculated at the genera level, is 1.7; this suggests less heterogeneity, with evenness or equitability measured at 0.78, signifying an assemblage moving away from equal distribution.

Table 11.3. Faunal Analysis Results with NISP by Site, Context, Screen Size, and Taxon Total Across Sites.

		CA-SCR-7 Column 1	CA-SCR-7 Column 2	CA-SCR-7 Column 3	CA-SCR-7 Column 4	CA-SCR-7 Column 5	CA-SCR-7 Column 6	CA-SCR-7 Column 7
	Screen size	>2 mm	>2 mm	>2 mm	>2 mm	>2 mm	>2 mm	>2 mm
Taxon	**Common name**							
SALT WATER FISHES								
Elasmobranchiomorphi			2		8			2
Rajidae	Skates				5	1		10
Raja sp.	Skate	1			19	7		12
Myliobatidae	Eagle rays							
Myliobatis californica	Bat ray		1					
Squalus suckleyi	Spiny dogfish							1
Lamnidae	Mackerel sharks							
Isurus sp.	Mako							
Actinopterygii	Ray-finned fishes	59	47	5	442	62	2	212
Engraulidae	Anchovies							
Engraulis mordax	Northern anchovy	1	2					
Clupeidae	Herrings	5	2	1	1		2	2
Clupea pallasii	Pacific herring				1			
Sardinops sagax	Pacific sardine							
Osmeridae	Smelts							
Spirinchus starksi	Night smelt							
Gadidae	Cods							
Microgadus proximus	Pacific tomcod	1	3					

CA-SCR-7 (Auger 2) >2 mm	CA-SCR-7 (Auger 3) >2 mm	CA-SCR-10 >2 mm	CA-SCR-10 >3.2 mm	CA-SCR-10 (Auger) >2 mm	CA-SCR-14 >2 mm	CA-SCR-14 >3.2 mm	CA-SCR-15 >2 mm	CA-SCR-15 >3.2 mm	CA-SCR-123/38 >2 mm	Taxon Total
1										**13**
		3								**19**
5		1			2					**47**
										1
1										**2**
						1				**1**
64	1	195	10	1	104	14	56	4	14	**1292**
		17			45		1		3	**69**
1		44	1		8	4				**71**
		1								**2**
		6		1					1	**8**
		2								**2**
					1					**1**
										4

Table 11.3. Continued.

	Screen size	CA-SCR-7 Column 1 >2 mm	CA-SCR-7 Column 2 >2 mm	CA-SCR-7 Column 3 >2 mm	CA-SCR-7 Column 4 >2 mm	CA-SCR-7 Column 5 >2 mm	CA-SCR-7 Column 6 >2 mm	CA-SCR-7 Column 7 >2 mm
Batrachoididae	Toadfishes							
Porichthys notatus	Plainfin midshipman							1
Atherinopsidae	New World silversides	2					1	1
Atherinopsis californiensis	Jacksmelt							1
Scorpaenidae	Scorpionfishes							
Sebastes sp.	Rockfishes	1			16	7		12
Hexagrammidae	Greenlings	6	5	1	8	8	4	19
Hexagrammos sp.	Greenlings				4	1		9
Ophiodon elongatus	Lingcod	1			9	1		1
Cottidae	Sculpins	4	8	1	2		1	1
Artedius sp.		1						
Clinocottus sp.		1	3	1	1			
Oligocottus sp.		1						
Scorpaenichthys marmoratus	Cabezon	1	2		4			2
Sciaenidae	Drums and croakers							
Genyonemus lineatus	White croaker							
Embiotocidae	Surfperches	20	21	3	24	6	1	19
Amphistichus sp.								
Damalichthys vacca	Pile perch	3	1		1			
Rhacochilus toxotes	Rubberlip seaperch							

CA-SCR-7 (Auger 2)	CA-SCR-7 (Auger 3)	CA-SCR-10	CA-SCR-10	CA-SCR-10 (Auger)	CA-SCR-14	CA-SCR-14	CA-SCR-15	CA-SCR-15	CA-SCR-123/38	Taxon Total
>2 mm	>2 mm	>2 mm	>3.2 mm	>2 mm	>2 mm	>3.2 mm	>2 mm	>3.2 mm	>2 mm	
		5	1							7
		12								16
										1
4		4	3		2	8	2	1	3	63
6		4			6	7	8	2	1	85
		2	1	1		1	2			21
1		1			2					16
		1			1		2			21
		1								2
1							1			8
										1
		1	1		2	5	2	1		21
						1				1
9	1	31	3		19	12	15	1		185
1		1					1	1		4
					6	1	3			15
			1							1

Table 11.3. Continued.

	Screen size	CA-SCR-7 Column 1 >2 mm	CA-SCR-7 Column 2 >2 mm	CA-SCR-7 Column 3 >2 mm	CA-SCR-7 Column 4 >2 mm	CA-SCR-7 Column 5 >2 mm	CA-SCR-7 Column 6 >2 mm	CA-SCR-7 Column 7 >2 mm
Stichaeidae	Pricklebacks							
Cebidichthys violaceus	Monkeyface prickleback			1				
Xiphister sp.	Black or Rock prickleback	1						
Clinidae	Kelp blennies							
Gibbonsia sp.	Spotted, Striped, or Crevice kelpfish	1			1			
Pleuronectiformes		1	1			1		1
Platichthys stellatus	Starry flounder							
FRESH OR SALT WATER FISHES								
Salmonidae	Trouts and salmons							
Oncorhynchus sp.		1	4			1	2	
Oncorhynchus mykiss	Rainbow trout							
Gasterosteidae	Sticklebacks							
Gasterosteus aculeatus	Threespine stickleback		1					
Cottidae	Sculpins							
Leptocottus armatus	Pacific staghorn sculpin		3		1			
Embiotocidae	Surfperches							
Cymatogaster aggregata	Shiner perch							
Total		112	106	13	547	95	13	306

	CA-SCR-7 (Auger 2)	CA-SCR-7 (Auger 3)	CA-SCR-10	CA-SCR-10	CA-SCR-10 (Auger)	CA-SCR-14	CA-SCR-14	CA-SCR-15	CA-SCR-15	CA-SCR-123/38	Taxon Total
	>2 mm	>2 mm	>2 mm	>3.2 mm	>2 mm	>2 mm	>3.2 mm	>2 mm	>3.2 mm	>2 mm	
						1					1
			1	1		1	2				6
			1								2
			1								1
											2
											4
	1										1
	1		1				1				11
										1	1
			1		1	1					4
	2										6
	2										2
	100	2	337	22	4	201	53	97	10	23	2041

Table 11.4. Results of Fish Analysis for CA-SCR-7 from the >2 mm Recovery Method with NISP by Context.

Taxon	Common name	CA-SCR-7 Column 1	CA-SCR-7 Column 2	CA-SCR-7 Column 3	CA-SCR-7 Column 4	CA-SCR-7 Column 5	CA-SCR-7 Column 6	CA-SCR-7 Column 7
	Screen size	>2 mm	>2 mm	>2 mm	>2 mm	>2 mm	>2 mm	>2 mm
SALT WATER FISHES								
Elasmobranchiomorphi			2		8			2
Rajidae	Skates				5	1		10
Raja sp.	Skate	1			19	7		12
Myliobatidae	Eagle rays							
Myliobatis californica	Bat ray		1					
Squalidae	Dogfish sharks							
Squalus suckleyi	Spiny dogfish							1
Actinopterygii	Ray-finned fishes	59	47	5	442	62	2	212
Engraulidae	Anchovies							
Engraulis mordax	Northern anchovy	1	2					
Clupeidae	Herrings	5	2	1	1		2	2
Clupea pallasii	Pacific herring				1			
Gadidae	Cods							
Microgadus proximus	Pacific tomcod	1	3					
Batrachoididae	Toadfishes							
Porichthys notatus	Plainfin midshipman							1

CA-SCR-7 (Auger 2)	CA-SCR-7 (Auger 3)	Total	Total w/o RFF	Relative Abundance w/o RFF
>2 mm	>2 mm			
1		13	13	0.03
		16	16	0.04
5		44	44	0.11
		1	1	0.00
1		2	2	0.01
64	1	894		
		3	3	0.01
1		14	14	0.04
		1	1	0.00
		4	4	0.01
		1	1	0.00

Table 11.4. Continued.

Taxon	Common name	Screen size	CA-SCR-7 Column 1 >2 mm	CA-SCR-7 Column 2 >2 mm	CA-SCR-7 Column 3 >2 mm	CA-SCR-7 Column 4 >2 mm	CA-SCR-7 Column 5 >2 mm	CA-SCR-7 Column 6 >2 mm	CA-SCR-7 Column 7 >2 mm
Atherinopsidae	New World silversides		2					1	1
Atherinopsis californiensis	Jacksmelt								1
Scorpaenidae	Scorpionfishes								
Sebastes sp.	Rockfishes		1			16	7		12
Hexagrammidae	Greenlings		6	5	1	8	8	4	19
Hexagrammos sp.	Greenlings					4	1		9
Ophiodon elongatus	Lingcod		1			9	1		1
Cottidae	Sculpins		4	8	1	2		1	1
Artedius sp.			1						
Clinocottus sp.			1	3	1	1			
Oligocottus sp.			1						
Scorpaenichthys marmoratus	Cabezon		1	2		4			2
Embiotocidae	Surfperches		20	21	3	24	6	1	19
Amphistichus sp.									
Damalichthys vacca	Pile perch		3	1		1			
Stichaeidae	Pricklebacks								
Cebidichthys violaceus	Monkeyface prickleback				1				

CA-SCR-7 (Auger 2) >2 mm	CA-SCR-7 (Auger 3) >2 mm	Total	Total w/o RFF	Relative Abundance w/o RFF
		4	4	0.01
		1	1	0.00
4		40	40	0.10
6		57	57	0.14
		14	14	0.04
1		13	13	0.03
		17	17	0.04
		1	1	0.00
1		7	7	0.02
		1	1	0.00
		9	9	0.02
9	1	104	104	0.26
1		1	1	0.00
		5	5	0.01
		1	1	0.00

Table 11.4. Continued.

Taxon	Common name	CA-SCR-7 Column 1 >2 mm	CA-SCR-7 Column 2 >2 mm	CA-SCR-7 Column 3 >2 mm	CA-SCR-7 Column 4 >2 mm	CA-SCR-7 Column 5 >2 mm	CA-SCR-7 Column 6 >2 mm	CA-SCR-7 Column 7 >2 mm
	Screen size							
Xiphister sp.	Black or Rock prickleback	1						
Clinidae	Kelp blennies							
Gibbonsia sp.	Spotted, Striped, or Crevice kelpfish	1			1			
Pleuronectiformes		1	1			1		1
Platichthys stellatus	Starry flounder							
FRESH OR SALT WATER FISHES								
Salmonidae	Trouts and salmons							
Oncorhynchus sp.		1	4			1	2	
Gasterosteidae	Sticklebacks							
Gasterosteus aculeatus	Threespine stickleback		1					
Cottidae	Sculpins							
Leptocottus armatus	Pacific staghorn sculpin		3		1			
Embiotocidae	Surfperches							
Cymatogaster aggregata	Shiner perch						.	
Total		112	106	13	547	95	13	306

CA-SCR-7 (Auger 2)	CA-SCR-7 (Auger 3)	Total	Total w/o RFF	Relative Abundance w/o RFF
>2 mm	>2 mm			
		1	1	0.00
		2	2	0.01
		4	4	0.01
1		1	1	0.00
1		9	9	0.02
		1	1	0.00
2		6	6	0.02
2		2	2	0.01
100	2	1294	400	

Table 11.5. Results of Fish Analysis for CA-SCR-10 from the >2 mm Recovery Method with NISP by Context.

Taxon	Common name	CA-SCR-10	CA-SCR-10 (Auger)	Total	Total w/o RFF	Relative Abundance w/o RFF
	Screen size	>2 mm	>2 mm			
SALT WATER FISHES						
Rajidae	Skates	3		**3**	3	**0.02**
Raja sp.	Skate	1		**1**	1	**0.01**
Actinopterygii	Ray-finned fishes	195	1	**196**		
Engraulidae	Anchovies					
Engraulis mordax	Northern anchovy	17		**17**	17	**0.11**
Clupeidae	Herrings	44		**44**	45	**0.29**
Clupea pallasii	Pacific herring	1		**1**	1	**0.01**
Sardinops sagax	Pacific sardine	6	1	**7**	7	**0.04**
Osmeridae	Smelts	2		**2**	2	**0.01**
Batrachoididae	Toadfishes					
Porichthys notatus	Plainfin midshipman	5		**5**	6	**0.04**
Atherinopsidae	New World silversides	12		**12**	12	**0.08**
Scorpaenidae	Scorpionfishes					
Sebastes sp.	Rockfishes	4		**4**	7	**0.04**
Hexagrammidae	Greenlings	4		**4**	4	**0.03**
Hexagrammos sp.	Greenlings	2	1	**3**	4	**0.03**
Ophiodon elongatus	Lingcod	1		**1**	1	**0.01**

Table 11.5. Continued.

Taxon	Common name	Screen size	CA-SCR-10	CA-SCR-10 (Auger)	Total	Total w/o RFF	Relative Abundance w/o RFF
			>2 mm	>2 mm			
Cottidae	Sculpins		1		1	1	0.01
Artedius sp.			1		1	1	0.01
Scorpaenichthys marmoratus	Cabezon		1		1	2	0.01
Embiotocidae	Surfperches		31		31	34	0.22
Amphistichus sp.			1		1	1	0.01
Rhacochilus toxotes	Rubberlip seaperch				0	1	0.01
Stichaeidae	Pricklebacks						
Cebidichthys violaceus	Monkeyface prickleback		1		1	2	0.01
Xiphister sp.	Black or Rock prickleback		1		1	1	0.01
Clinidae	Kelp blennies		1		1	1	0.01
Fresh or Salt Water Fishes							
Salmonidae	Trouts and salmons						
Oncorhynchus sp.			1		1	1	0.01
Gasterosteidae	Sticklebacks						
Gasterosteus aculeatus	Threespine stickleback		1	1	2	2	0.01
Total			337	4	**341**	157	

Table 11.6. Results of Fish Analysis for CA-SCR-10 from the >3.2 mm Recovery Method with NISP.

Taxon	Common name	CA-SCR-10	Total w/o RFF	Relative Abundance w/o RFF	Total w/o RFF	Relative Abundance w/o RFF
	Screen size	**>3.2 mm**				
SALT WATER FISHES						
Actinopterygii	Ray-finned fishes	10			34	0.22
Clupeidae	Herrings	1	1	0.08	1	0.01
Batrachoididae	Toadfishes				1	0.01
Porichthys notatus	Plainfin midshipman	1	1	0.08		
Scorpaenidae	Scorpionfishes				2	0.01
Sebastes sp.	Rockfishes	3	3	0.25	1	0.01
Hexagrammidae	Greenlings				1	0.01
Hexagrammos sp.	Greenlings	1	1	0.08		
Cottidae	Sculpins					
Scorpaenichthys marmoratus	Cabezon	1	1	0.08		
Embiotocidae	Surfperches	3	3	0.25	1	0.01
Rhacochilus toxotes	Rubberlip seaperch	1	1	0.08		
Stichaeidae	Pricklebacks				2	0.01
Cebidichthys violaceus	Monkeyface prickleback	1	1	0.08	157	
Total		**22**	**12**			

CA-SCR-14. For CA-SCR-14, the quantification and analyses for the site were separated by screen size and recovery method. I divided those samples recovered through dry screening with >3.2 mm mesh sieves and those recovered through flotation methods with >2 mm sieves. At CA-SCR-14, the most abundant organisms recovered through flotation and with 2 mm sieves are Northern anchovies at 46% (Table 11.7). Surfperches comprise 26% of the assemblage and include pile perch. Together, Northern anchovies and surfperches make up 72% of the >2 mm assemblage. In total, 8 genera make up the remaining 28% of the site assemblage. The density of fish remains recovered and identified within the assemblage to at least the class level is 4.1 NISP/l. The Shannon-Weiner Index for the assemblage, calculated at the genera level, is 1.1; this suggests less heterogeneity, with evenness or equitability measured at 0.45, signifying an assemblage away from equal distribution.

For the dry-screened materials at CA-SCR-14, the most abundant organisms recovered with >3.2 mm sieves are surfperches at 33%, including pile perch (Table 11.8). Also, the assemblage comprises greenlings at 21%, which may include kelp, rock, and masked greenling. The assemblage also includes rockfishes at 21%. Together, surfperches, greenlings, and rockfishes make up 75% of the site assemblage. Then, 5 genera, including trout and salmon, makos, cabezon, monkeyface pricklebacks, and white croaker, make up the remaining 25% of the site assemblage (Table 11.8). The density of fish remains recovered and identified within the >3.2 mm assemblage to at least the class level is 0.2 NISP/l. The Shannon-Weiner Index for the assemblage, calculated at the genera level, is 1.7; this suggests less heterogeneity, with evenness or equitability measured at 0.75, signifying an assemblage moving closer to equal distribution.

CA-SCR-15. In the quantification and analyses of materials for CA-SCR-15, I divided those samples recovered through dry screening with >3.2 mm mesh sieves from those recovered through flotation methods with >2 mm sieves. At CA-SCR-15, the most abundant organisms recovered through flotation and with >2 mm sieves are surfperches at 46%, including pile perch and barred, calico, or redtail surfperch (Table 11.9). The assemblage includes greenlings at 21%, which may include kelp, rock, and masked greenling. Herrings make up 10% of the site total. Together, surfperches, greenlings, and herrings comprise 77% of the CA-SCR-15 site assemblage. Then, 4 genera make up the remaining 23% of the site. The density of fish remains recovered and identified within the assemblage to at least the class level is 1.3 NISP/l. The Shannon-Weiner Index for the assemblage, calculated at the genera level, is 1.9; this suggests less heterogeneity, with evenness or equitability measured at 0.83, signifying an assemblage closer to equal distribution.

For the dry-screen materials at SCR-15, the most abundant organisms recovered with >3.2 mm sieves are greenlings at 33% (Table 11.10). Surfperches also make up 33% of the assemblage, including barred, calico, or redtail surfperch. Greenlings and surfperches make up 66% of the site assemblage. Lastly, rockfishes and cabezon make up the remaining 34% of the site assemblage. The density of fish remains recovered and identified within the >3.2 mm assemblage to at least the class level is 0.04 NISP/l. The Shannon-Weiner Index for the assemblage, calculated at the genera level, is 1.1, suggesting less heterogeneity, with evenness or equitability measured at 0.67. These data signify an assemblage that is moving away from equal distribution.

Table 11.7. Results of Fish Analysis for CA-SCR-14 from the >2 mm Recovery Method with NISP.

Taxon	Common name	CA-SCR-14	Total w/o RFF	Relative Abundance w/o RFF
	Screen size	>2 mm		
SALT WATER FISHES				
Rajidae	Skates			
Raja sp.	Skate	2	**2**	**0.02**
Actinopterygii	Ray-finned fishes	104		
Engraulidae	Anchovies			
Engraulis mordax	Northern anchovy	45	**45**	**0.46**
Clupeidae	Herrings	8	**8**	**0.08**
Osmeridae	Smelts			
Spirinchus starksi	Night smelt	1	**1**	**0.01**
Scorpaenidae	Scorpionfishes			
Sebastes sp.	Rockfishes	2	**2**	**0.02**
Hexagrammidae	Greenlings	6	**6**	**0.06**
Ophiodon elongatus	Lingcod	2	**2**	**0.02**
Cottidae	Sculpins	1	**1**	**0.01**
Scorpaenichthys marmoratus	Cabezon	2	**2**	**0.02**
Embiotocidae	Surfperches	19	**19**	**0.20**
Damalichthys vacca	Pile perch	6	**6**	**0.06**
Stichaeidae	Pricklebacks	1	**1**	**0.01**
Cebidichthys violaceus	Monkeyface prickleback	1	**1**	**0.01**

Table 11.7. Continued.

Taxon	Common name	CA-SCR-14	Total w/o RFF	Relative Abundance w/o RFF
	Screen size	>2 mm		
FRESH OR SALT WATER FISHES				
Gasterosteidae	Sticklebacks			
Gasterosteus aculeatus	Threespine stickleback	1	1	0.01
Total		**201**	**97**	

Table 11.8. Results of Fish Analysis for CA-SCR-14 from the >3.2 mm Recovery Method with NISP.

Taxon	Common name	CA-SCR-14	Total w/o RFF	Relative Abundance w/o RFF
	Screen size	>3.2 mm		
SALT WATER FISHES				
Lamnidae	Mackerel sharks			
Isurus sp.	Mako	1	1	0.03
Actinopterygii	Ray-finned fishes	14		
Scorpaenidae	Scorpionfishes			
Sebastes sp.	Rockfishes	8	8	0.21
Hexagrammidae	Greenlings	7	7	0.18
Hexagrammos sp.	Greenlings	1	1	0.03
Cottidae	Sculpins			
Scorpaenichthys marmoratus	Cabezon	5	5	0.13
Sciaenidae	Drums and croakers			
Genyonemus lineatus	White croaker	1	1	0.03
Embiotocidae	Surfperches	12	12	0.31
Damalichthys vacca	Pile perch	1	1	0.03
Stichaeidae	Pricklebacks			
Cebidichthys violaceus	Monkeyface prickleback	2	2	0.05

Table 11.8. Continued.

Taxon	Common name	CA-SCR-14	Total w/o RFF	Relative Abundance w/o RFF
	Screen size	>3.2 mm		
FRESH OR SALT WATER FISHES				
Salmonidae	Trouts and salmons			
Oncorhynchus sp.		1	1	**0.03**
Total		**53**	**39**	

Table 11.9. Results of Fish Analysis for CA-SCR-15 from the >2 mm Recovery Method with NISP.

Taxon	Common name	CA-SCR-15	Total w/o RFF	Relative Abundance w/o RFF
	Screen size	>2 mm		
SALT WATER FISHES				
Actinopterygii	Ray-finned fishes	56		
Engraulidae	Anchovies			
Engraulis mordax	Northern anchovy	1	1	0.02
Clupeidae	Herrings	4	4	0.10
Scorpaenidae	Scorpionfishes			
Sebastes sp.	Rockfishes	2	2	0.05
Hexagrammidae	Greenlings	8	8	0.20
Hexagrammos sp.	Greenlings	2	2	0.05
Cottidae	Sculpins	2	2	0.05
Clinocottus sp.		1	1	0.02
Scorpaenichthys marmoratus	Cabezon	2	2	0.05
Embiotocidae	Surfperches	15	15	0.37
Amphistichus sp.		1	1	0.02
Damalichthys vacca	Pile perch	3	3	0.07
Total		**97**	**41**	

Table 11.10. Results of Fish Analysis for CA-SCR-15 from the >3.2 Recovery Method with NISP.

Taxon	Common name	CA-SCR-15	Total w/o RFF	Relative Abundance w/o RFF
	Screen size	**>3.2 mm**		
SALT WATER FISHES				
Actinopterygii	Ray-finned fishes	4		
Scorpaenidae	Scorpionfishes			
Sebastes sp.	Rockfishes	1	1	0.17
Hexagrammidae	Greenlings	2	2	0.33
Cottidae	Sculpins			
Scorpaenichthys marmoratus	Cabezon	1	1	0.17
Embiotocidae	Surfperches	1	1	0.17
Amphistichus sp.		1	1	0.17
Total		**10**	**6**	

Discussion

Diachronic Shifts in Fish Relative Abundances

The analysis of fish remains and radiocarbon dates for the four sites studied from the Santa Cruz coast suggests that our excavations encountered deposits dating from ~6,800 cal BP to the historic era. Diachronic examination of the results of the Shannon-Weiner Index measuring diversity and evenness suggests that the fish derived from fishing practices became less diverse through time or that there was a general shift from broad-based fisheries to more specialized fishing practices in the >2 mm assemblages (Figure 11.2). However, at CA-SCR-15, we see a slight increase in the Shannon-Weiner Index. These data reflect the fact that the diversity index considers the number and quantity of organisms in the calculation. Therefore, the higher values observed at CA-SCR-7, CA-SCR-10, and CA-SCR-15 reflect communities with many species, each with few individuals represented. Conversely, the Shannon-Weiner Index is lower at CA-SCR-14, suggesting an assemblage with fewer organisms and that few organisms make up the bulk of abundance within the assemblage.

For instance, at CA-SCR-7, which dates from 6,740–6,660 cal BP to 4,240–4,090 cal BP, 74% of the site assemblage recovered with >2 mm sieves is comprised of surfperches, greenlings, skates, and rockfishes. The remaining 26% of the site assemblage is made of 18 genera. The Shannon-Weiner Index for the site was 2.4, suggesting an assemblage with greater heterogeneity or diversity of genera, and an evenness or equitability measured at 0.85. The density of fish remains in the site equals 3.5 NISP/l. Therefore, it appears that the fishery of CA-SCR-7 was broad-based and that these practices were in place for over 2,500 years.

In the case of CA-SCR-10, the site deposits analyzed dated from 5,850–5,650 cal BP to 730–670 cal BP. The >2 mm assemblage suggests that herrings, surfperches, and New

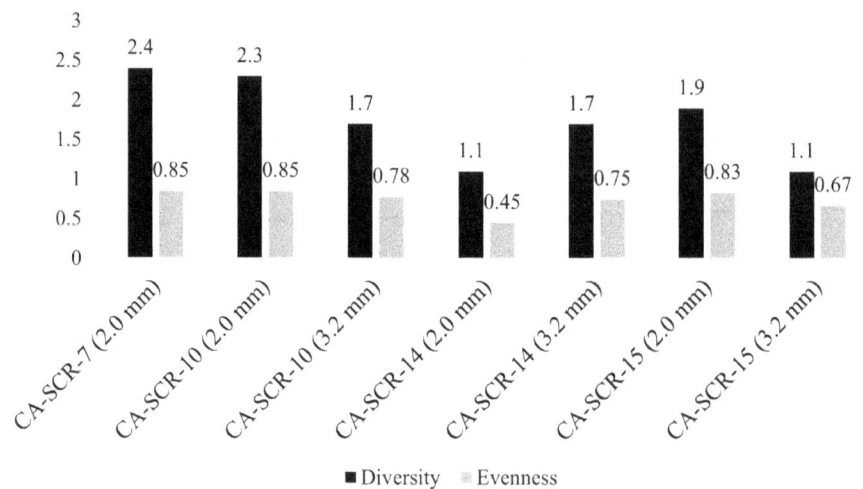

Figure 11.2. Shannon-Weiner Index for Diversity and Evenness by Site and Recovery Method.

World silversides amount to 75% of the site assemblage, with 11 genera making up the remaining 25%. However, the density of fish remains recovered per liter is 8.2 NISP/l, which is more than twice the amount recovered from CA-SCR-7; this suggests more fish usage at site CA-SCR-10. Further, at CA-SCR-10, the Shannon-Weiner Index calculated at the genera level is 2.3, suggesting greater heterogeneity. The evenness or equitability measurement at 0.85 signifies an assemblage closer to equal distribution.

For site CA-SCR-14, which dates from 940–800 cal BP to 260–30 cal BP, the >2 mm assemblage is primarily comprised of Northern anchovies and surfperches. They include 72% of fish remains by relative abundance; and 8 genera make up the remaining 28% of the site assemblage. The density of fish remains recovered is 4.1 NISP/l, suggesting a decrease in fish remains compared to CA-SCR-10 but similar to the fish densities at CA-SCR-7. The Shannon-Weiner Index for the assemblage, calculated at the level of genera, is 1.1, which suggests less heterogeneity, with evenness or equitability measured at 0.45, signifying an assemblage farther from equal distribution.

Lastly, at CA-SCR-15, which dates from 1,180–1,000 cal BP to 470–310 cal BP, the >2 mm assemblage is made up of surfperches, greenlings, and herrings. They comprise 77% of the assemblage by relative abundance; and 4 genera make up the remaining 23% of the site assemblage. The density of fish remains recovered and identified is 1.3 NISP/l. The Shannon-Weiner Index for the assemblage, calculated at the level of genera, is 1.9. This suggests less heterogeneity, with evenness or equitability measured at 0.83, signifying an assemblage closer to equal distribution.

Before deriving interpretations of the significance of these patterns and their implications for regional history, cultural practices, or other factors, it is essential to ensure that the results are not correlated with issues such as sample size (Grayson 1984). As Grayson (1984) noted, small sample sizes are unreliable for statistical inferences, such as relative abundance, given that changes in relative abundances of organisms may be a reflection of sample size, excavation, and analytical biases rather than a reflection of cultural, technological, or environmental variation. In essence, differences in relative abundances may reflect sample sizes rather than other circumstances.

To test if the changing relative abundances of organisms are a reflection of cultural, environmental, or technological variation, or a reflection of sample size, I conducted a Spearman rank correlation or Spearman's rho in R version 3.5.0 to test if the NISP counts of select taxa are correlated with sample size (see [a]R, Table 11.11 for reproducible R code). I began by exploring the relationship between surfperch NISP and the NISP of all other fishes. I hypothesized that there would be a positive correlation between sample size and surfperch NISP. Stated otherwise, it is expected that if the abundance of surfperch is correlated with sample size, surfperch NISP will increase in larger assemblages. The opposite would be true if there were a negative correlation, meaning that surfperch NISP would decrease as sample size increases.

The test results suggest a strong correlation between surfperch NISP and sample size ($r_s = 1$, $p < 0.05$). Therefore, it appears that surfperch NISP and sample size are associated; in other words, as sample size increases, surfperch NISP also increases. As a result, it is problematic to derive interpretations of changing relative abundances for the sites, as the information could tell us about changing NISP or sample size per site rather than other factors.

Table 11.11. Surfperch NISP by Site, their Relative Abundance, Rank, the NISP of All Other Fishes, and Fish Rank.

Site	Embiotocidae NISP	Embiotocidae Relative Abundance	Embiotocidae Rank	NISP of All Other Fishes	Fish Rank
CA-SCR-7	110	0.28	1	290	1
CA-SCR-10	36	0.23	2	121	2
CA-SCR-14	25	0.26	3	72	3
CA-SCR-15	19	0.46	4	22	4

[a]R Code for Spearman's rho test:
    Embiotocidae <- c(110, 36, 25, 19)
    Other_Fishes <- c(290, 121, 72, 22)
    cor.test(Embiotocidae, Other_Fishes, exact = TRUE, conf.level = 0.95, method = "spearman", alternative = "greater")

To test if the patterns observed with the surfperches are consistent across other organisms in the assemblage, I conducted Spearman's rho in R to test if the NISP counts of rockfishes are correlated with sample size (see [a]R, Table 11.12 for reproducible R code). The test results suggest a strong correlation between rockfish NISP and sample size (r_s = .95, p< 0.05). Therefore, like the surfperch NISP, rockfish NISP and sample size are associated. As a result, it is problematic to derive interpretations of changing relative abundances for the sites, as the information could tell us about changing NISP or sample size per site.

Table 11.12. Rockfish NISP by Site, their Relative Abundance, Rank, the NISP of All Other Fishes, and Fish Rank.

Site	sp. NISP	sp. Relative Abundance	sp. Rank	NISP of All Other Fishes	Fish Rank
CA-SCR-7	40	0.10	1	360	1
CA-SCR-10	7	0.04	2	150	2
CA-SCR-14	2	0.02	3	95	3
CA-SCR-15	2	0.05	4	39	4

[a]R Code for Spearman's rho test:
    Rockfish <- c(40, 7, 2, 2)
    Other_Fishes_R <- c(360, 150, 95, 39)
    cor.test(Rockfish, Other_Fishes_R, exact = TRUE, method = "spearman", alternative = "greater")

In an attempt to see if inferences regarding changes in the focus of fisheries between sites were possible, given the results of the Spearman's rho tests, I conducted a chi-square (χ^2) test to measure if there were statistically significant differences in the NISP values of fishes that serve as proxies for net-based and mass capture fishing practices (i.e., Northern anchovies, herrings, Pacific tomcod, and smelts) and hook-and-line fishing (i.e., rockfishes, greenlings, lingcod, and cabezon). These tests were conducted to broadly identify technological changes or variation in the focus of the fisheries across the four sites.

In these analyses, I excluded New World silversides and surfperches from consideration as previous research suggests these organisms can be captured using both technologies (Bertrando and McKenzie 2012; Sanchez et al. 2018). The χ^2 test was conducted in JMP version 7.0.1 based on the NISP values for net-based and mass capture fishes. The results suggest that the NISP values for these organisms do not meet expectations at CA-SCR-10 and CA-SCR-14 ($p < 0.05$), suggesting that the fish remains from these two sites are primarily derived from mass capture and net-based fishing rather than hook-and-line techniques and technologies, (Figure 11.3) and (Table 11.13). It appears that CA-SCR-7 and CA-SCR-15 represent fisheries heavily reflective of hook-and-line rather than net-based fishing.

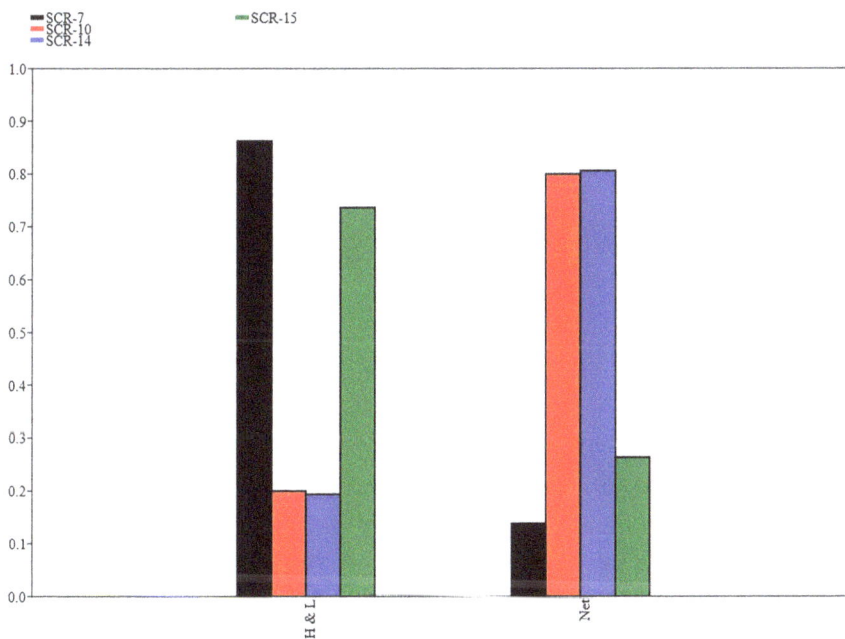

Figure 11.3. Results of the χ^2 Test of Hook-and-Line (H & N) and Net Based Fishing.

Table 11.13. Results of the χ^2 Test by Site and Capture Method.

	Count Total % Expected Cell Chi^2	Hook-and-Line	Net Mass Capture
CA-SCR-10	18 5.36 49.0179 19.6277	72 21.43 40.9821 23.4763	90 26.79
CA-SCR-14	13 3.87 36.4911 15.1223	54 16.07 30.5089 18.0875	67 19.94
CA-SCR-15	14 4.17 10.3482 1.2887	5 1.49 8.65179 1.5414	19 5.65
CA-SCR-7	138 41.07 87.1429 29.6806	22 6.55 72.8571 35.5003	160 47.62
Total	**183** **54.46**	**153** **45.54**	**336**

Sampling Biases in Santa Cruz Coast Archaeology

As previously mentioned, in conducting this study, I was interested in exploring the effects of recovery and sampling biases on the archaeology of fisheries along the Santa Cruz coast. To account for these biases, paired bulk sediment samples and dry-screened >3.2 mm materials were collected from the majority of the sites, except for CA-SCR-7 where my work focused on the collection of bulk sediment samples from column and auger samples to minimize impacts on the site. In addition, dry screening at CA-SCR-7 only occurred in highly disturbed contexts without paired >2 mm flotation samples. Therefore, the dry-screened samples from CA-SCR-7 are excluded in the analysis.

As indicated in the first section of this chapter, recovery rates for fishes along the Santa Cruz coast from previous excavations of sites sampled in this study (CA-SCR-7 and CA-SCR-10) as well as other sites (CA-SCR-9, CA-SCR-35, CA-SCR-44, CA-SCR-60/130) suggest significantly low fish bone recovery rates and density values per liter sampled (Table 11.1 and 11.2). For instance, at CA-SCR-7, fish bone density per liter was 0.0001 NISP/l. At CA-SCR-10, 8,000 liters of sediment were sampled during previous excavations, and no fish remains were recovered or reported, suggesting a lack of fishing in the local economies.

To visualize the density of fish bone recovered per liter excavated, I plotted the NISP/l in the paired samples from CA-SCR-10, CA-SCR-14, and CA-SCR-15 (Figure 11.4). These data demonstrate that the use of >3.2 mm recovery methods along the Santa Cruz coast is significantly biasing the recovery of fish remains from these sites, which is supported by the findings from earlier research phases (Table 11.1). For example, the >3.2 mm dry-screen recovery method at CA-SCR-10 and CA-SCR-14 yielded only 0.2 fish remains per liter, while at CA-SCR-15, it was only 0.04 fish remains. In contrast, the >2 mm heavy fraction materials from CA-SCR-10 include 8.2 NISP/l, CA-SCR-14 at 4.1 NISP/l, and CA-SCR-15 at 1.3 NISP/l.

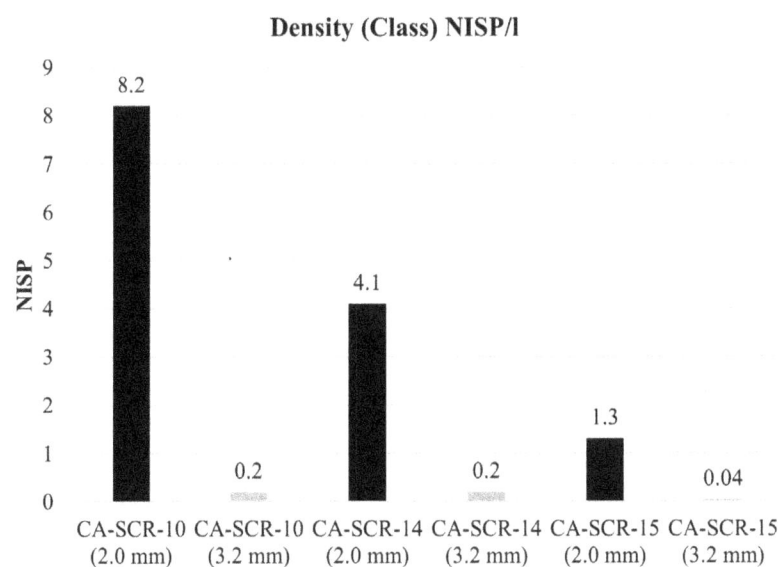

Figure 11.4. Density (Class) NISP/l for >3.2 mm and >2 mm Recovery Methods at CA-SCR-10, CA-SCR-14, and CA-SCR-15.

Conclusion

The archaeology of Santa Cruz County fisheries remains significantly understudied compared to other portions of the coastline and interior (Jones et al. 2007). While large-scale archaeological excavations within Santa Cruz County have occurred, most suffer from two serious issues. First, excavations were conducted with coarse-grained recovery methods—≥6.4 mm and 3.2 mm mesh sieves—affecting the representation of materials. Second, most excavated sites have never been systematically analyzed or formally published. Both factors affect our understanding of ancient and historical human-environmental relationships within Santa Cruz County.

In terms of ancient and historic era fisheries, there has been a dearth of data regarding human-fish relationships along the Santa Cruz coastline and the interior. This lack of information originates from the two biases reviewed above and a lack of fisheries-driven research, with some exceptions (Gobalet and Jones 1995). The scarcity of fisheries-related studies along the Santa Cruz coast affects our understanding of both human subsistence practices and ancient and historic fish biogeography, as well as the relevancy of archaeological data in Tribal cultural revitalization practices and fisheries conservation.

The findings of this case study suggest that the dearth of information regarding ancient and historic fisheries along the Santa Cruz coast is likely a result of significant recovery biases introduced through the standardization of >6.4 mm and >3.2 mm mesh sieves in local and regional archaeology, as well as the lack of fine-grained analyses necessitated for the recovery of minute materials (Casteel 1970, 1972, 1976; Colley 1990; Fitch 1969; Gobalet 1989; Grayson 1984; Moss et al. 2017; Sanchez et al. 2018). These findings have relevance for Santa Cruz archaeology and the archaeological community, more broadly.

The results of the fish analysis for the coastline and interior suggest that from 6,740–6,660 cal BP to 4,240–4,090 cal BP at CA-SCR-7, the fishery was diverse, with a variety of fishes represented. At CA-SCR-7, no single organism dominates the assemblage, with the primary technology in fish acquisition likely being hook-and-line fishing. In contrast, at CA-SCR-10, where I analyzed fish remains from deposits dating from 5,850–5,650 cal BP to 730–670 cal BP, the fishery was predominantly comprised of fish derived from net-based and mass capture fishing, but the fish remains within the site were still near to equal distribution. However, the assemblage from CA-SCR-10 is derived from the 730–670 cal BP contexts, with no fish remains included in this analysis from the 5,850–5,650 cal BP contexts (i.e., auger samples). Therefore, these data support other studies highlighting Late Period fisheries intensification of mass capture fishes throughout the Central California coast (Boone 2012; DeGeorgey 2016; Sanchez 2020; Sanchez, Gobalet et al. 2018).

By the Late Period at CA-SCR-14, which dates from 940–800 cal BP to 260–30 cal BP, it appears that a net-based mass capture fishery dominated by Northern anchovy is in place by this time, similar to fishery studies elsewhere in California (Boone 2012; Sanchez et al. 2018; Tushingham et al. 2013). These fish of marine origin were transported inland to CA-SCR-14 from the coast. At the nearby site CA-SCR-15, which dates from 1,180–1,000 cal BP to 470–310 cal BP, we see evidence of a hook-and-line fishery focused primarily on marine organisms. There is limited evidence at CA-SCR-14 and CA-SCR-15 for anadromous fishes and euryhaline fishes, but we find no definitive evidence of the use of freshwater fishes.

The significant biases in previous fisheries studies have relevance for our collaborators, including the AMTB and California State Parks. As the Amah Mutsun are revitalizing Indigenous knowledge and cultural practices, the lack of material evidence from previous excavations cannot inform contemporary cultural revitalization. For instance, our study is the first within Santa Cruz County to identify Northern anchovy net-based fishing practices, as evidenced by faunal remains at CA-SCR-10 and CA-SCR-14. Based on the biases in previous analyses, these data would not be available to Tribal collaborators unless fine-grained field and laboratory analyses are conducted. From the perspective of California State Parks, the results of our field studies can inform future archaeological permitting by providing required field recovery methods, sampling strategies, and reporting procedures for fieldwork.

References

Balée, William L.

1992   People of the Fallow: A Historical Ecology of Foraging in Lowland South America. In *Conservation of Neotropical Forests: Working from Traditional Resource Use*, edited by Kent Hubbard Redford and Christine Padoch, pp. 35–57. Columbia University Press, New York.

1998   *Advances in Historical Ecology.* The Historical Ecology Series. Columbia University Press, New York.

2006   The Research Program of Historical Ecology. *Annual Review of Anthropology* 35: 75–98.

2010   Contingent Diversity on Anthropic Landscapes. *Diversity* 2(2): 163–181.

2018   Brief Review of Historical Ecology. *Les nouvelles de l'archéologie* (152): 7–10.

Barker, Graeme

1975   To Sieve or Not to Sieve. *Antiquity* 49(193): 61–63.

Bertrando, Ethan B., and Dustin K. McKenzie

2012   Identifying Fishing Techniques from the Skeletal Remains of Fish. In *Exploring Methods of Faunal Analysis: Insights from California Archaeology*, edited by Michael A. Glassow and Terry L. Joslin, pp. 169–186. Perspectives in California Archaeology. Cotsen Institute of Archaeology University of California, Los Angeles.

Binford, Lewis R.

1964   A Consideration of Archaeological Research Design. *American Antiquity* 29(4): 425–441.

Boone, Cristie

2012   Integrating Zooarchaeology and Modeling: Trans-Holocene Fishing in Monterey Bay, California. PhD Dissertation, Department of Anthropology, University of California, Santa Cruz.

Breschini, Gary S., and Trudy Haversat

2000   *Archaeological Data Recovery at CA-SCR-44, at the Site of the Lakeview Middle School, Watsonville, Santa Cruz County, California.* Coyote Press Archives of California Prehistory 49. Coyote Press, Salinas.

Cannon, Aubrey

2000   Assessing Variability in Northwest Coast Salmon and Herring Fisheries: Bucket-Auger Sampling of Shell Midden Sites on the Central Coast of British Columbia. *Journal of Archaeological Science* 27(8): 725–737.

Cartier, Robert

1989   Scotts Valley Chronology and Temporal Stratigraphy. *Proceedings of the Society for California Archaeology* 2(8): 81–111.

Casteel, Richard W.

1970   Core and Column Sampling. *American Antiquity* 35(4): 465–467.

1972   Some Biases in the Recovery of Archaeological Faunal Remains. *Proceedings of the Prehistoric Society* 38: 382–388.

1976   Comparison of Column and Whole Unit Samples for Recovering Fish Remains. *World Archaeology* 8(2): 192–196.

Colley, Sarah M.

1990   The Analysis and Interpretation of Archaeological Fish Remains. In *Archaeological Method and Theory*, edited by Michael B. Schiffer, vol. 2, pp. 207–253. University of Arizona Press, Tucson.

Culleton, Brendan J., Robert H. Gargett, and Thomas L. Jackson

2005   *Data Recovery Excavations at CA-SCR-60/130 for the Pajaro Valley Water Management Agency Local Water Supply and Distribution Project.* Pacific Legacy Incorporated, Santa Cruz, California.

DeGeorgey, Alex

2016   *Archaeological Excavation of the Stege Mound (CA-CCO-297), A Late Period Shell Mound Located on the San Francisco Bayshore.* Alta Archaeological Consulting, Santa Rosa, California.

Driver, Jonathan C.

2011   Identification, Classification and Zooarchaeology. *Ethnobiology letters* 2: 19–39.

Faith, Tyler J., and Andrew Du

2017   The Measurement of Taxonomic Evenness in Zooarchaeology. *Archaeological and Anthropological Sciences* 10(6): 1419–1428.

Fitch, John E.

1969   Fish Remains, Primarily Otoliths, From a Ventura, California, Chumash Village Site (VEN-3). *Memoirs of the Southern California Academy of Sciences* 8: 56–71.

Gifford, Diane P., and Francine Marshall

1984   *Analysis of the Archaeological Assemblage from CA-SCR-35 Santa Cruz County, California.* Coyote Press, Salinas.

Gobalet, Kenneth W.

1989   Remains of Tiny Fish from a Late Prehistoric Pomo Site near Clear Lake, California. *Journal of California and Great Basin Anthropology* 11(2): 231–239.

2001   A Critique of Faunal Analysis: Inconsistency among Experts in Blind Tests. *Journal of Archaeological Science* 28(4): 377–386.

Gobalet, Kenneth W., and Terry L. Jones
1995   Prehistoric Native American Fisheries of the Central California Coast. *Transactions of the American Fisheries Society* 124(6): 813–823.

Gobalet, Kenneth W., Peter D. Schulz, Thomas A. Wake, and Nelson Siefkin
2004   Archaeological Perspectives on Native American Fisheries of California, with Emphasis on Steelhead and Salmon. *Transactions of the American Fisheries Society* 133(4): 801–833.

Gonzalez, Sara L.
2016   Indigenous Values and Methods in Archaeological Practice: Low-Impact Archaeology Through the Kashaya Pomo Interpretive Trail Project. *American Antiquity* 81(3): 533–549.

Gonzalez, Sara L., Darren Modzelewski, Lee M. Panich, and Tsim D. Schneider
2006   Archaeology for the Seventh Generation. *The American Indian Quarterly* 30(3): 388–415.

Grayson, Donald K.
1984   *Quantitative Zooarchaeology: Topics in the Analysis of Archaeological Faunas.* Studies in Archaeological Science. Academic Press, Orlando.

Harper, David A. T.
1999   *Numerical Palaeobiology: Computer-Based Modelling and Analysis of Fossils and their Distributions.* John Wiley & Sons, New York.

Hylkema, Mark G.
1991   Prehistoric Native American Adaptations along the Central California Coast of San Mateo and Santa Cruz Counties. MA Thesis, Department of Anthropology, San Jose State University, San Jose, California.
2002   Tidal Marsh, Oak Woodlands, and Cultural Florescence in the Southern San Francisco Bay Region. In *Catalysts to Complexity: Late Holocene Societies of the California Coast,* edited by Jon M. Erlandson and Terry L. Jones, pp. 233–262. Cotsen Institute of Archaeology, University of California, Los Angeles.

Jones, Deborah A., and William R. Hildebrandt
1990   *Archaeological Excavation at Sand Hill Bluff: Portions of a Prehistoric Site CA-SCR-7, Santa Cruz County, California.* Far Western Anthropological Research Group. Davis, California.
1994   *Archaeological Investigations at Sites CA-SCR-10, CA-SCR-17, CA-SCR-304, and CA-SCR-38/123 for the North Coast Treated Water Main Project, Santa Cruz County, California.* Far Western Anthropological Research Group. Davis, California.

Jones, Terry L., Nathan E. Stevens, Deborah A. Jones, Richard T. Fitzgerald, and Mark G. Hylkema

    2007   The Central Coast: A Midlatitude Milieu. In *California Prehistory: Colonization, Culture, and Complexity*, edited by Terry L. Jones and Kathryn A. Klar, pp. 125–146. AltaMira Press, Lanham, Maryland.

Langenwalter, Paul E., II

    2000   Vertebrate Animal Remains Recovered from CA-SCR-44 near Watsonville, Santa Cruz County, California. In *Archaeological Data Recovery at CA-SCR-44, at the Site of the Lakeview Middle School, Watsonville, Santa Cruz County, California*, edited by Gary S. Breschini and Trudy Haversat, pp. 153–193. Coyote Press Archives of California Prehistory 49. Coyote Press, Salinas, California.

Lightfoot, Kent G.

    2008   Collaborative Research Programs: Implications for the Practice of North American Archaeology. In *Collaborating at the Trowel's Edge: Teaching and Learning in Indigenous Archaeology*, edited by Stephen W. Silliman, pp. 211–227. University of Arizona Press, Tucson.

Lyman, Lee R.

    2008   *Quantitative Paleozoology*. Cambridge Manuals in Archaeology. Cambridge University Press, New York.

Magurran, Anne E.

    1988   *Ecological Diversity and its Measurement*. Princeton University Press, Princeton.

Meighan, Clement W.

    1950   *Observations on the Efficiency of Shovel Archaeology*. Reports of the University of California Archaeological Survey 7(4). University of California, Berkeley.

Moratto, Michael J.

    1984   *California Archaeology*. New World Archaeological Record. Academic Press, Orlando.

Moss, Madonna L., Rick Minor, and Kyla Page-Botelho

    2017   Native American Fisheries of the Southern Oregon Coast: Fine Fraction Needed to Find Forage Fish. *Journal of California and Great Basin Anthropology* 37(2): 169–182.

Nims, Reno, Diane Gifford-Gonzalez, Mark Hylkema, and Kara Potenzone

    2016   The CA-SCR-9 Archaeofauna: Insights into Prey Choice, Seasonality, and Processing. *Journal of California and Great Basin Anthropology* 36(2): 243–267.

Page, Lawrence M., Héctor Espinosa-Pérez, Lloyd T. Findley, Carter R. Gilbert, Robert N. Lea, Nicholas E. Mandrak, Richard L. Mayden, and Joseph S. Nelson

2013   *Common and Scientific Names of Fishes from the United States, Canada, and Mexico.* American Fisheries Society Special Publication 34. American Fisheries Society, Bethesda, Maryland.

Payne, Sebastian

1972   Partial Recovery and Sample Bias: The Results of Some Sieving Experiments. In *Papers in Economic Prehistory*, edited by E. S. Higgs, pp. 49–64. Cambridge University Press, New York.

Pearsall, Deborah M.

2000   *Paleoethnobotany: A Handbook of Procedures.* Academic Press, San Diego, California.

Ross, Anne, and Ryan Duffy

2000   Fine Mesh Screening of Midden Material and the Recovery of Fish Bone: The Development of Flotation and Deflocculation Techniques for an Efficient and Effective Procedure. *Geoarchaeology* 15(1): 21–41.

Sanchez, Gabriel M.

2020   Indigenous Stewardship of Marine and Estuarine Fisheries?: Reconstructing the Ancient Size of Pacific Herring through Linear Regression Models. *Journal of Archaeological Science: Reports* 29: 102061.

Sanchez, Gabriel M., Jon M. Erlandson, and Nicholas Tripcevich

2017   Quantifying the Association of Chipped Stone Crescents with Wetlands and Paleoshorelines of Western North America. *North American Archaeologist* 38(2): 107–137.

Sanchez, Gabriel M., Kenneth W. Gobalet, Roberta Jewett, Rob Q. Cuthrell, Michael Grone, Paul M. Engel, and Kent G. Lightfoot

2018   The Historical Ecology of Central California Coast Fishing: Perspectives from Point Reyes National Seashore. *Journal of Archaeological Science* 100: 1–15.

Sanchez, Gabriel M., Torben C. Rick, Brendan J. Culleton, Douglas J. Kennett, Michael Buckley, Jon M. Erlandson, and Robert L. Losey

2018   Radiocarbon Dating Legacy Collections: A Bayesian Analysis of High-Precision AMS 14C Dates from the Par-Tee Site, Oregon. *Journal of Archaeological Science: Reports* 21: 833–848.

Shaffer, Brian S.

1992   Quarter-Inch Screening: Understanding Biases in Recovery of Vertebrate Faunal Remains. *American Antiquity* 57(1): 129–136.

Sweeney, Francoise M. G.

    1986   Vertebrate Remains from CA-SCR-35, Santa Cruz County, California. In *Papers on California Prehistory*, edited by Gary S. Breschini and Trudy Haversat, pp. 15–56. Coyote Press Archives of California Prehistory. Coyote Press, Salinas, California.

Thomas, David H.

    1969   Great Basin Hunting Patterns: A Quantitative Method for Treating Faunal Remains. *American Antiquity* 34(4): 392–401.

Tushingham, Shannon, and Jennifer Bencze

    2013   Macro and Micro Scale Signatures of Hunter-Gatherer Organization at the Coastal Sites of Point St. George, Northwestern Alta California. *California Archaeology* 5(1): 37–77.

Tushingham, Shannon, Amy M. Spurling, and Timothy R. Carpenter

    2013   The Sweetwater Site: Archaeological Recognition of Surf Fishing and Temporary Smelt Camps on the North Coast of California. *Journal of California and Great Basin Anthropology* 33(1): 25–37.

Wheeler, Alwyne, and Andrew K. G. Jones

    1989   *Fishes.* Cambridge Manuals in Archaeology. Cambridge University Press, New York.

CHAPTER 12

Archaeofaunal Evidence for Landscape Management in Cotoni-Quiroste: Assessing Rodent Species Abundance as a Proxy

DIANE GIFFORD-GONZALEZ

In collaboration with the Amah Mutsun Tribal Band (AMTB), our research group has worked for over a decade to assess the evidence for fire-based landscape management, as well as marine resource management, in the Central California coast. We initially found multiple lines of evidence for Native cultural burning in the Quiroste Valley Cultural Preserve, near Point Año Nuevo, in southern San Mateo County (Lightfoot et al. 2013: 64). The large site excavated in the Quiroste Valley, SMA-113, is quite probably the settlement described by Crespí as containing plank houses and a large ceremonial building when visited by members of the Portolá Expedition in October 1769 (Crespí 2001), which was called *Metenne* by its Quiroste inhabitants. Radiocarbon determinations date around 1000–1300 cal CE (1000–700 cal BP), placing it in the Late Middle to Late Period, according to regional chronologies. Smaller samples were also collected for flotation analysis from previously excavated special-purpose and residential sites on Point Año Nuevo, also within the homeland of the Quiroste people. As detailed in this monograph, the project more recently expanded its scope to northern Santa Cruz County, focusing on a zone south of the present-day village of Davenport and north of the city of Santa Cruz. This region was known to its resident Awaswas-speaking Native peoples as Cotoni (Milliken et al. 2009).

In the original Quiroste Valley research, the species abundances of archaeofaunal rodents appeared to support conclusions drawn from other lines of evidence: at the time of habitation, the site lay amidst open coastal prairies. The reconstructed vegetation contrasts with that around CA-SMA-113 in recent times, dominated by northern coastal scrub shrublands and Douglas fir (*Pseudotsuga menziesii*) forest. Open country-adapted voles were numerous in the site's rodent archaeofauna, and closed habitat rodents were rare. One exception was the modest occurrence of woodrats, known from many parts of Native California as favored prey species (Lightfoot et al. 2009). Subsequent, low-impact investigations in Cotoni produced lower numbers of rodents, with some inferred not to derive from archaeological components (see *Other Methodological Considerations*). This new evidence, combined with a re-examination of the archaeological sites from the Año Nuevo area earlier, prompted a rethinking of rodent species abundances as a reliable proxy for vegetation, an exploration of what thermal alterations to rodent bone might contribute to such a study, and different conclusions about how to use rodent remains as an index of fire-assisted landscape management.

The first section of this chapter lays the groundwork for discussing how and why considerable caution is required when using rodent species abundances as a proxy for vegetation. It begins by discussing proxies as a form of analogy, and what makes some proxies stronger than others. In the process, it discusses the problem of equifinality when using patterning such as species abundances to infer causation. Finally, it reflects on what our field experience and analyses suggest about the site types and retrieval methods that are best for the recovery of evidence relevant to Indigenous landscape management practices.

One may ask why expend time discussing a secondary proxy that may prove to be an ambiguous indicator of causal factors. At the very least, it cautions others interested in using the same lines of evidence about potential problems of equifinality. However, the chapter also suggests tactics for strengthening rodent relative abundances as an environmental proxy.

Proxies for Past Processes

In archaeology, as in other historical sciences such as geology, researchers cannot directly observe the processes they want to study. A recommended approach in these situations is to assess multiple, independent lines of evidence for the operation of past processes. If most of these lines of evidence point to the same causal process or processes, the assertion that such processes produced the evidence in the past is strengthened (for archaeologically oriented discussions, see Gifford-Gonzalez 1991; Lyman 1987). When investigating Native cultural burning on the Central Coast of California, our team juxtaposed multiple lines of archaeological and paleoenvironmental evidence. These included direct evidence for fire occurrence and vegetation communities. We reasoned that, if most of these lines of evidence point toward plant communities that diverged substantially from natural patterns of each in the same locales today, this strongly suggests deliberate Native management of past landscapes; and if direct evidence of episodic fire events coincides in time with these, this management was probably done using fire.

In science, such lines of evidence are often called proxies. In the best case, each proxy of an unobservable past process was generated by a single, specifiable cause. Fire scars on long-lived trees, such as redwoods, are proxies because only a certain intensity of flames acting on the outer tissues of a tree trunk can produce them. Using this proxy is strongly warranted because inferring the action of fire from such distinctive scars rests on a *relational analogy*, an analogy based on a determinative relationship between a specified process––exposure to fire––and its observable product––the scar developed on the tree's tissues (Gifford-Gonzalez 1991; Wylie 1985). A proxy is thus strongest if based on a relational analogy testifying to a single, specific generative process. Our research project also used proxies for ancient vegetation communities: preserved pollen, seeds, nut hulls, stems, and burned wood, also grounded in relational analogies, many of which are discussed in Chapter 8.

In zooarchaeology, many such analogies were experimentally defined from the 1980s onward. One example was that of a sharp stone edge moving across a bone surface, leaving distinctive marks of its passage, which provided strong evidence for use of stone tools in large animal butchery by human ancestors nearly two million years ago (Bunn 1981; Potts and Shipman 1981). However, it soon emerged that at least two different *actors* and *contexts of production* could produce this trace: either an ancient hominin actor, using a stone tool while cutting up an animal, or a hoofed animal—another actor—trampling bones against angular stones on a land surface (Behrensmeyer et al. 1986; Fiorillo 1989). Each type of evidence had the same immediate cause—sharp stones moving with force across a bone surface—but if one is actually interested in ancient human actors and their behavioral contexts, such marks on the bone are an ambiguous proxy. Patterning initially supposed to be evidence for hominin behavior was *equifinal* with that produced by hoofed animals' trampling, compromising inferences from the proxy.

In response to this problem, zooarchaeologists proposed ways to clarify the causal context and actor by bringing in other lines of evidence, including the texture of the bones' sedimentary matrix—did it contain sharp-edged stones, for example—and the placement of the marks of anatomical elements—did they follow the logic of butchery actions (for a review of the discussion, see Gifford-Gonzalez 2018a: 281–300). This chapter will return to the theme of equifinality after its data presentation.

Rodents as Indirect Proxies for Vegetation

Direct proxies for fire incidence and vegetation communities are strong because they are based upon multiple relational analogies. Due to their adaptations to local food sources and limited foraging ranges, rodents have long been used to infer the immediate habitats around archaeological sites, using the contemporary requirements of identified species to infer past local vegetation and rainfall (e.g., Holbrook 1977). Rodent taxa traditionally have been identified using morphological traits, especially those of mandibles and teeth, with morphometrics to distinguish closely related species. More recently, two other analytic tools have become available for distinguishing species that use different habitat types: 1) bone stable isotope assays reflecting open vs. closed habitats, and 2) aDNA analysis to identify species and aspects of their genomes. We were unable to work with enough rodent specimens to enable a stable isotope study of open vs. closed habitats (Vicky Oelze, personal communication 2020). Chapter 13 addresses the aDNA investigations with voles' remains from project study sites as a distinct line of evidence. This chapter also presents the results of an earlier aDNA study of CA-SMA-113 rodents by Peter Heintzmann, then of the University of California, Santa Cruz (UC Santa Cruz), now of the Arctic University Museum of Norway, Tromsø.

Rodents and Site Formation

Rodents present some complications to their use as proxies for local vegetation because they may enter an archaeological site through several different pathways, at least one of which might not reflect the immediate environs of the site.

1. Most critically for using rodents as an environmental proxy, people may selectively target some rodent species in habitats other than that around the site where they are found.

2. A few rodent species can enter archaeological deposits through their burrowing behavior.

3. Rodent species' relative abundances in an archaeofaunal sample are influenced by various species' size-based, differential durability under human handling, as well as destructive effects of other postmortem processes—or taphonomy.

4. Some species' adaptations make them more or less prone to be a kind of "bycatch" of cultural burning—that is, prone to being killed during managed fires, then expediently collected for the cookpot.

5. Exploring each of these factors provides context for interpreting our archaeological samples and is outlined below. Appendix 12.1 details those species likely to enter archaeological sites via each of these site formation processes.

Thermal Alterations to Rodent Bones

Given the possibility that Native landscape maintenance with fire could lead to the deaths of rodent species and to their incorporation into archaeological deposits, the effects of fire on rodent bodies and bones were explored in this study to see whether possible bycatch deaths could be distinguished from culinary processing. Little research has as yet been published to clarify the possible pathways leading to patterns of thermal alteration to rodent bones.

Understanding such evidence gives some knowledge of bone's responses to heat. Fairgrieve (2008) and Holden and colleagues (1995) stress that the key factor in thermally induced color and structural changes to bone is the actual temperature reached by the bone itself. As bone fails to disperse heat, it builds up temperatures higher than the air around it. When it reaches around 285°C (545°F), its internal collagen starts to oxidize, and with increasing temperatures, it oxidizes completely. This is higher than standard oven settings for cooking, but during the application of lower levels of dry heat during roasting, exposed bone portions can accumulate enough heat to initiate these changes (Gifford-Gonzalez 2018a: 321, Table 15.2).

Experiments have shown that collagen oxidation produces a regular sequence of color changes: pale brown → dark brown → black → gray → white, though the precise temperatures at which the color shifts occur may vary (Holden et al. 1995; Shipman et al. 1984a). If bone continues to accumulate heat to about 600°C, the crystalline hydroxyapatite component of bone can go into flux and assume a new, ceramic-like texture. These structural changes are independent of bone color shifts, but, because most specimens reaching such high temperatures have already moved into a white color range, vitrified pieces are usually light gray or white, though not all white pieces are vitrified.

Some culinary techniques do not produce color shifts in bone, such as whole carcasses or carcass segments being wrapped in leaves, grasses, or hides and placed in pits for moist-roasting, thus shielding bone elements from direct exposure to ambient temperatures (Wandsnider 1997). Boiling also does not alter bone color. If rodent bodies were boiled using pot-boiling technologies typical of Central California, as suggested by Thoms (2008), their remains would probably lack both color shifts and cutmarks, though they might display marks of human chewing or digestive acid modifications (Gifford-Gonzalez 2018a: 237, 326–327). Despite decades of research seeking to discern indicators of boiling, until recently no definitive indicator of boiling was defined that was not equifinal with the effects of post-depositional diagenesis. Heating alters hydroxyapatite crystal size and changes collagen fiber morphology, but uncooked bone buried in the ground for even a decade or two develops the same traces (Shipman et al. 1984b; Stiner et al. 1995; Taylor et al. 1995). Bosch and colleagues (2011) report that SEM with energy-dispersive X-ray spectroscopy can discern changes to bone surface structure in experimentally boiled bone, but this technique has not been widely replicated.

Thermally induced color changes to bone can occur either during culinary processing or in post-depositional heating of discarded bones in a sedimentary matrix. Direct exposure to dry heat causes color changes on bone ends or protuberances exposed directly to ambient temperature, as with turkey wing tips or drumstick ends roasting in an oven, while bone sections covered by muscle do not undergo color changes at normal cooking temperatures, roughly 165–250°C (325–475°F). Roasting thus produces a partially color-altered specimen, with the color-shifted sections reflecting direct exposure to ambient heat (Holden et al. 1995).

Human forensic and zooarchaeological experiments show that bones lying in deposits 5–10 cm below a hearth can, depending upon the duration of their exposure and the nature of the matrix, undergo color shifts up to complete "carbonization" (Bennett 1999; Stiner et al. 1995). This is especially likely in repeatedly occupied, spatially constrained rock shelters or caves (Stiner et al. 1995). Post-culinary burning can also develop in open sites, for example, if skeletal elements are disposed of by placing them in their roasting pits holding still-hot embers or stones (Gifford-Gonzalez and Parham 2008). Such color shifts normally are more uniform over an entire specimen, since these have already been stripped of protective tissues.

Because this monograph deals with rodents and cultural burning, a third pathway to thermally induced color changes in bone can be considered: modification during or after fire-related death. Modern wildfire and prescribed burn research indicates that most rodents killed are not badly burned, but, regardless of the cause of death, their bodies are temporarily subject to high temperatures. Forensic anthropology has formulated general expectations regarding thermal modifications to skeletons subject to high heat that are relevant here. Bone segments thinly covered by skin and tendons are more likely to display color changes when exposed to high heat than are skeletal segments more heavily covered with muscle. Bones of the cranium and mandible, including teeth, and the distal extremities of the fore- and hind limbs fall into the former category for rodents, as they do for humans (Medina et al. 2012).

To summarize, several contexts of thermal stress can cause color changes to archaeological rodent bone:

1. During roasting, or during shorter-term fire exposure intended to singe off hair. These color changes would be partial and mostly on the less protected skull and lower extremities.
2. During heating of sediments in which discarded bones lay. These changes are likely to be uniform over an entire specimen, due to the lack of adhering soft tissues.
3. During exposure to short-term cultural burns that killed the animal and subjected its body to high temperatures. The anatomical pattern of modifications would be very similar to that inflicted on whole rodent bodies by singeing or roasting.
4. Causal contexts 1 and 3 above appear to be equifinal in their predicted effects on rodent bones. Therefore, partial thermal color alterations to bone specimens cannot unambiguously testify to either culinary preparation or to rodents' being the bycatch of landscape maintenance via burning.

Archaeofaunal Evidence and Fire-Based Land Management: Materials and Methods

Intersite comparisons work best when retrieval and analytic methods are comparable enough to reduce the possibility that the perceived patterning is an artifact of disparate recovery methods. Although these sites differ considerably in the volumes of sediments processed, recovery methods for the four Cotoni archaeological sites discussed here—CA-SCR-7, CA-SCR-10, CA-SCR-14, CA-SCR-15 (Sanchez 2020)—are outlined in Chapter 4, and recovery of the small samples from CA-SMA-18/18A, CA-SMA-19, CA-SMA-218 on Point Año Nuevo were comparable to these.

Sediments from the University of California, Berkeley's (UC Berkeley), 22 1 m² excavation units at CA-SMA-113 were screened through ¼–⅛" mesh screens in 2007 and in 2008–2009, through ⅛" mesh. UC Berkeley and Cabrillo College both took samples for flotation, and the UC Berkeley project excavated features in natural levels to collect their distinctive sediment samples. After 2007, these and other UC Berkeley column samples were processed by flotation, producing light and heavy fractions. The latter was passed through nested geologic screens (>4 mm, 2–4 mm, and 1–2 mm) in the field and at the California Archaeology Laboratory at UC Berkeley. Heavy-fraction samples included small fish, rodent, and small bird remains, as well as lithic micro-flakes and small mollusk shell fragments.

Quiroste Phase of Zooarchaeological Research

The vertebrate samples from all sites were initially sorted in the California Archaeology Laboratory at UC Berkeley (Chapter 4). From 2010 to 2012, with some later additions, the mammals and birds from CA-SMA-113 were analyzed by Gifford-Gonzalez (DGG), and its fish fauna was analyzed by Cristie Boone of Albion Environmental. The smaller CA-SMA-18/18A, CA-SMA-19, and CA-SMA-218 samples were identified by DGG in the UC Santa Cruz Zooarchaeology Lab (Gifford-Gonzalez 2018b) in subsequent years. Thomas J. Banghart aided in definitive identifications of rare CA-SMA-113 bird specimens in 2018–19 and during a 2019 visit to the Museum of Vertebrate Zoology at UC Berkeley.

Cotoni Phase of Research

As part of the second phase of research into the timing and emergence of landscape management in the Cotoni region, Gabriel Sanchez (GMS) undertook zooarchaeological analyses on the preponderance of the precolonial excavated sites listed earlier. DGG analyzed the fauna from CA-SCR-123/38, two overlapping precolonial sites from Middle and Late Periods, underlying the post-Mission Period Bolcoff Adobe, built in 1839–1841. Upon excavation, this site was deemed too mixed by rodent activity to yield a clear signal from the two precolonial sites below the historic deposits. Because it does not definitively testify to precolonial land management, the CA-SCR-123/38 fauna will not be discussed further in this chapter.

DGG examined all specimens for bone surface modifications with a 10x binocular loupe under a high-intensity lamp, clarifying the nature of some marks with a 20x light microscope as necessary. Further details on data collection and recording methods can be found in Nims and colleagues (2016). Data were recorded in a FileMaker Pro® database.

GMS identified mammal, bird, amphibian, reptile, and fish remains from CA-SCR-7, CA-SCR-10, CA-SCR-14, and CA-SCR-15 in the California Archaeology Laboratory at UC Berkeley using reference materials at the California Academy of Science, with advice from Dr. Kenneth Gobalet for fishes, and the Museum of Vertebrate Zoology at UC Berkeley, with consultations with Professor James L. Patton and Dr. Chris Conroy for rodents. Data on taxonomic identifications and bone surface modifications were recorded in an Excel® database.

When DGG was analyzing the CA-SMA-18, CA-SMA-113, and Año Nuevo rodent faunas, UC Santa Cruz had no complete postcranial comparative specimens for the *Peromyscus* species likely to be in the area, namely, *P. maniculatus*, *P. californicus*, *P. boylii*, and *P. truei*. Given this, she chose to assign 26 deer mouse cranial and postcranial specimens only to their genus, "*Peromyscus* sp." Peter Heintzmann's later aDNA analysis of a sub-sample of three "*Peromyscus* sp." specimens assigned all to *P. maniculatus* (Table 12.1, see **SMA-113: aDNA Results on Varied Species**).

By 2020, when the importance of *P. maniculatus* as an early recolonizer of burned areas became apparent, the COVID-19 lockdown of UC Santa Cruz and UC Berkeley facilities prevented DGG from taking the balance of the specimens to consult with the same experts with whom GMS had consulted in 2018–2019. The findings regarding *Peromyscus* at CA-SMA-113 and CA-SMA-18 thus remain pending and are open to further updates. In addition to the species-identifiable cranial and postcranial rodent elements from CA-SMA-113 listed in Table 12.2.b, 237 specimens from the site were so fragmentary as to be assigned to an "Indeterminate Rodent" category.

Specimens from the Quiroste-Cotoni project are being prepared in the California Archaeology Laboratory at UC Berkeley for curation with California State Parks.

Comparative Datasets: Archaeological and Zoological

Our zooarchaeological analysis used both an archaeological comparative dataset from a Point Año Nuevo site previously analyzed by DGG and colleagues and a modern, habitat-specific live-trapping enumeration of species from a recent zoological and isotopic ecology research project.

Archaeological Comparative Dataset
Gifford-Gonzalez and colleagues (2013) included a comparative dataset of rodent archaeofauna from CA-SMA-18, a late Middle Period site in a northern coastal scrub habitat on the southern side of Point Año Nuevo itself. Radiocarbon dates on charcoal, single-shell mussels, and two northern fur seal specimens, reported by Hildebrandt and colleagues (2006), range from 700 to 800 CE (1300–1200 cal BP), and additional dates run by this project and reported by Cuthrell in Chapter 8 expand the temporal range of the site to 550–850 CE (800–1100 BP), placing it close to the earliest dates for CA-SMA-113. The 2004 excavation of CA-SMA-18 was a "rapid recovery" operation, intended to swiftly salvage archaeological materials from the destructive effects of invasive northern elephant seals, *Mirounga angustirostris*, as well as to stabilize the remaining sediments from coastal erosion. Two north-south trench lines, consisting of five 1x2 m excavation units, were placed on the eastern and western sides of the deposit, and an east-west trench of the same dimensions was then excavated between them. All were dug in arbitrary, 20 cm levels. Three additional units were dug along the exposed erosion face to stabilize this cut.

Sediments were screened using ¼" or ⅛" screens, depending upon the nature of the units and side of the extant, ⅛" screened sample. Additionally, four 20x20 cm column samples, and two flotation samples from two "fire-affected rock" features were taken. The heavy fraction from flotation samples was processed through graduated geological screens, yielding elements of very small fishes and marine shells, but no rodent remains (Hildebrandt et al. 2006: 8).

Added to this CA-SMA-18 rodent sample for this chapter were small samples from the Lightfoot project's test pits and flotation program for previously excavated Point Año Nuevo sites, labeled CA-SMA-18 and CA-SMA-18A, analyzed by DGG. Recovery and processing were as described for flotation samples in Chapter 4 of this volume.

Of the rapid recovery vertebrate fauna, Jeanne C. Geary, then of San José State University (CSU San José), identified bird specimens, Kenneth Gobalet and Jereme Gaeta, then of California State University, Bakersfield (CSU Bakersfield), identified fish specimens, while DGG analyzed the mammal assemblage, with help from undergraduate students on the rodent fauna (Gifford-Gonzalez et al. 2006). It should be noted that in Gifford-Gonzalez et al. 2006: Table 12.1, the valley pocket gopher, *T. bottae* specimens were erroneously assigned to the northern pocket gopher, *T. talpoides*, based on a mislabeled reference specimen. This chapter's Table 12.3 corrects this error and recounts the rodents from the SMA-18 archaeofauna's FileMaker® database, taking into account the addition of four new species-identifiable specimens from the Lightfoot samples.

In addition to the species-identifiable cranial and postcranial rodent elements listed in Table 12.3, 70 specimens were so fragmentary that they were assigned to an "Indeterminate Rodent" category and are noted there, including the later Far Western flotation and 2 specimens from the later sampling and flotation program of the Lightfoot group at UC Berkeley.

The original rapid recovery sample from CA-SMA-18 is in the Department of Parks' Santa Cruz District Office curation facility, 303 Big Trees Park Road, Felton, California. Later samples are being prepared for deposit with State Parks by the Lightfoot California Archaeology Laboratory at UC Berkeley.

Contemporary Habitat Rodent Dataset

Rodent analyses from the sites discussed here faced another interpretive dilemma: while archaeofaunal rodent abundances appeared to diverge markedly from what would be expected of the closed habitats characterizing the present-day Quiroste Valley, did any contemporary research document rodent species abundances in closed and more open coastal scrub habitats nearer the ocean? Around the time results of the Quiroste Valley investigations were being written up, Rachel Brown Reid, then a doctoral student in the Department of Earth and Planetary Sciences at UC Santa Cruz, was conducting a micro-mammal live-trapping program in closed habitats in the next drainage south of Quiroste Valley's Whitehouse Creek, about 2.5 km away. Reid's live-trapping program was part of her dissertation's diachronic research on coyote isotopic ecology (Reid 2014). Reid also laid out trapping transects in coastal willows and coyote brush in Point Año Nuevo State Park, about 9 km south of Quiroste Valley, where current or recent cultivation has not impacted the land.

All of Reid's live-trapping was done in localities not subject to deliberate burning for around a century. Reid followed standard zoological methods for censusing and tagging captured individuals so as not to double-count capture-prone individuals; further details can be found in her 2014 dissertation. We collaborated with Reid, using her data as a comparative dataset to refine our expectations about the abundances of rodent species in fire-suppressed, inland, and coastal vegetation (Gifford-Gonzalez et al. 2013). The Reid data are reported in Table 12.2.

Rodents from Cotoni's Sand Hill Bluff Site

With the permission of GMS, the relatively large aggregate rodent sample from the Sand Hill Bluff Site, CA-SCR-7, was initially explored as another comparative dataset (Tables 12.3.a and 12.3.b). Because CA-SCR-7 was known from previous radiocarbon dates and the recovery of elements from the early Holocene flightless duck, *Chendytes lawi*, the present project undertook many additional radiometric dates from two components in Locus 1, the tall dune where two vertically distinct archaeological deposits had previously been defined. The Lower Midden Stratum lies at the base of the dune and a distinctive Upper Midden Stratum at the top of the eroding dune, both of which are summarized in Chapter 4. From these new radiocarbon assays, the lower component dates to 4800–3600 cal BCE and the younger to 2700–2200 cal BCE (Lightfoot et al. 2022).

Other aspects of methods applied in the analysis are included in Appendix 12.2.

Archaeofaunal Evidence and Fire-Based Land Management: Results

Fine and colleagues (Chapter 13, this volume) present a comprehensive approach to the history of *Microtus californicus* in our region, using modern and ancient DNA, some from archaeological samples recovered during the research reported here. An earlier aDNA analysis of specimens from CA-SMA-113 is reported below.

SMA-113: aDNA Results on Varied Species

Table 12.1 presents Peter Heintzmann's earlier aDNA analysis of a sample of 24 non-thermally altered rodent specimens from CA-SMA-113. This earlier aDNA analysis aimed to assess fragmentary cranial and postcranial elements, which, based on morphology, were attributed to the two cricetid species, *Microtus californicus* and genus *Peromyscus*. The larger, morphologically distinctive woodrat specimens were excluded from aDNA analysis. Heintzmann's analysis corrected four morphological identifications of postcranial specimens from "*Peromyscus* sp." to *Microtus californicus*, unbalancing a hoped-for proportionate representation of analyzed specimens relative to that in the overall sample. The three remaining "*Peromyscus* sp." were assigned to *P. maniculatus*. The aDNA analysis also corrected one thoracic vertebral specimen originally identified as *Microtus* to *Dipodomys* sp. At the time of the analysis, genetic reference material for *D. venustus*, the Santa Cruz kangaroo mouse, did not exist (P. Heintzmann, personal communication, 2021). It is therefore prudent to label this specimen *Dipodomys* sp., though its geographic provenience suggests the specimen is probably attributable to *D. venustus*, today an endangered species in Santa Cruz County.

Table 12.1. Ancient DNA Species Identifications from 24 Non-thermally Altered Samples Submitted to Peter Heintzmann (Formerly of the University of California, Santa Cruz, now of Arctic University Museum of Norway, Tromsø) in 2012. Of the original 24 morphologically based identifications by Gifford-Gonzalez or her students, four "*Peromyscus* sp." were identified by aDNA as *Microtus californicus*, and one "*Microtus californicus*" identification was changed to *Dipodomys* sp. All remaining "*Peromyscus* sp." were identified as *P. maniculatus*.

Linnaean Name	Common Name	NISP	Site NISP	Site %
Microtus californicus	California vole	20	114	48.5
Peromyscus maniculatus	North American deer mouse	3	6	2.6
Dipodomys sp.	Kangaroo rat	1	1	0.4
Total		**24**		

Reid's Live-Trapping Results for Four Habitats

Table 12.2 presents Reid's live-trapped rodent counts of individuals by species. Because each captive animal was marked and if recaptured not counted a second time, the counts represent independent individuals. Reid was able to identify the three species of *Peromyscus* from variations in their sizes and coat patterns. Pocket gophers were absent from Reid's live-trapping data in all environments, making her data commensurate with the NISP of the archaeofaunal data from which *T. bottae* had been excluded on methodological grounds. Given Reid's focus on possible prey of nocturnally foraging coyotes and her nocturnal sampling tactics, the absence of other diurnal species such as tree squirrels, *Sciurus griseus*, and California ground squirrels, *Otospermophilus beecheyi*, in her live-trapping program is not surprising, despite their undoubted presence in some habitats and their appearance in the archaeofaunas.

Table 12.2. Reid's Live-trapped Rodent Census for Four Habitat Zones in the Año Nuevo Area. See text for details on species representation. See text for details on species representation. Excluded by Reid: *Tamias merriami* Merriam's chipmunk, 3 individuals in Inland Forest.

Linnaean Name	Common Name	Inland Forest N	Inland Forest %	Inland Brush N	Inland Brush %	Coast Brush N	Coast Brush %	Coas Willow N	Coast Willow %
Sciuridae indet.	Indeterminate squirrels	0	0.0	0	0.0	0	0.0	0	0.0
Sciurus griseus	Western gray squirrel	0	0.0	0	0.0	0	0.0	0	0.0
Otospermosphilus beecheyi	California ground squirrel	0	0.0	0	0.0	0	0.0	0	0.0
Dipodomys cf. venustus	Kangaroo rat sp. indet.	0	0.0	0	0.0	0	0.0	0	0.0
Chaetopodis californicus	California pocket mouse	0	0.0	0	0.0	0	0.0	0	0.0
Peromyscus sp.	Deer mice species indet.	0	0.0	0	0.0	0	0.0	0	0.0
Peromyscus maniculatus	North American deer mouse	0	0.0	19	25.7	23	50.0	81	73.6
Peromyscus californicus	California mouse	42	53.2	2	2.7	0	0.0	19	17.3
Peromyscus boylii	Brush mouse	27	34.2	19	25.7	0	0.0	0	0.0
Microtus californicus	California vole	0	0.0	0	0.0	10	21.7	0	0.0
Reithrodontomys megalotis	Harvest mouse	0	0.0	15	20.3	13	28.3	10	9.1
Neotoma fuscipes	Dusky-footed wood rat	10	12.7	19	25.7	0	0.0	0	0.0
	Total	**79**	**100.0**	**74**	**100.0**	**46**	**100.0**	**110**	**100.0**

Table 12.2 shows that, in the inland forest, toward which the present-day Quiroste Valley vegetation had recently been succeeding, only California mice, *P. californicus,* brush mice, *P. boylii,* and woodrats, *N. fuscipes,* were trapped, as might be expected from their habitat preferences and nocturnal habits. In the inland coyote brush samples, *P. californicus* and *P. boylii* were again dominant, with some representation of *P. maniculatus,* as well as of *N. fuscipes.* An abundance of voles, *Microtus californicus,* was trapped in coastal coyote brush, along with *P. maniculatus* only. Coastal willow is a habitat where *P. maniculatus* thrives, along with some *P. californicus.* Significantly for archaeofaunal interpretation, no woodrats were trapped in either coastal habitat, and California voles were trapped *only* in coastal brush.

In terms of diversity and evenness, three of the live-trapping habitat samples—inland forest, inland brush, and coastal brush—reflect roughly similar diversity statistics and evenness close to 1, that is, approaching complete evenness. The coastal willow habitat, dominated by *P. maniculatus,* has lower diversity and less evenness (Appendix 12.3, Table 1).

Archaeofaunal Species Abundances

Table 12.3.a presents relative abundance data for the total rodent faunas from CA-SMA-18, including the later small flotation samples noted earlier, CA-SMA-113, as well as CA-SCR-7, with the understanding that the latter is simply presented for informational purposes, given the non-archaeological nature of the CA-SCR-7 rodent sample. As noted in **Methods**, valley pocket gopher NISP were excluded from statistical analyses here, and NISP and recalculated frequencies are given in Table 12.3.b. The species in both sites were predominantly voles, deer mice, and woodrats and show a "simplified" species profile, with relatively low species diversity and lack of evenness: California voles, comprising 45–53% of NISP, a lower proportion of deer mice, which, when species identifications were determined, were identified as *P. maniculatus,* plus 21–34% woodrats. CA-SCR-7's extreme proportion of voles, 83%, differs highly significantly at p<0.001 from the other two sites, perhaps unsurprisingly given its non-archaeological origins. NMDS does place the Sand Hill Bluff sample's archaeofaunal composition closer to those from CA-SMA-18 and CA-SMA-113 than to Reid's four live-trapping sample localities (Figure 12.1). The CA-SCR-7 sample will not be discussed further.

Figure 12.1. NMDS Distribution of Rodents from Reid's Four Habitat Live-trapping Samples, Counted as the Number of Individuals, plus CA-SMA-18, CA-SMA-113, and CA-SCR-7 Archaeofaunas counted as NISP. CA-SCR-7 was treated as a whole sample. Stress value: 8.179637e^{-05}.

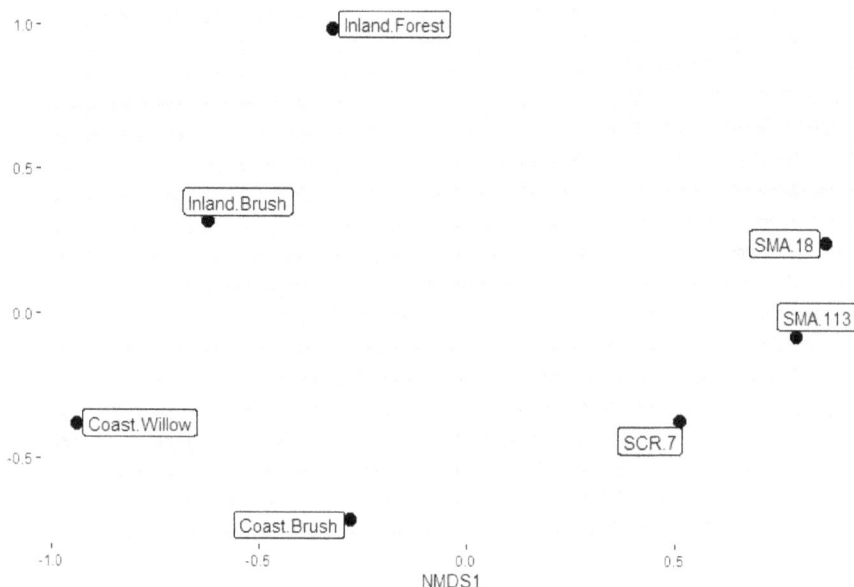

Table 12.3.a. Identifiable Rodent Taxa from CA-SMA-113, CA-SMA-18, and, for Information only, CA-SCR-7, Including Thomomys bottae. Some species, though not found in all sites, are included in the table to make them commensurate with those in Reid's (Table 12.2) live-trapping data. Specimens so fragmentary they could only be assigned to the Order Rodentia are also listed.

Taxon		SMA-18 NISP	SMA-18 %	SMA-113 NISP	SMA-113 %	SCR-7 NISP	SCR-7 %
Linnaean Name	**Common Name**						
Sciuridae indet.	Indeterminate squirrels	6	3.5	10	2.7	0	0.0
Sciurus griseus	Western gray squirrel	0	0.0	0	0.0	0	0.0
Otospermophilus beecheyi	California ground squirrel	0	0.0	5	1.4	2	0.9
Dipodomys sp.	Kangaroo rat	0	0.0	1	0.3	1	0.5
Chaetodipus californicus	California pocket mouse	1	0.6	1	0.3	0	0.0
Peromyscus sp.	Deer mice species indet.	8	4.7	27	7.4	0	0.0
Peromyscus maniculatus	North American deer mouse	0	0.0	3	0.8	11	5.1
Peromyscus californicus	California (deer) mouse	0	0.0	0	0.0	0	0.0
Peromyscus boylii	Brush (deer) mouse	0	0.0	0	0.0	0	0.0
Microtus californicus	California vole	33	19.3	92	25.1	99	45.6
Reithrodontomys megalotis	Harvest mouse	0	0.0	0	0.0	0	0.0
Neotoma fuscipes	Dusky-footed wood rat	26	15.2	36	9.8	6	2.8
Thomomys bottae	California pocket gopher	97	56.7	191	52.2	98	45.2
Subtotal		**171**	**100.0**	**366**	**100.0**	**217**	**100.0**
Indeterminate rodent		38		322		86	
Total		**209**		**588**		**303**	

Table 12.3.b. Identifiable Rodent Taxa from CA-SMA-113, CA-SMA-18, and, for Information Only, CA-SCR-7, excluding Thomomys bottae. Some species, though not found in all sites, are included in the table to make them commensurate with those in Reid's (Table 12.2) live-trapping data. Specimens so fragmentary they could only be assigned to the Order Rodentia are also listed.

Taxon		SMA-18 NISP	SMA-18 %	SMA-113 NISP	SMA-113 %	SCR-7 NISP	SCR-7 %
Linnaean Name	**Common Name**						
Sciuridae indet.	Indeterminate squirrels	6	8.1	10	5.7	0	0.0
Sciurus griseus	Western gray squirrel	0	0.0	0	0.0	0	0.0
Otospermophilus beecheyi	California ground squirrel	0	0.0	5	2.9	2	1.7
Dipodomys sp.	Kangaroo rat	0	0.0	1	0.6	1	0.8
Chaetodipus californicus	California pocket mouse	1	1.4	1	0.6	0	0.0
Peromyscus sp.	Deer mice species indet.	8	10.8	27	15.4	0	0.0
Peromyscus maniculatus	North American deer mouse	0	0.0	3	1.7	11	9.2
Peromyscus californicus	California (deer) mouse	0	0.0	0	0.0	0	0.0
Peromyscus boylii	Brush (deer) mouse	0	0.0	0	0.0	0	0.0
Microtus californicus	California vole	33	44.6	92	52.6	99	83.2
Reithrodontomys megalotis	Harvest mouse	0	0.0	0	0.0	0	0.0
Neotoma fuscipes	Dusky-footed wood rat	26	35.1	36	20.6	6	5.0
Subtotal		**74**	**100.0**	**175**	**100.0**	**119**	**100.0**
Indeterminate rodent		38		322		86	
Total		**112**		**497**		**205**	

Recalling that NMDS stress values of <0.2 are acceptable stopping points for iterative comparisons, this dataset's NMDS stress value of 8.179637e^{-05} reflects strong goodness of fit. The vole-dominant profile, with deer mice definitely or likely attributable to *P. maniculatus*, thus reflects two early recolonizing rodent species, plus woodrats, a species often targeted as human food.

However, the *nature* of that disturbance does not logically follow from the species relative abundance data. From the perspective on reliable proxies laid out at the beginning of this chapter, no clear chain of causality links rodent species abundances *only* with vegetation disturbance by humans––cultural burning, or other activities––because a broader array of non-human vegetation disturbances has not been excluded (see **Discussion**). These facts prompted exploration of the independent line of archaeofaunal evidence: thermal alteration to the bone specimens themselves, to assess whether patterns in these could refine the link between the archaeofaunal evidence and Native burning.

Species-Specific Thermal Modifications

Tables 12.4 and 12.5 summarize by-species and by-body segment thermal modifications for all taxonomic groups, including valley pocket gophers, to discern differences, if any, in human culinary handling or exposure to heat as bycatch. Thermal modifications were explored via Pairwise Fisher's exact tests, first between the *same taxonomic groups* from CA-SMA-18 and CA-SMA-113 and then via intrasite comparisons of thermal modifications in *similarly sized taxonomic groups* (Bochenski et al. 2009). These latter comparisons assumed that, barring significant anatomical differences, people would probably have handled like-sized species in similar ways. Larger rodents such as squirrels and woodrats might have been individually roasted, while smaller ones might have been gutted, skinned, and put in the cookpot with other plant and animal foods. Variables used in the Fisher's exact tests were the totals of thermally altered vs. non-altered NISP for each taxon, which collapses the body segment-specific information displayed in Tables 12.4 and 12.5. However, it was deemed that strong patterning displayed by statistical tests could then be further explored by returning to the segment-specific data to discern the sources of the contrasts.

Table 12.4. CA-SMA-18: By-species Ratios of Thermally Altered Specimens, Subdivided into Elements of Skull, Axial, and Extremity Body Segments. The "Axial" segment includes scapula and innominate elements, for reasons outlined in the text.

Species	Skull		Axial*		Extremities		Total	Overall
	NISP	Ratio	NISP	Ratio	NISP	Ratio	NISP	Ratio
California ground squirrel + squirrel indet.								
Altered	2	0.400	1	1.000	0	0.000	3	0.500
Unaltered	3	0.600	0	0.000	0	0.000	3	0.500
	5		**1**		**0**		**6**	
Kangaroo rat								
Altered	0	0.000	0	0.000	0	0.000	0	0.000
Unaltered	0	0.000	1	1.000	0	0.000	1	1.000
	0		**1**		**0**		**1**	
Harvest mouse								
Altered	0	0.000	0	0.000	0	0.000	0	0.000
Unaltered	0	0.000	0	0.000	1	1.000	1	1.000
	0		**0**		**1**		**1**	
North American deer mouse + deer mouse indet.								
Altered	0	0.083	0	0.059	0	0.267	0	0.000
Unaltered	2	0.917	0	0.941	6	0.733	8	1.000
							8	
California vole								
Altered	6	0.333	1	0.250	0	0.000	7	0.212
Unaltered	12	0.667	3	0.750	11	1.000	26	0.788
							33	

Table 12.4. Continued.

Species	Skull		Axial*		Extremities		Total	Overall
	NISP	Ratio	NISP	Ratio	NISP	Ratio	NISP	Ratio
Dusky-footed wood rat								
Altered	2	0.143	1	0.250	0	0.000	3	0.115
Unaltered	12	0.857	3	0.750	8	1.000	23	0.885
							50	
Valley pocket gopher								
Altered	15	0.181	0	0.000	5	0.455	20	0.206
Unaltered	68	0.819	3	1.000	6	0.545	77	0.794
							97	
Indeterminate rodent								
Altered	6	0.222	3	0.600	3	0.500	12	0.316
Unaltered	21	0.778	2	0.400	3	0.500	26	0.684
Total							**234**	

Table 12.5. CA-SMA-113: By-species Ratios of Thermally Altered Specimens, Subdivided into Elements of Skull, Axial, and Extremity Body Segments. * The "Axial" segment includes scapula and innominate elements, for reasons outlined in the text.

Species	Skull		Axial*		Extremities		All	
	NISP	Ratio	NISP	Ratio	NISP	Ratio	NISP	Ratio
California ground squirrel + squirrel indet.								
Altered	1	0.250	0	0.000	2	0.571	3	0.200
Unaltered	3	0.750	3	1.000	6	0.429	12	0.800
	4		**3**		**8**		**15**	
Kangaroo rat								
Altered	0	0.000	0	0.000	0	0.000	0	0.000
Unaltered	0	0.000	1	1.000	0	0.000	1	1.000
	0		**1**		**0**		**1**	
Harvest mouse								
Altered	0	0.000	0	0.000	0	0.000	0	0.000
Unaltered	0	0.000	0	0.000	1	1.000	1	1.000
	0		**0**		**1**		**1**	
North American deer mouse + deer mouse indet.								
Altered	0	0.000	0	0.000	0	0.000	0	0.000
Unaltered	11	1.000	16	0.941	5	0.733	30	1.000
	11		**16**		**5**		**30**	
California vole								
Altered	18	0.236	0	0.000	3	0.214	21	0.217
Unaltered	54	0.764	6	1.000	11	0.786	71	0.783
	72		**6**		**14**		**92**	

Table 12.5. Continued.

Species	Skull		Axial*		Extremities		All	
	NISP	Ratio	NISP	Ratio	NISP	Ratio	NISP	Ratio
Dusky-footed wood rat								
Altered	1	0.091	0	0.000	5	0.312	6	0.280
Unaltered	10	0.909	9	1.000	11	0.688	30	0.720
	11		**9**		16		**36**	
Valley pocket gopher								
Altered	10	0.066	1	0.083	10	0.233	21	0.105
Unaltered	126	0.934	11	0.917	33	0.767	170	0.895
.	**136**		**12**		43		**191**	
Indeterminate rodent								
Altered	78	0.650	28	0.538	42	0.400		
Unaltered	42	0.350	24	0.462	63	0.600		
	112		**52**		**107**		**271**	
Total							**688**	

Intersite Comparisons of Thermal Modifications

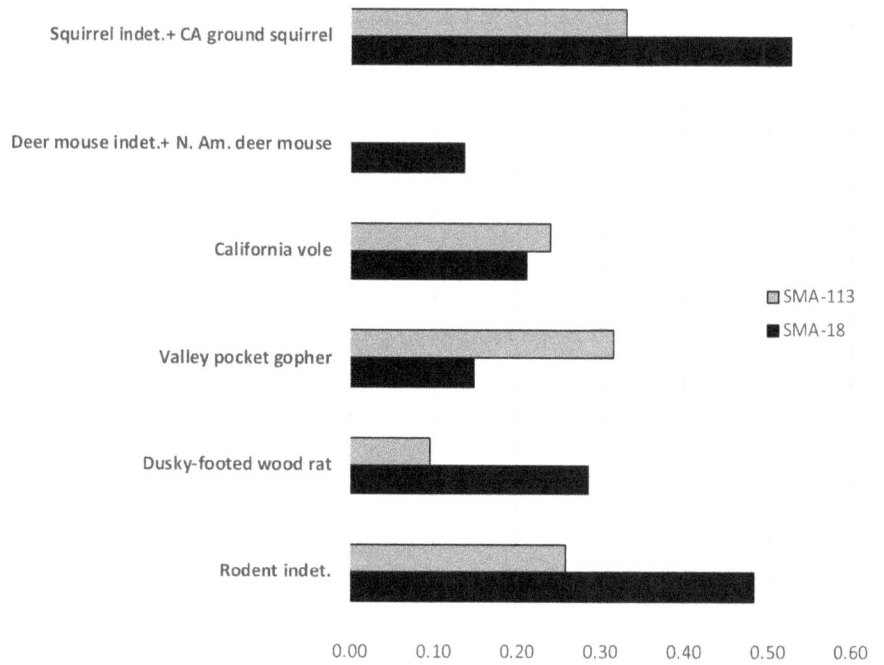

Figure 12.2. CA-SMA-18 and CA-SMA-113: Ratios of Thermally Altered Bone by the Taxonomic Group for the Two Sites.

Table 12.6. CA-SMA-18 and CA-SMA-113: Values of Pairwise Two-tailed Fisher's Exact Test for Between-site, By-taxon Comparisons of Thermal Alteration to Bone.

	SMA-113					
SMA-18	Squirrels	Wood rats	Pocket gophers	Voles	Deer mice	Rodent indet.
Squirrels						
Wood rats	0.2906					
Pocket gophers		0.7222				
Voles			**0.0293**			
Deer mice				1.0000		
Rodent indet.					**0.0015**	

Table 12.6 shows that no statistically significant differences among thermal alterations to squirrels, voles, and woodrats were discerned between CA-SMA-18 and CA-SMA-113, but valley pocket gophers at the two sites did diverge statistically significantly at the p≥.05 level of significance, with SMA-18's gophers showing higher rates of burning. The only other Fisher's exact value at the p≥.05 in the intersite comparison was that of the "Rodent indet." category. This is most likely attributable to the very different recovery methods at the two sites (see **Discussion**).

Intrasite Comparisons of Thermal Modifications

Results of the comparison of roughly like-sized taxonomic groups represented by NISP ≥20 at each site:

CA-SMA-18. Table 12.7 shows that, of the few taxa with ≥20 specimens, none differed from one another at the p=.05 level of significance. Although valley pocket gopher specimens showed more thermally induced color shifts than those at CA-SMA-113, they shared rates of thermal alteration with CA-SMA-18 voles—both around 20%, more than most other of the more numerous taxa at the site. For gophers, this rate is remarkable, given that the species' NISP in the deposits were quite probably "inflated" by unaltered elements from later, natural deaths in burrows (see **Discussion**).

CA-SMA-113. Table 12.8 shows that squirrels, voles, and woodrats display rates of thermal modifications to bones of 17–21%, with skull parts and bones of the extremities showing the highest incidence of modification. Given the likelihood of valley pocket gophers' "inflating" unaltered specimen NISP for their taxon by dying in CA-SMA-113 deposits after the settlement was abandoned, their 11% rate of thermal alteration might reflect a "dilution" of an originally higher rate. The much higher rate of thermal color changes in the "nonidentifiable" rodent specimens sets this category apart from the balance of specimens for largely taphonomic reasons (see **Discussion).**

Table 12.7. CA-SMA-18: Values of Pairwise Two-tailed Fisher's Exact test for Within Site, Taxon-to-Taxon Comparisons of Thermal Alteration to Bone.

Taxon	Wood rats	Pocket gophers	Voles	Rodent indet.
Wood rats		0.4004	0.4880	0.0777
Pocket gophers			1.0000	0.1851
Voles				0.4228
Rodent indet.				

Table 12.8. CA-SMA-113: Values of Pairwise Two-tailed Fisher's Exact test for Within Site, Taxon-to-Taxon Comparisons of Thermal Alteration to Bone.

Taxon	Squirrels	Wood rats	Pocket gophers	Voles	Deer Mice	Rodent indet.
Squirrels		1.0000	0.2258	1.0000	**0.0321**	**0.0027**
Wood rats			0.2658	0.6293	**0.0280**	**<0.00001**
Pocket gophers				**0.0169**	0.0835	**<0.00001**
Voles					**0.0033**	**<0.00001**
Deer mice						**<0.00001**
Rodent indet.						

At both CA-SMA-18 and CA-SMA-113, voles show more traces of some kind of exposure to fire than do most other taxa, slightly over 20% for both samples (Tables 12.4, 12.5). This suggests that more voles entered the human food chain, perhaps as bycatch from land maintenance fires (Cuthrell 2013), or perhaps as commensals killed in the settlements and added to boiled meals (CA-SMA-18, CA-SMA-113). By contrast, no deer mouse specimens from either site showed traces of thermal modification (see **Discussion).** The larger rodent sample from CA-SMA-113, where multiple reliable proxies and a historical account testify to grassland maintenance by Native burning, sheds further light on the question of human involvement. About 11% of 191 valley pocket gopher specimens displayed color shifts, nearly all whole-element. Two of three thermally altered squirrel bones displayed partial color shifts, but of the 22% of 92 vole specimens showing color shifts, all were whole-specimen. The 17% of woodrat specimens displaying color shifts were also whole-specimen (but see **Discussion).**

Archaeofaunal Evidence and Fire-Based Land Management: Discussion

The following sections discuss the problem of rodent species abundances as a proxy for land management using fire, outlining alternative scenarios for the production of similar patterns in species abundance, whether and how thermally induced color changes to specimens can shed light on their use by humans, evidence for targeted predation at CA-SMA-18 and CA-SMA-113, and finally, a proposed means to coping with the equifinality produced by nonhuman and human sources of habitat disturbance—enough to make them somewhat useful proxies.

Alterative Scenarios for Species Abundance Patterning

CA-SMA-18 lies in northern coastal scrub vegetation in which both California voles and North American deer mice naturally thrive; and it is possible that some of the rodent elements at CA-SMA-18 are from the natural background of rodent deaths incorporated into sandy sediments, as the analysts of CA-SCR-7 concluded about their sample. However, CA-SMA-18's squirrels, woodrats, voles, and pocket gophers all display thermal alteration rates of 11–21% of NISP, while none of CA-SCR-7's did. This begs the question of how some individuals of these species became entangled with fire-wielding people. Given the site's lack of strong evidence for vegetation transformation by Native cultural burning at the site (see *Rodents as a Proxy for Land Management Using Fire*, below), these are less likely to have been bycatch resulting from repeated burns.

Recalling that bones in sediments superheated by fires above them tend to be more evenly altered, but partial vs. total color shifts are not an infallible proxy for deliberate human handling vs. accidental exposure under hearths, this presents another potential case of equifinality. Squirrels, woodrats, and voles displayed whole-specimen color shifts, perhaps reflecting only that their remains were under human hearths or ovens. This would be parallel to Cuthrell's (2020) observation that some charred macrobotanical remains recovered from sites may have been parts of the soil seedbank that were superheated by fires above them.

A hint, albeit a negative one, exists that some of the whole-element heat-altered specimens may reflect more deliberate human handling than simply an accident of building a hearth atop sediments containing rodent bones. In marked contrast with similarly sized and smaller rodent taxa, none of the rather small samples of *Peromyscus* specimens from CA-SMA-18 (NISP=8) nor CA-SMA-113 (NISP=36) specimens display any thermal alterations (Tables 12.4, 12.5, and 12.8). This suggests that deer mice were incorporated into site deposits neither via human culinary processing nor via hearth use in the same ways as were other rodents. They may represent only the natural background of deer mice deaths, apparently unaffected by combustion features. Some refuse-disposal practices, such as dumping hot embers and ashes together with meal refuse into pits, could produce the same whole-element color shifts in by-products of meals. CA-SMA-113 did yield relatively small ash dumps containing heat-altered faunal and botanical remains (Cuthrell 2013). Although the precise details of human engagement with rodents are not yet resolved, their remains suggest that processes other than below-hearth heating, such as refuse-disposal practices, more likely account for whole-element color changes.

Higher rates of thermally induced color changes among CA-SMA-113's nonidentifiable rodent specimens arise from burned bone's reduced resilience to taphonomic stresses. Stiner and colleagues (1995) noted that thermally altered archaeofaunal specimens from Hayonim Cave, Israel, were collagen-depleted compared to other specimens, and hence more liable to break. Higher breakage rates in thermally altered bones simultaneously inflate the NISP counts of heat-altered specimens and reduce their taxonomic identifiability. "Rodent indet." specimens, especially those not identifiable to element or taxon, display higher rates of thermal modification than do taxonomically and anatomically identifiable ones. The rapid recovery salvage of CA-SMA-18 probably did not retrieve small bits of burned bone from the sandy sediments, so these are not represented in that site's archaeofaunal sample, accounting for the strong divergence of the nonidentifiable category between the two sites (Table 12.6).

The low rates of gray squirrel occurrence in the CA-SMA-113 sample, a site within walking distance of mixed woodlands where they usually live, is of interest in terms of cultural burning practices. As noted earlier, even small-scale cultural burns of oaks and other nut-bearing trees' understory—intended to reduce fungal and larval infestations of the year's crop (Anderson 2013: 145–148)—may reduce the availability of underground fungal foods. This, in turn, would limit foods critical to tree squirrels' survival and reproduction. Thus, gray squirrels may have been less accessible to hunters in oak woodlands managed by cultural burning practices.

Targeted Predation at CA-SMA-18 and CA-SMA-113

As this chapter's section on rodents and site formation indicated, members of the Rodentia are among the few groups that might find their way into archaeological deposits without human agency. Thermally induced bone color shifts support some form of human involvement in their deposition. In contrast to the expedient predation noted above, targeted predation involves the intentional taking of prey. In the case of woodrats slumbering in their highly visible nests, such predation might be embedded, *sensu* Binford (1979), in journeys to the coast or brushy habitats to harvest other foods, fuel, or other useful materials. Woodrat predation probably could be accomplished by anyone out of early childhood with a heavy stick, especially on days when hunting deer or other large mammals came to naught, as it often does (Bliege Bird and Bird 2008; O'Connell et al. 2002). One 230 gm woodrat in a cookpot would yield at least 125 kcal, of which 7.45% would be fat (Kaufman et al. 1975). Given these factors, woodrats might well be overrepresented in archaeofaunas, relative to their actual abundances in the immediate environment of an archaeological site (see Lightfoot et al. 2009).

Close inspection of rodent elements for traces of human culinary handling, such as cutmarks or color shifts on skeletal parts from roasting (Medina et al. 2012), can help in assessing whether a species' numbers may have been enhanced by targeted human predation (Blong et al. 2020: Figure 2; Schmitt and Juell 1994; Schmitt and Lupo 1995). Rates of thermal modifications on *N. fuscipes* bones in the CA-SMA-18 sample are low, about 11%, half that of voles and pocket gophers, which display 20–21% thermal alteration to specimen, and no woodrat specimens bore cutmarks. However, as stated in the introduction to thermal modifications, animals destined to be boiled may not show much in the way of cutmarks or color shifts. Possibly, the lack of success in extracting aDNA from an "unburned" sample of woodrat specimens submitted to her lab (B. Shapiro, personal communication, 2019) could be due to the specimens having been boiled during their taphonomic histories. Without cutmarks, which would probably not be the case for animals simmered in a stew, this evidence is presently equivocal regarding human use.

CA-SMA-18, located among stabilized dunes on Point Año Nuevo itself where Reid's sampling of long-undisturbed habitats yielded no *N. fuscipes*, has a substantial proportion of woodrats, accounting for 35% of rodent NISP (Table 12.3b). The high frequency of woodrats in this "rapid recovery" sample might result from recovery bias because their larger bones are more evident to screeners than are those of much smaller voles or deer mice. However, Cuthrell (2020) suggested that woodrat-friendly habitats with stands of Monterey pine may have been closer to CA-SMA-18 during the time the site was created than documented in recent historical times.

At CA-SMA-113, woodrats comprise 20.6% of the identifiable rodent sample, slightly less than their proportion, 25.7%, in Reid's Inland Brush habitats, but more than the 12.7% in her Inland Forest sample (Tables 12.2 and 12.3.b). Specimens also have a relatively low, 16.7%, rate of thermal color shifts, in comparison with voles' 21.7%. Two possibilities for woodrats' presence in such proportions exist: first, they could reflect the existence of a closed habitat so close to the village that they might be bycatch of cultural burning. Given Lee and Tietje's (2005) report of the low rates of woodrat deaths in prescribed burns noted above, this is less likely than the second option—that they were occasional targeted prey, either by adults or by children—given their size, accessibility, and ease of dispatch.

Although they were excluded from some statistical assessments of site archaeofaunas, Valley pocket gophers from CA-SMA-18 and CA-SMA-113 offer an interesting contrast in signs of processing by people. At CA-SMA-18, over 20% of the 97 *T. bottae* specimens showed thermally induced color shifts, inflicted primarily on the skull parts and extremities, as would be found on the body segments most liable to change color with roasting (Table 12.4). CA-SMA-113 *T. bottae* specimens that display color shifts did so at about half the rate of those at CA-SMA-18. Specimens from CA-SMA-113 showed whole-element color shifts, suggesting a greater likelihood that the heating of gopher bones in the sedimentary matrix contributed to the thermal modifications. Given that gophers can readily flee fires by escaping into their deep burrows, they are less likely to be bycatch of burning at CA-SMA-113. About one-third of the 17 specimens of heat-altered CA-SMA-18 gophers showed partial color changes, a hallmark of roasting.

Gophers are remarkably curious about activities outside their burrows, and, though they may present opportunities for predation when they stick their heads up to check these, they can readily flee predators by withdrawing into their tunnel systems. However, they are susceptible to being tricked in various ways. The simple act of scooping out the soil "plug" at the mouth of an active burrow can provoke a gopher to come out to close the hole, where they can then be trapped by a snare set at the mouth of the burrow, a tactic used to trap the species today. In hunting groups documented in recent times, capture and cooking, usually by roasting of relatively harmless, small prey is done by children, while adults go about their tasks (Bird and Bleige Bird 2017 [2005]). Although we cannot be certain which actors targeted the CA-SMA-18 gophers, it is highly likely that someone did so and then roasted them, as was done with ground squirrels.

Rodents as a Proxy for Land Management Using Fire

For a few years, it appeared that a faunal profile typical of reiterated land management using fire had been identified. Our original research site, CA-SMA-113 in the Quiroste Valley, had a rodent archaeofauna with a preponderance of voles, with traces of *P. maniculatus* among the indeterminate deer mice, and kangaroo rat. North American deer mice are among the earliest post-fire recolonizers, and voles were especially abundant after grass and forbs regrew. Solitary kangaroo rats and Western harvest mice are also typical of post-fire recovery but would have been much less likely than the above species to enter into archaeological deposits in any numbers, given their solitary habits. The coastal site of CA-SMA-18 shared this species profile, as well as the seemingly contradictory representation of closed country-adapted woodrats in some numbers (Table 12.3.b), and the sample of over 100 specimens from CA-SCR-7 was similar.

In 2013, we inferred that the environs of Metenne were not the closed landscape it has been recently, but rather an open grassland to which such rodent species are adapted. Our conclusion was strongly supported by several lines of paleoethnobotanical and charcoal evidence, as well as by the Portolá Expedition's historical records. Except for woodrats, no strong signals existed within the archaeofauna for the forest or brush habitats near the site during its formation. Buoyed by this convergence of evidence, as the lead analyst of mammal fauna, I did not thoroughly consider whether archaeofaunal rodent abundances were a strong proxy, based upon relational analogies.

From late 2020 through 2021, doubts emerged about rodents as a proxy for cultural burning, based upon both botanical and faunal evidence from more sites. Cuthrell's 2021 unpublished report (see also Chapter 8) on Cotoni-Quiroste paleoethnobotany found that coastal sites on Año Nuevo Point (CA-SMA-18, CA-SMA-19, CA-SMA-216, CA-SMA-218) and at CA-SMA-7 lacked the macrobotanical evidence for seed processing such as that recovered from inland sites like CA-SMA-113, CA-SCR-10, CA-SCR-14, and CA-SCR-15. Moreover, when opal phytolith evidence was comparatively analyzed across these contrasting habitat types, compared to inland sites, coastal sites display low proportions of opal phytoliths. Because relatively high proportions of opal phytoliths in and around sites has been deemed a strong indicator of the existence of grasslands, the much lower proportions in coast sites are interpreted by Cuthrell as reflecting those of grass species typical of the mixture of northern coastal scrub, grassland, and coastal strand vegetation habitats in the coastal zone. Cuthrell has stressed that absence of evidence sufficient to address the question of anthropogenic burning in this zone does not "prove" it was absent (R. Cuthrell, personal communication, 2022).

The paleobotanical evidence recovered from these sites did, however, call into question facile equivalency between high vole frequencies with extensive grasslands. The coastal vegetation attested by macrobotanical evidence at CA-SCR-18 is the only habitat where voles were live-trapped by Reid. However, their remains occur in higher relative proportions in the CA-SMA-18 sample than did individuals in Reid's live-trap sample.

The second line of evidence that suggested high frequencies of voles may not always be tied to land clearance by Native burning was our project's vole researchers' conclusion, cited earlier, that many of the CA-SCR-7 voles were not associated with archaeological deposits (Chapter 13, this volume). Cuthrell's Chapter 8 cites several lines of paleoethnobotanical evidence that this coastal site, like those at Año Nuevo Point, lay in the northern coastal scrub/grass/strand vegetation favored by voles and North American deer mice.

Given that they are secondary proxies for vegetation, rodent archaeofaunal abundances are less reliable indicators of human use of fire than are multiple lines of paleobotanical evidence, especially when these are combined with proxies for fire frequency. This is because *any* form of vegetation disturbance, not simply human ones such as landscape burning, can invite booms in early recolonizing rodent species such as voles, North American deer mice, and kangaroo rats. This chapter cannot definitively identify the causes of high recolonizing rodent frequencies, especially based on such a small sample of sites.

One consideration is the disturbance of vegetation communities around campsites by human trampling and the process of vegetation and microfaunal regrowth after abandonment of a campsite. Trampling has been extensively documented in the modern conservation literature (e.g., Bogucki et al. 1975; Cole 2004; Cole and Monz 2003; Davies et al. 2022; Kerbiriou et al. 2008). Most of the coastal sites documented by the project to date are inferred to be special-purpose sites, occupied for relatively short spans, and perhaps not by entire households. These are all on sandy substrates and most among dunes that were stable during the sites' occupation, and surrounded by relatively fragile vegetation. CA-SMA-18, with its greater diversity of artifactual and faunal evidence (Hildebrandt et al. 2006), may reflect repeated occupations of somewhat longer time spans. It is likely that human foot traffic, as well as enrichment of a periphery around each camp with nitrogen from human waste, provided environments of disturbance and regrowth that amplified the dynamic natural processes affecting coastal vegetation communities (see Cuthrell, Chapter 8).

More systematic research is also needed on how much microfaunal bone "background" exists in the loose sandy sediments characterizing the coastal zone, which is hinted at by the CA-SCR-7 vole data discussed in Chapter 13. Given voles' high reproductive rates, their remains may significantly accumulate in sandy deposits, thereby inflating the numbers of specimens recovered during archaeological sampling.

In sum, the problem of equifinal rodent species abundance patterns parallels the case of equifinality between cut marks and pseudo-cut marks presented early in this chapter. As in that case, contextual evidence can be used to narrow the options. One can propose a statement of the *contexts* in which rodent archaeofaunas *could* reinforce the evidence already established by botanical evidence, namely:

> *If* found in areas that *support closed, brushy vegetation or forest* without human intervention, peaks in vole and *P. maniculatu*s further support inferences from paleoethnobotanical evidence and direct proxies of fire for open prairie maintained by human disturbance using fire (viz. CA-SMA-113).

Secondary proxies such as rodents are just that because the best estimate of paleoenvironmental conditions depends upon the paleobotanical evidence, to which rodent abundances only serve as a support in well-defined cases. Nonetheless, rodent faunas and their modifications can offer insights into other Native interactions with various rodent species at a fine-grained level.

Where to Find and How to Recover Evidence for Fire-Based Landscape Management

Researchers interested in recovering the best samples of paleoenvironmental evidence with which to investigate Native landscape maintenance using fire must consider whether some kinds of archaeological sites offer better samples for this than others. In the process of trying various sampling protocols, our research has further defined the sample size range for obtaining evidence adequate to research landscape management.

Types of Sites

Residential Sites

Even if they were occupied only during a specific span in a seasonal round of movement, residential sites are the logical foci for harvesting local, managed landscapes and wild harvesting. In most cases, people consider the distance that they are willing to transport masses of harvested seeds, nuts, and other bulk foods from where they are harvested to where these are stored, processed, and consumed. Except in extraordinary cases (e.g., Hildebrandt et al. 2009), those tasked with gathering and transporting foods tend to minimize these distances. It follows that decisions in locating residential sites involve a consideration of areas intended to be burned and managed, balanced against other key considerations, such as proximity of water, fuel, and food species not managed by burning, and providing for the safety of very young children. Time and energy expenditures, especially for those tasked with much of household and camp maintenance, are common-sensical but can be assessed within a Central Place foraging model (Bettinger et al. 1997). Repeated seasonal occupations of such places by numerous households year after year build up evidence of many activities, yielding more densely concentrated, and thus more recoverable and interpretable, lines of evidence for fire-based landscape modification.

CA-SMA-113, with its clear signature of grassland maintenance in the Quiroste Valley and evidence of a diversity of activities, is such a site. In Cotoni's open oak-madrone woodlands, CA-SCR-14 and CA-SCR-15 display similar phytolith densities to grasslands maintained over a substantial time, and they probably are residential in the sense used here (Cuthrell, Chapter 8).

Special-Purpose Sites

Other places where people regularly obtain specific resources need not be chosen based on all the factors listed above. These localities are often called "special-purpose sites" by archaeologists, and in the Quiroste-Cotoni region, they may include stone quarries for tools, spots good for catching fishes, or places where land-breeding marine mammals are killed and butchered. Some of these places, especially along shorelines, may be uncomfortable for longer visits due to tidal action or high onshore winds, or outright dangerous places for children. They are unlikely to be loci of longer-term, group residences. This is especially true if they offer access to rare, spatially restricted resources—such as the outcrop of Monterey Bay banded chert at Año Nuevo Point, the only one for many miles along the greater Monterey Bay region, best accessed at low tide. These may be visited repeatedly for short stays, sometimes by a subset of the social group. Regardless of resources gathered, special-purpose localities tend to be only intermittently used and bear evidence of only a few activities. From the paleoethnobotanical evidence (Cuthrell, Chapter 8), it seems increasingly the case that coastal sites are unlikely to yield evidence of extensive, fire-based vegetation management, although they may be important sites for obtaining a clear picture of other kinds of focused management and harvesting, as discussed by Grone in Chapter 9. Moreover, the people at such localities as CA-SMA-218—where banded chert was formed into tradeable bifaces and marine mammals were butchered—appear to have provisioned themselves at least in part from storable seeds and nuts from inland sources (Cuthrell, Chapter 8). This is also true of CA-SMA-19, a site from the same period as

Metenne but dominated by evidence of Olivella shell harvesting and the first stages of processing these for bead manufacture (Mark Hylkema, personal communication, 2022).

Sites Between Two Ends of the Spectrum

Some sites can be expected to fall along a continuum between large residential camps—out of which numerous activities by diverse actors are logistically organized (Binford 1980)—and small, special-purpose camps, which are focused on only a few activities. The latter might comprise a few household groups and be devoted to seasonally harvesting and preserving local resources—for example, ethnohistoric sources testify to acorn-harvesting and deer-hunting camps in the early autumn in the Sierra Nevada foothills (Anderson 2013). Sites can also change their nature over time, from special-purpose to residential or vice versa. CA-SCR-10, within an easy walk to the shoreline, may have been such an encampment, used seasonally over a very long period, as suggested from its dates—but with more residential groups present and evidence of more plant food processing, and later, for some landscape management, than the above-mentioned sites (Cuthrell, Chapter 8).

Retrieving Samples Adequate for Eco-archaeological Research

Another consideration is how to define the optimal methods for obtaining adequate macrobotanical, marine invertebrate and vertebrate, and terrestrial faunal evidence from archaeological sites. I appreciate, in both senses of the word, the ethos of doing low-impact archaeology using ground-penetrating radar (Nelson, Chapter 5) and smaller excavation units, thus coinciding with Tribal wishes to minimize site disturbance, given the risks of encountering ancestral remains. At the same time, researchers, whether Indigenous or settler in their origins, must present convincing, evidence-based assertions about the lives and activities of precolonial Native peoples based on the evidence they recover. Paleoethnobotanists, zooarchaeologists, and archaeomalacologists rely on sample sizes adequate to make statistically based arguments for the meaning of patterns in their data, and the "adequacy" of a sample can vary with the type of evidence. So, a balance must be struck between these two needs, but no *a priori* formula exists for how to obtain such a sample. Our project has explored this issue and has some findings that may be useful to others. I have worked with faunal remains from the large CA-SMA-113 excavation and from much smaller flotation samples from Point Año Nuevo sites. I also reviewed the rodent data of the Cotoni site samples recovered by augering, column sampling, and small excavations. After these experiences, I have reflected on the trade-offs involved in each approach.

Excavations at CA-SMA-113 were extensive, using both ¼" and 1/8" dry screening for lateral excavations, along with a comprehensive flotation sampling strategy that involved processing a minimum of five liters of soil from every excavated context (e.g., arbitrary level, natural stratigraphic level, feature), as well as collecting multiple samples from the same interpretive context (Cuthrell 2013). It is estimated that around 13,428.6 liters of earth were dry screened, and another 1,525 liters were processed using flotation (R. Cuthrell, personal communication, 2021). The vertebrate faunal sample from CA-SMA-113 had a total specimen count of 19,460. Fully 42% of these are listed as "nonidentifiable," small fragments without diagnostic landmarks, which could only be assigned to general zoological classes—fishes, reptiles, birds, mammals—and sometimes to size groups within those classes—e.g., very small birds. This is typical of many Central Coast faunas. Winnowing down the 12,910-specimen mammal sample to those identifiable to the zoological family, genus, or species level gives a total NISP of 1.471. Inspecting Table 12.9, one sees quite a low specimen count for rodents, despite the amount of soil moved overall, with clear indications that flotation achieves much higher rates of retrieval.

Table 12.9 also includes comparable data from CA-SMA-18, where the dry-screen rate of retrieval was even lower than at CA-SMA-113, expectable in light of the rapid retrieval approach to recovery there. The four small-scale site samples in Cotoni moved a total of 2,423 liters of sediments (Sanchez 2019), recovering a total of 452 identifiable rodent specimens from all sites, with another 132 unidentifiable fragments (Table 12.9). This much more restricted excavation strategy, with its emphasis on sampling for flotation and close sorts of the heavy fraction for macrobotanicals and smaller bird, mammal, and fish bones, produced roughly comparable rodent retrieval rates, as did recovery techniques at CA-SMA-113.

Table 12.9. Comparison of the Number of Rodent Specimens (NSP) Recovered, Liters of Site Sediment Processed, and Rodent Recovery Rates per Liter for CA-SMA-18 (Hildebrandt et al. 2006: Table 3), CA-SMA-113 (Rodents: Present Study Tabulation, Sediment Processed (Cuthrell 2013: 640–648, Table 1.2 and Personal Communication, 2021) and Cotoni Sites of CA-SCR-7, CA-SCR-10, CA-SCR-14, CA-SCR-15 (Sanchez 2020).

Site	Column, Auger			Dry Screen		
	Rodent NSP	Liters	Rodent:Liter	Rodent NSP	Liters	Rodent:Liter
CA-SMA-18	---	----	----	112	14,100.0	0.008
CA-SMA-113	533	2025.5	0.263	155	13,428.6	0.012
CA-SCR-7, CA-SCR-10, CA-SCR-14, CA-SCR-15	584	2423.0	0.241	----	----	----

However, in contrast to CA-SMA-113, which boasted an abundance of larger mammal specimens, from rabbit-sized to elk- and fur-seal-sized, the Cotoni sites yielded only smaller shells, small fish bones, and micromammals. For answering questions about overall ecology and landscape and seascape management, this may be a good enough trade-off between research goals and Tribal sensibilities. However, the diminution of faunal returns is extreme and sheds no light on Native relations with larger terrestrial and marine birds, mammals, and fishes. Although I expected to be the only specialist raising questions about sample size, Cuthrell (Chapter 8) has raised concerns over whether the resulting paleoethnobotanical samples from the Cotoni sites are sufficient to draw dependable inferences about vegetation history. Thus, this eco-archaeological research program for paleoenvironmental sampling may have explored both the upper size limits at CA-SMA-113 and its lower ones at CA-SCR-7, CA-SCR-10, CA-SCR-14, and CA-SCR-15.

From my point of view, to ask and answer questions about the everyday lives of ancestors—men, women, and children—in these landscapes and along these shores, the lowest-impact approaches explored by the Cotoni phase of the project may not be optimal. One option for avoiding sample size problems, while at the same time respecting further disturbance of ancestral remains, is to make sure that *every* relevant, previously excavated museum collection, many obtained in a time when relations between Native Californians and non-Native archaeologists were very different (Schneider 2021), is fully analyzed. Many such site excavations used ½" or ¼" screens and did no flotation to retrieve very small plant, animal, shell, and lithic objects; thus, such museum collections cannot address certain questions raised in this monograph, for example, what three to five lines of paleoethnobotanical evidence can say about ancient vegetation. However, as was long-recommended in California archaeology (Hester et al. 1975), column samples of soil were frequently excavated and stored by these fieldworkers and saved for more sophisticated, future analyses. These samples now can be processed using flotation to retrieve some such detailed data. With Tribes' and other stakeholders' consent, new auger or column samples from previously excavated sites also could be taken, as Cuthrell did for Año Nuevo site samples. Most of these collections have also retrospectively been subject to repatriation according to the Native American Graves Protection and Repatriation Act (NAGPRA) or could be, at the request of a Tribe, thereby making them more amenable to collaborative analyses.

Conclusion

Rodent species abundances were initially thought to display a distinctive signature of iterative Native landscape management using fire, with early recolonizing seed-eaters dominating the assemblages, thus adding one more of several lines of biological evidence for Native vegetation maintenance. However, a critical assessment of rodents from Quiroste-Cotoni revealed that three sites with such profiles were located in areas seldom, if ever, manipulated by burning. This, in turn, has led to the conclusion that, because disparate environmental processes can produce equifinal species abundance profiles, rodent relative frequencies are not as reliable a proxy as are the actual paleoethnobotanical remains. I hope that this chapter's discussion of the nature of strong proxies and problems of equifinality can be of use to other researchers, as can this chapter's analysis of thermally altered bone evidence, in teasing out the situation of their samples.

Residential sites are probably the optimal locales for further exploring Native cultural burning due to their proximity to maintained grasslands and their density of evidence. The low-impact approach for evidence retrieval used to sample the Cotoni sites probably represents the lowest possible limits of recovery for adequate samples. While respectfully acknowledging the need for a trade-off between Tribal reluctance to disturb the ancestors and adequate sample recovery, I suggest that many questions that cannot be explored with such small samples could, however, be investigated with existing collections from earlier excavations. These can be augmented as needed by flotation-based studies of their column samples or by limited column or auger sampling of previously excavated sites, as Cuthrell did with Año Nuevo Point sites, given Tribal and other stakeholders' consent. By working with older collections, recovered in times before collaborative research between archaeological researchers and Tribes seeking to relearn ancestral ways, today's archaeologists—whether Native or settler in descent—can partially redeem archaeological predecessors' disregard for Tribal wishes and protocols and disturbance of Tribal ancestors.

Acknowledgments

The faunal analysis reported here was supported by subawards from NSF Archaeology BCS-0320168 Kent Lightfoot, PI; NSF Archaeology BCS-0912162, Kent Lightfoot and Paul Fine, Pis; the Gordon and Betty Moore Foundation, Rosemary Gillespie, Kent Lightfoot, and David Ackerly Pis. I am deeply grateful to Amah Mutsun Chair Valentín Lopez, the Amah Mutsun elders, and other members of the Tribe for their continued interest and advice on this project. I thank my colleagues Thomas J. Banghart, Cristie Boone, Mark Hylkema, Kent Lightfoot, Rachel Brown Reid, and Dave Schmitt for their advice and support. I especially wish to acknowledge and thank Rob Cuthrell for his responses to my many questions, his close reading of a draft of this chapter, and the detailed and thoughtful content of his dissertation and later writings. Gabriel Sanchez graciously shared the microfaunal data from the precolonial Cotoni site samples with me, which I have condensed into Table 12.2. Peter Heintzmann generously supplied his earlier, unpublished aDNA data on CA-SMA-113 rodents, and I thank him. A special thanks goes out to Dr. Vicky Oelze who explored the possibility of stable isotope research with selected elements from the study areas. I am deeply grateful to the zooarchaeology lab interns who helped with the final phases of this project: Danielle Aparicio, Nathalie Chávez, Justin Colón, María Contreras Vega, and Ashley Navas, as well as to earlier interns and employees who worked on sorting and identifying CA-SMA-113 small mammal fauna and managing its imposing database, who included Basia Atkinson-Barr, Ben Curry, Emily Zimmerman Gilstrap, and Patrick O'Meara. Maria Viteri (Ecology and Evolutionary Biology, Stanford University) ran the NMDS analysis from the number of individuals on Reid's four live-trapping data and NISP for the three sites discussed in the chapter, and I am deeply indebted to her for her advice and discussions of this method. Visiting Assistant Professor Claudia Wehrhahn Cortes (Department of Statistics, UCSC) further advised me on NMDS as an appropriate tool for analyzing these data and on the use of the "r" statistical package for it. All errors of opinion or judgment are my own.

References

Anderson, M. Kat
2013   *Tending the Wild: Native American Knowledge and the Management of California's Natural Resources.* University of California Press, Berkeley.

Ayinde, Kayode, and A. O. Abidoye
2010   Simplified Freeman-Tukey Test Statistics for Testing Probabilities in Contingency Tables. *Science World Journal* 2.

Bandoli, James H.
1987   Activity and Plural Occupancy of Burrows in Botta's Pocket Gopher *Thomomys bottae. The American Midland Naturalist* 118(1): 10–14.

Behrensmeyer, Anna K., Kathleen D. Gordon, and Glenn T. Yanagi
1986   Trampling as a Cause of Bone Surface Damage and Pseudo-Cutmarks. *Nature* 319: 768–771.

Bennett, Joanne L.
1999   Thermal Alteration of Buried Bone. *Journal of Archaeological Science* 26(1): 1–8.

Bettinger, Robert L., Ripan Malhi, and Helen McCarthy
1997   Central Place Models of Acorn and Mussel Processing. *Journal of Archaeological Science* 24: 887–899.

Binford, Lewis R.
1979   Organization and Formation Processes: Looking at Curated Technologies. *Journal of Anthropological Research* 35(3): 255–273.
1980   Willow Smoke and Dogs' Tails: Hunter-Gatherer Settlement Systems and Archaeology. *American Antiquity* 45: 4–20.

Bird, Douglas W., and Rebecca Bliege Bird
2017 [2005]   Mardu Children's Hunting Strategies in the Western Desert, Australia. In *Hunter-Gatherer Childhoods: Evolutionary, Developmental and Cultural Perspectives,* edited by B. S. Hewlett and M. E. Lamb, pp. 129–146. Routledge, New York.

Bliege Bird, Rebecca, and Douglas W. Bird
2008   Why Women Hunt: Risk and Contemporary Foraging in a Western Desert Aboriginal Community. *Current Anthropology* 9(4): 655–693.

Blong, John C., Martin E. Adams, Gabriel Sanchez, Dennis L. Jenkins, Ian D. Bull, and Lisa-Marie Shillito

2020   Younger Dryas and Early Holocene Subsistence in the Northern Great Basin: Multiproxy Analysis of Coprolites from the Paisley Caves, Oregon, USA. *Archaeological and Anthropological Sciences* 12(9): 224.

Bocek, Barbara

1986   Rodent Ecology and Burrowing Behavior: Predicted Effects on Archaeological Site Formation. *American Antiquity* 51: 589–603.

1992   The Jasper Ridge Reexcavation Experiment: Rates of Artifact Mixing by Rodents. *American Antiquity* 57 (Clairmont): 261–269.

Bochenski, Zbigniew M., Teresa Tomek, Risto Tornberg, and Krzysztof Wertz

2009   Distinguishing Nonhuman Predation on Birds: Pattern of Damage Done by the White-Tailed Eagle *Haliaetus albicilla*, with Comments on the Punctures Made by the Golden Eagle *Aquila chrysaetos. Journal of Archaeological Science* 36: 122–129.

Bogucki, Donald J., John L. Malanchuk, and T. E. Schenck

1975   Impact of Short-Term Camping on Ground-Level Vegetation. *Journal of Soil and Water Conservation* 30 (Robinson, et al.): 231–232.

Bonadio, Christopher

2000   *Neotoma fuscipes* (On-Line). In *Animal Diversity Web*, vol. 2013. University of Michigan Museum of Zoology, Ann Arbor.

Bosch, Pedro, Inmaculada Alemán, Carlos Moreno-Castilla and Miguel Botella

2011   Boiled Versus Unboiled: A Study on Neolithic and Contemporary Human Bones. *Journal of Archaeological Science* 38(10):2561–2570.

Brylski, Phillip, and Jay Harris

1990   *California's Wildlife Volume III.* California Department of Fish and Game, Sacramento, California.

Bunn, Henry T.

1981   Archaeological Evidence for Meat-Eating by Plio-Pleistocene Hominids from Koobi Fora and Olduvai Gorge. *Nature* 291: 574–577.

Carraway, Leslie N., and B. J. Verts

1994   *Sciurus griseus. Mammalian Species* 474: 1–7.

Chew, Robert M., Bernard B. Butterworth, and Richard Grechman

1959   The Effects of Fire on the Small Mammal Populations of Chaparral. *Journal of Mammalogy* 40(2): 253.

Cole, David N.
　　2004　Impacts of Hiking and Camping on Soils and Vegetation: A Review. *Environmental Impacts of Ecotourism* 41: 60.

Cole, David N., and Christopher A. Monz
　　2003　Impacts of Camping on Vegetation: Response and Recovery Following Acute and Chronic Disturbance. *Environmental Management* 32(6): 693–705.

Crespí, Juan
　　2001　*A Description of Distant Roads: Original Journals of the First Expedition into California, 1769–1770.* Translated by Alan K. Brown. University of San Diego State University Press, San Diego, California.

Cuthrell, Rob Q.
　　2013　An Eco-Archaeological Study of Late Holocene Indigenous Foodways and Landscape Management Practices at Quiroste Valley Cultural Preserve, San Mateo County, California. PhD Dissertation, Department of Anthropology, University of California, Berkeley.
　　2021　Archaeobotanical Research at Eight Archaeological Sites West of the Santa Cruz Mountains: Implications for Subsistence Practices and Anthropogenic Burning In *The Study of Indigenous Landscape and Seascape Stewardship Practices on the Santa Cruz Coast: A Collaborative Eco-Archaeological Approach*, edited by Kent G. Lightfoot, Rob Q. Cuthrell, Mark G. Hylkema, Valentin Lopez, Diane Gifford-Gonzalez, Roberta A. Jewett, Michael A. Grone, Gabriel M. Sanchez, Peter A. Nelson, Alec J. Apodaca, Ariadna Gonzalez, Kathryn Field, and Alexii Sigona. Report Prepared for California Department of Parks and Recreation, Santa Cruz District, Archaeological Research Facility, University of California, Berkeley.

Davies, Benjamin, Mitchell J. Power, David R. Braun, Matthew J. Douglass, Stella G. Mosher, Lynne J. Quick, Irene Esteban, Judith Sealy, John Parkington, and J. Tyler Faith
　　2022　Fire and Human Management of Late Holocene Ecosystems in Southern Africa. *Quaternary Science Reviews* 289: 107600.

Erlandson, Jon M.
　　1984　A Case Study in Faunalturbation: Delineating the Effects of the Burrowing Pocket Gopher on the Distribution of Archaeological Materials. *American Antiquity* 49: 785–790.

Fairgrieve, Scott I.
　　2008　*Forensic Cremation: Recovery and Analysis.* CRC Press, Boca Raton.

Fiorillo, Anthony R.

 1989　An Experimental Study of Trampling: Implications for the Fossil Record. In *Bone Modification*, edited by R. Bonnichsen and M. Sorg, pp. 61-72. University of Maine Center for the Study of Early Man, Orono.

Gifford-Gonzalez, Diane

 1991　Bones Are Not Enough: Analogues, Knowledge, and Interpretive Strategies in Zooarchaeology. *Journal of Anthropological Archaeology* 10: 215–254.

 2018a　*An Introduction to Zooarchaeology.* Springer Publishing, New York.

 2018b　*Mammal, Bird, and Reptile Flotation Sample Specimens from Four Sites in Año Nuevo State Park.* University of California, Santa Cruz.

Gifford-Gonzalez, D., and J. Parham

 2008　The Fauna from Adrar Bous and Surrounding Areas. In *Adrar Bous: The Archaeology of a Granitic Ring Complex in Central Sahara*, edited by D. G. E. Gifford-Gonzalez, J. D. Clark, E. Agrilla, D. C. Crader, A. B. Smith, and M. A. J. Williamson, pp. 313–353. Royal Africa Museum Annales in Archaeology, Tervuren, Belgium.

Gifford-Gonzalez, Diane, Cristie M. Boone, and Rachel E. B. Reid

 2013　The Fauna from Quiroste: Insights into Indigenous Foodways, Culture, and Land Modification. *California Archaeology* 5 (Clairmont): 291–317.

Gifford-Gonzalez, Diane, Kenneth Gobalet, Jeremy Gaeta, and Jean C. Geary

 2006　The Faunal Sample from CA-SMA-18: Environment, Subsistence, Taphonomy, Historical Ecology. In *Archaeological Investigations at CA-SMA-18: A Study of Prehistoric Adaptations at Año Nuevo State Reserve, Report Prepared for California Department of Parks and Recreation, Santa Cruz District*, edited by M. Hylkema, W. Hildebrandt, J. Farquhar, D. Gifford-Gonzalez and K. Gobalet, pp. 29–50. Far Western Anthropological Research Group, Davis, California.

Grzimek, Bernhard

 1990　*Grzimek's Encyclopedia of Mammals. Volume Three.* McGraw-Hill Publishing, New York.

Hester, Thomas R., Robert F. Heizer, and John A. Graham

 1975　*Field Methods in Archaeology.* 5th ed. Mayfield, Palo Alto.

Hildebrandt, William R., Jennifer Farquhar, and Mark G. Hylkema (editors)

 2006　*Archaeological Investigations at CA-SMA-18: A Study of Prehistoric Adaptations at Año Nuevo State Reserve.* Far Western Anthropological Research Group, Davis, California.

Hildebrandt, William R., Jeffrey Rosenthal, and Glenn Gmoser
2009   Shellfish Transport, Caloric Return Rates, and Prehistoric Feasting on the
Laguna De Santa Rosa, Alta California. *California Archaeology* 1(1): 55–78.

Holbrook, Sally J.
1977   Rodent Faunal Turnover and Prehistoric Community Stability in
Northwestern New Mexico. *The American Naturalist* 111(982): 1195–1208.

Holden, J. L., P. P. Phaky, and J. G. Clement
1995   Scanning Electron Microscope Observations of Incinerated Human
Femoral Bone: A Case Study. *Forensic Science International* 74: 17–28.

Howard, W. E., R. L. Fenner, and H. E. Childs, Jr.
1959   Wildlife Survival in Brush Burns. *Journal of Range Management*
12: 230–234.

Jameson, Everett W., Jr., and Hans J. Peeters
1988   Brush Mouse (*Peromyscus boylii*). In *California Mammals*, edited by E.
W. Jameson, Jr. and H. J. Peeters, pp. 301–302, 376. California Natural History
Guides. Vol. 52. University of California Press, Berkeley.

Johnson, Brent E., Rand R. Evett, Kent G. Lightfoot, and Charles J. Stiplen
2010   Exploring the Traditional Use of Fire in the Coastal Mountains of Central
California. In *JFSP Research Project Reports*. vol. 74. DigitalCommons@University
of Nebraska, Lincoln.

Kalcounis-Rueppell, Matina C., and John S. Millar
2002   Partitioning of Space, Food, and Time by Syntopic *Peromyscus boylii* and *P.
californicus. Journal of Mammalogy* 83 (Clairmont): 614–625.

Kalcounis-Rueppell, Matina C., and Tracey R. Spoon
2009   *Peromyscus boylii* (Rodentia: Cricetidae). *Mammalian Species* 838: 1–14.

Kaufman, Donald W., Glennis A. Kaufman, and James G. Wiener
1975   Energy Equivalents for Sixteen Species of Xeric Rodents. *Journal of
Mammalogy* 56(4): 946–949.

Kerbiriou, Christian, Isabelle Leviol, Frédéric Jiguet, and Romain Julliard
2008   The Impact of Human Frequentation on Coastal Vegetation in a Biosphere
Reserve. *Journal of Environmental Management* 88(4): 715–728.

Kroeber, Alfred L.
1925   *Handbook of the Indians of California.* Bulletin No. 78. Bureau of American
Ethnology, Smithsonian Institution, Washington, DC.

Lawrence, George E.

1966   Ecology of Vertebrate Animals in Relation to Chaparral Fire in the Sierra Nevada Foothills. *Ecology* 47 (Clairmont): 278–291.

Lee, Derek E. and William D. Tietje

2005   Dusky-Footed Woodrat Demography and Prescribed Fire in a California Oak Woodland. *Journal of Wildlife Management* 69(3):1211-1220.

Lightfoot, Kent G., and Otis Parrish (editors)

2009   *California Indians and Their Environment: An Introduction.* University of California Press, Berkeley and Los Angeles.

Lightfoot, Kent G., Rob Q. Cuthrell, Cristie M. Boone, Roger Byrne, Andreas S. Chavez, Laurel Collins, Alicia Cowart, Rand R. Evett, Paul V. A. Fine, Diane Gifford-Gonzalez, Mark G. Hylkema, Valentin Lopez, Tracy M. Misiewicz, and Rachel E. B. Reid

2013   Anthropogenic Burning on the Central California Coast in Late Holocene and Early Historical Times: Findings, Implications, and Future Directions. *California Archaeology* 5: 371–390.

Lightfoot, Kent G., Rob Q. Cuthrell, Mark G. Hylkema, Valentin Lopez, Diane Gifford-Gonzalez, Roberta A. Jewett, Michael A. Grone, Gabriel M. Sanchez, Peter A. Nelson, Alec J. Apodaca, Ariadna Gonzalez, Kathryn Field, Jordan Brown, Alexii Sigona, and Paul V. A. Fine

2021   The Eco-Archaeological Investigation of Indigenous Stewardship Practices on the Santa Cruz Coast. *Journal of California and Great Basin Anthropology* 41 (Clairmont): 187–205.

Linsdale, Jean M.

1946   *The California Ground Squirrel.* University of California Press, Berkeley and Los Angeles.

Linsdale, Jean M., and Lloyd P. Tevis

1951   *The Dusky-Footed Woodrat.* University of California Press, Berkeley.

Lyman, R. Lee

1987   Archaeofaunas and Butchery Studies: A Taphonomic Perspective. *Advances in Archaeological Method and Theory* 10: 249–337.

McDonald's

2017–2021   Mcdonald's Online Nutrition Calculator, vol. 2021.

Medina, Matías E., Pablo Teta, and Diego Rivero
2012 Burning Damage and Small-Mammal Human Consumption in Quebrada Del Real 1 (Cordoba, Argentina): An Experimental Approach. *Journal of Archaeological Science* 39(3): 737–743.

Merritt, Walter W.
1974 Factors Influencing the Local Distribution of *Peromyscus californicus* in Northern California. *Journal of Mammalogy* 55(1): 102–114.

Meserve, P.
1977 Three-Dimensional Home Ranges of Cricetid Rodents. *Journal of Mammalogy* 58: 549–558.

Milliken, Randall, Laurence H. Shoup, and Beverly Ortiz
2009 *Ohlone/Costanoan Indians of the San Francisco Peninsula and Their Neighbors, Yesterday and Today.* Report for the National Park Service, Cultural Resources and Museum Management Division, Golden Gate National Recreation Area, San Francisco, California.

National Park Service
2005 Final Environmental Impact Statement for a Fire Management Plan, Santa Monica Mountains National Recreation Area, California. US Department of Interior, National Park Service, Calabasas, California.

O'Connell, James F., Kristen Hawkes, Karen D. Lupo, and Nicholas G. Blurton Jones
2002 Male Strategies and Plio-Pleistocene Archaeology. *Journal of Human Evolution* 43(6): 831–872.

Potts, Richard B., and Pat Shipman
1981 Cutmarks Made by Stone Tools on Bones from Olduvai Gorge, Tanzania. *Nature* 291: 577–580.

Quinn, Ronald D.
1990 Habitat Preferences and Distribution of Mammals in California Chaparral. vol. Research Paper PSW-202. Pacific Southwest Research Station, U.S. Department of Agriculture, Forest Service, Berkeley.

Reid, Rachel E. B.
2014 Dietary Ecology of Coastal Coyotes (*Canis latrans*): Marine-Terrestrial Linkages from the Holocene to Present, Earth and Planetary Sciences, University of California, Santa Cruz, Santa Cruz, California.

Reid, Rachel E. B., Eli N. Greenwald, Yiwei Wang, and Christopher C. Wilmers
2013 Dietary Niche Partitioning by Sympatric *Peromyscus boylii* and *P. californicus* in a Mixed Evergreen Forest. *Journal of Mammalogy* 94(6): 1248–1257.

Roberts, Susan L., Douglas A. Kelt, Jan W. van Wagtendonk, A. Keith Miles, and Marc D. Meyer

> 2015   Effects of Fire on Small Mammal Communities in Frequent-Fire Forests in California. *Journal of Mammalogy* 96(1): 107–119.

Sanchez, Gabriel

> 2019   The Historical Ecology and Ancient Fisheries of the Central California Coast: Insights from Point Reyes National Seashore and the Santa Cruz Coast, Anthropology, University of California, Berkeley.
>
> 2020   *Report to National Science Foundation Archaeology Research Grant BCS-0320168: Middle and to Late Holocene Faunal Remains from the Santa Cruz Coast.* University of California, Berkeley.

Schmitt, Dave N., and Kenneth E. Juell

> 1994   Toward the Identification of Carnivore Scatological Faunal Accumulations in Archaeological Contexts. *Journal of Archaeological Science* 21: 249–262.

Schmitt, Dave N., and Karen D. Lupo

> 1995   On Mammalian Taphonomy, Taxonomic Diversity, and Measuring Subsistence Data in Zooarchaeology. *American Antiquity* 60(3): 496–514.

Schneider, Tsim D.

> 2021   *The Archaeology of Refuge and Recourse. Coast Miwok Resilience and Indigenous Hinterlands in Colonial California.* University of Arizona Press, Tucson.

Shaffer, Brian S.

> 1992   Interpretation of Gopher Remains from Southwestern Archaeological Assemblages. *American Antiquity* 57(4): 683–691.

Shipman, Pat, Giraud Foster, and Margaret Schoeninger

> 1984a   Burnt Bones and Teeth: An Experimental Study of Color, Morphology, Crystal Structure and Shrinkage. *Journal of Archaeological Science* 11(4): 307–325.

Smith, Jennifer E., Douglas J. Long, Imani D. Russell, Kate Lee Newcomb, and Valeska D. Muñoz

> 2016   *Otospermophilus Beecheyi* (Rodentia: Sciuridae). *Mammalian Species* 48(939): 91–108.

Stienecker, Walter E., and Bruce M. Browning

> 1970   Food Habits of the Western Gray Squirrel. *California Fish and Game* 56: 36–48.

Stiner, Mary C., Steven L. Kuhn, Stephen Weiner, and Ofer Bar-Yosef
    1995   Differential Burning, Recrystallization, and Fragmentation of Archaeological Bone. *Journal of Archaeological Science* 22: 223–237.

Taylor, R. E., P. E. Hare, and T. D. White
    1995   Geochemical Criteria for Thermal Alteration of Bone. *Journal of Archaeological Science* 22: 115–119.

Thoms, Alston V.
    2008   The Fire Stones Carry: Ethnographic Records and Archaeological Expectations for Hot-Rock Cookery in Western North America. *Journal of Anthropological Archaeology* 27:443–460.

Ugan, Andrew, and Jason Bright
    2001   Measuring Foraging Efficiency with Archaeological Faunas: The Relationship between Relative Abundance Indices and Foraging Returns. *Journal of Archaeological Science* 28(12): 1309–1321.

Wandsnider, LuAnn
    1997   The Roasted and the Boiled: Food Composition and Heat Treatment with Special Emphasis on Pit-Hearth Cooking. *Journal of Anthropological Archaeology* 16: 1–48.

Wylie, Alison
    1985   The Reaction against Analogy. *Advances in Archaeological Method and Theory* 8: 63–111.

Young, Tanner M.
    2014   Biodiversity Calculator, vol. 2019. Al Young Studios.

CHAPTER 13

Ancient DNA from Voles from Archaeological Sites in Central California Reveals Population Continuity over the Last 6,000 Years: Implications for Past Land Management Practices

PAUL V. A. FINE, CHRIS J. CONROY, CAMERON SHARD MILNE, BETH SHAPIRO, DIANE GIFFORD-GONZALEZ, GABRIEL M. SANCHEZ, AND KENT G. LIGHTFOOT

How do we learn about the distant past, from the time of hundreds or even thousands of years ago? In many parts of the world, we have oral histories, passed down from our ancestors. On the California coast, some of these stories were lost in the genocide that occurred after European settlers arrived in the eighteenth century (see Chapter 2). Native Californian survivors of this genocide are now looking to academic research as another source for stories about the past. Archaeology, paleontology, and biology all construct narratives about the past, and in this chapter, we try to weave together threads from these fields to learn about the Central California coast over the past several thousand years, in large part to provide information for Native Californians who are working to recover their stories and lands.

Ethnographic records from coastal Tribes and Spanish accounts from the eighteenth century describe coastal prairies covering large areas of the California coast between Monterey and San Francisco Bay. These accounts were recently corroborated by a multidisciplinary research project including archaeological evidence, fire scars, phytoliths, and pollen records around Quiroste Valley in San Mateo County over the past 1,000 years (Lightfoot et al. 2013). However, people have been living on the California coast for substantially longer periods of time. Have they been using fire to maintain coastal prairie and grasslands over the past 5,000 years or even longer? How extensive was this landscape management? How far back in time and over what geographic scale can we reconstruct past landscapes using tools from archaeology and biology?

To investigate human management of coastal California landscapes, an interdisciplinary team of researchers has been sampling archaeological sites over a wide range of time periods on the Santa Cruz coast (Lightfoot et al. 2021). Radiocarbon dated archaeological sites yielded copious plant and animal material. Some of this material (especially animal remains) can contain sufficient amounts of DNA to piece together sequence fragments that can be used for phylogenetic analysis. By matching the DNA sequences of known genes (for example, the cytochrome *b* gene of the mitochondrial genome), it is possible to compare the sequences of ancient organisms to present-day organisms living in the same area. With enough genetic information, one could in theory estimate population size and whether the populations have changed in size over time (Fine et al. 2013). To test whether people were living in open habitats (as opposed to forested landscapes), we selected the California vole (*Microtus californicus*), a species that is associated with prairies and grasslands and whose bones and teeth are abundant in the archaeological record across most sites and time periods (see Chapter 12, this volume).

In addition to being abundant in the archaeological record in this project, California voles are currently extremely abundant in California, including in coastal habitats. Although they are most abundant in open habitats, they can also be found in other kinds of habitats at low densities. They live in burrows and make tunnels above and below ground in grasslands. They eat seeds, leaves, and roots of grasses and forbs. Their home ranges are small, usually 10x10 m in size, and they usually do not disperse further than 50 m in their short lives (generally, their life span is only a few months) (Cudworth and Koprowski 2010). Populations experience frequent booms and busts within and among years. They are known to disperse across unfavorable habitats (including swimming) (Conroy and Neuwald 2008; Cudworth and Koprowski 2010). Voles and other small rodents probably were a minor component of Central California coast peoples' diets during the past millennia, which is why they are likely found in so many archaeological sites (Gifford-Gonzalez et al. 2013; Chapter 12, this volume; Blong et al. 2020).

We asked two main questions:

1. Do extant vole populations in the Santa Cruz area represent the same populations that occurred in the area at the times of six Central Coast (Table 13.1) archaeological sites over the past 6,000 years?
2. Fire suppression causing shrub and tree encroachment and the success of invasive plants has radically transformed coastal prairie ecosystems over the past 250 years; how have these habitat changes affected California vole population size and geographic distribution?

Methods

Archaeological Samples

We collected 23 vole bone and tooth samples from six archaeological sites along California's Central Coast (Table 13.1). Most of the vole bones were found at the sites in clusters, suggesting that they were deposited naturally—often likely from one individual given the representation of non-repetitive skeletal elements. None of the vole bones were burned or showed cut marks or other evidence of human consumption, such as acid etching from digestion or bone deformation from mastication (Blong et al. 2020). None of the voles were found in hearths or other discrete features and instead were found in the shell midden, but this does not necessarily indicate deposition by humans. Therefore, we interpret these voles as naturally deposited specimens, although radiocarbon dating evidence suggests that these voles are ancient rather than modem intrusive specimens.

Table 13.1. Site Information, aDNA Analysis Results, Zooarchaeological Species Identification, Element, and Site Age Derived from Calibrated Radiocarbon Dates on Charred Botanicals.

Catalog #	Site	aDNA	Species	Element	Site Age
100027	CA-SCR-35	Analyzed	*Microtus californicus*	Mandible	390 BCE–820 CE*
100028	CA-SCR-35	Low endogenous	*Microtus californicus*	Mandible	390 BCE–820 CE*
401390	CA-SCR-35	Unanalyzed	*Microtus californicus*	Mandible	390 BCE–820 CE*
401391	CA-SCR-35	Analyzed	*Microtus californicus*	Mandible	390 BCE–820 CE*
007-0007-05-05	CA-SCR-7	Analyzed	*Microtus californicus*	Mandible	3635–3520 BCE
007-0012-05-05	CA-SCR-7	Analyzed	*Microtus californicus*	Tooth	3945–3710 BCE
007-0013-05-05	CA-SCR-7	Unanalyzed	*Microtus californicus*	Mandible	---
007-0018-05-05	CA-SCR-7	Analyzed	*Microtus californicus*	Tooth fragment	3905–3655 BCE
007-0040-05-05	CA-SCR-7	Unanalyzed	*Microtus californicus*	Molar	4795–4710 BCE
010-0030-02-05	CA-SCR-10	Analyzed	*Microtus californicus*	Mandible	770–880 CE
010-0035-05-05	CA-SCR-10	Analyzed	*Microtus californicus*	Crania	---
014-0002-02-05	CA-SCR-14	Analyzed	*Microtus californicus*	Tibia	---

*Represents radiocarbon dates for site occupation not context.

---Denotes no radiocarbon dating conducted for context.

Table 13.1. Continued

Catalog #	Site	aDNA	Species	Element	Site Age
014-0007-05-05	CA-SCR-14	Unanalyzed	*Microtus*	Molar	1695–1920 CE
015-0004-02-05	CA-SCR-15	Analyzed	*Microtus californicus*	Mandible	1490–1640 CE
038-0001-05-05	CA-SCR-38/123	Unanalyzed	*Microtus californicus*	Tooth fragment	---
038-0003-05-05	CA-SCR-38/123	Analyzed	*Microtus californicus*	Mandible	---
038-0012-05	CA-SCR-38/123	Low endogenous	*Microtus californicus*	Mandible	---
04-0001-103	CA-SMA-218	Low endogenous	*Microtus californicus*	Tooth fragment	410–200 BCE*
04-0002-103	CA-SMA-218	Low endogenous	*Microtus californicus*	Bone fragment	410–200 BCE*
05-0001-003	CA-SMA-18	Unanalyzed	*Microtus californicus*	Molar	670–880 CE*
05-0001-103	CA-SMA-18	Low endogenous	*Microtus californicus*	Tooth fragment	670–880 CE*
05-0003-103	CA-SMA-18	Unanalyzed	*Microtus californicus*	Tooth fragment	670–880 CE*
06-0005-003	CA-SMA-19	Analyzed	*Microtus californicus*	Ulna	1470–1660 CE*

*Represents radiocarbon dates for site occupation not context.
---Denotes no radiocarbon dating conducted for context.

Radiocarbon Dating

For sites CA-SCR-7, CA-SCR-10, CA-SCR-14, and CA-SCR-15, we obtained direct radiocarbon dates of vole remains analyzed in this study; and in other instances, we radiocarbon-dated charred botanical remains found in the same excavation level as the voles. Vole remains submitted for radiocarbon dating were identified by author Gabriel Sanchez. Rob Cuthrell identified terrestrial paleoethnobotanical remains for radiocarbon dating (see Chapters 6 and 8). Rhytidome and parenchymous tissue of terrestrial vegetation were selected to avoid biases or the "old wood" effect (Ashmore 1999; Stuiver et al. 1986; Schiffer 1986). Specimens for radiocarbon dating were selected from light fraction materials >1 mm in size.

Bone and charred botanicals selected for radiocarbon dating were analyzed at the W. M. Keck Carbon Cycle Accelerator Mass Spectrometry Laboratory, University of California, Irvine (UCIAMS). Bone collagen was extracted and purified using the modified Longin method with ultrafiltration (Brown et al. 1988; Hoggarth et al. 2014). Organic samples (200–400 mg) were demineralized for 24–36 h in 0.5 N HCl at 5°C, followed by a brief (< 1 h) alkali bath in 0.1 N NaOH at room temperature to remove humates. The pseudomorph was rinsed to neutrality in multiple changes of 18.2 MΩ H$_2$O, and then gelatinized for 10 h at 60°C in 0.01 N HCl. Gelatin solution was pipetted into precleaned Centriprep® 30 ultrafilters (retaining>30 kDa molecular weight gelatin) and centrifuged three times for 20 minutes, diluted with 18.2 MΩ H₂O, and centrifuged three more times for 20 minutes to desalt the solution. More detailed ultrafilter cleaning methods are described by McClure and colleagues (2010). Ultrafiltered collagen was lyophilized and weighed to determine the percent yield as a first evaluation of the degree of bone collagen preservation. All $\delta^{13}C$ and $\delta^{15}N$ values were measured to a precision of <0.1‰ and <0.2‰, respectively, on aliquots of ultrafiltered collagen, using a Fisons NA1500NC elemental analyzer/Finnigan Delta Plus isotope ratio mass spectrometer. Sample quality was evaluated by % crude gelatin yield, %C, %N, and C:N ratios before AMS radiocarbon dating. C:N ratios for the samples ranged from 3.28 to 3.31, indicating good collagen preservation and fitting within the threshold, as advocated by DeNiro (1985) (2.9–3.6) and van Klinken (1999) (3.1–3.5). Radiocarbon samples (~2.5 mg) were combusted for 3 hours at 900°C in vacuum sealed quartz tubes with CuO wire and Ag wire. Sample CO_2 was reduced to graphite at 550°C using H2 and a Fe catalyst, with reaction water drawn off with $Mg(ClO_4)_2$ (Santos et al. 2004). Graphite samples were pressed into targets in Al cathodes and loaded on the target wheel for AMS analysis. Radiocarbon ages were corrected for mass-dependent fractionation with measured δ13C values on the AMS (Stuiver and Polach 1977) and compared with samples of ^{14}C-free whale bone and mammoth bone. Conventional radiocarbon dates were calibrated in OxCal 4.4 using the IntCal20 calibration curve (Bronk Ramsey 1995; Reimer et al. 2020). (See Tables 13.1 and 13.2.)

Table 13.2. Conventional and Calibrated AMS ^{14}C Dates on Vole Bone from the Central California Coast.

Sample ID	Taxon	Element	UCIAMS #	δ¹³C (‰, VPDB)	δ¹⁵N (‰, Atm N2)	C/N	Conventional ¹⁴C Age BP	cal BCE/CE (95.4% CI)	Associated Charred Botanical cal BCE/CE (95.4% CI)
007-0007-05-05	*Microtus californicus*	Mandible	210898	---	---	---	4340 ± 25	3020 2895 BCE	3635–3520 BCE
007-0040-05-05	*Microtus californicus*	Molar	210899	-22.9	4.3	3.28	5815 ± 20	4775–4555 BCE	4795–4710 BCE
010-0030-02-05	*Microtus californicus*	Mandible	210900	-20.8	7.1	3.31	1390 ± 20	605–670 CE	770–880 CE
015-0004-02-05	*Microtus californicus*	Mandible	210901	---	---	---	Modern	Modern	1490–1640 CE

Samples labeled "Modern" contain excess 14C, probably from mid-twentieth century atmospheric thermonuclear weapons tests. Note for Table 13.2: Samples labeled "Modern" contain excess 14C, probably from mid-twentieth century atmospheric thermonuclear weapons tests.

Bayesian Statistical Modeling: Testing the Contemporaneity of Vole Remains and Charred Botanicals

The construction and modeling of archaeological chronologies through Bayesian approaches incorporates prior information about the archaeological site(s) and regional cultural histories, emphasizing the context, provenience, relative dating, and stratigraphic relationships of samples (Bayliss and Ramsey 2004; Bayliss et al. 2007). The primary goal in the current research is to provide a reliable and precise chronological model to guide the interpretation of vole deposition and the formation of shell midden, and to suggest when interpretations of environmental data derived from vole presence (i.e., presence of open prairies) can reliably be used to infer human agency in landscape management. In this analysis, radiocarbon dates were calibrated using the IntCal20 Northern Hemisphere calibration curve and Bayesian models developed and tested in OxCal 4.4 (Bronk Ramsey 1995; Reimer et al. 2020).

OxCal chronological modeling calculates an individual agreement (A) index for each dated item or sample and an index for the model (Amodel), which is a measure of the agreement between the model and the observed data (Bronk Ramsey 1995). An overall agreement (Overall) index for the model is also determined, calculated from the individual agreement indices (Bayliss et al. 2007; Bronk Ramsey 1995). The combination of Bayesian analysis and chronological modeling of six new AMS radiocarbon dates for human plant use and vole presence along the California coast provide an exceptional opportunity to test if these samples are contemporaneous or if the voles may be intrusive. We combined paired charred plant and vole radiocarbon samples in OxCal and used the Combine command. Given that the dates are not from the sample or the same object, we did not apply the R_Combine command but instead applied the Combine command to test and combine probability distributions with a chi-square test.

The results of the Bayesian modeling suggest that only radiocarbon samples from CA-SCR-7 context 007-0040 and Column 1 Level 8 (70–80 cm BD) approach a statistically significant relationship, suggesting that they may be contemporaneous with a combination agreement index (Acomb) of 60.6% (Appendix 13.1, Table 1). However, the additional samples from CA-SCR-7 and those from the CA-SCR-10 agreement indices are well below the 60% threshold, suggesting that these are not contemporaneous and that one or more dates are problematic (Bayliss et al. 2007). For example, only sample UCIAMS# 197229 has an agreement index above zero (57.2%), while the rest of the samples fail outright (0%). Therefore, although the vole remains from CA-SCR-7 and CA-SCR-10 are ancient, they should not be interpreted to be a result of human occupation, as radiocarbon evidence and Bayesian modeling suggest that human and vole presence may not be contemporaneous. In addition, the sample from CA-SCR-15 suggests the radiocarbon-dated vole is modern, with deposition likely after the mid-twentieth century, when atmospheric thermonuclear weapons tests were conducted. The charred botanical remains from that level dated from 1490 to 1640 CE; thus, these findings of intrusive vole remains at other sites are not surprising.

Ancient DNA Analyses

We performed DNA extraction and library preparation in the dedicated ancient DNA processing facility at the University of California, Santa Cruz (UC Santa Cruz), Paleogenomics Lab following protocols developed for ancient DNA analysis (Fulton and Shapiro 2019). For each sample, we collected 7–80 mg of bone or tooth powder, from which we extracted ancient DNA following Dabney and colleagues (2013). We quantified DNA using a Qubit 4.0 Fluorimeter, revealing DNA yields ranging from 0.214 to 6.32 ng/uL.

We converted each extract into dual-indexed Illumina sequencing libraries using a single-stranded library preparation protocol optimized for short fragments (Kapp et al. 2021). To assess preservation, we sequenced each library to a target depth of 500,000 reads on an Illumina MiSeq 2x75 run and performed quality control as follows: we used SeqPrep2 to remove adapter sequences, merge paired-end reads (overlap=15), and filter sequences shorter than 30bp; we removed low complexity reads using the DUST algorithm with PRINSEQ-lite using a complexity threshold of 7 (Schmieder and Edwards 2011); we removed duplicate reads using Samtools v1.19 (Li et al. 2009). After filtering, we mapped the remaining reads to the prairie vole (*Microtus ochrogaster*) nuclear genome using BWA (Li and Durbin 2009) and to the California vole mitochondrial genome using mapping iterative assembler (MIA; Green et al. 2008). To assess the authenticity of the ancient DNA, we used mapDamage2 (Jonsson et al. 2013) to detect ancient DNA damage in mapped reads. Of 23 samples, 10 showed signs of ancient DNA damage and had endogenous contents >2% (average: 20.87%). For these, we performed hybridization-based capture enrichment using a custom myBaits (Arbor Biosciences) set that includes the mitochondrial genome of 76 fauna species, including *Microtus kikuchii* and 9 other species in the order *Rodentia* (described in Kirillova et al. 2017), following the myBaits v4 protocol. We sequenced captured libraries on Illumina NextSeq 2x150 to recover a target 10X mitochondrial coverage. For an additional six samples, we performed shotgun sequencing to a target mitochondrial coverage of ~10X using an Illumina HiSeqX and 2x75 v3 chemistry. We performed filtering as above, and mapped recovered reads to the California vole genome using MIA. To create consensus mitochondrial genomes, we called bases that were present at >3X coverage and for which there was >67% base consensus. The resulting mitochondrial genomes had an average of coverage of 35.64X, with 24-1154 Ns after filtering.

Present-Day Vole Sampling from the Santa Cruz Area

To generate mitochondrial genomes from modern California vole populations, we collected ear punches from nine California voles (three from each site) from three sites in Central California: Fort Ord Natural Reserve (FONR), North Marshall Field—UC Santa Cruz (NMF), and Younger Lagoon Reserve (YLR). We cleaned the surface of the collected tissue prior to DNA extraction using the Qiagen DNeasy Blood and Tissue Kit according to the manufacturer's instructions. We sheared the DNA using a Bioruptor (Diagenode) and converted the extracts into dual-indexed Illumina sequencing libraries using the NEBNext Ultra II kit (New England Biosciences) following the manufacturers' instructions. We sequenced each library to a target coverage of ~10X, and performed data filtering, mitochondrial genome assembly, and base-calling as above. We aligned the assembled ancient and modern vole genomes using MUSCLE (Edgar 2004) and visually inspected the alignment for consistency.

Haplotype Network

We obtained published (Conroy and Neuwald 2008, Lin et al. 2020) and unpublished (Christopher J. Conroy, personal communication, 2021) mitochondrial cytochrome *b* (cyt b) data for "Northern clade" *Microtus californicus* (sensu Conroy and Neuwald 2008). These sequences come from vole populations that range from the border of California and Oregon, south to Santa Barbara County. Together with the modern samples from the Santa Cruz area mentioned above, and the 11 ancient DNA samples from the archaeological sites, we compiled a dataset of 440 sequences for *cyt b*. Of these sequences, 56 were deemed to have undefined states and were masked, so we removed these from our analysis and ended up with a data file with 384 sequences. We used the program PopArt (Leigh and Bryant 2015) to simultaneously collapse the data file to unique sequences and build a Minimum Spanning Network with a default of Epsilon = 0 (Bandelt et al. 1999). The resulting network and logged collapsed sequences were used to identify haplotypes by a letter code, e.g., A, B, C. These haplotypes were mapped in QGIS (v. 2.8.2; QGIS. org 2021). For various reasons, the haplotypes that were derived include sequences that differed by single base differences. Nearly every sequence differed from others by 1 base pair. We also used a Maximum Parsimony analysis in PAUP* v4.0 b10 (Swofford 2002) to display all differences among sequences.

Results

Vole bones and teeth from 11 archaeological samples yielded mitochondrial aDNA sequence data from the cyt *b* gene of sufficient length and quality to compare to contemporary samples. These ranged from modern to 3945–3710 BCE (Table 13.1).

We found that ancient voles from the sites on the Santa Cruz coast had nearly identical sequences (with one exception) to the contemporary voles sampled from the same region. One haplotype was a singleton with one base-pair change from the remaining 10 samples (Figure 13.1). The parsimony analysis showed that the study area sequences are most closely related to the present-day populations sequenced from Jasper Ridge (Adams and Hadly 2010), indicating there is some geographic structure to the mtDNA data (Appendix 13.2; Figure 1).

Collapsing the sequences in PopArt to a manageable number of haplotypes is fraught with challenges, as our data are not exactly overlapping. The sequences collapsed into haplotype A are similar, divergent only by a few base pairs, suggesting a rapid spread and large population across this range. Haplotype A matched the most common haplotype reported in a recent survey of California voles (Conroy and Neuwald 2008). This haplotype corresponded to the most dominant and widespread of northern California vole haplotypes and is found from Santa Barbara to Siskiyou county, in the Coast Ranges, the Central Valley, and the Sierra Nevada foothills (Figure 13.2). The area covered by

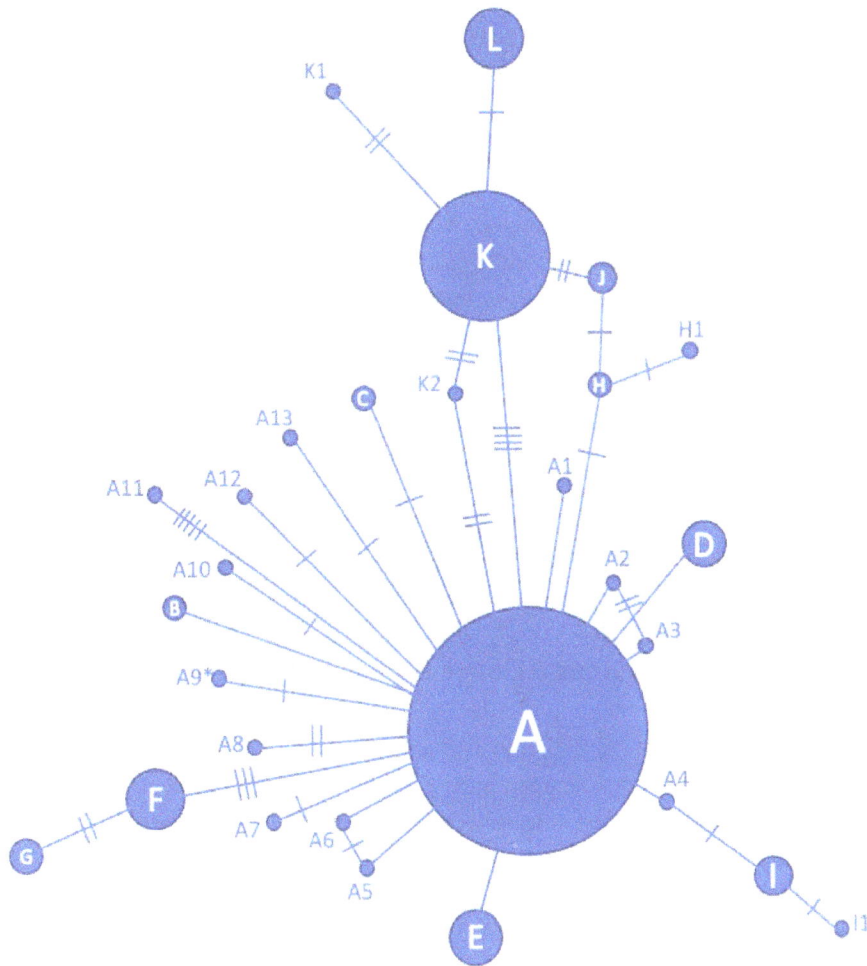

Figure 13.1. Haplotype network showing the relationship among Northern California vole haplotypes. Ancient samples are all haplotype A plus A9*. The relative size of each circle corresponds to the number of individual voles in each haplotype group. Lines without crossbars mean there is one base pair difference between circles; each crossbar represents one additional base pair difference.

this lumped haplotype matches the estimates of where there was extensive grassland and coastal prairie, and despite the large amount of change in habitat use and species composition, California voles have maintained their large population sizes over time.

Discussion

Our main finding is that the 11 vole bones recovered from six archaeological sites spanning the past 6,000 years represent very similar haplotypes (i.e., populations) to those of voles found today in three locations around Santa Cruz. The Santa Cruz area haplotypes are very similar to those recovered across most of lowland California, suggesting that northern California voles represent a large, single panmictic population that has spread its range throughout open habitats in California over the last several thousand years (Lin et al. 2020). If vole habitat had been discontinuous and rare during this time period, we would expect more geographic structure (i.e., a diversity of different haplotypes that corresponded to different geographic localities). For example, southern Californian *Microtus californicus* populations show geographic structure, and this corresponds with the high variation in climate and topography found along the southern California coast,

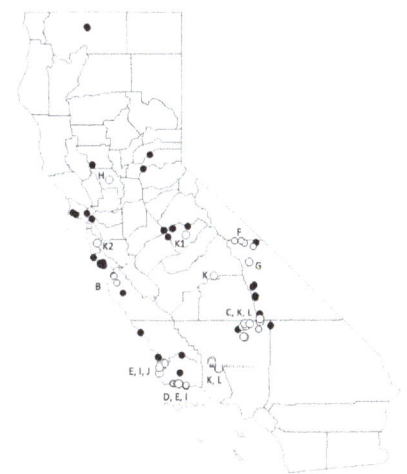

Figure 13.2. Map of the vole populations from the complete dataset. Populations of haplotype A are represented by black circles; the white circles are other haplotypes and are labeled on the map. See Figure 1 and Appendix 13.2, Figure 1 for relationships among haplotypes.

causing fragmenting of vole populations over time (Conroy and Neuwald 2008). Voles have maintained a contiguous and continuous population over this length of time, likely due to their good dispersal ability and capacity to use a wide variety of habitats.

Our results are consistent with the idea that it is very unlikely that coastal prairie and grasslands were rare habitats between 2,000 and 6,000 years ago on the Central California coast. Much more probable is that open habitats, including coastal prairie and native grasslands, were a major component of coastal California. A recent study quantifying phytoliths (silica bodies found in leaves) in the soil columns across California grasslands found evidence of long-term (hundreds of years and more) presence of native grass and forbs over much of the Central California coast (Evett and Bartholome 2013). These open habitats must be maintained by regular fire regimes or periodic disturbance; otherwise, they become closed shrublands or Douglas-fir forests (Evett and Bartholome 2013; Lightfoot et al. 2013). Lightning strikes are exceedingly rare in coastal California (Anderson et al. 2006), thus the best explanation for how such large and contiguous areas of coastal prairie and grasslands were maintained to support a large, panmictic vole population is that people must have been managing their open landscapes with frequent fire.

Comparing Ancient and Contemporary Vole Population Genetics

This is not the first study to compare ancient DNA of voles with modern populations. Hadly and colleagues (2004) investigated a chronosequence of *Microtus montanus* (Montane vole) bones found in a cave in Wyoming that spanned the past 3,000 years and compared the population-level signatures (haplotypes and microsatellites) with contemporary voles sampled in the same area. They found substantial change in haplotype diversity over time; and using a large sample of fossils together with genetic data from the gopher *Thomomys talpoides*, they were able to show changes in population size and morphological traits in both voles and gophers over time. They interpreted the changes to the large fluctuations in climate that occurred in Wyoming over the past 3,000 years, driving immigration of new voles to the cave area and/or selection for new, locally adapted phenotypes (Hadly et al. 2004). It is interesting to compare our findings to this study, given that we do not find corresponding change in haplotypes over the same span of time, but our sample size was much smaller, and our time resolution not as discrete. Another important difference between the studies is that the climate velocity of coastal California during this period would likely be much less pronounced than in the interior of the continent (Dobrowski et al. 2013).

Voles' Response to Habitat Transformation

After European colonization, fire suppression and the import of cattle and sheep transformed the landscape, as did the introduction of annual grasses and European weeds such as star thistle. The plant species composition of open habitats has changed radically during this period. There has been almost a complete turnover in dominant species, with forb and native grass communities across the state becoming replaced by non-native annual grasses (Evett and Bartolome 2013). Although California voles do not thrive in heavily grazed pastures, their populations can recover after ranchlands are left fallow (Harper et al. 2020). Thus, although the introduction of cattle since Spanish colonization has radically transformed California's lowland landscapes, and during this period heavy grazing would have often reduced and even extirpated voles from some of their previous range, spatial variation in overgrazed and fallow ranchlands likely maintained healthy source populations of voles such that we still see abundant populations across the state that share the same haplotypes (Adams and Hadly 2010; Conroy and Neuwald 2008). The implication is that over the past 250 years, California voles have been able to quickly adapt to new foods and new habitat structure (Cudworth and Koprowski 2010; Adams and Hadly 2010).

Future Studies

In our study, we were able to extract DNA from only 11 ancient voles, limiting our ability to estimate within-population genetic diversity or changes in population size over time. Other population genetics studies in voles with larger sample sizes have been able to estimate how vole population sizes have responded to climate change or habitat modification (Hadly et al. 2004; Adams and Hadly 2010), and these approaches could be applied to our archaeological sites with future work and the sampling of more vole bones. Moreover, with greater investment in new aDNA techniques, it could be possible to investigate SNPs or other genomic data to bring to bear on understanding how the ancient vole samples were related to contemporary populations in much more detail and how the populations may have changed in size over time. Additional taxa that are common in the archaeological record would also be important to investigate, including other organisms more specialized to open habitats than voles and comparison organisms more specialized to forested environments (i.e., woodrats).

Acknowledgements

The work would not have been possible without the assistance, hard work, and support from members of the Amah Mutsun Tribal Band, the Amah Mutsun Land Trust, and the Native Stewardship Corps. Our collaborative, eco-archaeological project was funded by the National Science Foundation (BCS-1523648), the Research Institute for Humanity and Nature, and the UC Berkeley Class of 1960 Chair in Undergraduate Education. The mitochondrial cytochrome b data that we used comes partially from sequences on Genbank as well as nearly 20 years of field and lab work by C. J. Conroy and many others. Funding came from the MVZ, James L. Patton, C. J. Conroy, the Bureau of Land Management, the Department of Defense—US Navy, UC Davis (Janet Foley), California Department of Fish & Wildlife. Finally, we would like to thank the organizers of the 2020–2021 California Archaeological Association conference.

References

Adams, Rachel. I., and Elizabeth A. Hadly

2010   High Levels of Gene Flow in the California Vole (*Microtus californicus*) are Consistent Across Spatial Scales. *Western North American Naturalist* 70: 296–311.

Anderson, M. Kat

2006   The Use of Fire by Native Americans in California. In *Fire in California's Ecosystems*, edited by Neil Sugihara, Jan van Wagtendonk, J. Fites-Kaufman, Kevin Shaffer, and Andre Thode, pp. 417–430. University of California Press, Berkeley.

Ashmore, Patrick J.

1999   Radiocarbon Dating: Avoiding Errors by Avoiding Mixed Samples. *Antiquity* 73(279): 124–130.

Bandelt, Hans-Jurgen, Peter Forster, and Arne Röhl.

1999   Median-joining Networks for Inferring Intraspecific Phylogenies. *Molecular Biology and Evolution* 16(1): 37-48.

Bayliss, Alex, Christopher Bronk Ramsey, Johannes van der Plicht, and Alasdair Whittle

2007   Bradshaw and Bayes: Towards a Timetable for the Neolithic. *Cambridge Archaeological Journal* 17(1): 1–28.

Bayliss, Alex, and Christopher Bronk Ramsey

2004   Pragmatic Bayesians: A Decade of Integrating Radiocarbon Dates into Chronological Models. In *Tools for Constructing Chronologies*, edited by Caitlin E. Buck and Andrew R. Millard, pp. 25–41. Lecture Notes in Statistics 177. Springer, London.

Blong, John C., Martin E. Adams, Gabriel Sanchez, Dennis L. Jenkins, Ian D. Bull, and Lisa-Marie Shillito

2020   Younger Dryas and Early Holocene Subsistence in the Northern Great Basin: Multiproxy Analysis of Coprolites from the Paisley Caves, Oregon, USA. *Archaeological and Anthropological Sciences* 12(9): 1–29.

Bronk Ramsey, Christopher

1995   Radiocarbon Calibration and Analysis of Stratigraphy: The OxCal Program. *Radiocarbon* 37(2): 425–430.

Brown, Thomas A., Earle D. Nelson, John S. Vogel, and John R. Southon

1988   Improved Collagen Extraction by Modified Longin Method. *Radiocarbon* 30(2): 171–177.

Conroy, Christopher J., and Jennifer L. Neuwald

2008   Phylogeographic Study of the California Vole, *Microtus californicus. Journal of Mammalogy* 89: 755–767.

Cudworth, Nichole L., and John L. Koprowski

2010   *Microtus californicus* (Rodentia: Cricetidae). *Mammalian Species* 42: 230–243.

Dabney, Jesse, Michael Knapp, Isabelle Glocke, Marie-Theres Gansauge, Antje Weihmann, Birgit Nickel, Cristina Valdiosera, Nuria García, Svante Pääbo, Juan-Luis Arsuaga, and Matthias Meyer

2013   Complete Mitochondrial Genome Sequence of a Middle Pleistocene Cave Bear Reconstructed from Ultrashort DNA Fragments. *Proceedings of the National Academy of Sciences* 110: 15758–15763.

DeNiro, Michael J.

1985   Postmortem Preservation and Alteration of In Vivo Bone Collagen Isotope Ratios in Relation to Palaeodietary Reconstruction. *Nature* 317(6040): 806–809.

Dobrowski, Solomon Z., John Abatzoglou, Alan K. Swanson, Jonathan A. Greenberg, Alison R. Mynsberge, Zachary A. Holden, and Michael K. Schwartz

2013   The Climate Velocity of the Contiguous United States During the 20th Century. *Global Change Biology* 19: 241–251.

Edgar, Robert C.

2004   MUSCLE: Multiple Sequence Alignment with High Accuracy and High Throughput. *Nucleic Acids Research* 32: 1792–1797.

Evett, Rand R., and James W. Bartolome

2013   Phytolith Evidence for the Extent and Nature of Prehistoric Californian Grasslands. *The Holocene* 23: 1644–1649.

Fine, Paul V. A., Tracy M. Misiewicz, Andreas S. Chavez, and Rob Q. Cuthrell

2013   Population Genetic Structure of California Hazelnut, an Important Food Source for People in Quiroste Valley in the Late Holocene. *California Archaeology* 5: 353–370.

Fulton, Tara L., and Beth Shapiro

2019   Setting up an Ancient DNA Laboratory. In *Ancient DNA*, edited by Beth Shapiro, Axel Barlow, Peter D. Heintzman, Michael Hofreiter, Johanna L.A. Paijmans, and André E.R. Soares, pp. 1–13. Humana Press, New York.

Gifford-Gonzalez, Diane, Cristie M. Boone, and Rachel E. Reid
    2013   The Fauna from Quiroste: Insights into Indigenous Foodways, Culture, and Land Modification. *California Archaeology* 5: 291–317.

Green, Richard E., Anna-Sapfo Malaspinas, Johannes Krause, Adrian W. Briggs, Philip L. F. Johnson, Caroline Uhler, Matthias Meyer, Jeffrey M. Good, Tomislav Maricic, Udo Stenzel, Kay Prüfer, Michael Siebauer, Hernán A. Burbano, Michael Ronan, Jonathan M. Rothberg, Michael Egholm, Pavao Rudan, Dejana Brajković, Željko Kućan, Ivan Gušić, Mårten Wikström, Liisa Laakkonen, Janet Kelso, Montgomery Slatkin, and Svante Pääbo
    2008   A Complete Neandertal Mitochondrial Genome Sequence Determined by High-throughput Sequencing. *Cell* 134: 416–426.

Hadly, Elizabeth A., U. M. A. Ramakrishnan, Yvonne L. Chan, Marcel van Tuinen, Kim O'Keefe, Paula A. Spaeth, Chris J. Conroy, and Craig Moritz
    2004   Genetic Response to Climatic Change: Insights from Ancient DNA and Phylochronology. *PLoS Biology* 2(10): e290.

Harper, Alan, Anny Peralta-García, Jorge H. Valdez-Villavicencio, Scott Tremor, and Chris J. Conroy
    2020   Current Distribution of the California Vole (*Microtus californicus*) in Baja California, Mexico. *Western North American Naturalist* 80: 194–203.

Hoggarth, Julie A., Brendan J. Culleton, Jaime J. Awe, and Douglas J. Kennett
    2014   Questioning Postclassic Continuity at Baking Pot, Belize, Using Direct AMS 14C Dating of Human Burials. *Radiocarbon* 56(3): 1057–1075. DOI:10.2458/56.18100.

Jónsson, Hákon, Aurélien Ginolhac, Mikkel Schubert, Philip L. F. Johnson, and Ludovic Orlando
    2013   MapDamage2.0: Fast Approximate Bayesian Estimates of Ancient DNA Damage Parameters. *Bioinformatics* 29: 1682–1684.

Kapp, Joshua D., Richard E. Green, and Beth Shapiro
    2021   A Fast and Efficient Single-stranded Genomic Library Preparation Method Optimized for Ancient DNA. *Journal of Heredity* 112: 241–249.

Kirillova, Irina V., Olga F. Chernova, Jan Van der Made, Vladimir V. Kukarskih, Beth Shapiro, Johannes van der Plicht, Fedor K. Shidlovskiy, Peter D. Heintzman, Thijs Van Kolfschoten, and Oksana G. Zanina
    2017   Discovery of the Skull of *Stephanorhinus kirchbergensis* (Jäger, 1839) Above the Arctic Circle. *Quaternary Research* 88: 537–550.

van Klinken, Gert J.

    1999   Bone Collagen Quality Indicators for Palaeodietary and Radiocarbon Measurements. *Journal of Archaeological Science* 26(6): 687–695.

Leigh, Jessica W., and David Bryant

    2015   Popart: Full-Feature Software for Haplotype Network Construction. *Methods in Ecology and Evolution* 6: 1110–1116.

Li, Heng, and Richard Durbin

    2009   Fast and Accurate Short Read Alignment with Burrows—Wheeler Transform. *Bioinformatics* 25 (14): 1754–1760.

Li, Heng, Bob Handsaker, Alec Wysoker, Tim Fennell, Jue Ruan, Nils Homer, Gabor Marth, Goncalo Abecasis, and Richard Durbin

    2009   1000 Genome Project Data Processing Subgroup, The Sequence Alignment/Map Format and SAMtools. *Bioinformatics* 25 (16): 2078–2079. https://doi.org/10.1093/bioinformatics/btp352

Lightfoot, Kent G., Rob Q. Cuthrell, Cristie M. Boone, Roger Byrne, Andreas S. Chavez, Laurel Collins, Alicia Cowart, Rand R. Evett, Paul V. A. Fine, Diane Gifford-Gonzalez, and Mark G. Hylkema

    2013   Anthropogenic Burning on the Central California Coast in Late Holocene and Early Historical Times: Findings, Implications, and Future Directions. *California Archaeology* 5: 371–390.

Lightfoot, Kent G., Valentin Lopez, Mark G. Hylkema, Rob Q. Cuthrell, Michael A. Grone, Gabriel A. Sanchez, Peter A. Nelson, Roberta A. Jewett, and Diane Gifford-Gonzalez

    2021   *The Study of Indigenous Landscape and Seascape Stewardship Practices on the Santa Cruz Coast: A Collaborative Eco-Archaeological Approach.* Report Prepared for California Department of Parks and Recreation, Santa Cruz District, Archaeological Research Facility, University of California, Berkeley.

Lin, Dana, Eileen A. Lacey, Bryan H. Bach, Ke Bi, Christopher J. Conroy, Anton Suvorov, and Rauri C. K Bowie

    2020   Gut Microbial Diversity Across a Contact Zone for California Voles: Implications for Lineage Divergence of Hosts and Mitonuclear Mismatch in the Assembly of the Mammalian Gut Microbiome. *Molecular Ecology* 29: 1873–1889.

McClure, Sarah B., Oreto Garcia Puchol, and Brendan J. Culleton

    2010   AMS Dating of Human Bone from Cova De La Pastora: New Evidence of Ritual Continuity in the Prehistory of Eastern Spain. *Radiocarbon* 52: 25–32. QGIS.org

    2021   QGIS Geographic Information System. A Free and Open Source Geographic Information System. QGIS Association. http://www.qgis.org.

Reimer, Paula J., William E. N. Austin, Edouard Bard, Alex Bayliss, Paul G. Blackwell, Christopher Bronk Ramsey, Martin Butzin, Hai Cheng, R. Lawrence Edwards, Michael Friedrich, Pieter M. Grootes, Thomas P. Guilderson, Irka Hajdas, Timothy J. Heaton, Alan G. Hogg, Konrad A. Hughen, Bernd Kromer, Sturt W. Manning, Raimund Muscheler, Jonathan G. Palmer, Charlotte Pearson, Johannes van der Plicht, Ron W. Reimer, David A. Richards, E. Marian Scott, John R. Southon, Christian S. M. Turney, Lukas Wacker, Florian Adolphi, Ulf Büntgen, Manuela Capano, Simon M. Fahrni, Alexandra Fogtmann-Schulz, Ronny Friedrich, Peter Köhler, Sabrina Kudsk, Fusa Miyake, Jesper Olsen, Frederick Reinig, Minoru Sakamoto, Adam Sookdeo, and Sahra Talamo
    2020   The IntCal20 Northern Hemisphere Radiocarbon Age Calibration Curve (0–55 cal kBP). *Radiocarbon* 62: 1–33.

Santos, Guaciara M., John R. Southon, Kevin C. Druffel-Rodriguez, Sheila Griffin, and Maya Mazon
    2004   Magnesium Perchlorate as an Alternative Water Trap in AMS Graphite Sample Preparation: A report on Sample Preparation at KCCAMS at the University of California, Irvine. *Radiocarbon* 46(1): 165–173.

Schiffer, Michael B.
    1986   Radiocarbon Dating and the "Old Wood" Problem: The Case of the Hohokam Chronology. *Journal of Archaeological Science* 13(1): 13–30.

Schmieder, Robert, and Robert Edwards
    2011   Quality Control and Preprocessing of Metagenomic Datasets. *Bioinformatics* 27: 863–864.

Stuiver, Minze, and Henry Polach
    1977   Reporting of ^{14}C Data. *Radiocarbon* 19(3): 355–363.

Stuiver, Minze, Gordon W. Pearson, and Tom Braziunas
    1986   Radiocarbon Age Calibration of Marine Samples Back to 9000 cal yr BP. *Radiocarbon* 28(2): 980–1021.

Swofford, David L.
    2002   PAUP* Version 4.0 b10. Phylogenetic Analysis Using Parsimony (* and Other Methods). Sinauer, Sunderland.

CHAPTER 14

"Returning to the Path of our Ancestors": Using Collaborative Eco-Archaeology to Support Contemporary Indigenous Landscape and Seascape Stewardship

ALEXII SIGONA, ALEC J. APODACA, AND VALENTIN LOPEZ

The Relevance of Collaborative Eco-Archaeology

Collaborative eco-archaeology is an effective method for recovering dormant practices of Indigenous landscape and seascape stewardship. In recent years, this approach to archaeology has helped the Amah Mutsun Tribal Band (AMTB) better understand historical stewardship practices that may be re-implemented today to restore local habitats and the health of Indigenous terrestrial and marine species. While research that occurred over a decade ago at Quiroste Valley Cultural Preserve and Pinnacles National Park found that AMTB ancestors were important stewards of coastal prairies and many other plants on the landscape (Cuthrell 2013a, 2013b; Lightfoot et al. 2013), recent work along the Santa Cruz coast has indicated that seascapes may have been stewarded as well. The continuing eco-archaeological research on stewardship reifies our presence as important stewards of the land and the coast.

Past relationships between Native communities and archaeologists were tumultuous (Lightfoot 2008; McManamon 1999). Despite such tensions, in 2007, the Tribe collaborated on a research project to better understand the long-term history of Indigenous natural resource management using archaeological records from the coastal area of Santa Cruz (see Lopez et al., Chapter 3, this volume). Each decision regarding archaeological research design and methodology, such as excavation procedures and other sampling strategies, required consent from AMTB leadership. Other agreements involved promoting participation of Tribe members in research, centering interests of the Tribe, and supporting future stewardship efforts of the community. The collaboration between the AMTB and archaeologists is a continuous process from the initial phase of research design, real-time discoveries during fieldwork, and laboratory analysis of materials. These aspects culminate in jointly interpreting the results of the research and implementing the findings in ongoing natural resource management plans (see Sanchez et al. 2021).

Collaborative research contributes to a growing body of scholarship and projects recognizing descendant communities as important contributors to archaeological research (see Morgan 2010; Silliman and Ferguson 2010; Watkins 2012). Collaborative practices in California have demonstrated the ability of archaeology to support the broad goals of California Native groups (see Gonzalez 2016; Lightfoot and Lopez 2013, Nelson 2020; Panich 2007; Schneider 2015). For example, it can support Indigenous communities by reforming education curriculum, developing site stewardship programs, and fostering the revitalization of Traditional Ecological Knowledge (TEK). Eco-archaeology makes use of different datasets and emergent fields of analysis to better understand the legacies of Indigenous stewardship practices and ecosystems.

This chapter is intended for Native peoples, archaeologists, and natural resource managers interested in how collaborative environmental archaeology can lead to an alternative approach to natural resource management today. AMTB members and archaeologists have worked cooperatively toward mutual goals that promote Indigenous people as leaders in land and ocean stewardship. To reflect the outcomes of this work, a few examples of the longevity of eco-archaeology outcomes are briefly discussed.

TEK and Eco-Archaeology

Archeological scholarship incorporating Traditional Knowledge (TK), Traditional Ecological Knowledge (TEK), and Local Traditional Knowledge (LTK) is not novel. Research in other parts of the world, mainly in the Pacific Northwest of North America, has integrated the perspectives of TK in interdisciplinary research and resource management (Berkes 2007). Furthermore, TK has also been incorporated to concepts of "human ecodynamics" or historical ecology (Fitzhugh et al. 2019). However, we argue the reach of TK can go much further than supplemental value to the broader research (Fitzhugh et al. 2019: 1087). Fitzhugh and colleagues also note how its use validates LTK and informs bi-lateral knowledge production. We argue that TK should not be relegated to having just supplemental value or being rendered valid by academic researchers, but rather, collaborative research should support the needs of Indigenous collaborators to build knowledge systems important for self-determination (Whyte 2018).

We demonstrate how combining TK and Scientific Ecological Knowledge (SEK) can support the recovery of TEK and provide valuable insight that supports AMTB's governance systems. TEK can be defined as "a cumulative body of knowledge, practice, and belief, evolving by adaptive processes and handed down through generations by cultural transmission, about the relationship of living beings (including humans) with one another and with their environment" (Berkes 2007: 8). Fikret Berkes understands TK to have four interrelated levels: (1) localized ecological knowledge, (2) resource management system, (3) social institution, and (4) worldview (Berkes 2007). Each level lies within and informs the successive level, all informing the greater Indigenous worldview. The levels of TK suggest that Indigenous communities who lack specific localized knowledge due to histories of removal, colonization, forced assimilation, or other factors, still maintain TEK within their worldviews.

The recovery of TEK, a key component within TK, through eco-archaeology informs localized knowledge and resource management. In turn, broader TK worldviews facilitate how such knowledge is put into practice. Berkes' concept of interrelated levels allows for a better understanding of the symbiosis between TK and SEK, in which a TK worldview can inform SEK—in this case eco-archaeology—to guide the recovery of more localized TEK. Indigenous Studies scholar Gregory Cajete (Tewa) sees TEK as fitting within his concept of Native Science, highlighting reciprocal relationships with non-human relatives and advocating for a shift in practices of science to account for "Indigenous relationship to land, plants, animals, community, self, cosmos, spirit, and the creative animating processes of life" (Cajete et al. 2018: 15). In this case, SEK informs local ecological components of TEK.

The restoration of localized knowledge relies on opportunities for Indigenous communities to engage with local ecosystems, within or outside of ancestral territory. Potawatomi botanist Robin Wall Kimmerer reminds us to forefront reciprocal relationships when discussing Indigenous environmental restoration. Kimmerer conceptualizes the process of restoration of land and culture as *reciprocal restoration* (Kimmerer 2011). Kimmerer notes the "positive feedback" characteristics when restoring land and culture. In a similar pattern, Kari Norgaard notes that North American colonialism and genocide have been "assaults" on Indigenous ecological relationships. She notes a similar positive feedback loop between diminishing ecological conditions and cultural connection to land: "the destruction of these relationships has a circular effect with both cultural and ecological impacts reinforcing one another" (Norgaard 2019: 109). As shown below, collaborative eco-archaeology informs TEK and supports AMTB revitalization of reciprocal restoration, allowing for a beneficial positive feedback cycle. Similarly, integrative cultural resource management simultaneously facilitates and informs AMTB reciprocal restoration.

The Amah Mutsun Tribal Band and Collaborative Research

The AMTB are driven by a cultural obligation to steward ancestral landscapes (Amah Mutsun Tribal Band, 2021). For many Native American Tribes, including AMTB, natural resources are understood to be relatives and land tending practices animate important kinship relations (Anderson 2005; Kimmerer 2013). As a community without a land base, AMTB stewardship requires collaborating with land holding entities, often state and federal government agencies and private conservation organizations, within ancestral territory. It has resulted in the formation of the Amah Mutsun Land Trust (AMLT) as the Tribe's vehicle for "returning to the lands, knowledge, and practices of our ancestors" (Amah Mutsun Land Trust, 2021). The AMLT is part of a growing movement of Native American land trusts that espouse Indigenous collaborative land management practices (Middleton 2014; Diver 2016).

Today, the land trust hosts multiple programs, employs a number of Tribe members, and reaches dozens of households through youth programming and educational opportunities. The core program of the AMLT is the Native Stewardship Corps (NSC), which is a workforce of young adult Tribe members that carry out a range of natural resource management and conservation projects (see Chapter 3). For example, the NSC is engaged in forestry fuel reduction projects, ecological restoration of grasslands, and the implementation of cultural burning, when possible. As of December 2020, the NSC alongside State Parks has removed 2,500 Douglas-fir trees and burned 175 burn piles in Quiroste Valley Cultural Preserve. The NSC is also on track to propagate and outplant 90,000 native grassland plants in field beds, which will go on to produce an additional 30,000 to be outplanted at the restoration sites in Quiroste Valley Cultural Preserve.

The AMLT is on the frontline helping other researchers, natural resource managers, and Indigenous communities study and combat the detrimental effects of human-induced climate change, exotic species invasions, and other challenges. Ongoing eco-archaeology projects that create and continue to inform stewardship opportunities for the AMTB community are useful for AMLT objectives. Collaborations with UC Berkeley archaeologists extend beyond research projects to include technical support involving regular meetings and field visits, as well as infrastructural support, such as sharing lab facilities and equipment. Sacred site protection is another area involving support from this collaboration. AMTB's relationship with archaeology is often the "ace-in-sleeve" because it is one of the best methods to provide long-term historical ecological context when making decisions regarding deciding appropriate species and habitat restoration decisions, as well as designing cultural learning activities. The relationship has also gone at the pace of the AMTB since trust building is a slow but important process (see Chapter 3). We turn to several examples of AMLT programming actively incorporating knowledge generated from past eco-archaeology projects that continue to produce practical outcomes.

Although some have cautioned about the problems that may occur when archaeology is designed to support disenfranchised communities (McGhee 2008), we emphasize that the findings of research intended to serve community interests can have lasting positive effects on resource stewardship and produce tangible, practical outcomes for diverse stakeholders. Our findings are informative not only for California archaeology but for the discipline as a whole, ranging from compliance-based CRM projects to other proactive forms of cultural resource studies that occur before the planning phase of public land development (Cuthrell 2017, 2018, 2019; Apodaca and Cuthrell 2020).

AMTB-led eco-archaeology restores important relationships between the Tribe and ancestral lands in diverse ways. AMTB members have gained physical access to important cultural sites through our collaboration on eco-archaeological projects along the Santa Cruz coast involving land managers, UC Berkeley researchers, and the AMLT. AMLT communicates the research to the broader community through newsletters and cultural learning days. Furthermore, the AMLT has created a Tribe-centered professional archaeological resource management program, which involves AMTB members in archaeology more than ever before. First, we detail AMTB's archaeological collaborations and approach to stewardship of archaeological resources to contextualize ongoing work. We also rely on the scholarship of fellow members of our research team who provide a richer and more nuanced analysis of collaborations and research findings in other chapters of this volume. This chapter discusses the relevance of collaborative eco-archaeology and the current stewardship, cultural programming, and educational goals designed by and for AMTB members.

Amah Mutsun Land Trust's Integrative Approach to Indigenous Cultural Resource Management

For the AMTB and many other Native American Tribes, biological and abiotic natural resources that were used traditionally for cultural purposes for foods, medicines, and crafting materials are essential for contemporary cultural practitioners and cultural revitalization efforts. Documenting and stewarding such resources in the traditional territory of the AMTB is an integral part of the AMLT's mission to protect, document, and take care of our ancestral places for future generations (Amah Mutsun Land Trust 2021). Providing resource management organizations with relevant information about the location and condition of sensitive cultural resources as early in the planning phase as possible is an efficient way to avoid adverse impacts to such resources, saving all parties involved time, money, and headaches. Harm to cultural resources can also negatively affect the AMTB community and can disrupt the afterlife of ancestors (see Chapter 3, this volume).

The AMTB believes that in order to adequately protect and make informed decisions regarding the stewardship of archaeological sites and other culturally sensitive resources (e.g., natural springs, ethnobotanical taxa and vegetation types, wildlife areas, caves, rock outcroppings, minerals, viewsheds), we must attain adequate information about these places. An important outgrowth of the recent eco-archaeology along the Santa Cruz coast is the development of a proactive and integrative approach to cultural and natural resource management that simultaneously collects systematic and opportunistic data in the field. AMTB stewards are directly involved in collecting information regarding culturally relevant archaeological, botanical, and other abiotic resources. Documenting and stewarding such resources in the traditional territory of the AMTB is an integral part of the AMLT's mission and serves three purposes.

First, documenting archaeological sites makes it possible to assess and monitor their condition to determine whether natural and/or anthropogenic processes are negatively affecting Indigenous resources. Archaeological sites can be adversely affected through artificial processes, such as earth movement, archaeological excavations, looting, and so on, as well as through natural processes such as erosion. This first purpose has become increasingly crucial considering the rapid development of certain areas and the ongoing effects of sea level rise and more severe storms (Newland 2012). Our ancestral sites have been facing ongoing destruction for several centuries; thus, protecting all archaeological sites remains a priority issue for the AMTB.

Second, recording ancestral resource locations is a necessary step for the Amah Mutsun when considering appropriate places for resource stewardship and harvesting culturally important plants and animals. In many cases, we have observed that archaeological sites contain important ethnobotanical taxa and vegetation types that may be the result of many generations of careful plant stewardship. In other cases, archaeological sites have been completely stripped of native flora and replaced with exotic invasive species, such as jubata grass (*Cortaderia jubata*), poison hemlock (*Conium maculatum*), and other herbaceous exotic species. Archaeological resources are not considered separate from the living landscape (e.g., plants, animals) and are approached as one comprehensive unit of resources that requires case-by-case stewardship plans.

Finally, recording archaeological sites and culturally relevant botanical and abiotic resources allows the Tribe to affirm connections to their ancestral lands and co-create the narratives that are disseminated in public interpretive centers, youth education, and other spaces where Tribal histories are taught to the public. The Tribe has worked to change perceptions of their ancestors as nomadic hunter-gatherers to peoples occupying a specific region with a long-standing relationship with plants and animals. In certain cases, public interpretive programs often gloss over the thousands of years of Indigenous history, which are overshadowed by information focusing on post-Contact time periods. "Reimagining" the ways which we produce archaeological knowledge begins with fore-fronting Indigenous perspectives regarding how we think about cultural resources and how to steward them (Atalay 2006, 2012; Schneider and Hayes 2020; Yellowhorn 2000).

Since 2016, the AMLT has partnered with the Bureau of Land Management (BLM) to proactively conduct Indigenous-led cultural resource surveys of the Cotoni-Coast Dairies (CCD) National Monument. This approach to surveying CCD for Indigenous cultural resources was built upon the experiences the AMTB has gained with eco-archaeological methods, which proved useful for evaluating the condition of existing sites while also making it possible to detect unrecorded sites and other undocumented cultural resources. Limited areas of the property have been surveyed for archaeological sites in the past (Edwards and Carr-Simpson 2006a, 2006b) and therefore presented a unique opportunity to use CCD as a testing ground for the AMLT's proactive and integrative approach to cultural resource assessment. By 2020, the AMLT had systematically surveyed more than 700 acres of grassland, riparian forests, and woodlands that resulted in the recordation of 14 previously unidentified Indigenous archaeological sites. Furthermore, hundreds of acres of culturally significant vegetation types were mapped, and the location of over a hundred ethnobotanical plants and other natural resources were recorded over four seasons by AMTB Native Stewards and archaeologists.

Altogether, the integrative survey has demonstrated that two extensive areas within the CCD property contain associations of sensitive Indigenous cultural resources and culturally significant natural resources located in proximity to each other. The AMLT considers such areas to be "culturally significant landscapes," a type of traditional cultural property meriting special management considerations to preserve and revitalize the sensitive resources, rare qualities, and/or associations between landscape attributes contributing to the significance of the resource. The AMLT documents culturally significant landscapes when several cultural and natural resource components occur in a defined area. Some watersheds on the property contain several archaeological sites in proximity to mature stands of nut food trees, such as California bay (*Umbellularia californica*), California buckeye (*Aesculus californica*), hazel (*Corylus cornuta* subsp. *Californica*), and coast live oak (*Quercus agrifolia*). Perennial streams that sustain a permanent population of steelhead (*Oncorhynchus mykiss*) and intermittent Coho salmon (*Oncorynchus kisutch*) are also considerable components of cultural landscapes. Also, several notable stands of ethnobotanical plants such as five finger fern (*Adiantum aleuticum*), paniceled bulrush (*Scirpus microcarpus*), and cow parsnip (*Heracleum maximum*) are found in association with archaeological sites. Through this project, the AMTB has been actively contributing to research on Indigenous settlement patterns, the locations of non-archaeological cultural resources, and the history of Indigenous habitation in the CCD area.

At the core of the approach is the AMLT's NSC, who are directly involved in collecting systematic data regarding culturally relevant archaeological, botanical, and other abiotic resources. In broadening the conception of cultural resources beyond archaeological sites to also include all components of the cultural landscape—for example, stands of ethnobotanical plants or viewsheds that contain important landmarks, such as mountain peaks and oceans—the Native Stewards employed an integrated survey methodology when assisting organizations with cultural resource management. To this end, Native Stewards surveyed for archaeological and other cultural resources in "areas of potential impact" and other locations that are known to exhibit high levels of Indigenous cultural resource sensitivity. Stewards also focused on systematically surveying areas that were predicted to have a relatively high probability of containing Indigenous cultural resources (e.g., alluvial flats near perennial streams). In locations where previously unrecorded Indigenous archaeological sites were detected, additional survey was carried out to define site boundaries and collect data on site constituents. The catch-and-release survey method, which collects, records, and returns materials to sample locations immediately after documentation in the field, does not result in collection or retention of artifacts or ecofacts.

Survey methods documented the following landscape components: a) non-biological Native American cultural resources, such as archaeological sites, caves, and springs; b) vegetation type, based on general classes (e.g., "conifer forest," "exotic grassland"); and c) ethnobotanical resources significant to the AMTB, including plants that were traditionally used for food, crafting, ceremony, medicine, and so on. This approach is based on the understanding that for the AMTB, natural resources used for traditional cultural purposes also constitute cultural resources. Each type of data can be recorded concurrently during fieldwork; thus, this integrative approach to cultural resource assessment is also more efficient than carrying out cultural resource and natural resource surveys separately.

Native Stewards employ a systematic surface collection grid across the landscape to collect soil samples (2–4 liters) that are screened with 1/8" mesh, and all cultural materials are counted, weighed, and returned to the unit. While surface collection units usually occur at 10–25-meter intervals on gently sloping terrain within 100 meters of a perennial water source, surface units can also occur at 50 m intervals in areas that are more sloped and further away from water (see Cuthrell 2019 for details about stratified survey methods). The standard density data calculated from the surface collection units are then used to delineate site boundaries and areas with artifact concentrations.

The collaboration with the BLM at CCD has facilitated access for Native Stewards, seasonal youth (under the age of 18), and Native Steward Interns (seasonally employed high school and college age AMTB members) to ancestral lands for the first time and allowed for several weeks of hands-on experience in documenting and evaluating the conditions of cultural resources, be they an archaeological site, isolated artifact find, or stand of mature hazel. More importantly, these field experiences also serve as a fundamental point of interaction to reconnect with ancestral places and resources and as an essential part of the relearning process. The integrative cultural resource program is a key step for future site stewardship at Cotoni-Coast Dairies, as well as for other cultural landscapes within the AMTB's ancestral territory. It provides geospatial and quantitative information that is of special use to the ATMB but is also valuable for land planners engaged in open space stewardship and development. The ethnobotanical and vegetation geospatial data that results from the survey also adds immense value as a cumulative database to be used by Tribe members when planning plant stewardship and gathering activities. Native Stewards have already utilized these ethnobotanical location maps to harvested willow (*Salix* sp.) for traditional structures and the construction of sweat lodges.

A next step for the Amah Mutsun's approach to cultural resource management is exploring ways to regularly steward archaeological sites and increase cooperation between stakeholders (Padon 2012). This problem is well-known amongst archaeologists. The AMLT is currently experimenting with alternative, integrative methods of site stewardship. Recent proactive cultural resource assessments highlight how Tribe members have documented more than a dozen previously unrecorded Indigenous archaeological sites and have updated the existing site records of the same number. Yet an issue remains: how can the AMLT continuously monitor these ancestral areas after they are first recorded or updated? How do we increase the incentive to systematically monitor the condition of archaeological sites on a regular basis when other projects and goals (e.g., wildfire management, other post-disaster work) often supersede the priority of archaeological site monitoring?

Certain Indigenous archaeological sites may be suitable for continuous monitoring if a "two-for-one" outcome can be achieved in regard to planning, scheduling, and logistics. For example, an archaeological site situated in the interior with a currently denuded surface and exposed matrix could be a prime candidate for the reintroduction of Native plants that were once common in that locality. The propagation of locally appropriate vegetation, such as ethnobotanical woody vegetation—e.g., Yerba Santa (*Eriodictyon californicum*), carpet-forming manzanita (*Arctostaphylos* sp.)—on slopes and herbaceous vegetation—e.g., California oatgrass (*Danthonia californica*) and soaproot (*Chlorogolum pomeridianum*)—on relatively flat surfaces could mitigate surface erosion at the archaeological site, while simultaneously providing a resource patch where Amah Mutsun stewards can tend, gather, and conduct opportunistic visual inspection of the condition of archaeological sites.

A similar approach may be suitable for sites adjacent to the shoreline that are under threat from sea-level rise and increased rates of erosion from severe storm and flooding events. Coastal archaeological middens in sand dune environments can also be propagated with important, salt-tolerant taxa, such as native dunegrass (*Leymus mollis / Elymus mollis*) (Pickart 2008), maritime brome (*Bromus maritimus*), or other crafting taxa from Cyperaceae and/or Juncacaeae. By setting up locations with accessible, valuable food, medicinal, and crafting taxa near or in denuded archaeological sites, the Tribe provides incentives to regularly steward and tend ancestral sites on more than occasional visits to keep track of natural disturbances, such as storms, or artificial disturbances, such as looting.

The experiences gained from collaborative eco-archaeology over the years has helped the AMLT provide an important and timely service to partner agencies who are committed to more holistic cultural resource management approaches. The AMLT's ongoing integrative surveys of Indigenous cultural resources at CCD is a prime example of how initial collaborative archaeology efforts can proliferate over time and evolve into a survey program that contributes to archaeological method and theory. The AMLT is currently involved in several integrative surveys on the San Vicente Redwoods property in the Santa Cruz Mountains. Native Stewards are central to these projects, and their expertise in the many AMLT programs are invaluable for educating the next generation of stewards at the AMLT Youth Camp.

Native Steward Cultural Days

The NSC partake in weekly cultural days as part of their work schedule. Cultural days involve archaeological site visits, plant harvesting, marine resource education, and crafting. Marine stewardship days, led by UC Berkeley archaeologists, are a recent highlight of the Stewardship Corps programming. Stewards and youth interns are taught about intertidal resources, traditional foods, and about how sea-level rise impacts coastal archaeology sites. Archaeological site visits where stewards identify ethnobotanical resources and learn about research findings also contribute to cultural day planning. Recent harvesting activities undertaken by Stewards within Quiroste Valley have utilized the findings of eco-archaeological research that showed tarweed seeds (*Madia sativa*) to be a common food in the past (Figure 14.1). Employing this information, Native Stewards have now harvested tarweed for the past two years. Stewards are able to use their knowledge informed by regular interactions with ancestral landscapes to discuss priorities for resource management projects. The NSC includes many elements of stewardship, both spiritual and physical (see Lopez et al., Chapter 3).

Figure 14.1. Culturally Important Plant Tarweed (*Madia sativa*) Growing in Quiroste Valley Preserve. Photo by authors.

The NSC also uses cultural days for fire training. All stewards are qualified as National Wildlife Coordinating Group type-2 wildland firefighters (Figure 14.2). In fall 2020, following the CZU-Lightning Complex fires in the Santa Cruz Mountains, the NSC joined California State Parks in a prescribed grassland fire in Pescadero, California (Hagemann 2020). Integrating the broader objectives of prescribed burning with Indigenous-directed cultural burning has been a unique way that Amah Mutsun has rekindled relationships with ancestral lands. The impetus behind returning stewards to the land as certified wildland firefighters was the strong evidence for many centuries of cultural burning in the Quiroste Valley, resulting in an anthropogenic fire regime that promoted valuable grassland prairies vegetation during the Late Holocene (Cuthrell 2013a). Such findings provide one way for how the Tribe can return to the path of their ancestors. The Native Stewards are emerging leaders as cultural practitioners informed by eco-archaeology. The benefits of collaborations also extend to many other Tribe members, particularly through youth educational programming.

Figure 14.2. Amah Mutsun Land Trust Native Stewards Participating in a Prescribed Burn at Cascade Field, San Mateo County. Photo by Guadalupe Delgado.

Amah Mutsun Youth Camp

The annual AMLT Youth Coastal Stewardship camp has been a central program since 2016. The camp brings together AMTB members between the ages of 5 and 17 from as far away as Long Beach, California, and Las Vegas, Nevada, to camp on the Central California coast. Youth Camp is a community highlight, and many generations of the community—children, parents, grandparents—camp out along the coast and engage in programs including marine stewardship, language lessons, plant tending, and ethnobotanical hikes. Many of these programs are either informed by archaeological findings or led by archaeology research associates.

The 2020 youth camp was hosted virtually with 62 youth participants participating from the safety of their homes due to the COVID-19 pandemic. Camp was made possible by a virtual camp box hand-delivered to the residences of camp participants. These boxes contained a host of materials such as traditional foods, craft materials, seed starter sets, Mutsun language guides, shells, and other educational materials (Figure 14.3). Food items such as chia seeds, native black walnut, hazelnut, sea palm, elderberry, huckleberry, elk, manzanita berries, and salmon were chosen following consultation with archaeology research associates. Research Associate Michael Grone and others have hosted the ocean stewardship day for the past three youth camps, relying on findings from past collaborative research including traditional seaweed harvesting, seasonality of mussel harvesting, and use of shell materials for various purposes such as foods, currency, and jewelry. The 2020 camp was unable to visit the coastal environment as done in past sessions, but youth were able to sample sea palm (*Postelsia palmaeformis*) distributed in camp boxes (Grone 2020).

Figure 14.3. Huckleberries Provided by AMLT Youth Camp Boxes. Photo by Michelle Glowa.

During virtual sessions, archaeological research informed presentations concerning coastal stewardship, cultural fire management practices, medicine plant tending practices and uses, and ethnobotanical lessons.

The youth camp cultivates the next generation of AMLT land stewards, leaders, and cultural practitioners. By informing programming via eco-archaeology, youth can observe how ecological science and traditional knowledge can be integrated. A central focus of AMTB leadership is to have the community return to their path of their ancestors, through an informed educational curriculum.

Conclusion

From the onset, eco-archaeology has supported the interests of the Amah Mutsun community; and now, over a decade later, it continues to do so in diverse ways. Collaborative eco-archaeology has provided Tribe members with jobs, training, and opportunities to connect with their culture. It has informed TEK and supports the return of ecological conditions integral to AMTB's cultural revitalization. TEK is not stagnant, and the recovery of past practices has informed contemporary TEK practices within changing ecological conditions. Collaborations not only validate traditional knowledge systems but also support AMTB's self-determination and are built on strong relationships involving mutual trust. Findings have supported a myriad of stewardship objectives, as discussed in other chapters of this volume, and have provided a foundation capable of supporting interests in prescribed and cultural fire, marine resource stewardship, youth education, and novel integrative ethnobotanical and archaeological site stewardship. Recent developments within the AMLT demonstrate the relevance of collaborative archaeology and indicate opportunities for mutually beneficial collaborations, which provide opportunities for research to directly support contemporary Indigenous eco-cultural revitalization.

References

Amah Mutsun Land Trust

    2021   Our History and Vision. Electronic document.
    https://www.amahmutsunlandtrust.org/our-vision, accessed January 30, 2021.

Amah Mutsun Tribal Band

    2021   Culture. Electronic document, https://www.amahmutsun.org/culture,
    accessed January 30, 2021.

Anderson, M. Kat

    2005   *Tending the Wild: Native American Knowledge and the Management of
    California's Natural Resources.* University of California Press, Berkeley.

Apodaca, Alec, and Rob Cuthrell

    2020   *Results of Integrative Indigenous Cultural Resource Survey in the Cotoni-Coast
    Dairies Unit, California Coastal National Monument, Santa Cruz County, California.*
    Report prepared by Amah Mutsun Land Trust for Bureau of Land Management,
    Central Coast Field Office. On file at the Bureau of Land Management, Central
    Coast Field Office, Marina, California.

Atalay, Sonya

    2006   Indigenous Archaeology as Decolonizing Practice. *American Indian
    Quarterly* 30: 280–310.
    2012   *Community-Based Archaeology: Research with, by, and for Indigenous and Local
    Communities.* 1st ed. University of California Press, Berkeley.

Berkes, Fikret

    2007   Community-based Conservation in a Globalized World. *Proceedings
    of the National Academy of Sciences* 104(39): 15188–15193. DOI:10.1073/
    pnas.0702098104.

Cajete, Gregory, Melissa K. Nelson, and Daniel Shilling

    2018   Native Science and Sustaining Indigenous Communities. In *Traditional
    Ecological Knowledge: Learning from Indigenous Practices for Environmental
    Sustainability,* edited by Melissa Nelson and Daniel Shilling, pp. 15–26. Cambridge
    University Press, Cambridge.

Cuthrell, Rob Q.

    2013a   Archaeobotanical Evidence for Indigenous Burning Practices and
    Foodways at CA-SMA-113. *California Archaeology* 5(2): 265–290. DOI:10.1179/19
    47461X13Z.00000000015.
    2013b   An Eco-Archaeological Study of Late Holocene Indigenous Foodways and
    Landscape Management Practices at Quiroste Valley Cultural Preserve, San Mateo

County, California. PhD Dissertation, Department of Anthropology, University of California, Berkeley.

2017   *Results of Native American Cultural Resource Survey of 35 Hectares in Lower Laguna Creek Watershed, Santa Cruz County, California.* Report prepared by Amah Mutsun Land Trust for Bureau of Land Management, Central Coast Field Office. On file at Bureau of Land Management, Central Coast Field Office, Marina, California.

2018   *Results of Integrative Indigenous Cultural Resource Survey in the Yellow Bank and Laguna Creek Watersheds, Cotoni-Coast Dairies Unit, California Coastal National Monument, Santa Cruz County, California.* Report prepared by Amah Mutsun Land Trust for Bureau of Land Management, Central Coast Field Office. On file at Bureau of Land Management, Central Coast Field Office, Marina, California.

2019   *Results of Integrative Indigenous Cultural Resource Survey in the Cotoni-Coast Dairies Unit, California Coastal National Monument, Santa Cruz County, California.* Report prepared by Amah Mutsun Land Trust for Bureau of Land Management, Central Coast Field Office. On file at the Bureau of Land Management, Central Coast Field Office, Marina, California.

Diver, Sibyl, Mehana Vaughan, Merrill Baker-Médard, and Heather Lukacs

2019   Recognizing "Reciprocal Relations" to Restore Community Access to Land and Water. *International Journal of the Commons* 13(1): 400–429. DOI:10.18352/ijc.881.

Edwards, Rob, and Charr Simpson-Smith

2006a   *Archaeological Reconnaissance and Cultural Resource Site Report Completed of a Five Acre Portion of the Coast Dairies Ranch: Laguna Creek, Santa Cruz County, California.* Report prepared by Cabrillo College Archaeological Technology Program for Bureau of Land Management, Central Coast Field Office. On file at Bureau of Land Management, Central Coast Field Office, Marina, California.

2006b   *Archaeological Reconnaissance of a Ninety-Five Acre Portion of the Coast Dairies Ranch: Warrenella Road Survey, Santa Cruz County, California.* Report prepared by Cabrillo College Archaeological Technology Program for Bureau of Land Management, Central Coast Field Office. On file at Bureau of Land Management, Central Coast Field Office, Marina, California.

Fitzhugh, Ben, Virginia L. Butler, Kristine M. Bovym Michael A. Etnie

2019   Human Ecodynamics: A Perspective for the Study of Long-term Change in Socioecological Systems. *Journal of Archaeological Science: Reports* 23. DOI:10.1016/j.jasrep.2018.03.016, accessed February 17, 2021.

Gonzalez, Sara L.

2016   Indigenous Values and Methods in Archaeological Practice: Low-impact Archaeology Through the Kashaya Pomo Interpretive Trail Project. *American Antiquity* 81(3): 533–549. DOI:10.7183/0002-7316.81.3.533.

Grone, Michael A.

2020   Of Molluscs and Middens: Historical Ecology of Indigenous Shoreline Stewardship along the Central Coast of California. PhD Dissertation, Department of Anthropology, University of California, Berkeley.

Hagemann, Hannah

2020   *Amah Mutsun Tribal Band reignites cultural burning.* Santa Cruz Sentinel. November 25, 2020, https://www.santacruzsentinel.com/2020/11/25/amah-mutsun-tribal-band-reignites-cultural-burning/.

Kimmerer, Robin

2011   Restoration and Reciprocity: The Contributions of Traditional Ecological Knowledge. In *Human dimensions of ecological restoration*, pp. 257–276. Island Press, Washington, DC.
2013   *Braiding Sweetgrass: Indigenous Wisdom, Scientific Knowledge, and the Teachings of Plants.* Milkweed Editions, Minneapolis, Minnesota.

Lightfoot, Kent G.

2008   Collaborative Research Programs: Implications for the Practice of North American Archaeology. In *Collaborating at the Trowel's Edge: Teaching and Learning in Indigenous Archaeology*, edited by Stephen W. Silliman, pp. 211–227. University of Arizona Press, Tucson, Arizona.

Lightfoot, Kent G., and Valentin Lopez

2013   The Study of Indigenous Management Practices in California: An Introduction. *California Archaeology* 5(2): 209–219. DOI:10.1179/1947461X13Z.00000000011.

Lightfoot, Kent G., Rob Q. Cuthrell, Chuck J. Striplen, and Mark G. Hylkema

2013   Rethinking the Study of Landscape Management Practices Among Hunter-Gatherers in North America. *American Antiquity* 78(2): 285–301.

McGhee, Robert

2008   Aboriginalism and the Problems of Indigenous Archaeology. *American Antiquity* 73(4): 579–597. DOI:10.2307/25470519.

McManamon, Francis P.

1999   The Protection of Archaeological Resources in the United States: Reconciling Preservation with Contemporary Society. In *Cultural Resource Management in Contemporary Society: Perspectives on Managing and Presenting the Past*, edited by A. Hatton, and Francis P. McManamon, pp. 104–131. Taylor and Francis, New York.

Middleton, Beth Rose
2011   *Trust in the Land: New Directions in Tribal Conservation.* University of Arizona Press, Tucson.

Morgan, David W.
2010   Descendant Communities, Heritage Resource Law, and Heritage Areas. In *Cultural Heritage Management: A Global Perspective*, edited by Phyllis Mauch Messenger, and George S. Smith, pp. 199–211. University Press of Florida, Gainesville.

Nelson, Peter A.
2020   Refusing Settler Epistemologies and Maintaining an Indigenous Future for Tolay Lake, Sonoma County, California. *The American Indian Quarterly* 44(2): 221–242.

Newland, Michael
2012   The Potential Effects of Climate Change on Cultural Resources within Point Reyes National Seashore, Marin County, California. Final Report, On File, Point Reyes National Seashore, 1 Bear Valley Road, Point Reyes Station, CA 94956.

Norgaard, Kari Marie
2019   *Salmon and Acorns Feed Our People: Colonialism, Nature, and Social Action.* Rutgers University Press, Rutgers.

Padon, Beth
2012   Public Partnership in Site Preservation: The California Archaeological Site Stewardship Program. On the Archaeological Institute of America (AIA) website.

Panich, Lee
2007   Collaborative Archaeology South of the Border. *News from Native California* 20(4): 12–15.

Sanchez, Gabriel M., Michael A. Grone, Alec J. Apodaca, R. Scott Byram, Valentin Lopez, and Roberta A. Jewett
2021   Sensing the Past: Perspectives on Collaborative Archaeology and Ground Penetrating Radar Techniques from Coastal California. *Remote Sensing* 13(2): 285. DOI:10.3390/rs13020285.

Schneider, Tsim D.
2015   Placing Refuge and the Archaeology of Indigenous Hinterlands in Colonial California. *American Antiquity* 80(4): 695–713.

Schneider, Tsim D., and Katherine Hayes

2020   Epistemic Colonialism: Is it Possible to Decolonize Archaeology? *The American Indian Quarterly* 44(2): 127–148.

Silliman, Stephen W., and Thomas J. Ferguson

2010   Consultation and Collaboration with Descendant Communities. In *Voices in American Archaeology*, edited by Wendy Ashmore, Dorothy T. Lippert, and Barbara J. Mills, pp. 48–72. SAA Press, Washington, DC.

Watkins, Joe

2012   Public Archaeology and Indigenous Archaeology: Intersections and Divergences from a Native American Perspective. In *The Oxford Handbook of Public Archaeology*, edited by Robin Skeates, Carol McDavid, and John Carman, pp. 1–15. Oxford University Press, Oxford.

Whyte, Kyle

2018   What Do Indigenous Knowledges Do for Indigenous Peoples? In *Traditional Ecological Knowledge: Learning from Indigenous Practices for Environmental Sustainability*, edited by Daniel Shilling and Melissa K. Nelson, pp. 57–82. New Directions in Sustainability and Society. Cambridge University Press, Cambridge.

Yellowhorn, Eldon

2000   Indians, Archaeology, and the Changing World. In *Ethics in American Archaeology*, edited by Mark J. Lynott, and Alison Wylie, pp. 126–137. Society for American Archaeology, Washington, DC.

CHAPTER 15

The Findings from the Eco-Archaeological Study of the Central California Coast: New Insights on the Timing, Development, Scale, and Relevance of Indigenous Stewardship Practices

KENT G. LIGHTFOOT, VALENTIN LOPEZ, MARK G. HYLKEMA, ROB Q. CUTHRELL, MICHAEL A. GRONE, GABRIEL M. SANCHEZ, PETER A. NELSON, ROBERTA A. JEWETT, DIANE GIFFORD-GONZALEZ, PAUL V. A. FINE, ALEC J. APODACA, ALEXII SIGONA, JORDAN F. BROWN, ARIADNA GONZALEZ, AND KATHRYN FIELD

The purpose of this final chapter is to synthesize the results of our collaborative, eco-archaeological study that examined evidence for Indigenous landscape and seascape stewardship practices on the greater Santa Cruz coast. As presented in Chapters 1, 4, and 5, scholars from the Amah Mutsun Tribal Band (AMTB), the Amah Mutsun Land Trust (AMLT), California State Parks, and the University of California, Berkeley (UC Berkeley), and the University of California, Santa Cruz (UC Santa Cruz), employed a low-impact, fine-grained methodology designed to minimize impacts to ancestral places, while emphasizing the detailed investigation of pertinent artifacts, archaeobotanical remains, and zooarchaeological specimens. We begin this chapter by summarizing the results of the field and laboratory research at CA-SCR-7, CA-SCR-10, CA-SCR-14, and CA-SCR-15, as detailed in previous chapters. We then address the four primary goals for the project outlined in Chapter 2 that involve the timing, development, geographic scale, and contemporary relevance of Indigenous landscape and seascape stewardship practices on the greater Santa Cruz coast over the last 7,000 years. We evaluate these goals employing the new findings from the four sites, as well as by incorporating previous findings from other sites in Santa Cruz and San Mateo Counties. In the final section, we conclude with four major observations concerning our collaborative, low-impact, eco-archaeological study.

Four Santa Cruz Sites: Summary of Findings

The following discussion is based on the findings from both the surface and subsurface investigations conducted at CA-SCR-7, CA-SCR-10, CA-SCR-14, and CA-SCR-15. To facilitate the comparison of the sites, we compiled (Table 15.1) the densities for the various archaeological data sets documented in the surface units and flotation samples from auger (AU), column (CU), and excavation units (EU). This table includes the densities for total artifacts (n/m^2), lithic artifacts (n/m^2), faunal elements (n/m^2), and shellfish (g/m^2) from the surface units, as well as the densities for the lithic artifacts (n/l), macrobotanical specimens (identified species, edible nut, and edible seed [n/l]), shellfish remains (g/l), mammal/bird/reptile faunal elements (NISP/l), and fish faunal elements (NISP/l) from subsurface contexts.

Table 15.1. Densities of Archaeological Materials from Surface and Subsurface Contexts for the Four Santa Cruz Sites. The Subsurface Densities Are Calculated from the Flotation Samples. NA (i.e., Not Applicable) Indicates Cases Where Samples Were Not Collected and/or Analyzed. Note that the Macrobotanical Densities for CA-SCR-7 Were Calculated from the Values Recorded for Components A and B in Chapter 8. Subsurface Densities for CA-SCR-10 Are from EU 1 Only.

Site	Surface Artifact Denisty (n/m2)	Surface Lithic Density (n/m2)	Surface Vertebrate Faunal Density (n/m2)	Surface Shell Density (g/m2)	Flotation Lithic Density (n/l)	Flotation Macrobotanical Remains Identified Specimens (n/l)	Flotation Macrobotanical Remains Edible Nut (n/l)	Flotation Macrobotanical Remains Edible Seed (n/l)	Flotation Shell Density (g/l)	Flotation Mammal, Bird, Reptile Density (NISP/l)	Flotation Fish Density (NISP/l)
CA-SCR-7 Locus 1	NA	NA	NA	NA	8.53	0.8	0.24	0.49	87.4	1.17	3.5
CA-SCR-7 Locus 4	9.59	9.33	0.36	1.65	NA	NA	NA	NA	NA	NA	NA
CA-SCR-10	3.68	3.62	0.16	NA	6.52	58.81	5.81	48.22	52.4	1.25	8.2
CA-SCR-14	17.12	16.12	8.8	85.12	1.98	86.61	18.27	51.97	38.26	0.6	4.1
CA-SCR-15	62.04	62.04	7.08	124.04	4.9	60.39	10.91	43.43	34.27	0.39	1.3

CA-SCR-7

Sand Hill Bluff, one of the most notable archaeological sites in Central California, is renowned for its imposing size, complex site structure, antiquity, and evidence for the extinct flightless duck, *Chendytes lawi*. The archaeological complex, covering an estimated 8.3 hectares at the mouth of Laguna Creek, is comprised of four separate loci. Locus 1 consists of an impressive sand dune mound rising 10.6 m above the coastal terrace that contains archaeological strata. Locus 2 is a midden deposit situated to the southeast that has been heavily impacted by the construction of a defunct abalone farm. Locus 3 is defined by a smaller dune and archaeological complex to the north of Locus 1, while Locus 4 consists of an extensive scatter of artifacts found in adjacent former agricultural fields. Our fieldwork involved the excavation of three auger units and seven column samples in Locus 1 and the investigation of 76 surface units systematically placed across Locus 4.

The study provides further refinement and clarification about the chronology and site structure of Locus 1. As outlined in Chapter 4, previous researchers had defined two distinct archaeological strata in Locus 1: a Lower Midden Stratum at the base of the sand dune and a distinctive Upper Midden Stratum at the top of the eroding sand dune. We employed a field strategy that involved the placement of CU 1 and CU 2 in the Upper Midden, CU 3 in Lower Midden, and AU 1–3 and CU 4–7 in areas in between these two known strata. The geophysical survey of Locus 1 provided excellent information for the placement of these additional subsurface units (see Chapter 5). Our purpose was to obtain fine-grained samples from the upper and lower deposits and to evaluate if other extant archaeological deposits could be found in the internal area of the mound. Our findings indicate that the sand dune complex is interlaced with additional archaeological strata that exist above the Lower Midden and below the Upper Midden. The geophysical survey detected at least three separate midden strata within the sand dune complex (Chapter 5). The 27 radiocarbon dates from these various strata indicate two chronological periods are represented in Locus 1: Component A from ca. 4800 to 3600 cal BCE and Component B from ca. 2700 to 2200 cal BCE (see Chapter 6).

Our findings support and enhance the interpretation presented by Hildebrandt and colleagues (2007: 3) concerning the growth of the sand dune and associated archaeological deposits over time. They proposed that people first began using the marine terrace by at least 6000 BP for hunting-gathering-fishing activities. Our study suggests that the earliest occupation of the terrace may extend back almost 7,000 years. The mound complex grew extensively during the period of 4800–3600 cal BCE, as evidenced by the artifact-bearing strata intertwined within the sand dune deposits that were sampled in AU 1–3, CU 3, CU 4, CU 5, and CU 7, and the lower levels of CU 1 and CU 2. Hildebrandt and colleagues (2007) believe the rapid growth of the mound complex resulted from increased sediment loads in Laguna Creek and the formation of extensive beaches in the nearby environs, which supplied wind-blown sand that was transported onto the marine terrace. The development of the sand dune complex appears to have slowed down about 5,600 years ago. Our sampling of the upper deposits suggests that this later component of the site was used from about 2700 to 2200 cal BCE as indicated by the radiocarbon dates in the upper levels of CU 1, CU 2, and CU 6. Our sampling of the Upper Midden Stratum indicates that the deposits from the near surface to about 60 cm below surface date primarily to

Component B, and that deposits found about 60–80 cm below surface are associated with the earlier Component A occupation.

We interpret these data as suggesting that people did not continuously occupy the Sand Hill Bluff from roughly 7,000 to 4,000 years ago, but rather that they used this place in fits and starts, depending on their seasonal rounds and the progression and movement of the sand dune complex. It appears that there were times when a rapid buildup of windblown sand took place (see Chapter 5). During these spurts of mound growth, Native peoples probably made little use of the dune area proper. But these episodes of mound growth were then followed by periods of dune stasis when people resided on the dune complex—as evidenced by the archaeological strata we detected in the mound. We also note that there was also a roughly 1,000-year interval between Component A and Component B when people may not have inhabited the mound proper. It is not clear from our study whether any further occupation of Locus 1 took place after 2700–2200 BCE.

The analysis of the archaeological materials from the flotation of sediments from the three auger and seven column units provided the following insights about what people were doing when they resided at CA-SCR-7. The site yielded the highest density of lithic artifacts (8.53 lithics/l) recovered from the four sites (Table 15.1). There is considerable evidence for the knapping of Monterey chert that involved the reduction of cores and creation of flakes as expedient tools (see Chapter 7). There is also evidence, as discussed in Jones and Hildebrandt (1990), for the production of bifaces and other formal chipped stone tools. We found little evidence for ground stone artifacts, such as mortars, pestles, and milling stones that may have been employed in the processing of plant foods, a finding also supported by Jones and Hildebrandt's (1990) study. There was also a paucity of worked bone, antler, and shell artifacts. We found little evidence for temporal changes in the lithic assemblages for Component A (ca. 4800–3600 cal BCE) and Component B (ca. 2700–2200 cal BCE) beyond a slight increase in obsidian and other lithic raw materials (Chapter 7).

CA-SCR-7 is notable for the relative scarcity of archaeobotanical remains (Chapter 8). We recovered low densities of identified specimen (.8 n/l), edible nut (.24 n/l) and edible seed (.49 n/l) from the flotation samples. These density values were significantly lower than from any of the other sites examined (Table 15.1). It is possible that the paucity of macrobotanical remains may be the product of taphonomic processes and the considerable antiquity of the site. However, the paucity of ground stone implements suggests this may have more to do with people not engaging in intensive plant processing or wood burning while residing in the dune complex. There may have been some processing and consumption of certain plant foods, such as tanoak (*Notholithocarpus densiflorus*) and oak (*Quercus* sp.) acorns, California bay (*Umbellularia californica*) nuts, manzanita (*Arctostaphylos* sp.) seeds, and the roasting of soaproot (*Chlorogalum* sp.), but their presence is pretty limited. The findings indicated minimal differences between Components A and B, as they both displayed consistently low densities of macrobotanical remains. We did find that Component A displayed a higher wood charcoal density than Component B (0.18 g/l and 0.13 g/l, respectively (see Table 8.1 in Chapter 8). Components A and B had similar densities of edible small seeds (0.43 n/l and 0.55 n/l, respectively), while edible nut density by count in Component A (0.13 n/l) was only about a third of that observed in Component B (0.35 n/l). The anthracological analysis found that redwood

(*Sequoia sempervirens*) or probable redwood ("cf. *Sequoia*") comprised 74.4% of the entire assemblage and 91.3% of identified specimens (see Table 8.2 in Chapter 8). Only two other taxa were identified: alder (*Alnus* sp.) and California sagebrush (*Artemisia californica*). We believe that the primary source of firewood for the people at Sand Hill Bluff was probably redwood driftwood.

We recorded the highest density of invertebrate remains from the flotation samples (87 g/l) at CA-SCR-7 (Table 15.1). California mussel (*Mytilus californianus*) and acorn barnacles (*Balanus* spp.) were ubiquitous, and the most common species found throughout all samples and levels. The overall size of the mussels at Sand Hill Bluff was larger than those measured from the other sites, and there is evidence that both plucking and stripping methods of harvests were employed (Chapter 9). The findings from the isotopic analysis suggested mussel harvesting took place from spring through fall (Chapter 10). We found in the smaller size fractions (4–8mm, 2–4mm) a much greater diversity of intertidal taxa represented, especially limpets, leaf barnacles, chitons, urchins, and turban snails (Chapter 9). These findings suggest broad-spectrum and intensive harvesting of shoreline resources. Our fine-grained analysis of the flotation samples documented the common presence of *Lottia insessa,* as well as *Littorina* spp. and *Lacuna* spp.—gastropods that are associated with Feather Boa Kelp (*Egregia menziesii*) and surfgrass (*Phyllospadix scouleri*), respectively. Thus, it appears that an important activity of the Sand Hill Bluff inhabitants was the harvesting of kelp and surfgrass, presumably for food and crafting purposes.

We identified a moderate density (1.17 NISP/l) of mammal, bird, and reptile remains from the flotation samples (Table 15.1). A diverse range of pelagic and shore waterbirds were represented, including ducks (*Anas* sp.), a Canada goose (*Branta canadensis*), a brown pelican (*Pelecanus occidentalis*), a red-throated loon (*Gavia stellate*), common murres (*Uria aalge*), and a sandpiper (Scolopacidae). We recorded five elements of the extinct, flightless duck, *Chendytes lawi*, in CU 1 (Sanchez 2021). The presence of waterbirds was strikingly different from the other sites where we identified none. We also recorded an assortment of marine and terrestrial mammals (Sanchez 2021). The faunal analysis detected the presence of indeterminate seals (Pinnipedia), indeterminate sea lions/fur seals (Otariidae), northern fur seals (*Callorhinus ursinus*), harbor seals (*Phoca vitulina*), and a sea otter (*Enhydra lutris*). We also found multiple elements of large terrestrial mammals, such as indeterminate deer (Cervidae), indeterminate elk (*Cervus* sp.), and deer (*Odocoilius* sp.). Smaller mammals were also well-represented, including black-tailed jack rabbits (*Lepus californicus*), brush rabbits (*Sylvilagus bachmani*), and rodents. The latter included the California ground squirrel (*Otospermophilus beecheyi*), California vole (*Microtus californicus*), dusky-footed wood rat (*Neotoma fuscipes*), deer mouse (*Peromyscus maniculatus*), Botta's pocket gopher (*Thomomys bottae*), a shrew mole (*Neurotrichus gibbsii*), a kangaroo rat (*Dipodomys* sp.), and a number of unidentified Rodentia (see Chapter 12). Some elements of unidentified scaled reptiles (Squamata) and snakes (Serpentes) were also detected in CA-SCR-7 (Sanchez 2021).

We detected only a modest density of fish remains (3.5 NISP/l) (Table 15.1). Many of the identified fish remains (28%) were surfperches, including pile perch (*Damalichthys vacca*), shiner perch (*Cymatogaster aggregate*), and other surfperches (e.g., barred, calico, or redtail surfperch) (Chapter 11). Greenlings were also common with lingcod (*Ophiodon elongatus*) and other taxa comprising 21% of the identified NISP. Other common taxa included skates (*Raja* sp.) and rockfishes (Sebastes sp.) at about 15% and 10% of the identified fish assemblage, respectively. The remainder of the saltwater fish assemblage at CA-SCR-7 was composed of 18 genera (see Chapter 11).

The survey of Locus 4 involving the investigation of 76 surface units provided insights about what people were doing in the hinterland of Locus 1. Similar to our findings from Locus 1, we detected a relatively high density of lithic artifacts in the survey units (9.33 lithics/m^2), comprised mostly of Monterey chert flakes and debitage (Chapter 4). We detected few ground stone artifacts and fire-cracked rocks in the surface units. We recorded low densities of faunal remains and shellfish remains in the survey work (.36 faunal elements/ m^2, 1.65 shellfish g/ m^2, respectively, in Table 15.1). The primary activities conducted by Native peoples in the hinterland appears to have been the knapping of Monterey chert and the deposition of some bifaces and projectile points, possibly during hunting forays.

In summary, our investigation of CA-SCR-7, in combination with other previous work, suggests that Locus 1 was used by Native peoples primarily between about 7,000 to 4,000 years ago. The inhabitants appear to have resided here intermittently, depending on the dynamics of the local environment and the deposition of the sand dune. We suggest that there were probably times during the rapid growth of the sand dune when people did not live on the mound complex, as evidenced by deposits of almost pure sand that we excavated. However, there were also times when local conditions favored the occupation of the mound, as marked by the interlacing archaeological strata that we detected in the mound proper with our geophysical survey and field testing.

Our findings indicate that local groups used the mound complex as a base for undertaking a broad spectrum of activities involving the harvesting and processing of coastal and terrestrial resources. Native peoples harvested a diverse range of coastal resources that included the gathering of mussel, acorn barnacles, and other shellfish, the hunting of waterbirds and marine mammals, and the fishing for intertidal species such as surfperches, greenlings, and rockfish. Small schooling fish, such as northern anchovy (*Engraulis mordax*) and herring (Clupeidae), made up about 5% (n=18) of the identified fish specimen.

Our fine-grained analysis found little evidence for the Sand Hill bluff inhabitants gathering or processing terrestrial plant foods. Yet the ubiquity of small gastropods recovered throughout the deposits indicates that Native inhabitants actively collected and processed kelp and seagrass, probably for use as food and as crafting material, a point that Jones and Hildebrandt (1990: 74) hypothesized in their earlier study. It is possible that some of the sharp-edged flakes knapped at the site were used for processing marine plant foods.

The broad-spectrum economy practiced at Sand Hill Bluff also involved the hunting of terrestrial game in the nearby hinterland. The small sample size of our excavation precludes any discussion about the extent to which this activity took place, but the presence of small and large mammals, such as deer, at the site supports this interpretation. Hylkema's (2021) study of bifaces that had been previously amassed by artifact collectors from the nearby agricultural fields supports the idea that hunting was taking place in the hinterland of CA-SCR-7. While the dating of subsurface materials from EU 1 in Locus 4 was ambiguous, Hylkema's analysis of the collectors' assemblages suggests a Middle Holocene age for many of the projectile points.

A relatively open environment that may have attracted ungulates also supports the possibility that people residing at Sand Hill Bluff may have been hunting game near the dune complex. The sandy soils associated with the remnant dune system probably supported coastal strand vegetation. The phytolith analysis of the soils from Locus 4 displayed moderate proportional phytolith weight (ca. 0.4–0.8%), but relatively low densities of grass phytoliths (ca. 27k–181k n/g), with three of the four samples analyzed containing density values of only ca. ≤100k n/g (Chapter 8). Phytoliths in these soils may have originated primarily from sedges (Cyperaceae) rather than grasses.

The rodents found at Sand Hill Bluff, which presumably resided in the local area, support further evidence for an open, coastal strand environment. Of the identified rodents, the majority consisted of California voles (n=99) and Botta's pocket gopher (n=98). In contrast, we only identified six elements of the dusky-footed wood rat (see Table 12.3.a in Chapter 12). While various taphonomic factors may have played a role, these results suggest that the nearby Sand Hill Bluff environs in Middle Holocene times may have been characterized by a relatively open environment conducive to California voles in contrast to the closed, brushy habitats favored by the wood rats (see Chapter 12; and Gifford-Gonzalez et al. 2013: 309–313).

CA-SCR-10

This impressive site covers an extensive area of ca. 400x250 m along a terrace overlooking Baldwin Creek, a short distance to the southeast of CA-SCR-7. Much of the site was under agricultural production when it was first recorded in 1950, and it remained under cultivation for Brussels sprouts when we initiated our fieldwork in 2016 and 2017. However, after our study, the site was taken out of agricultural production to protect and preserve the extant site. Despite many years of plowing, our field team observed a mounded area in the center of the site that rose slightly above the relatively flat periphery. During a brief interval in October 2016 when the Brussel sprout field was fallow, we initiated a study of this central area that involved a geophysical survey, the recording of archaeological materials from 38 surface units, and the excavation of three augers. The purpose of this work was to better define the archaeological context of the central, low-mounded area. Our field investigation also involved the excavation of EU 1, a 1x1 m unit placed along the southeastern edge of the agricultural field. We placed EU 1 near an excavation unit (Unit 2) where, in 2011, Cabrillo College and California State Parks had detected a dense assemblage of shell, vertebrate remains, and artifacts to a depth of 1.7 meters below surface. The purpose of excavating EU 1 was to obtain fine-grained samples that would enhance and build upon the findings from Unit 2 (see Chapter 4).

The geophysical survey of the central area of the site indicated intact deposits and the possibility of features, such as a potential house floor with a central hearth pit (Chapter 5). Unfortunately, the field-testing of the central area proved disappointing. We found that recent tractor work, combined with the effects of heavy rain, had produced a compact surface that was not conducive to our surface investigation. The solid surface was not suitable to our standard practice of collecting near-surface sediments and screening them through 3.2 mm sieves. Consequently, we recorded archaeological materials from only the compact surface. We believe that the results of the surface survey are reflected in this change in methodology. The lowest densities for lithic artifacts (3.62 n/m^2) and vertebrate faunal remains (.16 n/m^2) were recorded at CA-SCR-10 (Table 15.1). The lithics included debitage, flakes, and fire-cracked rocks. We observed mussel, barnacle, clam, and abalone shells in the surface units. The three auger units detected cultural materials to a depth of about 60–80 cm below surface. The relatively shallow depth of the auger units was surprising given the much deeper archeological deposits unearthed on the site's periphery in Unit 2 and EU 1. We recovered a low density (2.4 n/l) of lithic artifacts from AU 1–AU 3 that consisted of Monterey chert flakes and debitage (see Table 7.3). Three radiocarbon dates from the basal levels of AU 1, AU 2, and AU 3 returned dates from ca. 3800 to 3200 cal BCE (Chapter 6).

The excavation of EU 1 detected archaeological materials to a depth of 1.44 m below surface. Unfortunately, we observed evidence of recent historical disturbance to a depth of ca. 1 m below surface, as evidenced by plastic wrappers and other contemporary materials. The geophysical investigation of this area discovered evidence for horizontal layers of disturbance, probably from seasonal plowing and flattening of the agricultural field (Chapter 5). We did uncover intact archaeological deposits at a depth of about 112 cm to 144 cm below surface. Here we unearthed a series of cultural features, including shell concentrations, clusters of fire-cracked rocks, and ash lenses. The 11 radiocarbon dates for EU 1 support these observations. The dates for the upper levels were erratic, ranging from ca. 810 to 795 cal BCE for Level 4, ca. 1225 to 1270 cal CE for Level 9, and ca. 875 to 970 cal CE for Level 12. These data provide further evidence for the disturbed nature of these upper deposits that involved the mixing of older and younger-aged materials. In contrast, the lower levels of EU 1 (Levels 14, 16, 18, 20, 22, 23 and 24) that contained archaeological features revealed a tight range of dates from ca. 680 to 880 cal CE (see Chapter 6).

We recovered a moderate density of lithic artifacts (6.52 n/l) from the flotation samples for EU 1 that was slightly lower than that from CA-SCR-7, but higher than the other sites (Table 15.1). The flotation samples yielded flake shatter, angular shatter, complete flakes, proximal flakes, and a core made primarily from Monterey chert (Chapter 7). The dry-screen samples augmented these findings with the addition of bifaces, a uniface, a hammerstone, a pestle fragment, and three shell beads and debitage. It appears that some of the larger artifacts (ground stone materials) and formal artifacts have been systematically removed over the years from the surface of the field by agricultural workers. We observed a number of large ground stone and fire-cracked lithic objects on the site's periphery that field workers probably removed from the Brussels sprout field. Near EU 1, we point provenienced cores, fire-cracked rocks, hammerstones, and a pestle on the site surface. The previous excavations undertaken by Jones and Hildebrandt (1994) on the western periphery of the site also suggest that a relatively diverse lithic assemblage is associated with CA-SCR-10. While this area was disturbed by agricultural and road-building activities, their excavation of 8 cubic meters from ten 1x1 m units revealed flakes and debitage, as well as 3 projectile points, 19 bifaces, 3 drills, 16 flake tools, 10 cores, 3 core tools, 1 pitted cobble, 5 cobble tools, 3 battered cobbles, 3 hand stones, and 1 pestle fragment.

The archaeobotanical study of the lower levels of the site produced a relatively robust assemblage with an identified specimen density of 58.8 n/l, an edible nut density of

5.8 n/l, and an edible seed density of 48.2 n/l (Table 15.1). Edible nuts included hazelnut (*Corylus cornuta* var. *californica*), tanoak (*Notholithocarpus densiflorus*), oak (*Quercus* sp.), and California bay (*Umbellularia californica*) (Chapter 8). The high density of grass seeds in the flotation samples strongly indicates processing of grass seeds for food. In addition, the findings suggest that manzanita and soaproot may have been harvested by Native peoples. The analysis of wood charcoal samples revealed a rich assemblage comprised of 18 taxa (Chapter 8). The most common woods identified included redwood and probable redwood ("cf. *Sequoia*"), willow (*Salix* sp.), alder (*Alnus* sp.), and California bay. People probably harvested these woods from the nearby riparian corridor of Baldwin Creek and from habitats farther inland and at higher elevations. Coyote brush (*Baccharis pilularis*) was also recovered, which may have been used for medicinal and ceremonial purposes. The phytolith study for CA-SCR-10 revealed a moderate phytolith content ranging from 0.5 to 1.0%, with grass short cell density values between 86 and 208k n/g (Chapter 8). Phytolith evidence for long-term grasslands was only slightly stronger at CA-SCR-10 than at CA-SCR-7.

The invertebrate faunal analysis revealed a relatively high density of shellfish remains in EU 1 (52.4 g/l; Table 15.1). California mussel (*Mytilus californianus*) and acorn barnacles (*Balanus* spp.) comprised 92% and 5% of the assemblage by weight, respectively (Chapter 9). The intact lower deposits contained well-preserved concentrations of shellfish associated with charcoal deposits and a hearth feature. Leaf barnacle (*Pollicipes polymerus*) and black turban snail (*Tegula funebralis*), commonly associated with mussel beds, were found throughout the unit. We recovered the seaweed limpet (*Lottia insessa*) in the >4 mm mesh from the lower levels of the unit, and seaweed limpets *Littorina* spp. and *Lacuna* spp. in the 2–4 mm mesh. These findings suggest that people harvested kelp and surfgrass from nearby intertidal waters and conveyed these resources to CA-SCR-10 (Chapter 9).

The vertebrate faunal analysis documented a moderate density (1.25 n/l) of mammal, bird, and reptile remains in the flotation samples (Sanchez 2021). While a few indeterminate bird remains were recorded in the flotation and dry-screen samples, a striking difference compared to CA-SCR-7 was the paucity of identified pelagic and shore birds. The combination of the flotation and dry-screen samples from CA-SCR-10 also revealed a paucity of marine mammals in contrast to Sand Hill Bluff. A few medium and large mammals were identified, including two *Odocoileus* sp. in the flotation samples and one medium mammal, and one *Odocoileus* sp. in the dry-screen samples, However, the majority of the terrestrial faunal remains from both the flotation and dry-screen samples consisted of small mammals, including a black-tailed jackrabbit, brush rabbits, a striped skunk (*Mephitis mephitis*), various rodents (California vole, dusky-footed wood rate, a deer mouse, a broad faced mole, and Botta's pocket gopher), and several snake elements.

We detected the highest density of fish remains in our study from the flotation samples at CA-SCR-10 (8.2 NISP/l; Table 15.1). The most abundant fishes identified in the flotation samples were herrings, including Pacific herring (*Clupea pallasii*) and Pacific sardine (*Sardinops sagax*), which comprised 34% of the fish assemblage (Chapter 11). Surfperches made up 23% of the assemblage, while the northern anchovy (*Engraulis mordax*) and New World silversides embraced another 11% and 8% of the identified fish, respectively (Chapter 11). Thus, the fine-grained flotation analysis employing >2 mm mesh found that fully 75% of the fish assemblage consisted of small to medium size schooling fish, while the remaining 25% of the fish were dispersed across 11 genera. The dry-screen fishes recovered with >3.2 mm sieves were dominated by surfperches, such as rubberlip seaperches (*Rhacochilus toxotes*), that made up 33% of the dry-screen assemblage followed by rockfishes at 25%. Herrings, greenlings, plainfin midshipman (*Porichthys notatus*), cabezon (*Scorpaenichthys marmoratus*), and monkeyface pricklebacks (*Cebidichthys violaceus*) comprised the remaining 42% in relatively equal numbers (Chapter 11).

In summary, CA-SCR-10 is characterized by a complex occupational history. People resided in the central area of the site from at least ca. 3800 to 3200 cal BCE. It also appears that the periphery of the site was occupied many centuries later, about 680–880 cal CE. The decades of agricultural production have adversely impacted this ancestral place. We believe the mounded space that once comprised the central area has been repeatedly pushed and flattened over the years. Plowing and grading have no doubt greatly reduced the configuration of the mound and probably accounts for the relatively shallow depth of archaeological deposits in the central area (60–80 cm below surface). The leveling of the central area has also resulted in the pushing of cultural materials to the periphery of the site, a process that appears to have buried intact younger deposits with older materials transported from the former mound. This interpretation is supported by the findings of the geophysical survey (Chapter 5), the mixed stratigraphy and evidence of recent historical disturbances in the upper levels of EU 1, along with the reverse stratigraphy found in Unit 2 where the upper level (60–70 cm) dated to 4425–4205 cal BP, and the lowest level (170–180 cm) dated to 3545–2775 cal BP.

While the geophysical survey of the central area of the site suggests the possible existence of intact deposits and features, the constrained surface survey and limited auger units produced a paucity of materials. Further surface and subsurface investigations involving much larger sample sizes need to be done in this area to obtain a better picture of the lifeways of people at CA-SCR-10 ca. 3800–3200 cal BCE. Our investigation of the lower, intact levels of EU 1, along with the previous work reported in Jones and Hildebrandt (1994), did provide insights about what the inhabitants of CA-SCR-10 were doing about 680–880 cal CE. They knapped locally available Monterey chert cobbles into expedient flakes that were used as tools. They may have also been producing bifaces. The Native community was oriented toward both the seascape and the landscape. Like the Sand Hill Bluff denizens, they collected mussels, acorn barnacles, kelp, seagrass and a smattering of other shellfish species. Unlike the former, there is little evidence at this time for the hunting of waterbirds and marine mammals. Furthermore, the generalized fishing pattern found at CA-SCR-7 had now been transformed to one focusing on the harvesting of primarily small- and medium-sized schooling fish, along with some rockfishes. While the presence of large- and small-sized mammals suggest that the hunting of terrestrial game took place, the sample size is too small to provide any details.

In contrast to Sand Hill Bluff, the inhabitants of CA-SCR-10 were actively involved in the gathering and use of terrestrial plants, particularly grasses and to a lesser extent, nuts. The presence of ground stone artifacts, including hand stones and pestle fragments, support the idea that terrestrial plants were processed here. The detection of fire-cracked rocks suggests that cooking activities took place here as well. While the phytolith evidence for long-term grasslands in the nearby environs was somewhat ambiguous, it was stronger than at CA-SCR-7. While the sample sizes are low, there was slightly more evidence for

California voles than dusky-footed woodrats at CA-SCR-10, a finding that suggests open areas probably existed near the site (Chapter 12). We believe these findings indicate that people were probably instigating frequent cultural burns to maintain open grasslands and other habitats at least 1,200 years ago.

CA-SCR-14

Situated in the uplands 2.0 km from the coast on the southeastern bank of Laguna Creek, this site measured about 77 m (E/W) by 35 m (N/S) in size. We observed a rich midden deposit on relatively flat ground that extended down the slope of the creek bank. Field crews initiated a geophysical survey, the systematic investigation of 35 surface units, the examination of five auger units, and the excavation of two 50x50 cm units (EU 1, EU 2). The geophysical work revealed anomalies that might represent intact subsurface cultural features, such as hearths or earth ovens (Chapter 5). The surface investigation recorded a relatively high density of lithic artifacts (16.12 n/m²; Table 15.1) comprised of primarily Monterey chert flakes and debitage pieces, one possible pestle fragment, and fire-cracked rocks (Chapter 4). Some recent Euro-American manufactured materials were also found. The surface investigation documented the highest density of vertebrate faunal remains (8.8 n/m²; Table 15.1), which included both mammal and fish elements. Field crews also recorded a relatively high density (85.12 g/m²; Table 15.1) of surface shellfish remains that included a diverse array of taxa. The spatial distribution of the lithic artifacts, vertebrate faunal elements, and shellfish remains suggests three major loci of cultural materials: one concentration in the northcentral area, another in the southcentral area, and a third trailing down the slope to Laguna Creek on the west side of the site.

The placement of the two excavation units was based on the findings from the geophysical work, the surface survey, and the auger units (Chapters 4 and 5). Excavation crews unearthed an intact fire-cracked rock feature in EU 1 that may represent the remains of a cooking area or some other related activity. The excavation of EU 2 revealed an area probably used for dumping refuse on the periphery of the site as evidenced by multiple, small discrete deposits of ashy, ecofact-rich materials. The archaeological deposits unearthed in both units were fairly shallow (45–55 cm below surface). The eight radiocarbon dates for the two units indicate a Late Holocene occupation ca. 1000–1700 cal CE, and more conservatively, with the removal of dates from potentially disturbed near-surface contexts, about 1000–1510 cal CE (Chapter 6).

We recovered a relatively low density of lithic artifacts (1.98 n/l) in the flotation samples that consisted primarily of Monterey chert debitage and flakes and one pestle fragment (Table 15.1). The much larger volume of sediments passed through the dry-screen mesh produced essentially the same results—primarily flakes and debitage and one pestle fragment. No worked bone or shell tools were recovered (Chapter 7).

The archaeobotanical assemblage from CA-SCR-14 was robust, with the greatest concentration of plant remains found in EU 2 where inhabitants appear to have been disposing of refuse on the edge of the site (Chapter 8). We recorded the highest densities of identified specimens (86.61 n/l), edible nut (18.27 n/l), and edible seed (51.97 n/) at this site (Table 15.1). Edible nuts were dominated by hazelnut (*Corylus cornuta* var. *californica*), followed by tanoak (*Notholithocarpus densiflorus*), and lesser amounts of oak (*Quercus* sp.)

and California bay (*Umbellularia californica*). The high edible seed density indicates the processing of terrestrial plants for food, particularly grasses, tarweed (*Madia* sp.), panicled bulrush, and clover. The densities of fiddleneck (*Amsinckia* sp.) and phacelia (*Phacelia* sp.) in the assemblage could indicate decreased woody vegetation cover at the time of site occupation, as both genera contain fire-followers that thrive in grasslands. People may have used them for medicinal purposes. The dominant charcoal specimens recovered from EU 2 were redwood and probable redwood ("cf. *Sequoia*"), with some Douglas fir, California lilac/ceanothus, oak, and pine (*Pinus* sp.) (Chapter 8). The phytolith study from five samples collected near CA-SCR-14 and CA-SCR-15 presented the strongest evidence for long term grasslands among sites analyzed in our study. Four of the samples collected from this area were relatively consistent in phytolith data, with phytolith content between 0.8% and 1.2% and grass short cell density ranging from 150k n/g to 193k n/g (Chapter 8). One of the five samples was anomalously low in phytolith content and grass short cell density, which may indicate nearby forested habitats that were less subjected to frequent fires as suggested by the presence of the Douglas fir.

The invertebrate faunal study yielded a relatively high density (38.26 g/l) of shell-fish remains (Table 15.1). Similar to the archaeobotanical findings, EU 2 produced a more robust assemblage than EU 1 (Chapter 9). While California mussel and barnacles were again the most abundant species represented, other taxa such as limpets, chitons, urchins, whelks, and turban snails were present in nearly all levels. Purple sea urchin, (*Strongylocentrotus purpuratus*) was ubiquitous in EU 2. The seaweed limpet (*Lottia insessa*) was recovered in the >4 mm mesh in both units, as well as in the 2–4 mm sieve along with *Littorina* spp. and *Lacuna* spp. The metric analysis of mussels in Chapter 9 indicates that the average size of the mussels from CA-SCR-14 is smaller than CA-SCR-7. The finding of a range of small- to medium-sized shellfish at CA-SCR-14, which remained the same size or increased slightly in size over the occupation span of the site, is consistent with expectations of a stripping method of harvest and resource stability. The isotopic analysis of the mussels suggests people employed the stripping method to harvest mussels primarily in the winter months (Chapter 10).

We recovered a relatively low density (.6 NISP/l; Table 15.1) of mammal, bird, and reptile remains in the flotation sample (Sanchez 2021). While no bird remains were identified, the flotation samples revealed one sea otter (*Enhydra lutris*) element, and the dry-screen samples produced one sea otter and one harbor seal (*Phoca vitulina*). Terrestrial game was present, including one *Odocoileus* sp. and one brush rabbit in the flotation samples, and one Cervidae, four *Odocoileus* sp., and two brush rabbit elements in the dry-screen samples. The majority of the terrestrial faunal remains from both the flotation and dry-screen samples were rodents, including a western gray squirrel (*Sciurus griseus*), California voles, a dusky-footed wood rat, a deer mouse, and Botta's pocket gopher, and two snake elements.

We documented a moderate density (4.1 NISP/l) of fish remains in the flotation sample (Table 15.1). The most abundant of the identified fishes in the >2mm mesh were Northern anchovies and surfperches that comprised 46% and 26% of the flotation assemblage, respectively (Chapter 11). The remaining 28% of the assemblage consisted of eight genera. Interestingly, the dry-screen samples passed through the 3.2 mm mesh failed to detect any of the smaller Northern anchovy remains, but did recover surfperches at 33%, greenlings at 21%, which may include kelp, rock, and masked greenling, as well as rockfishes at 21%. The remaining 25% of the dry-screen elements consisted of a few trout and salmon, makos, cabezon, monkeyface pricklebacks, and white croaker (Chapter 11).

In summary, Native peoples residing along Laguna Creek at CA-SCR-14 in Late Holocene times (ca. 1000–1510 CE) enjoyed the bounty of both intertidal and terrestrial resources. The findings from both the surface investigation and excavation of two units indicate that they knapped Monterey chert, produced expedient flake tools, processed terrestrial and maritime plants on site, and created thermal features with fire-cracked rocks that may have served as hearths or ovens. The site's residents gathered plant foods from nearby environs, as indicated by the abundance of edible seeds and nuts. The diversity and abundance of intertidal resources suggest that they were intensively harvesting and transporting shellfish, kelp, and surfgrass from the coast to CA-SCR-14 for processing as food and crafting materials. While the sample size is small, there is evidence that people hunted some terrestrial game. A significant finding from the flotation samples was the relative abundance of small- and medium-sized schooling fish, greenlings, and rockfish. The abundance of grasses, tarweed, clover, and other plant resources, in combination with the phytolith findings, strongly suggests the existence of nearby coastal prairies that people probably maintained through frequent cultural burning. Indigenous stewardship practices appear to have produced a patchy landscape mosaic with many fire-adapted plants (e.g., California lilac/Ceanothus, hazelnut) that probably included not only grasslands but also neighboring woodlands and forests populated with tanoak, oak, California bay, and redwoods, along with less frequently burned habitats containing Douglas fir.

CA-SCR-15

A nearby neighbor of CA-SCR-14, this large midden site sits above Laguna Creek on a ridge that contains a low knoll connected to an extensive grassy field where we observed cultural materials covering an area of ca. 130 m (E-W) by 70 m (N-S). Here, we undertook a geophysical survey, the systematic investigation of 44 surface units, and the excavation of two 50x50 cm units (EU1, EU 3). The geophysical investigation revealed shallow, intact deposits and some anomalies that might be cooking or thermal features, as well as a potential small structure floor (Chapter 5). The surface investigation divulged the highest surface densities of lithics (62.04 n/m^2) and shellfish (124.04 g/m2) for the study, as well as the second highest surface density of vertebrate faunal remains (7.08 n/m^2) (Table 15.1). We recorded Monterey chert flakes and debitage pieces, ground stone fragments, many pieces of fire-cracked rock, and a shell bead fragment (Chapter 4). These cultural materials were found in large numbers on the western knoll and in the field area to the east. Notable artifacts included an unfinished projectile point and fire-cracked rock that may have derived from ground stone tools. The surface vertebrate faunal remains included mammal, bird, and fish elements (including one bat ray) dispersed on the western knoll and central area of the site. Field crews documented a diverse array of shellfish on the surface concentrated on the western knoll overlooking the creek, as well as impressive weights of shell in the central and eastern areas of the site (Chapter 4).

We located EU 1 in the grassy field east of the knoll in an area containing a high density of surface artifacts and geophysical anomalies. Field workers observed considerable evidence for bioturbation in the unit, but they did record a concentration of fire-crack rocks. Cultural remains were detected to a depth of 60–70 cm below surface. We placed EU 3 on the western knoll using the GPR and the findings from surface units that revealed high densities of surface artifacts and shellfish remains in the area. Field workers excavated a fire-crack rock feature and cultural materials to a depth of 60 cm below surface. EU 3 produced a much greater density of cultural materials than EU 1. We obtained nine radiocarbon dates from EU 1 and EU 3. Seven of these dates revealed a relatively tight age range from ca. 1050 to 1400 cal CE, while the other two were more questionable given the possibility of subsoil mixing and near-surface disturbances. It is possible that the site occupation could extend into the 1500s to early 1600s cal CE (see Chapter 6).

The excavation of EU 1 and EU 3 produced a modest density (4.9 n/l) of lithic artifacts with the majority recovered in the latter unit (Table 15.1). The lithic assemblage was comprised primarily of Monterey chert flakes and debitage along with a couple of cores unearthed in the dry-screen samples for EU 3 (Chapter 7). The flotation and dry-screen samples yielded no formal flaked stone tools, ground stone tools, or worked bone or shell from the two units. No Euro-American manufactured materials were found in the surface or subsurface samples.

The archaeobotanical investigation documented a robust macrobotanical assemblage characterized by a relatively high density of identified specimens (60.4 n/l), moderate edible nut density (10.9 n/l), and moderate edible seed density (43.4 n/l) (Table 15.1). When only EU 3 is considered, then the macrobotanical densities are quite impressive: identified specimens (113.1 n/l), edible nuts (20.4 n/l), and edible seeds (81.6 n/l) (Appendix 8.3). Most of the edible nuts consisted of hazelnut, followed distantly by tanoak, and a few oak and California bay. The densities of grasses were high enough to clearly indicate cultural use, including the processing of grass seeds for food, while densities of panicled bulrush (*Scirpus microcarpus*) and clover were also relatively high in EU 3 and probably indicate intentional cultural use (Chapter 8). Other edible plants observed in the assemblage included manzanita and potentially huckleberry that were probably transported to the site from outlying habitats. One native tobacco (*Nicotiana quadrivalvis* or *N. attenuata*) seed was identified from EU 3, raising the possibility that tobacco may have been actively cultivated in the area and/or opportunistically supported through cultural burning or acquired through trade (see Chapter 8).

The anthracological analysis from two flotation samples from EU 3 recorded a dominance of Redwood and probable redwood ("cf. *Sequoia*") along with lesser densities of California lilac/ceanothus, alder, and members of the honeysuckle family (Chapter 8). All of these would have been locally available to people residing at CA-SCR-15. The California lilac/ceanothus specimens are likely California lilac (*Ceanothus thyrsiflorus*) that appear to have thrived in the local environs. The presence of the California lilac and the paucity of fire-susceptible Douglas fir suggest that landscape fires were relatively frequent at the time of site occupation.

A moderate density of shellfish remains (34.27 g/l) was recovered from the flotation samples (Table 15.1). The density of shell was considerably higher in EU 3 (48.44 g/l) than in EU 1 (20.1 g/l) (Chapter 9). The shellfish assemblage was again dominated by California mussel and barnacles, while smaller intertidal species such as purple sea urchin (*Strongylocentrotus purpuratus*), *Chiton* spp., gooseneck barnacle (*Pollicipes polymerus*), black turban snail (*Tegula funebralis*), and Pacific littleneck clam (*Leukoma staminea*) were common in EU 3. The seaweed limpet (*Lottia insessa*) was recovered in the >4 mm mesh in both units, as well as in the 2–4 mm sieve along with *Littorina* spp. and *Lacuna* spp. (Chapter 9).

A low density (.39 NISP/l) of mammal, bird, and reptile remains was recorded for the flotation assemblage (Table 15.1). While we found no bird remains, we did record one seal (Pinnipedia), one northern fur seal (*Callorhinus ursinus*), and one harbor seal (*Phoca vituline*) in the flotation samples, and one harbor seal (*Phoca vituline*) and two sea otter (*Enhydra lutris*) elements in the dry-screen samples (Sanchez 2021). Terrestrial game was scarce, with one *Odocoileus* sp. and one rabbit/hare (Leporidae) from the flotation samples, and two *Odocoileus* sp. elements from the dry-screen samples. Like CA-SCR-14, most of the mammals in the flotation and dry-screen samples were identified as rodents, such as the California ground squirrel (*Otospermophilus beecheyi*), California voles, dusky-footed wood rats, a deer mouse, and Botta's pocket gopher, as well as one snake element (Chapter 12). Given the small sample size, we caution that these findings are only tentative at this time.

A low density (1.3 NISP/l) of fish remains was noted for the flotation samples from CA-SCR-15 (Table 15.1). The flotation assemblage recovered using >2mm mesh was dominated by surfperches at 46%, including pile perch and barred, calico, or redtail surfperch, followed by greenlings at 21%, which may include kelp, rock, and masked greenling (Chapter 11). Herrings comprised 10% of the flotation assemblage, while four other genera made up the remaining 23%. The dry-screen assemblage obtained with 3.2 mm mesh consisted of greenlings at 33%, surfperches at 33%, including barred, calico, or redtail surfperch, as well as rockfishes and cabezon that made up the remaining 34% (Chapter 11).

In summary, our investigation suggests that the Native community residing at CA-SCR-15 from about 1050 to 1400 cal CE (or a little later) was quite like its neighbors at CA-SCR-14. They implemented a variety of cultural practices that included knapping Monterey chert into useable flakes and possibly formal tools, harvesting shellfish, marine mammals, and fish from the coast, and gathering and processing terrestrial and maritime plants. The presence of terrestrial game, such as deer, at the site also suggests that hunting activities took place in nearby environs. However, the combination of the small sample sizes and paucity of specimens makes any interpretation about terrestrial hunting premature at this time. In contrast to CA-SCR-14, there is less evidence for the fishing of small schooling fish, such as herring, and more support for the fishing of mid-size schooling fish and rockfish from the intertidal zone. The inhabitants appear to have transported kelp, surfgrass, and other marine plants from the coast to the site, probably for use as both food and crafting materials. They gathered terrestrial nuts and seeds from nearby environs that were processed and cooked on site, possibly in ovens or baskets as indicated by the ubiquity of fire-cracked rocks. Like their neighbors at CA-SCR-14, the residents of this upland site appear to have initiated Indigenous stewardship practices that enhanced and maintained the growth of grasses, clover, hazelnut, and possibly Native tobacco. The presence of nearby coastal prairies and other habitats that would have been supported by frequent cultural burns is supported by the common occurrence of California lilac, the paucity of Douglas fir, and the phytolith findings that suggest the long-term maintenance of grasslands in the nearby hinterland.

Evaluation of Research Goals

Below, we employ the findings from our eco-archaeological investigation of CA-SCR-7, CA-SCR-10, CA-SCR-14, and CA-SCR-15, along with other sites in Santa Cruz and San Mateo Counties, to address the four major research goals of our project concerning the timing, development, geographic scale, and contemporary relevance of Indigenous landscape and seascape stewardship practices on the Central California coast, as introduced in Chapter 2. This discussion is an updated version of what we first presented in Lightfoot and colleagues (2021).

First Goal

The first goal is to investigate when people first initiated sustained cultural burning and seascape stewardship practices in the broader region. The findings from our previous NSF-funded project demonstrated that Indigenous people initiated cultural burns on the Central California coast from ca. 1000 to 1300 cal CE in Quiroste Valley. We argued that lightning ignitions alone (with fire return intervals of 50–100 years) were insufficient to sustain these coastal prairies. We showed that the long-term maintenance of coastal grasslands in Central California required a sub-decadal fire return interval with fires probably set at intervals of one to five years (Cuthrell 2013a, 2013b). NSF-funded research that we commenced elsewhere, discussed below, indicates evidence for both terrestrial and coastal stewardship practices in the Late Holocene. In undertaking our study of sites on the Santa Cruz coast with occupations dating to the Middle Holocene (6000–3000 BP) and Late Holocene (3000–500 BP) periods, we proposed to examine whether eco-archaeological evidence existed for landscape and seascape stewardship, when these practices may have been commenced, and how they compare to developments in Quiroste Valley and elsewhere.

When Did People First Initiate Sustained Cultural Burning and Seascape Stewardship on the Santa Cruz Coast?
Terrestrial Stewardship Practices. The results of the current study provide strong support that Indigenous people employed cultural burning to facilitate the creation and maintenance of productive grasslands west of the Santa Cruz Mountains from ca. 700 to 1500 CE. Furthermore, it appears they harvested plant and animal resources from patchy mosaics of biotic communities consisting of grassland, shrubland, woodland, and riparian resources. The earliest evidence for sustained grassland harvesting is from CA-SCR-10 in archaeological contexts dating to ca. 680–880 cal CE. Here, we detected high densities of grass seeds and hazelnuts, along with lesser quantities of other plant foods. Our investigation revealed a moderate phytolith content and some evidence for the development of long-term grasslands in the nearby environs. People gathered wood for fuels and raw materials from redwood, willow, alder, California bay, and other trees from the nearby riparian corridor of Baldwin Creek and from habitats farther inland and at higher elevations. Later evidence for the intensive gathering of coastal prairie resources is found at the upland sites of CA-SCR-14 and CA-SCR-15 from ca. 1000 to 1500 cal CE. At this time, people continued to gather grass seeds and hazelnuts, along

with tarweed, clover, panicled bulrush, tanoak, oak, and California bay. They may have also been involved in the cultivation or enhancement of Native tobacco. The findings from the phytolith study indicate the maintenance of long-term grasslands in the nearby environs of these upland sites. People obtained wood primarily from redwoods, California lilacs, oaks, and pines.

The findings from these three sites strongly suggest that Indigenous communities employed stewardship practices that facilitated the long-term upkeep of coastal prairies situated within a productive, patchy landscape mosaic that probably included not only grasslands but also neighboring woodlands, wetlands, and forests with fire-adapted species (e.g., hazelnut, California lilac, redwood). The detection of the fire-susceptible Douglas fir as a minor fuel source at CA-SCR-14 also suggests the presence in the upland hinterland of less frequently burned habitats. The convergence of multiple lines of evidence drawn from the archaeobotanical data (fire-enhanced food plants such as grasses, hazelnuts, tarweed, etc.), the anthracological data (fire-compatible plants gathered for fuel and raw materials such as redwood and California lilac), and the phytolith data (phytolith concentrations showing some evidence of nearby grasslands) is strong support for an anthropogenic fire regime from ca. 700 to 1500 cal CE. This fire regime would have been characterized by a much shorter fire interval (sub-decadal) than a natural fire regime based on lightning-ignited fires alone. We conclude that Native stewards tended frequent cultural burns over many generations on the Santa Cruz coast that greatly enhanced the quantity, diversity, and sustainability of fire-enhanced plants and animals in their territories.

While our findings indicate that Native peoples implemented cultural burning on the Santa Cruz coast at least 1,200 years ago, we found little evidence that the inhabitants of the earlier Sand Hill Bluff site (ca. 4800–2200 cal BCE) harvested and used resources from coastal grasslands. The paucity of archaeobotanical remains at this site compared to the others is striking. The Indigenous residents probably made use of some local terrestrial plant foods, such as tanoak, oak, California bay, manzanita, and soaproot, but their presence is spotty. The primary fuel used at the site appears to have been locally available redwood driftwood. Given the age and context of the site (e.g., sand dune matrix), the scarcity of macrobotanical remains may be the product of various taphonomic processes. But the paucity of ground stone implements suggests that the tending of terrestrial plants was not a focus of the site's inhabitants. Rather, it was the harvesting and processing of a diverse array of maritime resources, along with the hunting of some terrestrial game.

The paucity of grassland-associated plant foods at CA-SCR-7 does not negate the possibility that Indigenous cultural burning may have taken place earlier than 1,200 years ago on the Santa Cruz coast. The detailed analysis of archaeobotanical remains from the nine sites discussed in Chapter 8 shows a strong trend for coastally situated sites on the broader Santa Cruz coast to be associated with considerable evidence for maritime resource harvesting and little evidence for terrestrial plant food processing, consumption, or deposition. Where we find support for terrestrial plant tending, particularly those gathered from coastal prairies, is from inland sites (e.g., CA-SMA-113, CA-SCR-10, CA-SCR-14, CA-SCR-15) that are all Late Holocene in age. Thus, it appears that people who occupied oceanfront sites tended to maintain a focus on maritime resources, and when they shifted their residence to interior places on upland marine terraces and the foothills of the Santa Cruz Mountains, they continued to garner coastal resources as well as to harvest grassland-related and other terrestrial foods. We now recognize that coastally oriented CA-SCR-7 is not an ideal candidate for evaluating the possibility that sustained cultural burning took place in Middle Holocene or early Late Holocene time. A full evaluation of this question should be based on the study of additional inland-situated sites that predate 700 CE.

Coastal Stewardship Practices. Our investigation of the four Santa Cruz sites indicates some changes in Indigenous seascape-harvesting practices may have taken place from Middle to Late Holocene times. The residents of Sand Hill Bluff from 4800 to 2200 cal BCE employed a generalized, broad-spectrum maritime economy that involved the harvesting and processing of a diverse range of shellfish, fishes, pelagic and shore birds, marine mammals, kelp, and surfgrass. They gathered mussels in the spring through fall by both plucking and stripping them from patches that produced relatively large meat packages. They fished for intertidal species such as surfperches, greenlings, and rock fish, as well as a few small schooling fish probably captured in nets. In comparison, Native peoples who inhabited CA-SCR-10, CA-SCR-14, and CA-SCR-15 from about 700 to 1500 cal CE appear to have been more focused in the maritime resources they used. We found an increasing emphasis on small foraging fish along with surfperches and various rockfishes. A similar pattern was observed for CA-SMA-113 (Gifford-Gonzalez et al. 2013: 297–309).

While mussels continued to be the primary shellfish gathered through time on the Santa Cruz coast, our findings indicate increasing emphasis on the stripping method of harvest in Late Holocene times that allowed people to collect small- to medium-size shellfish in a sustainable manner. Native gatherers may have mass harvested mussels from specific coastal patches while allowing other patches to remain fallow for a season or two while the local mussel population rebounded and renewed itself. We believe this stewardship practice observed by at least 1000 cal CE at CA-SCR-10 and 1300 cal CE at CA-SMA-216 facilitated the intensive collecting of mussels while maintaining the sustainability of the shellfish population over time (Chapter 9).

Second Goal

The second goal is to investigate how local Indigenous communities may have modified stewardship practices over time. We are particularly interested in examining steward-ship practices during past episodes of climate change. Our research team scrutinized eco-archaeological evidence for Indigenous harvesting practices before and during the Medieval Climatic Anomaly (MCA) from about 900 to 1300 CE, as well as during the Little Ice Age (LIA) from about 1350 to 1850 CE, to explore how overall warmer and cooler temperatures may have structured fire regimes and Indigenous stewardship practices in Late Holocene times. As outlined in Chapter 2, if climate primarily influenced Indigenous harvesting practices in the MCA, then we would expect to see a decline in the frequency of fire-enhanced species in the archaeological record in the cooler LIA. On the other hand, if stewardship practices and cultural burning comprised significant components of the subsistence regimes of Indigenous people in the Late Holocene, then we would expect to see evidence for the tending and harvesting of fire-adapted species in both the MCA and LIA.

We are also interested in examining how Indigenous stewardship practices may have been influenced over time by sociopolitical transformations, technological innovations, and changes in human demography. We are particularly attentive to how the process of resource intensification, which may be associated with human population growth and technological innovations, may have modified Indigenous stewardship practices. Likewise, we consider the timing and development of seascape stewardship practices in the region in relation to climate change and societal transformations and how they may have overlapped with terrestrial stewardship activities.

Is There Evidence for Significant Changes in Anthropogenic Fire Regimes and Seascape Stewardship Practices Over Time That May Be Due to Climate Change, Sociopolitical Transformations, Technological Innovations, and Human Demography?

We found little correlation between the timing of anthropogenic burning on the Santa Cruz coast and major episodes of climate change. We recognize that our dating of cultural burning from 700 to 1500 cal CE is not fine-tuned enough to examine specific climatic events, but the findings suggest that people probably ignited fires slightly before, during, and after the MCA (ca. 900–1300 cal CE), when temperatures and drought conditions may have increased. It does appear that people facilitated the maintenance of the coastal prairies during this period of climatic warming. Yet our study also indicates that productive grasslands existed before the onslaught of the warming conditions of the MCA and continued to be enhanced during the early years of the LIA (ca. cal 1350–1850 CE), when cooler conditions prevailed. Thus, our study suggests that Indigenous people implemented a regime of frequent cultural burning under diverse climatic conditions and/or that climate change was more moderate here than elsewhere in California (Jones et al. 1999).

We also found little correlation between the timing of seascape stewardship practices involving the stripping harvest of mussels and major episodes of climate change. While our sample size of sites is undeniably small, we observed that the mass harvesting of mussels in a sustainable manner took place at CA-SCR-14 from 1000 to 1500 cal. CE and CA-SMA-216 from 1300 to 1640 cal. CE. Our findings indicate that this method of seascape stewardship took place both during the MCA and well into the cooler LIA. However, we note that Boone (2012) suggests that aquatic resources, such as from tidepools, may have been more reliable than some other terrestrial resources during the drought prone MCA. We recognize that our chronology may not be fine-tuned enough to see this difference in tidepool species that might be influenced by changing environmental conditions. No doubt, climatic fluctuations greatly influenced resource productivity and availability in both the sea and the land. Nevertheless, our findings suggest that Native peoples employed stewardship practices in times of both warming and cooling climatic fluctuations.

Our study suggests that the generalized broad-spectrum economy observed in Middle Holocene times at CA-SCR-7 transitioned to Late Holocene economies at CA-SCR-10, CA-SCR-14, CA-SCR-15, CA-SMA-113 that were more focused on smaller food packages from both the land and sea. We observed a greater emphasis over time on the harvesting of coastal prairie resources, the increasing use of small schooling fish, such as herring and anchovies, the mass harvesting of small and medium size mussels, and the continued harvesting of kelp and surfgrass. The first evidence for the intensive use of grass seeds and hazelnuts appeared at CA-SCR-10 from ca. 680 to 880 cal CE, when we also see evidence for the extensive harvesting of small foraging fish using nets and the continued use of marine plants presumably for food and raw materials. Moreover, in later times, when Indigenous people maintained coastal prairies around CA-SCR-14 and CA-SCR-15, this trend of harvesting forage fish continues (more so at the former site), along with the use of marine plants. Furthermore, there is also evidence in Late Holocene times for the intensive harvesting of mussels employing the stripping method that facilitated the mass collection of shellfish, while still sustaining the shellfish population over time.

The broader pattern that we see for the Late Holocene for the Santa Cruz coast is the implementation of various stewardships practices that enhanced the productivity and sustainability of small-food packages through frequent cultural burning and the selective harvesting of small- and medium-size forage fish and shellfish populations. While people may have initiated cultural burning at an earlier time, there appears to be a broad correlation between landscape stewardship practices and the more intensive use of the seascape in Late Holocene times on the Central California coast.

The increasing emphasis on small-food packages that we documented on the Central California coast in Late Holocene times appears to be part of a broader trend observed in many areas of California involving resource intensification—when people worked harder to increase the productivity of resource harvesting per unit of area. Much of the literature on resource intensification in California stresses its costs—that it can result in decreasing foraging efficiency and resource depression as people increasingly turned to the exploitation of costly, lower-ranked, small-food packages (Basgall 1987 Broughton 1994, 1999). Our research suggests that Indigenous people may have also implemented strategies of resource intensification that involved Indigenous stewardship practices designed to enhance the variety, quantity, and availability of small-food packages from both the land and sea. Our findings from Central California indicate that some Native peoples in Late Holocene times may have employed stewardship practices that allowed them to intensify the harvesting of coastal prairie resources, small schooling fish, and shellfish, while still maintaining the long-term productivity and sustainability of these populations. At this time, it is not clear why local communities initiated these strategies on the Santa Cruz coast, but we suspect there may be many potential reasons—climatic fluctuations, technological innovations, sociopolitical transformations, increasing sedentarism, trade, and/or human population growth. This is a research issue that we will consider in more detail with future work.

Third Goal

The third goal is to evaluate the geographic scale of Indigenous landscape and seascape stewardship practices on the Central California coast. As discussed in Chapter 2, the ongoing debate about the importance and pervasiveness of cultural burning in the past raises the important issue of how widespread its use was among Tribes. In expanding our study to include the Santa Cruz coast, we endeavored to evaluate whether cultural burning at Quiroste Valley was an isolated case or part of a broader pattern of landscape stewardship instituted by multiple Tribes in Central California.

Is There Evidence That People Initiated Anthropogenic Burning at a Regional Scale on the Central California Coast?

We found that cultural burning at Quiroste Valley in the homeland of the historic Quiroste Tribe (Hylkema and Cuthrell 2013: 225–228) was not an isolated case, but part of a broader pattern of landscape stewardship involving Indigenous tending of highly productive grasslands and other culturally modified habitats in coastal Central California. We believe that multiple Tribal groups employed frequent, low-intensity fires to enhance the diversity, quantity, and predictability of plant and animal resources available in their territories over many centuries. Our investigation of sites on the Santa Cruz coast in the homeland of the historic Cotoni Tribe indicates strong evidence that local groups sustained coastal grasslands through frequent cultural burning spanning back at least 1,200 years. Another eco-archaeological investigation farther north on the Central California coast suggests that sustained cultural burning and the maintenance of patches of coastal prairies extended back at least 600 years (Cuthrell 2020). Yet another eco-archaeological study undertaken by Peter Nelson (2017) in Coast Miwok territory north of the contemporary city of Petaluma also revealed strong evidence for grassland associated archaeobotanical remains, which indicates the long-term stewardship of productive prairies.

In addition, the analysis of phytolith remains and other ecological remains at Fort Ross State Historic Park and Salt Point State Park by Bicknell and colleagues (1993a, 1993b) suggests the Kashaya Pomo maintained an extensive swath of grassland in pre-colonial times. Russian eyewitness accounts of cultural burning by the Kashaya Pomo and other neighboring Tribes around the Ross settlement and coastal Central California further supports Indigenous tending of extensive grassland patches (Golovnin 1979: 168; Kotzebue 1830: 126–127; Lütke 1989: 257). Other phytolith studies discussed in Cuthrell (2013b; Chapter 8, this volume), Evett and Cuthrell (2013), and Evett and Bartholome (2013), further suggest the abundance of native grass and forb habitats over much of the Central California coast in ancient and early historic times.

The maintenance of extensive coastal prairies in coastal Central California is further reinforced by the findings of the aDNA study of California voles (*Microtus californicus*) sampled from both ancient and modern contexts, as detailed in Chapter 13. We found evidence for the genetic intermixture of vole populations over a broad swath of the Santa Cruz coast and beyond, which suggests this species had spread its range throughout open habitats in California over the last several thousand years. The results of the eco-archaeological research, the phytolith studies, and the aDNA research suggest that extensive patches of coastal grasslands extended along the Central California coast from Santa Cruz to Fort Ross in Late Holocene and Historic times. This finding supports the hypothesis proposed by Weiser and L'epofsky (2009) that wide-ranging anthropogenic coastal prairies existed along the Pacific Coast of North America, extending from southern British Columbia through Washington, Oregon, and into California in Late Holocene and early Historic times.

Fourth Goal

We commenced this volume (Chapter 1) by noting that eco-archaeological investigations undertaken in partnerships with Tribes and resource agencies can provide new perspectives about intimate interactions that Indigenous communities fostered with local environments over many centuries. We argued that fine-grained eco-archaeological studies complement other important sources (e.g., Tribal oral traditions/histories, ethnohistoric observations, ethnographic studies) in providing important information about how Native peoples stewarded their lands and waters to enhance the productivity, diversity, and sustainability of culturally important plants and animals in their traditional territories. These stewardship practices provide a deep pool of knowledge for rethinking how we might revitalize the lands and waters of Central California after more than two centuries of harmful historical impacts to local environments.

How Can Collaborative Eco-Archaeological Research Provide Useful Information for Tribes and Resource Agencies Today?
The AMTB, the AMLT, and California State Parks are working collaboratively in developing comprehensive plans for the co-management of some parklands on the Central California coast. They are interested in the re-implementation of stewardship practices that will enhance the richness and diversity of Native coastal and terrestrial species, provide places for Indigenous people to harvest foods, medicines, dance regalia, and crafting materials, and to reopen terrestrial landscapes with reduced fuel loads and lower wildfire risks. So how are the findings of our collaborative eco-archaeological investigation on the Central California coast being used by the AMTB, California State Parks, and other resource agencies? We outline below four ways in which lessons from the past, generated by eco-archaeological research, are providing useful information for both the Tribe and resource agencies in meeting their objectives.

1. Restoring Indigenous Knowledge. A major goal of the AMTB is to revive Indigenous knowledge and cultural practices in their Tribal lands after more than two centuries of brutal colonialism that devastated their Tribal organizations and communities, forced them from their lands, denied them access to resources in their lands and waters, and outlawed many of their stewardship practices, such as cultural burning. As outlined by Lopez and colleagues in Chapter 3, the tumultuous entanglements with Spanish, Mexican, and American colonists over many centuries disenfranchised many Tribal members from their lands and led to some lapses or gaps in their comprehension about the ways of their ancestors. The AMLT is actively combating this colonial legacy through the revitalization of their ceremonies, dances, spirituality, language, health, and cultural practices. Two cornerstones of this revitalization process involved the creation of the AMLT and the Native Stewardship Corps (NSC) to facilitate the restoration of knowledge of their ancestors and to revive stewardship practices in their Tribal lands. The relearning of the old ways is critical for the Tribe to fulfill its sacred obligation of healing Mother Earth, taking care of all living things, and restoring the health of a much-degraded environment (Chapter 3; and Lopez 2013). Our eco-archaeological work, undertaken in close partnership with the AMLT and NSC, provided important insights to the Tribe about the various stewardship practices employed by their ancestors to manage the plants and animals from both Tribal lands and waters. As discussed in Chapter 1, we are privileged to be able to give this information back to the Tribe so that they can choose how they will incorporate it back into their Traditional Ecological Knowledge (TEK) and employ it as part of their revitalization efforts.

2. Reviving Ancient Ways and Ecosystems. The findings of our eco-archaeological work described in this volume and elsewhere provided the AMTB with specific, diachronically derived information that may be incorporated into their TEK. This information concerns the nature of past anthropogenic fire regimes, the plants and animals that once thrived on the Santa Cruz coast, and the kinds of Indigenous stewardship practices employed to maintain these productive landscapes and seascapes. Based on this work, the Amah Mutsun and California State Parks are very interested in reviving patchy landscape mosaics that include swaths of coastal prairies and other prolific habitats that have largely disappeared from the region after years of fire suppression and environmental neglect. Much of this effort has focused on the co-management of the Quiroste Valley Cultural Preserve in Año Nuevo State Park. As discussed by Sigona and colleagues in Chapter 14, eco-archaeology serves as an "ace-in-sleeve" for these revival efforts in that it supplies crucial information for making decisions about what species and habitats to regenerate.

The findings of the eco-archaeological research have provided a blueprint for the AMLT and CSP for bringing back native plants and animals that once flourished in Quiroste Valley. The AMLT and NSC are now reopening patches of the landscape that involve fuel load reduction with the thinning of dense (sometimes almost impenetrable) thickets of vegetation, the creation of hundreds of burn piles, and the tending of remnant patches of culturally significant plants. Plans are in the works for cultural burns that will involve strategically placed low-intensity, broadcast fires to facilitate the opening of some patches. In close collaboration with California State Parks and Pie Ranch, the AMLT and NSC have created a Native Plant Propagation Program that involves the growing of culturally important plants in greenhouse facilities and outdoor fields at Cascade Ranch State Park. This propagation program is creating both seeds and seedlings of various grasses and forbs for the regeneration of coastal prairies and other habitats in the newly opened landscapes. As of spring 2024, the AMLT, NSC, and volunteers are tending 135,000 grassland plants in field beds for seed propagation and another 10,000 seedlings to be replanted in Quiroste Valley (for details of the plant propagation program, see Chapters 3 and 14). The AMLT is in the process of establishing 150–200 plots of culturally important plants in Quiroste Valley through its plant propagation program (Rob Cuthrell, personal communication, 2023). Notably, the eco-archaeological research served an important role in providing species lists of the culturally important plants being reintroduced into the revitalized Quiroste Valley landscape.

The AMLT is currently developing a Coastal Stewardship program building upon this research to restore and revitalize resource stewardship on the sea. The AMLT has partnered with the California Indian Environmental Alliance, Tolowa Dee-ni' Nation, Resighini Rancheria, Kashia Band of Pomo Indians, and Ecotrust to form a Tribal Marine Stewards Network. One project within this program will focus on assessing and stewarding ancient archaeological sites in coastal settings being affected by sea-level rise and coastal erosion. A second project of this program focuses on monitoring other important natural and cultural resources, such as kelp forests, rocky intertidal zones, and seagrass beds, which provide habitats for a diverse range of species and are essential for maintaining productive marine ecosystems. By monitoring these resources with aerial and underwater drones throughout the year, we can better understand the issues affecting them and contribute to efforts to protect and manage them. The Amah Mutsun, through their participation in the Coastal Stewardship Program, are applying TEK generated from Tribal oral traditions, the archaeological and ethnographic record, along with modern observations of marine ecology, to better steward and protect the coastal seashore of the Central California.

3. Synergetic Innovations. We have found that collaborative eco-archaeological research involving the AMLT, the NSC, resource agency specialists, and university faculty and students and other scholars produced an incredible synergy for rethinking how we revive moribund ecosystems and better protect and preserve important culturally significant landscapes. Sigona and colleagues, in Chapter 14, present two examples of how these partnerships served as incubators for the creation of innovative developments.

The first was the joint creation of the "integrated cultural resource survey" that arose out of our eco-archaeological work involving NSC and university personnel working together in the field. Our archaeological surveys began with the traditional objective of documenting archaeological sites but transformed into a broader examination of the landscape that involved not only the search for and recording of sites but also other culturally significant resources, including ethnobotanical taxa used for food, medicines and crafting material, natural springs, caves, rock outcroppings, wildlife areas, and notable minerals. We have found that the georeferencing of culturally sensitive archaeological, botanical, and abiotic resources as components of "culturally significant landscapes" can then be employed by the Tribe and resource agencies to make better-informed decisions about the placement of trails, parking lots, and development projects (see Chapter 14). This innovative approach also produces excellent information for making decisions about the revitalization of degraded environments. The integrated survey provides a systematic way for detecting and documenting remnant patches of culturally sensitive plants that may be revived via Tribal tending involving cultural burning and the infusion of new seedlings and/or seeds from the Native Plant Propagation Program. We feel that eco-archaeological surveys that document both Indigenous cultural resources and culturally significant natural resources yield important information that should be incorporated into the decision-making process of Tribes and resource agencies involved in the regeneration of any neglected landscape.

The second innovation concerns the preservation and protection of archaeological sites. We have found that many archaeological sites on the Central California coast are associated with culturally significant plants, which we suspect are remnants of patches maintained by Indigenous ancestors who occupied these places. As Chapter 14 outlines, these places provide a two-for-one incentive for the repeated monitoring of ancestral sites to evaluate their condition and to record recent evidence of looting or damage due to erosion and other natural factors. Remnant patches associated with sites may provide additional incentives for visiting these places not only to make observations about the state of the site, but also to tend the plants. In situations where erosion is impacting ancestral sites, culturally significant plants may be reintroduced to stabilize soils. Again, these places with concentrations of important cultural resources will provide incentives for the NSC to systematically visit them for stewardship objectives and to gather foods, medicines, and crafting materials, and, at the same time, to inspect the current state of the archaeological record.

4. Education and Access to Ancestral Lands. Our collaborative eco-archaeological research is contributing to educational programs that facilitate the relearning of cultural practices and empower Tribal access to its lands, sites, and resources that have been denied to them for many decades. Chapters 3 and 14 discuss the reciprocal learning that takes place during Native Stewards Cultural Days, which involve the NSC and archaeologists visiting archaeological sites, harvesting plant foods and medicines, identifying important maritime resources, and crafting Indigenous objects. Eco-archaeology also contributes to the Amah Mutsun Youth Camp when young Tribal members are exposed to many cultural practices, such as harvesting seaweed, shellfish, and fish, as evidenced in the local archaeological record and Amah Mutsun ethnobiological database. These events provide a fun space to engage in stewardship and directly apply the findings of archaeological research to community health and wellness.

Conclusion

We employed a collaborative, low-impact, eco-archaeological approach to study landscape and seascape stewardship practices implemented by Indigenous communities on the Santa Cruz coast in ancient and historic times. The primary purpose of our work was to recover dormant information about traditional ecological knowledge that we could give back to the Amah Mutsun in their quest to reimplement Indigenous stewardship practices and restore local environments after many decades of mistreatment under colonialism. The research was conducted by a team of scholars from the AMTB, the AMLT, California State Parks, UC Berkeley, and UC Santa Cruz. This chapter synthesized the findings from our investigation of four Santa Cruz sites (CA-SCR-7, CA-SCR-10, CA-SCR-14, and CA-SCR-15) that were occupied in Middle Holocene, Late Holocene, and early Historic times. We then employed these and other findings from our eco-archaeological investigations of coastal Central California to evaluate the four major goals for the project involving the timing, development, geographic scale, and contemporary relevance of Indigenous landscape and seascape stewardship practices on the greater Santa Cruz coast.

We conclude this chapter with four observations about our study concerning low-impact archaeology, resource intensification, coastal and inland sites, and the relevance of collaborative eco-archaeology to contemporary California.

Low-Impact Archaeology

Our low-impact, eco-archaeological approach as described in Chapter 1 is designed explicitly to minimize impacts to archaeological sites and to avoid disturbances to burials and other sacred Indigenous remains, while maximizing the recovery of useful information about cultural resources and Tribal histories. The methodology emphasizes the use of surface and near surface procedures, including site mapping, geophysical survey, and systematic surface collection, to obtain crucial information about site structure with minimal destruction of the archaeological record (see Chapters 4 and 5 for specific methods used in this project). This first phase of investigation is then followed by subsurface sampling if warranted and agreed upon by members of our collaborative team. If the decision to undertake subsurface sampling is made, then we collectively decide on the location, scale, and number of subsurface units. For those specific units we excavate, we employ an intensive recovery methodology of fine screening and bulk soil processing from which a broad suite of data classes can be recovered, including macro/micro artifacts, macro/micro faunal remains, macrobotanical remains, phytoliths, and radiocarbon samples.

So how did this low-impact, but highly precise archaeological program fare in the eco-archaeological investigation of the Santa Cruz sites?

Surface and Subsurface Comparisons. Overall, we feel the methodology worked well in providing basic information about site structure and associated archaeological remains. One of the great benefits of low-impact methodology is that considerable information can be collected relatively rapidly in the field. The entire field component of our project took place over a seven-week period. But the downside of this methodology is that most of the work takes place in the laboratory, where considerable time must be invested in the tedious analysis of fine-screened archaeological materials. It took a large laboratory crew several months just to process the archaeological materials from a relatively small volume of sediments, and then many more weeks for specialists to identify the artifacts, botanical, and faunal remains. We are pleased to see that the systematic surface collection involving the "catch-and-release" methodology produced results that, in many cases, were corroborated by the excavation units. That is, the findings from surface units that yielded various artifact classes and moderate-to-high densities of vertebrate and invertebrate faunal remains tended to be replicated in the excavation units. The major exceptions to this observation were CA-SCR-7—where the subsurface and surface units sampled different areas of the site (Locus 1, Locus 4, respectively)—and CA-SCR-10—where our standard methodology for surface recording had to be altered due to unforeseen conditions (hard-packed surface).

Eco-Archaeological Findings. Our low-impact, eco-archaeological investigation did provide important insights about the timing, development, and scale of Indigenous stewardship practices on the Santa Cruz coast. As outlined above, our results strongly suggest that Native peoples implemented landscape stewardship practices revolving around frequent cultural burning on the Santa Cruz coast more than 1,200 years ago, which greatly enhanced the productivity and diversity of various foods and other terrestrial resources in their territories. We also found evidence that the process of resource intensification in Late Holocene times involved not only small seeds, nuts, and other terrestrial resources but also the more intensive use of small-package marine resources, such as mussels, small schooling fish, and intertidal plants. We believe Indigenous stewardship practices were employed that maintained the viability of these terrestrial and marine resources over multiple generations of intensive harvesting.

Issues with Low-Impact Archaeology. Much can be learned about the spatial structure and associated cultural materials of ancestral places through detailed surface mapping, geophysical survey, and catch-and-release surface units. In some cases, that may be all that is needed to address Tribal, resource agency, and site stewardship questions. But when the decision is made to undertake excavation work, how much is enough? After completing our study, we now recognize several methodological shortcomings with our excavation program.

First, we experimented with the excavation of 50x50 cm units at CA-SCR-14 and CA-SCR-15. In retrospect, while they minimized impacts to the sites, the units were so small that they were difficult to dig below 50 cm in depth and it was tough to profile the walls. Furthermore, as discussed below, we examined only a tiny volume of the archaeological deposits from the sites. In the future, as Cuthrell suggests in Chapter 8, we would recommend larger size units on the order of .5x1 m or 1x1 m.

Second, we placed only one to two excavation units in most of the sites. While this plan also minimized impacts to the sites, the small sample sizes raise questions concerning our ability to characterize the archaeological remains from the entire site. Furthermore, the placement of units in marginal areas of sites, such as at CA-SCR-14 and CA-SCR-15, can have significant implications about how we interpret these places. As exemplified by our excavation at CA-SCR-10, we were able to document intact archaeological materials from ca. 680 to 880 cal CE, but our excavation strategy missed out on the opportunity for detailing intact deposits from earlier components of the site. As Cuthrell and Gifford-Gonzalez emphasize in Chapters 8 and 12, respectively, the small sample size also affects our ability to employ parametrical statistical tests and other kinds of analyses. In the future, we would recommend a larger sample of excavation units (e.g., at least 6–8 units) placed across the sites that are selected for subsurface investigation. This will provide a better understanding of the site structure and the range of variation in subsurface archaeological materials found across the site.

Third, we also experienced sampling biases in the kinds of materials we recovered given the small size and number of our excavation units (see Gifford-Gonzalez, Chapter 12). The relatively small volume of sediments that we excavated and processed for the sites appears to have precluded the recovery of larger and less common archaeological materials at sites (e.g., Grayson 1984, 1989). This potential sampling bias is an important concern for low-impact archaeological studies. The fine-grained methodology we employed appears to have been relatively successful in unearthing small-sized archaeological materials, such as charred seeds and nuts, fish bones, and fragments of shellfish, which allowed us to evaluate the stewardship research questions. But, in retrospect, we note the paucity of our recovery of larger-size archaeological materials in our subsurface assemblages. For example, we recovered few ground stone objects or larger-bodied terrestrial and marine mammal remains (that could be identified to species or genus) in our excavation units. Had we not detected ground stone materials in the surface collection at CA-SCR-15, then we would not have recognized that this class of artifacts existed at this site. Similarly, had we not had the results of the previous excavation work at CA-SCR-10 by Jones and Hildebrandt (1994), we would not have known about the existence of various classes of chipped stone and ground stone artifacts.

As Gifford-Gonzalez emphasizes in Chapter 12, the implementation of low-impact archaeology involves some challenging decisions and negotiations, particularly in working on sites protected in parklands or conservation lands. On one hand, in working closely with Tribal scholars and resource agencies, there needs to be an underlying ethical concern in preserving and protecting sites and for undertaking fieldwork that minimizes archaeological impacts to ancestral places. On the other hand, if decisions are made to undertake subsurface investigations to obtain important information about Tribal cultural practices and lifeways, particularly for sites that are now endangered by climate change and looting, then a sufficient number of subsurface units and volume of sediments need to be carefully examined to provide reliable sample sizes of diverse archaeological materials. Obviously, this decision will need to be made on a case-by-case basis.

As Gifford-Gonzalez notes in Chapter 12, fine-grained, low-impact, eco-archaeological work can be facilitated by incorporating previous excavations that used more coarse-grained methodologies that examined greater volumes of the site matrix. The findings from low-impact, fine-grained studies undertaken in combination with those from previous excavations can provide a more robust population of archaeological materials for generating interpretations. Our interpretations for CA-SCR-7 and CA-SCR-10 were facilitated greatly by previous broader-scale work (e.g., Jones and Hildebrandt 1990, 1994; Hildebrandt et al. 2007). Our point is that coordinated research using reports from previous relevant excavations and studies of museum assemblages from earlier excavations should play a significant role in low-impact archaeological studies. However, if such collections do not exist, then we advocate whenever possible in collaboration with Tribal partners for more extensive subsurface investigations than what we undertook in this project (e.g., ideally a sample of at least 6–8 excavation units measuring .5x1 m or 1x1 m).

Resource Intensification

An important outcome of our work on the Santa Cruz coast and elsewhere is the potential for developing a better understanding about how Indigenous people implemented strategies of resource intensification in Late Holocene times in Central California. Resource intensification is here defined as a process in which people expended more time and labor to procure necessary resources per unit of area. It typically involved people working harder to enhance the amount of food obtained within defined territories. A significant transformation observed by archaeologists for many Indigenous societies across California is the acceleration of resource intensification in Late Holocene times, particularly over the last 2,000 years or so. Some studies suggest that this process of resource intensification involved the anthropogenic degradation of local environments as larger, prime food packages were overharvested and people were forced to forage for smaller, more costly food packages (e.g., Broughton 1994, 2002). Our research suggests another potential scenario—that is, the process of resource intensification in some times and places may have involved Indigenous stewardship practices designed to enhance the diversity, quantity, and sustainability of both terrestrial and maritime resources in local places. Our findings on the Santa Cruz coast and elsewhere in Central California indicate that Native peoples appear to have incorporated stewardship practices that allowed them to intensify food harvests of small-food packages, while still maintaining the long-term viability of specific kinds of resources, such as coastal prairies, fisheries, and shellfish populations. The issue of resource intensification and how it was implemented on the Central California coast in Late Holocene times will be a significant area of future research by our collaborative research team.

Coastal and Inland Sites

The analysis of archaeobotanical remains at nine sites completed by Cuthrell in Chapter 8 detected an intriguing pattern: coastal sites exhibited little evidence for the use or processing of terrestrial foods, and it appears people using these places focused their attention on the harvesting of maritime related resources. In contrast, archaeobotanical evidence for terrestrial foods was found in sites situated in more interior locations (CA-SCR-10, CA-SCR-14, CA-SCR-15, and CA-SMA-113). This spatial pattern makes intuitive sense but will need to be further explored in the future. It also raises the issue of sampling biases in our findings. Our research currently suggests that Indigenous landscape stewardship practices involving the creation and maintenance of coastal prairies through cultural burning took place primarily in Late Holocene times. Yet our eco-archaeological research to date has focused primarily on Late Holocene sites located on or near the coast.

We emphasize that future investigations need to be undertaken on more Middle Holocene and Late Holocene coastal sites, but particularly on Middle and Late Holocene aged sites situated in the interior. We particularly need to initiate studies of sites in the interior that date prior to 700–800 cal CE. Here is where evidence for earlier terrestrial food processing and use may be found. Thus, in evaluating the question of when sustained cultural burning first took place in Central California, an important future direction for our collaborative research team will be undertaking additional studies of Middle and Late Holocene sites in *both* coastal and inland locations.

Contemporary Relevance of Collaborative Eco-Archaeology

We conclude the volume by emphasizing the significant role that eco-archaeological research undertaken in partnership with Tribes and resource agencies can play in contemporary California. As outlined in this volume, collaborative eco-archaeological research can provide a wealth of important information that Tribes and resource agencies can employ in making decisions about ecological restorations of lands and waters, improving the health of biological communities, enhancing the stewardship of terrestrial and coastal resources, restoring Indigenous knowledge that may have gone dormant through many centuries of colonialism, reducing the risks of catastrophic wildfires by lowering fuel loads through Indigenous cultural burning, and enhancing public education programs. We see a bright future ahead for collaborative eco-archaeology in California.

References

Basgall, Mark E.
> 1987   Resource Intensification among Hunter-Gatherers: Acorn Economies in Prehistoric California. *Research in Economic Anthropology* 9: 21–52.

Bicknell, Susan H., Amy T. Austin, Donna J. Bigg, and R. Parker Godar
> 1993a   *Fort Ross Historic State Park Prehistoric Vegetation Final Report.* Report of File at California Department of Parks and Recreation, Sacramento, California.

Bicknell, Susan H., R. Parker Godar, Donna J. Bigg, and Amy T. Austin
> 1993b   *Salt Point State Park Prehistoric Vegetation Final Report.* Report of File at California Department of Parks and Recreation, Sacramento, California.

Boone, Cristie M.
> 2012   Integrating Zooarchaeology and Modelling: Trans-Holocene Fishing in Monterey Bay, California. PhD Dissertation, Department of Anthropology, University of California, Santa Cruz.

Broughton, Jack M.
> 1994   Declines in Mammalian Foraging Efficiency during the Late Holocene, San Francisco Bay, California. *Journal of Anthropological Archaeology* 13(4): 371–401.
> 1999   *Resource Depression and Intensification during the Late Holocene, San Francisco Bay: Evidence from the Emeryville Shellmound Vertebrate Fauna.* Anthropological Records, Volume 32. University of California Press, Berkeley.
> 2002   Pristine Benchmarks and Indigenous Conservation? Implications from California Zooarchaeology. In *The Future from the Past: Archaeozoology in Wildlife Conservation and Heritage Management*, edited by Roel C. Lauwerier, and Plug Ina, pp. 6-18. Oxbow Books, Oxford, England.

Cuthrell, Rob Q.
> 2013a   Archaeobotanical Evidence for Indigenous Burning Practices and Foodways at CA-SMA-113. *California Archaeology* 5(2): 265–290.
> 2013b   An Eco-Archaeological Study of Late Holocene Indigenous Foodways and Landscape Management Practices at Quiroste Valley Cultural Preserve, San Mateo County, California. PhD Dissertation, Department of Anthropology, University of California, Berkeley.
> 2020   Analysis of Archaeobotanical and Phytolith Materials. In *Historical Ecology and Archaeology of Landscapes at Point Reyes National Seashore (HEALPR),* edited by Kent G. Lightfoot, Rob Q. Cuthrell, Peter A. Nelson, Michael A. Grone, Gabriel M. Sanchez, Roberta A. Jewett, Nick Tipon, Alec J. Apodaca, and Ariadna Gonzalez. Report Prepared for the National Park Service. Archaeological Research Facility, University of California, Berkeley.

Evett, Rand R., and James W. Bartolome
2013   Phytolith Evidence for the Extent and Nature of Prehistoric California Grasslands. *The Holocene* 23: 1644–1649.

Evett, Rand R., and Rob Q. Cuthrell
2013   Phytolith Evidence for a Grass-Dominated Prairie Landscape at Quiroste Valley on the Central Coast of California. *California Archaeology* 5(2): 319–335.

Gifford-Gonzalez, Diane, Cristie M. Boone, and Rachel E. B. Reid
2013   The Fauna From Quiroste: Insights into Indigenous Foodways, Culture, and Land Modification. *California Archaeology* 5(2): 291–317.

Golovnin, Vasilii M.
1979   *Around the World on the Kamchatka 1817–1819*. Translated by Ella L. Wiswell. The Hawaiian Historical Society and University Press of Hawaii, Honolulu.

Grayson, Donald K.
1984   *Quantitative Zooarchaeology: Topics in the Analysis of Archaeological Faunas*. Studies in Archaeological Science. Academic Press, Orlando, Florida.
1989   Minimum Numbers and Sample Size in Vertebrate Faunal Analysis. *American Antiquity* 43(1): 53–65.

Hildebrandt, William R., Deborah A. Jones, and Mark G. Hylkema
2007   *Sand Hill Bluff Site: CA-SCR-7*. National Register of Historic Places Registration Form. On file, California Department of Parks and Recreation, Sacramento, California.

Hylkema, Mark G.
2021   Middle Holocene Projectile Points from the Santa Cruz Coast of Northern Monterey Bay. Paper Presented at the 2021 Annual Meeting for the Society for California Archaeology, A Virtual Event. Paper Archived by the Society for California Archaeology.

Jones, Deborah A., and William R. Hildebrandt
1990   *Archaeological Excavation at Sand Hill Bluff: Portions of a Prehistoric Site CA-SCR-7, Santa Cruz County, California*. Far Western Anthropological Research Group, Davis, California.

Jones, Deborah A., and William R. Hildebrandt
1994   *Archaeological Investigations at Sites CA-SCR-10, CA-SCR-17, CA-SCR-304, and CA-SCR-38/123 for the North Coast Treated Water Main Project, Santa Cruz County, California*. Far Western Anthropological Research Group, Davis, California.

Jones, Terry L., Gary M. Brown, L. Mark Raab, Janet L. McVickar, W. Geoffrey Spaulding, Douglas J. Kennett, Andrew York, and Phillip L. Walker

    1999   Environmental Imperatives Reconsidered: Demographic Crises in Western North America during the Medieval Climatic Anomaly. *Current Anthropology* 40(2): 137–170.

Kotzebue, Otto Von

    1830   *A New Voyage Round the World, in the Years 1823, 24, 25, and 26.* Henry Colburn and Richard Bentley, London, England.

Lopez, Valentin

    2013   The Amah Mutsun Band: Reflections on Collaborative Archaeology. *California Archaeology* 5(2): 221–223.

Lütke, Fedor P.

    1989   September 4–28, 1818. From the Diary of Fedor P. Lütke during his Circumnavigation Aboard the Sloop Kamchatka, 1817–1819: Observations on California. In *The Russian American Colonies Three Centuries of Russian Eastward Expansion 1798–1867, volume 3: A Documentary Record*, edited by Basil Dmytryshyn, E. A. P. Crownhart-Vaughan, and Thomas Vaughan, pp. 257–285. Oregon Historical Society Press, Portland.

Nelson, Peter A.

    2017   Indigenous Archaeology at Tolay Lake: Responsive Research and the Empowered Tribal Management of a Sacred Landscape. PhD Dissertation, Department of Anthropology, University of California, Berkeley.

Sanchez, Gabriel M.

    2021   Middle and Late Holocene Faunal Remains from the Santa Cruz Coast. In *The Study of Indigenous Landscape and Seascape Stewardship Practices on the Santa Cruz Coast: A Collaborative Eco-Archaeological Approach*, edited by Kent G. Lightfoot, Rob Q. Cuthrell, Mark G. Hylkema, Valentin Lopez, Diane Gifford-Gonzalez, Roberta A. Jewett, Michael A. Grone, Gabriel M. Sanchez, Peter A. Nelson, Alec J. Apodaca, Ariadna Gonzalez, Kathryn Field, and Alexii Sigona. Report Prepared for California Department of Parks and Recreation, Santa Cruz District. Archaeological Research Facility, University of California, Berkeley.

Weiser, Andrea, and Dana Lepofsky

    2009   Ancient Land Use and Management of Ebey's Prairie, Whidbey Island, Washington. *Journal of Ethnobiology* 29(2): 184–212.

List of Contributors

Alec J. Apodaca Department of Anthropology, Archaeological Research Facility, University of California, Berkeley

Jordan F. Brown Department of Anthropology, Archaeological Research Facility, University of California, Berkeley

Chris J. Conroy Museum of Vertebrate Zoology, University of California, Berkeley

Rob Q. Cuthrell Archaeological Research Facility, University of California, Berkeley

Kathryn Field Archaeological Research Facility, University of California, Berkeley

Paul V. A. Fine Department of Integrative Biology, University of California, Berkeley

Ariadna Gonzalez Archaeological Research Facility, University of California, Berkeley

Diane Gifford-Gonzalez Anthropology Department, University of California, Santa Cruz

K. Michelee Glowa Consultant, Amah Mutsun Tribal Band, PhD

Michael A. Grone California Department of Parks and Recreation, Santa Cruz District

Mark G. Hylkema California Department of Parks and Recreation, Santa Cruz District

Roberta A. Jewett Archaeological Research Facility, University of California, Berkeley

Kent G. Lightfoot Department of Anthropology, Archaeological Research Facility, University of California, Berkeley

Valentin Lopez Chairman, Amah Mutsn Tribal Band

Cameron Shard Milne Department of Ecology and Evolutionary Biology, University of California, Santa Cruz

Peter A. Nelson Department of Environmental Science, Polity and Management, Ethnic Studies, University of California, Berkeley, Federated Indians of Graton Rancheria

Carolyn T. Rodriguez Amah Mutsun Tribal Band Member, MA

Gabriel M. Sanchez Department of Anthropology, University of Oregon

Beth Shapiro Department of Ecology and Evolutionary Biology and Howard Hughes Medical Institute, University of California, Santa Cruz

Alexii Sigona Amah Mutsun Tribal Band, Department of Environmental Science, Policy and Management, University of California, Berkeley

Acknowledgments

Our collaborative, eco-archaeological project was generously supported by the National Science Foundation (BCS-1523648), the Research Institute for Humanity and Nature (RIHN), the UC Berkeley Class of 1960 Chair in Undergraduate Education, and the Undergraduate Research Apprentice Program (URAP) at UC Berkeley. We appreciate greatly the tremendous contributions made by various members of the Amah Mutsun Tribal Band (AMTB), the Amah Mutsun Land Trust (AMLT), and the Native Stewardship Corps who made this a successful collaborative project. The synergetic program of research that took place on the Santa Cruz coast would not have been possible without the vision and leadership of Valentin Lopez, Chair of the AMTB. We are particularly indebted to Valentin Lopez, Eleanor Castro, and Ed Ketchum who played important roles in the planning and implementation of the research program. The participation of the Native Stewardship Corps during the 2016 and 2017 field seasons was greatly facilitated by the hard work of the AMLT staff and others—particularly, Ek Ong Kar Singh Khalsa, Sara French, Rob Cuthrell, Jay Sherf, Rick Flores, Reed Holderman, Diane Talbert, and Jeff Pace.

We are most thankful for the continued support of our collaborative research program by the Santa Cruz District, California State Parks. We thank Mark Hylkema, recently retired from California State Parks, for his inspiration, expertise, and skill in the inauguration of the collaborative program and for keeping it running for multiple years. Without his many efforts, continued advice, and contributions, the work would not have taken place. We are also indebted to many other people in the Santa Cruz District who made the work possible and provided camping facilities on the Santa Cruz coast, including Tim Reilly, Tim Hyland, Chris Spohrer, Portia Halbert, and many other staff who ensured contracting and field operations ran smoothly. Their continued support and collaboration resulted in a Memorandum of Understanding (MOU) between AMLT and California State Parks regarding shared efforts to reintroduce Indigenous land stewardship practices at several parks in the Santa Cruz District.

We thank Stephanie Mills for her ongoing interest in the project, her steadfast support for the collaborative research with the Amah Mutsun, and her hospitality and generosity that enabled much of our work to take place.

We were blessed to work with a wonderful team of AMTB members and UC Berkeley students who undertook fieldwork along the Santa Cruz coast in the summers of 2016 and 2017. The 2016 field team headquartered on State Park land near Davenport was directed by Mark Hylkema, Valentin Lopez, and Kent Lightfoot along with Rob Cuthrell, Roberta Jewett, Peter Nelson, Mike Grone, and Gabriel M. Sanchez. Nicholas Tripcevich provided much needed help, advice, and support in our use of geophysical and other sophisticated equipment from the Archaeological Research Facility (ARF) at UC Berkeley.

Eleanor Castro served as the camp's tribal elder and chef who prepared some amazing meals for us. We appreciate Ed Ketchum coming to the field to provide advice and to present tribal oral histories to the field participants. Five members of the Stewardship

Corps of the AMTB participated in all components of the fieldwork during the summer of 2016: these included Josh Higuera-Hood, Abran Lopez, Paul Lopez, Gabriel Pineida, and Nathan Vasquez. Alyssa Scott provided expertise in historical archaeology and Elizabeth Campos in conservation methods. Ten UC Berkeley undergraduate students participated in the summer 2016 field work, including Amanda Dobrov, Rachel Gordon, Michele Maybee, Trinity Miller, Mary Ellis Passey, Nick Shepetek, Kelsey Scott, Rosario Torres, Savannah Tucker, and Charles Woodward. Patty Buckley and Caleb Jewett also served as volunteers in the field.

The 2017 field team headquartered in the campground of Wilder Ranch State Park was directed by Mark Hylkema, Valentin Lopez, and Kent Lightfoot, along with Rob Cuthrell, Roberta Jewett, Mike Grone, and Gabriel M. Sanchez. Nicholas Tripcevich was once again a tremendous help in facilitating the use of field equipment from the ARF. Alyssa Scott and Thomas Banghart provided expertise in historical archaeology during the fieldwork. Eleanor Castro again contributed greatly as the camp's Tribal elder. She served as the camp's chef along with Lupe Delgado, and the two of them produced some real delicacies for the field team. Nine members of the Native Stewardship Corps of the AMTB participated in all components of the fieldwork during the summer of 2017: these included Lupe Delgado, Natalie Garcia, Josh Higuera-Hood, Ian Jirouard, Paul Lopez, Gabriel Pineida, Vanessa Sanchez, Nathan Vasquez, and Nathaniel Verdugo. Seven UC Berkeley undergraduate students participated in the fieldwork, including Erin Bridges, Ariadna Gonzalez, Mark Johnson, Miriam Lagunas, Sandra Martinez, Rosario Torres, and Leonardo Valdez Ordonez.

We are grateful for the wonderful team of faculty, staff, and students who participated in the analysis of the flotation samples and associated cultural materials in the California Archaeology Laboratory at UC Berkeley from 2016 to 2022. They continued to excel in their work despite major adversities involving the closing of the laboratory and campus during the COVID-19 pandemic throughout much of 2020 and 2021. Directed by Rob Cuthrell, Mike Grone, Gabriel M. Sanchez, and Kent Lightfoot, the team included Erin Bridges, Jacquie Cooper, Kathryn Field, Ariadna Gonzalez, Annes Kim, Miriam Lagunas, Cheyenne Laux, Michele Maybee, Pedro Mendoza, Trinity Miller, Katharine Nusbaum, Kasie Olmstead, Leonardo Valdez Ordonez, Paul Rigby, Alyssa Straub, Rosario Torres, Carola Rosenthal, and Michelle Vuong. Elizabeth Campos also directed a team consisting of Kathryn Field, Ariadna Gonzalez, and Carolina Gonzalez, who provided expertise in the curation and cataloging of cultural materials.

We appreciate greatly the continued support that the ARF has given in implementing collaborative eco-archaeology on the Santa Cruz coast and elsewhere in California. Their superb staff has been crucial in providing logistical support, advice, and encouragement throughout the field and laboratory research. We are particularly thankful for the many fine contributions of Director Christine Hastorf, Nicholas Tripcevich, Tomeko Wyrick, Sarah Kansa, and Kass Cazier. We are most grateful for the assistance and advice of Junko Habu, Chair of the Publications Committee of the ARF, in facilitating the production of this volume in the ARF Contributions Series. We appreciate the constructive comments from the anonymous peer review process of the volume that helped us strengthen and elaborate on several points concerning our collaborative, eco-archaeological approach.

We are indebted to Jessica Kaplan, Edward Zegarra, Morgane Leoni, Domenica Newell-Amato and the other fine people associated with Westwood Press for their excellent advice, assistance, and editorial and design expertise in the creation of this handsome volume for the ARF Contributions Series. Publication of this volume is made possible in part by the generous support from the Berkeley Research Impact Initiative (BRII), sponsored by the UC Berkeley Library.

www.ingramcontent.com/pod-product-compliance
Lightning Source LLC
Chambersburg PA
CBHW082126210326
41599CB00031B/5884